# FLUID DYNAMICS AT INTERFACES

Many of the significant issues in fluid dynamics occur at interfaces, that is, at the boundaries between differing fluids or between fluids and solids. Interfacial fluid dynamics is important in areas ranging from the flight of an aircraft, to the flow of blood in the heart, to chemical vapor deposition. The subject is also an area of active research and development, owing to improved analytical, experimental, and computational techniques.

This book describes current research and applications in interfacial fluid dynamics and stability. It is organized around five topics: Benard and thermocapillary instabilities, shear- and pressure-induced instabilities, waves and dispersions, multiphase systems, and complex flows. Chapters have been contributed by internationally recognized experts, both theoreticians and experimentalists.

Because of the range and importance of topics discussed, this book will interest a broad audience of graduate students, faculty, and researchers in mechanical, aerospace, materials, and chemical engineering, as well as in applied mathematics and physics.

Wei Shyy is professor and chairman of the Department of Aerospace Engineering, Mechanics and Engineering Science at the University of Florida, Gainesville.

Ranga Narayanan is a professor in the Department of Chemical Engineering at the University of Florida.

Professor C. S. Yih, 1918–1997

# FLUID DYNAMICS AT INTERFACES

*Edited by*

WEI SHYY

RANGA NARAYANAN

*University of Florida*

CAMBRIDGE
UNIVERSITY PRESS

PUBLISHED BY THE PRESS SYNDICATE OF THE UNIVERSITY OF CAMBRIDGE
The Pitt Building, Trumpington Street, Cambridge, United Kingdom

CAMBRIDGE UNIVERSITY PRESS
The Edinburgh Building, Cambridge CB2 2RU, UK   www.cup.cam.ac.uk
40 West 20th Street, New York, NY 10011-4211, USA   www.cup.org
10 Stamford Road, Oakleigh, Melbourne 3166, Australia
Ruiz de Alarcón 13, 28014 Madrid, Spain

First published 1999

Printed in the United States of America

*Typeface* Times Roman 10.25/12.5 pt.   *System* LaTeX $2_\varepsilon$   [TB]

*A catalog record for this book is available from the British Library.*

*Library of Congress Cataloging-in-Publication Data*
Fluid dynamics at interfaces / edited by Wei Shyy and Ranga Narayanan.
        p.      cm.
    Includes index.
    ISBN 0-521-64266-3 (hb)
    1. Interfaces (Physical sciences)   2. Fluid dynamics.   3. Surface
chemistry.   I. Shyy, W. (Wei)   II. Narayanan, Ranga.
    QC173.4.S94F54   1999
    532 – dc21                                      98-45448
                                                      CIP

ISBN 0 521 64266 3 hardback

# Contents

| | | |
|---|---|---|
| *Preface* | | *page* ix |
| *List of Contributors* | | xiii |

**Part I. Bénard and Thermocapillary Instabilities**    1

1   Nonlinear Dynamics of Thin Evaporating Liquid Films Subject
to Internal Heat Generation    3
ALEXANDER ORON

2   The Effect of Air Height on the Pattern Formation in Liquid–Air
Bilayer Convection    15
D. JOHNSON, R. NARAYANAN, AND P. C. DAUBY

3   The Third Type of Benard Convection Induced by Evaporation    31
WEN-JEI YANG

4   Waves Generated by Surface-Tension Gradients and Instability    43
M. G. VELARDE, A. YE. REDNIKOV, AND H. LINDE

5   Thermocapillary-Coriolis Instabilities    57
ABDELFATTAH ZEBIB AND CÉDRIC LE CUNFF

**Part II. Shear and Pressure Driven Instabilities**    71

6   Control of Instability in a Liquid Film Flow    73
S. P. LIN AND J. N. CHEN

7   Three-Dimensional Waves in Thin Liquid Films    85
A. A. NEPOMNYASHCHY

8   Modulation Wave Dynamics of Kinematic Interfacial Waves    99
H.-C. CHANG, E. A. DEMEKHIN, R. M. ROBERTS, AND Y. YE

9   Multifilm Flow Down an Inclined Plane: Simulations Based on the Lubrication
Approximation and Normal-Mode Decomposition of Linear Waves    112
C. POZRIKIDIS

10   Spatial Evolution of Interfacial Waves in Gas–Liquid Flows    129
M. J. McCREADY

11   The Shear Breakup of an Immiscible Fluid Interface    142
GRÉTAR TRYGGVASON AND SALIH OZEN UNVERDI

12   Two-Fluid-Layer Flow Stability    156
S. ÖZGEN, G. S. R. SARMA, G. DEGREZ, AND M. CARBONARO

**Part III. Waves and Dispersion**                                             169

13   On Modeling Unsteady Fully Nonlinear Dispersive Interfacial Waves         171
     THEODORE YAOTSU WU

14   Instabilities in the Coupled Equatorial Ocean–Atmosphere System           179
     HENK A. DIJKSTRA AND PAUL C. F. VAN DER VAART

15   Large-Amplitude Solitary Wave on a Pycnocline and Its Instability         198
     DANIEL T. VALENTINE, BRIAN C. BARR, AND TIMOTHY W. KAO

16   Stability and Pattern Selection in Parametrically Driven Surface Waves    211
     PEILONG CHEN AND JORGE VIÑALS

17   Deformation and Rupture in Confined, Thin Liquid Films Driven
     by Thermocapillarity                                                      221
     MARC K. SMITH AND DAVID R. VRANE

18   Linear and Nonlinear Waves in Flowing Water                               234
     CHIA-SHUN YIH AND WILLIAM W. SCHULTZ

19   Pinned-Edge Faraday Waves                                                 246
     DIANE M. HENDERSON AND JOHN W. MILES

20   Interfacial Shapes in the Steady Flow of a Highly Viscous Dispersed Phase 254
     DANIEL D. JOSEPH AND RUNYUAN BAI

**Part IV. Multiphase Systems**                                                263

21   Interaction between Fluid Flows and Flexible Structures                   265
     WEI SHYY, HENG-CHUAN KAN, H. S. UDAYKUMAR, AND ROGER TRAN-SON-TAY

22   Numerical Treatment of Moving Interfaces in Phase-Change Processes        278
     SURESH V. GARIMELLA AND JAMES E. SIMPSON

23   Accuracy and Convergence of Continuum Surface-Tension Models              294
     M. W. WILLIAMS, D. B. KOTHE, AND E. G. PUCKETT

24   Interaction of Convection and Solidification at Fluid–Solid Interfaces    306
     L. BÜHLER, A. EHRHARD, AND U. MÜLLER

25   Interfacial Motion of a Molten Layer Subject to Plasma Heating            320
     P. S. AYYASWAMY, S. S. SRIPADA, AND I. M. COHEN

26   The Fluid Mechanics of Premelted Liquid Films                             339
     M. G. WORSTER AND J. S. WETTLAUFER

27   Recent Advances in Lattice Boltzmann Methods                              352
     SHIYI CHEN, GARY D. DOOLEN, XIAOYI HE, XIAOBO NIE, AND RAOYANG ZHANG

28   Bubble Dynamics in Heterogeneous Boiling Heat Transfer                    364
     RENWEI MEI, JAMES F. KLAUSNER, AND GLEN THORNCROFT

**Part V. Complex Flows**                                                      379

29   Heat, Mass, and Momentum Exchanges between Outer Flow
     and Separation Bubble behind a Single Backward-facing Step with Gas
     Injection from One Duct Wall                                             381
     TONG-MIIN LIOU AND PO-WEN HWANG

30   A Moving Boundary Problem Arising from Stratigraphic Modeling             393
     J. MARR, J. B. SWENSON, AND V. R. VOLLER

31  Convection Generated by Lateral Heating of a Solute Gradient:
    Review and Extension                                                403
    C. F. CHEN

32  Heat Conduction from a Solid Particle and the Force on It in Stokes Flow
    in a Fluid with Position-Dependent Physical Properties              413
    ANDREAS ACRIVOS AND YONGGUANG WANG

33  Radiation-Affected Ignition Phenomena with Solid–Gas Interaction    427
    SEUNG-WOOK BAEK AND JAE HYUN PARK

34  Biomagnetic Fluid Dynamics                                          439
    YOUSEF HAIK, VINAY M. PAI, AND CHING-JEN CHEN

    The Man I Know: Chia-Shun Yih, July 25, 1918–April 25, 1997         453
    YUAN-CHENG FUNG

*Index*                                                                 459

# Preface

This volume is a collection of 34 articles on interfacial fluid dynamics, multiphase systems, and complex flows with relevance primarily to the transport processes at an interface. The oral presentations of these papers were delivered in Gainesville, Florida, in June of 1998, at a special symposium held in memory of the late Professor Chia-Shun Yih. Professor Yih pioneered the field of interfacial flows, and a considerable number of the papers in this volume are directly connected with his contributions. Many modern aspects of fluid dynamics problems that involve the above issues, such as phase changes at different scales, solitons, fluid–fluid and fluid–solid interactions with large interface deformation, and multiple transport mechanisms are addressed in this volume. In view of the collective scope of these articles, we have chosen the title *Fluid Dynamics at Interfaces*. The book is divided into five major parts, each having a focus. These parts are titled (i) Bénard and thermocapillary instabilities, (ii) shear- and pressure-induced instabilities, (iii) waves and dispersions, (iv) multiphase systems, and (v) complex flows.

The first part of the volume is focused primarily on Bénard and thermocapillary instabilities. Topics discussed include the nonlinear dynamics evolving from thin films that have internal heat generation in the presence and the absence of transverse temperature gradients, and the curious effects of coupling between phases in geometries of finite lateral extent are presented. Also included in this part are papers on convection induced by evaporation at an interface and on long-wave disturbance generation arising from Marangoni instabilities. Finally, a study explaining the onset of oscillatory motion in cylindrical bridges with thermocapillary and Coriolis forces is presented.

The second part of the volume deals with the flow of liquid films down an inclined plane and with the shear breakup of interfaces. The study of instabilities of inclined liquid films is related to the early work of Yih in 1963. In this connection there is a paper that shows the possible control of interface instability by the imposition of in-plane oscillations and another one that describes the weak nonlinear evolution of three-dimensional waves. In a different study, the slow modulation of the wave amplitudes by use of the evolutionary equation and a comparison with experiments is presented. A paper on instabilities in multifilm flows on an inclined plane is offered, with a systematic formulation of the theory of film thickness evolution, in the absence of inertia, then solved numerically under the lubrication approximation. Three papers in this category are concerned with shear-induced instability of the interface in horizontal flows and boundary-layer instabilities with one study having application to the deicing of airplane wings. In one of these papers the wave evolution for thin and thick layers is studied, while in another a numerical simulation of the Kelvin–Helmholtz instability for a multilayer is examined. When the density differences are made negligible, the simulations show that fingers of one fluid penetrate the other and when the density differences are large the waves of the heavy liquid grow into the lighter one, the sizes naturally depending on the Weber number.

In the third part of the book the main interests are waves and dispersions. Three papers have applications to oceanography. A study on nonlinear unsteady internal waves in layered fluids gives a careful theoretical development of the models that describe motion under the action of gravity and surface tension. The second is concerned with ocean–atmosphere interactions and the origin of oscillatory instabilities, while the third deals with a pycnocline and the numerical study of the instability of large-amplitude waves. In addition to these, there are four papers on the deformation and the rupture of thin films due to thermocapillarity, frequency, and mode-shape prediction in Faraday waves and the evolution of linear and nonlinear water waves. There is also a paper that gives an asymptotic analysis of a very viscous phase that is dispersed in a nearly inviscid phase.

The fourth part of this volume addresses topics on multiphase flows with an accent on interfacial dynamics. Several papers are concerned with the numerical methods of interface tracking with applications to deformation of viscous drops, interaction of air with flexible airfoils, the rising of a gas bubble in a fluid, and solidification fronts, showing dramatic pictorial representations of the physics. In another interesting study, kernels and convolution methods are incorporated into continuum surface-tension models with the objective of providing accurate representation of first- and second-order spatial derivatives, as these play a vital role in evaluation of functionals such as the unit normal and mean curvature. Another paper analyzes the motion of the interface under plasma heating with the plasma modeled as a continuum. In a different context, a paper exposes and analyzes thermal regelation, a process by which a solid migrates through another one in the presence of a thermal gradient with melting and resolidification, the melting being caused by intermolecular forces. Recent developments in lattice Boltzmann methods for simulating multiphase flows on different scales and analytical and experimental investigations of the onset and the development of bubble dynamics in boiling heat transfer also offer interesting insight into multiphase flow phenomena.

The last part of this volume is on complex flows. These papers deal with analytical, numerical, and empirical aspects of fluid dynamics involving multiple, competing physical mechanisms. Interactions between heat, momentum, and mass transfer in complex, recirculating flows are reviewed, with comprehensive data offered. In a numerical investigation, a new class of moving boundary problems arising from large-scale sediment transport and deposition is treated. Recent understanding of combined buoyancy- and solute-driven heat transfer, the double-diffusive flow, is summarized. In another article, a detailed analysis is presented to address the heat conduction and force surrounding a solid particle in Stokes flow with a special focus on spatially varying physical properties. Effects of external radiation on the ignition of a two-phase mixture of gas and solid particles are also explored. Finally, a paper on the interesting issues of blood flows in a magnetic field is presented.

The volume concludes with an article authored by Professor Yuan Cheng Fung, who offers a biographical sketch and personal reminisces of Chia-Shun Yih, a remarkable fluid physicist and a dignified person, one whose work will be remembered and memory cherished.

The symposium was part of the Thirteenth U.S. National Congress of Applied Mechanics. Professor Martin Eisenberg was the General Chair, Professor Ronald Adrian the Scientific Chair, and Professor Renwei Mei the Congress Secretary. We thank them for their support and encouragement during the organization of the symposium. We also appreciate the financial support received from the University of Florida's Office of Research and Technology, Department of Aerospace Engineering, Mechanics and Engineering Science, and Department of Chemical Engineering. A special thanks is due to Dr. Winfred Phillips, Dean of the College of Engineering of the University of Florida for arranging generous financial support from the College.

There were scientists from over all over the U.S. and nine different countries of outside the U.S. Many domestic and international organizations were supportive both morally and monetarily of our efforts. Among these were universities and organizations such as the Von Karman Institute, the Deutsche Forschungsanstalt für Luft und Raumfahrt, Germany, NATO, and the Society of Theoretical and Applied Mechanics of the Republic of China. To all of these we are indeed grateful. We owe gratitude in no small measure to Jan Rockey, who so effectively and patiently helped organize the various aspects of the symposium and the book. We thank Florence Padgett and Jackie Mahon of Cambridge University Press for their encouragement and advice.

Finally a word of thanks to the authors and the attendees. We appreciated the encouragement and support of Mrs. Shirley Yih, Professors and Mrs. Yuan Cheng Fung, Shan Fu Shen, and Theodore Yaotsu Wu. Their participation made the symposium a success and a fitting tribute to the memory of our esteemed colleague and distinguished scientist, Professor, Chia-Shun Yih.

# List of Contributors

**Andreas Acrivos**
The Levich Institute
City College of New York
New York, NY

**P. S. Ayyaswamy**
Department of Mechanical Engineering
and Applied Sciences
University of Pennsylvania
Philadelphia, PA

**Seung-Wook Baek**
Department of Aerospace Engineering
Korea Institute of Science and Technology
Korea

**Hsueh-Chia Chang**
Department of Chemical Engineering
University of Notre Dame
Notre Dame, IN

**Ching-Jen Chen**
Dean of FAMU-FSU College of Engineering
Department of Mechanical Engineering
FAMU-FSU College of Engineering
Tallahassee, FL

**Chuan F. Chen**
Department of Aerospace and Mechanical
Engineering
The University of Arizona
Tucson, AZ

**Shiyi Chen**
Center for Nonlinear Studies and Theoretical
Division
Los Alamos National Laboratory
Los Alamos, NM

**Henk Dijkstra**
Institut Meteorologie en Oceanografie
Universiteit Utrecht
Utrecht, The Netherlands

**Yuan Cheng Fung**
University of California–San Diego
Institute for Biomedical Engineering
La Jolla, CA

**Suresh V. Garimella**
University of Wisconsin
Mechanical Engineering Department
Milwaukee, WI

**Diane Henderson**
William G. Pritchard Fluid Mechanics
Laboratory
Department of Mathematics
Penn State University
University Park, PA

**Daniel D. Joseph**
Department of Aerospace Engineering
and Mechanics
University of Minnesota
Minneapolis, MN

**Douglas B. Kothe**
Theoretical Division
Material Science and Technology Division
Structure/Property Relations Group MST-8,
MS-G755
Los Alamos National Laboratory
Los Alamos, NM

**Sung Piau Lin**
Department of Mechanical and Aeronautical
Engineering

Clarkson University
Potsdam, NY

**Tong-Miin Liou**
Department of Power Mechanical
Engineering
National Tsing Hua University
Hsin Chu, Taiwan

**Mark J. McCready**
Department of Chemical Engineering
University of Notre Dame
Notre Dame, IN

**Renwei Mei**
Department of Aerospace Engineering
Mechanics and Engineering Science
University of Florida
Gainesville, FL

**Ulrich Mueller**
Institut fur Angewandte,
Thermo- und Fluiddynamik
Kernforschungszentrum Karlsruhe GmbH
Karlsruhe, Germany

**Ranga Narayanan**
Department of Chemical Engineering
University of Florida
Gainesville, FL

**Alexander A. Nepomnyashchy**
Department of Mathematics
Technion–Israel Institute of Technology
Haifa, Israel

**Alexander Oron**
Department of Mechanical Engineering
Technion–Israel Institute of Technology
Haifa, Israel

**Costas Pozrikidis**
Department of Applied Mechanics and
Engineering Sciences
University of California–San Diego
La Jolla, CA

**G. S. R. Sarma**
Von Karman Institute for Fluid Dynamics
Rhode-St-Genese, Belgium

**William W. Schultz**
Department of Mechanical Engineering
and Applied Mechanics

The University of Michigan
Ann Arbor, MI

**Wei Shyy**
Department of Aerospace Engineering,
Mechanics and Engineering Science
University of Florida
Gainesville, FL

**Marc K. Smith**
George W. Woodruff School of Mechanical
Engineering
Georgia Institute of Technology
Atlanta, GA

**Gretar Tryggvason**
Department of Mechanical Engineering
and Applied Mechanics
The University of Michigan
Ann Arbor, MI

**Daniel T. Valentine**
Department of Mechanical and Aeronautical
Engineering
Clarkson University
Postdam, NY

**Manuel G. Velarde**
Instituto Pluridisciplinar
Universidad Complutense De Madrid
Madrid, Spain

**Jorge Vinals**
Supercomputer Computations Research
Institute
Florida State University
Tallahassee, FL

**V. R. Voller**
St. Anthony Falls Laboratory
Department of Civil Engineering
University of Minnesota
Minneapolis, MN

**D. R. Vrane**
Becton Dickinson Immunocytometry
Systems
San Jose, CA

**M. G. Worster**
Department of Applied Mathematics
and Theoretical Physics
Institute of Theoretical Geophysics

University of Cambridge
Cambridge, England

**Theodore Y. Wu**
California Institute of Technology
Pasadena, CA

**Wen-Jei Yang**
Department of Mechanical Engineering
and Applied Mechanics

The University of Michigan
Ann Arbor, MI


**Abdelfattah M. G. Zebib**
Department of Mechanical and Aerospace
Engineering
Rutgers University
New Brunswick, NJ

# PART ONE

# Bénard and Thermocapillary Instabilities

# 1

# Nonlinear Dynamics of Thin Evaporating Liquid Films Subject to Internal Heat Generation

ALEXANDER ORON

## 1.1. Introduction

Thin liquid films are often encountered in various experimental settings and technological applications. In many situations these films are prone to different types of instability, being subjected to the influence of various factors. For instance, isothermal films can be unstable when their average thickness ranges within the action domain of long-range dispersion forces, when overlaid by a liquid bulk with a higher density, or when being deposited on a cylindrical surface. Nonisothermal liquid films are also prone to various kinds of instability, such as those driven by density stratification, surface shear stresses arising from temperature dependence of interfacial tension (the Marangoni effect), or when being subjected to evaporation and the effect of vapor recoil. These and many other instability mechanisms were discussed in the review by Oron et al. (1997).

The review paper by Oron et al. (1997) presented a unified mathematical theory that used the disparity of the length scales and discussed the asymptotic procedure of reduction of the governing equations to a simplified evolution equation or a set of evolution equations. As a result of this reduction, a general nonlinear evolution equation was obtained in the simplest case, and various particular cases were then considered.

Among many possible physical situations discussed in the review, the case of nonlinear behavior of thin liquid films with volumetric internal heat generation was not examined. Char and Chiang (1994) and Wilson (1997) undertook the linear stability analysis of Benard–Marangoni convection in a horizontal liquid layer with internal heat generation. Both works indicated that the effect owing to the presence of internal heat sources is always to destabilize the system. On the other hand, Oron and Peles (1998) performed a nonlinear analysis of a thin liquid nonvolatile film with internal heat generation and concluded that the latter effect is *stabilizing*. This stabilization was shown to be provided by creation of a thermocapillary stress that acts from the thicker to the thinner parts of the film. Moreover, the critical value of the parameter $Q$, defined in Section 2 and related to the intensity of the heat sources, above which the system is unconditionally stable, was specified as a function of the Biot number. This apparent controversy is resolved below.

The behavior of a volatile thin liquid film with internal heat generation is studied in this chapter. The nonlinear evolution equation describing the spatiotemporal behavior of the film is derived and numerically investigated. Two spatially uniform basic states are found in the presence and the absence of a transverse temperature gradient, and their linear stability properties are examined.

## 1.2. Problem Formulation and Derivation of the Evolution Equation

A thin incompressible volatile liquid film of density $\rho$, viscosity $\mu$, thermal conductivity $k$, and average thickness $d$ is considered on a solid horizontal plane maintained at the uniform temperature $\vartheta_\omega$. The free surface of the film with surface tension $\sigma$ is exposed to the ambient gas phase at the saturation vapor temperature $\vartheta_s$ and the saturation pressure $p_s$. Spatially uniform volumetric heat sources of intensity $\dot{q}$ are present within the liquid bulk. The surface tension $\sigma$ is assumed to be a linearly decreasing function of the temperature $\vartheta$:

$$\sigma = \sigma_r + \frac{\partial \sigma}{\partial \vartheta}(\vartheta - \vartheta_r), \tag{2.1}$$

where the subscript $r$ denotes the reference value of the corresponding property and $\partial\sigma/\partial\vartheta$ is negative.

The one-sided model introduced by Burelbach et al. (1988) is used here to describe the dynamics of a volatile film. This model is based on the assumption of the disparity between the physical properties of the liquid and the vapor, so that their dynamics is effectively decoupled.

The governing equations are, respectively, the mass conservation, the momentum, and the energy equations:

$$\nabla \cdot \mathbf{v} = 0, \tag{2.2a}$$

$$\rho[\partial_t \mathbf{v} + (\mathbf{v} \cdot \nabla)\mathbf{v}] = -\nabla p + \mu \nabla^2 \mathbf{v} - \rho \mathbf{g}, \tag{2.2b}$$

$$\rho c[\partial_t \vartheta + (\mathbf{v} \cdot \nabla)\vartheta] = k\nabla^2 \vartheta + \dot{q}, \tag{2.2c}$$

where $\mathbf{v} = (u, v, w)$ is the fluid velocity field, $p$ is the pressure, $\vartheta$ is the fluid temperature, $\mathbf{g}$ is gravity normal to the solid substrate, $c$ is the specific heat of the fluid, $t$ is time, and $\nabla \equiv (\partial_x, \partial_y, \partial_z)$. The coordinate system $x$, $y$, $z$ is chosen such that the solid substrate coincides with the $x$–$y$ plane and the $z$ axis is normal to it.

The boundary conditions are the no-slip, no-penetration, and the prescribed temperature, respectively, at the solid surface, and the balances of the normal and the tangential stresses, the continuity of the heat flux, the augmented kinematic condition, and the constitutive equation relating the temperature and the mass flux, respectively, at the film interface $z = h(x, y, t)$ (Burelbach et al. 1988):

at the rigid bottom $z = 0$:

$$\mathbf{v} = 0, \qquad \vartheta = \vartheta_w; \tag{2.3a}$$

at the interface $z = h$:

$$-(p - p_s) + 2\mu \mathbf{D} \cdot \mathbf{n} \cdot \mathbf{n} + 2\kappa\sigma = 0, \tag{2.3b}$$

$$2\mu \mathbf{D} \cdot \mathbf{n} \cdot \mathbf{t} = \nabla\sigma \cdot \mathbf{t}, \tag{2.3c}$$

$$-k\mathbf{n} \cdot \nabla\vartheta = jL, \tag{2.3d}$$

$$\partial_t h + \mathbf{v} \cdot \nabla h^* = j, \tag{2.3e}$$

$$\tilde{K}j = \vartheta_i - \vartheta_s, \tag{2.3f}$$

where $\mathbf{D}$ is the deviatoric stress tensor, $\kappa$ is the mean interfacial curvature, $L$ is the latent heat of evaporation, $\mathbf{n}$ and $\mathbf{t}$ are the normal and the tangential to the interface unit vectors, respectively, $j$ is the evaporative mass flux, $\vartheta_i$ is the temperature at the interface, and $h^* \equiv -z + h(x, y, t)$.

The parameter $\tilde{K}$ is given by Plesset and Prosperetti (1976) and Burelbach et al. (1988) as

$$\tilde{K} = \frac{\vartheta_s^{3/2}}{\alpha \rho_v L} \left( \frac{2\pi R_g}{M_w} \right)^{1/2},$$

where $\alpha$ is the accommodation coefficient, $\rho_v$ is the vapor density, $R_g$ is the universal gas constant, and $M_w$ is the molecular weight of the vapor. The effect of vapor recoil is neglected in Eqs. (2.3b) and (2.3d).

The methods of the long-wave theory of thin liquid films are applied here, as outlined by Oron et al. (1997), with a small expansion parameter $\epsilon$:

$$\epsilon \equiv \frac{d}{l} \ll 1,$$

where $l$ is a characteristic wavelength of interfacial disturbances. Equations (2.2) and (2.3) are normalized by the following dimensionless (denoted by the corresponding capital letters) variables:

$$(X, Y, Z, H) = \frac{1}{d}(\epsilon x, \epsilon y, z, h), \qquad T = \frac{\epsilon \mu t}{\rho d^2}, \qquad \Theta = \frac{\vartheta - \vartheta_s}{\vartheta_r}, \qquad J = \frac{dL_j}{k\vartheta_r}, \qquad (2.4a)$$

$$(U, V, W) = \frac{\rho d}{\mu}(u, v, \epsilon^{-1}w), \qquad P = \frac{(p - p_s)\epsilon \rho d^2}{\mu^2}, \qquad (2.4b)$$

where $\vartheta_r$ is the reference temperature chosen as $\vartheta_r \equiv \vartheta_w - \vartheta_s$ if $\vartheta_w > \vartheta_s$, and $\vartheta_r = \vartheta_w$ if $\vartheta_w = \vartheta_s$.

On substitution of the expansions

$$\{U, V, W, P, \Theta, J\} = \sum_{i=0}^{\infty} \epsilon^i \{U_i, V_i, W_i, P_i, \Theta_i, J_i\} \qquad (2.5)$$

into the dimensionless governing equations and boundary conditions, one obtains at each order a simplified set of equations to be solved.

At leading order in $\epsilon$, the heat and mass transfer problem is

$$\partial_z^2 \Theta_0 = -Q, \qquad (2.6a)$$

$$z = 0: \Theta_0 = \Theta_w, \qquad (2.6b)$$

$$z = H: \partial_z \Theta_0 + J_0 = 0, \qquad (2.6c)$$

$$z = H: K J_0 = \Theta_0, \qquad (2.6d)$$

where $\Theta_w = 1$ if $\vartheta_w \neq \vartheta_s$, $\Theta_w = 0$ if $\vartheta_w = \vartheta_s$, and

$$Q = \frac{\dot{q}d^2}{k\vartheta_r}, \qquad K = \frac{k\tilde{K}}{dL} \qquad (2.7)$$

are the dimensionless magnitude of distributed heat sources and the thermal resistance to the heat transfer at the interface, respectively.

The solution for Eqs. (2.6) is expressed by

$$\Theta_0 = \Theta_w \frac{H - Z + K}{H + K} + \frac{Q}{2}\left( \frac{H^2 + 2KH}{H + K}Z - Z^2 \right), \qquad (2.8a)$$

and therefore the interfacial temperature is given by

$$\Theta_i \equiv \Theta_0(Z = H) = \frac{K}{H + K}\left(\Theta_w + \frac{Q}{2}H^2\right). \tag{2.8b}$$

The dimensionless evaporative mass flux $J_0$ is thus found from Eq. (2.6$d$):

$$J_0 = \frac{1}{H + K}\left(\Theta_w + \frac{Q}{2}H^2\right). \tag{2.8c}$$

A general Eq. (2.85) with Eq. (2.28) and $\beta_0 = \tau_0 = 0$, both from the work Oron et al. (1997), is then used to obtain the evolution equation in the form

$$\partial_T H + \frac{E}{H + K}\left(\Theta_w + \frac{Q}{2}H^2\right) + \nabla_1 \cdot \left[KMH^2\frac{2\Theta_w - Q(H^2 + 2KH)}{4(H + K)^2}\nabla_1 H\right]$$

$$+ \nabla_1 \cdot \left[\frac{1}{3}\bar{C}^{-1}H^3\nabla_1\nabla_1^2 H\right] = 0, \tag{2.9}$$

where $\nabla_1 \equiv (\partial_X, \partial_Y)$ and

$$M = \frac{\epsilon\gamma\vartheta_r\rho d}{\mu^2}, \qquad \bar{C}^{-1} = \frac{\epsilon^3\sigma\rho d}{\mu^2}, \qquad E = \frac{\epsilon k\vartheta_r}{\mu L} \tag{2.10}$$

are, respectively, the modified Marangoni, capillary, and evaporation numbers, all unit order, and $\gamma = -d\sigma/d\vartheta > 0$.

The second term in Eq. (2.9) represents the effect of mass loss owing to evaporation, the third one is associated with the thermocapillary effect, while the last one describes the effect of capillary forces. The negative sign of the $Q$ term within the thermocapillary term of Eq. (2.9) shows that internal heat generation has a *stabilizing* impact on the evolution of the interface by means of the thermocapillary effect associated with it. This stabilization was considered by Oron and Peles (1998). In the derivation of Eq. (2.9) the effects of gravity, long-range van der Waals forces, and vapor recoil are neglected.

Equation (2.9) reduces in the case of no heat sources, $Q = 0$, and differential heating across the film, $\Theta_w = 1$, to the evolution equation that accounts for the effects of evaporation, capillarity, and thermocapillarity, e.g., Eq. (2.92) in the work of Oron et al. (1997) for $A = D^{-1} = 0$. Oron et al. (1996) showed that in the case of a nonzero thermal resistance of the solid substrate to heat conduction, the constant $K$ in the denominators of Eqs. (2.8$b$) and (2.8$c$) represents a sum of the latter and the thermal resistance to the interfacial phase transformation.

As found and discussed by Oron and Peles (1998), in the case of a nonvolatile film, internal heat generation stabilizes the interface by means of the Marangoni effect associated with it. This follows directly from the fact that if the temperature of the solid is equal to the ambient temperature, $\vartheta_w = \vartheta_s$, i.e., $\Theta_w = 0$, the interfacial temperature $\Theta_i$ of Eq. (2.8$b$) is an increasing function of the film thickness $H$. The same is valid in the volatile case for the evaporative mass flux $J_0$ of Eq. (2.8$c$). This is understood, as at leading order the heat transfer in the thin liquid film is one dimensional across it and the energy input from the heat sources in the thicker part of the film is greater than in its thinner part. Thus the temperature at the depression is lower than at the crest of the interface, and the thermocapillary stress drives the liquid into the depression, leading to equilibration of the interface. The same leads to an increase of the mass flux at the crest in comparison with that at the depression. This type of behavior promotes stabilization of the uniform state of the interface. All this is different from the standard case discussed in the literature, when heat sources are absent. In the latter case of $Q = 0$ and $\Theta_w = 1$, both the interfacial temperature and the mass flux decrease with $H$, and instability of the spatially uniform state of the interface is thus triggered.

Char and Chiang (1994) and Wilson (1997) studied the linear stability of the purely conductive state of a nonvolatile layer with a prescribed interfacial temperature subject to the Marangoni effect. In their case the basic state is such that the temperature at the film interface decreases with the distance from the solid surface, as in the case of no volumetric heat generation. This leads to the emergence of the thermocapillary stress produced by the differential heating across the layer that stems from the $\Theta_w$ component in Eq. (2.8$b$) and destabilizes the basic state. On the other hand, using the heat and mass transfer boundary conditions of Eqs. (2.3$d$) and (2.3$f$) here or the condition of continuity of the heat flux at the interface, Eq. (2$c$) in the work of Oron and Peles (1998), introduces a different mode associated with internal heat generation represented by the $Q$ component in Eq. (2.8$b$), which is found to be stabilizing. According to this, the evolution equation (2.9) contains two components associated with the Marangoni effect, the destabilizing $\Theta_w$ component and the stabilizing $Q$ component.

## 1.3. Analysis

### 1.3.1. *Basic States*

The evolution of an initially flat interface is considered first. In this case the capillary and the thermocapillary forces are neutralized and only evaporation is active. Therefore spatially uniform solutions $H(T)$ are sought for the evolution equation (2.9) that reduces to

$$\partial_T H + \frac{E}{H + K}\left(\Theta_w + \frac{Q}{2}H^2\right) = 0. \tag{3.1}$$

Direct integration of Eq. (3.1) along with the initial condition $H(T = 0) = 1$ yields the basic solution $H(T)$ in the implicit form of

$$\frac{K}{H} - \ln H = K + \frac{1}{2}QET \tag{3.2}$$

in the case of $\Theta_w = 0$, and

$$\frac{1}{Q}\ln\frac{Q+2}{QH^2+2} + \frac{\sqrt{2}K}{\sqrt{Q}}\left[\tan^{-1}\left(\frac{\sqrt{Q}}{\sqrt{2}}\right) - \tan^{-1}\left(\frac{H\sqrt{Q}}{\sqrt{2}}\right)\right] = ET \tag{3.3}$$

in the case of $\Theta_w = 1$.

In the former case a closed-form solution is admitted:

$$H(T) = \exp\left(-\frac{1}{2}QET\right) \tag{3.4}$$

when the equilibrium evaporation $K = 0$ takes place. It then follows from here that the film does not disappear in a finite time. When the total thermal resistance of the system is nonzero, i.e., $K \neq 0$, the film thinning is even slower than for $K = 0$, and therefore the film lives infinitely.

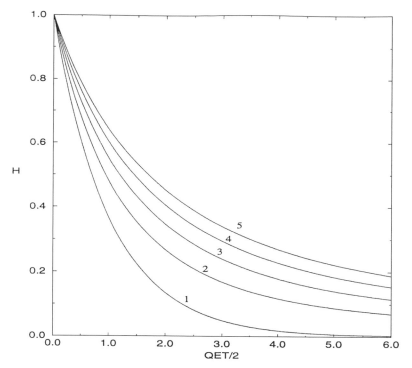

Figure 1.1. Temporal evolution of a spatially uniform interface for various values of the parameter $K$, as given by Eq. (3.2): curve 1, $K = 0$, as given by Eq. (3.4); curve 2, $K = 0.25$; curve 3, $K = 0.5$; curve 4, $K = 0.75$; curve 5, $K = 1$. The film thickness H decreases exponetially with time $T$ for $K = 0$. The film lifetime span is infinite for any $K$.

This is confirmed by calculation of solution (3.2) as a function of time $T$ and is shown in Fig. 1.1. The lifetime span of the film obviously increases with $K$.

If a nonzero temperature difference across the film is present, $\Theta_w = 1$, the film that evolves according to Eq. (3.3) disappears, i.e., becomes of a zero thickness H = 0, in a finite time $T_d$, where

$$T_d = E^{-1}\left[\frac{1}{Q}\ln\left(1 + \frac{Q}{2}\right) + \frac{\sqrt{2}K}{\sqrt{Q}}\tan^{-1}\left(\frac{\sqrt{Q}}{\sqrt{2}}\right)\right]. \qquad (3.5)$$

These solutions H($T$) are displayed in Fig. 1.2 for various $K$ and $Q$. As expected, an increase of the total thermal resistance of the system $K$ slows down the film thinning, when $Q$ is fixed. If the value of $K$ is fixed, an increase of $Q$ accelerates the film thinning. In the limiting case of a weak intensity of heat sources, $Q \ll 1$, solution (3.3) reduces to H($T$) $= -K + \sqrt{(1 + K)^2 - 2ET}$ (Burelbach et al. 1988), and the disappearance time $T_d$, as given by Eq. (3.5), reduces to the value

$$T_d^0 = \frac{1 + 2K}{2E}. \qquad (3.6)$$

In any case for fixed values of $K$ and $E$, internal heat generation accelerates the disappearance of the film, $T_d < T_d^0$, as displayed in Fig. 1.3.

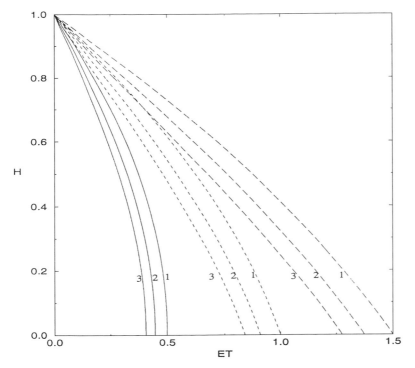

Figure 1.2. Temporal evolution of a spatially uniform interface for various values of the parameters $K$ and $Q$, as given by Eq. (3.3). The solid, dashed, and long-dashed curves correspond to $K = 0$, $K = 0.5$, and $K = 1$, respectively. The curves marked by 1, 2, and 3 correspond to $Q = 0$, 0.5, and 1, respectively.

### 1.3.2. *Stability of the Basic States*

Consider first the basic state given by Eq. (3.2) for $\Theta_w = 0$. The relevant evolution equation is obtained from Eq. (2.9) in the form

$$\partial_T H + \frac{EQH^2}{2(H+K)} + \frac{1}{3}\bar{C}^{-1}\nabla_1 \cdot \left(H^3\nabla_1\nabla_1^2 H\right) - \frac{QKM}{4}\nabla_1 \cdot \left[\frac{H^2(H^2 + 2KH)}{(H+K)^2}\nabla_1 H\right] = 0.$$

$$(3.7)$$

Introducing the disturbance of the base state $\mathsf{H}(T)$, Eq. (3.2), in the form

$$H(X, Y, T) = \mathsf{H}(\tau) + \alpha \exp[\omega(\tau) + i\mathbf{a} \cdot \mathbf{x}], \qquad \tau = ET, \tag{3.8}$$

into Eq. (3.7) yields, on linearization, the growth rate $\omega(\tau)$ of the perturbation:

$$\omega(\tau) = -E^{-1} \int_0^\tau \left[ \frac{Q}{2}\left(E + \frac{KMQ}{4}a^2\mathsf{H}^2\right)\frac{\mathsf{H}^2 + 2\mathsf{H}K}{(\mathsf{H} + K)^2} + \frac{\bar{C}^{-1}}{3}a^4\mathsf{H}^3 \right] d\tau. \tag{3.9}$$

Here $\alpha \ll 1$, $\mathbf{a} = a_x\mathbf{i}_x + a_y\mathbf{i}_y$ is the wave vector, $\mathbf{x} = X\mathbf{i}_x + Y\mathbf{i}_y$ and $a = |\mathbf{a}|$. It follows from Eq. (3.9) that the growth rate $\omega(\tau)$ is always negative; thus the base state $\mathsf{H}(T)$,

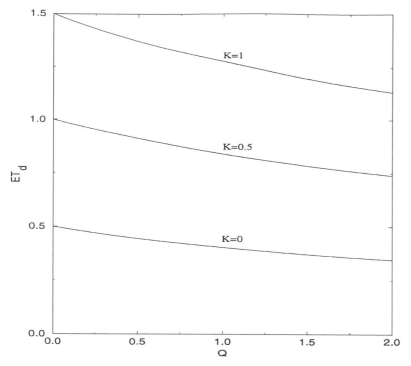

Figure 1.3. The film disappearance time $T_d$ of a spatially uniform interface given by
Eq. (3.5) as a function of $Q$ for various values of the parameter $K$.

Eq. (3.2), is linearly stable. This result is not surprising because, as mentioned above, in the case of $\Theta_\omega = 0$ the evaporative mass flux increases with H; therefore the mass loss is greater at the peaks of the interface and lower at its troughs. This indicates the tendency of the film to reduce the amplitude of the interfacial disturbance, even without capillary forces present at the interface.

Stability of the base state $H(T)$, as given by Eq. (3.3) for $\Theta_\omega = 1$, is studied next. Introducing disturbance (3.8) into Eq. (2.9) results, on linearization, in

$$\omega(\tau) = E^{-1} \int_0^\tau \left[ \frac{(E + \frac{KM}{2}a^2 H^2)}{(H + K)^2} \mathcal{Q}(H) - \frac{\bar{C}^{-1}}{3} a^4 H^3 \right] d\tau, \qquad (3.10)$$

where $\mathcal{Q}(H) = 1 - Q(H^2 + 2HK)/2$.

Two typical cases of the linear stability analysis of basic state (3.3) are presented in Fig. 1.4. First, in the case of $K = 0$, $Q = 1$, $\mathcal{E} \equiv 3E\bar{C} = 1$, and $\mathcal{M} \equiv KM\bar{C} = 0$, shown in Fig. 1.4(a), there is a finite band of unstable modes $a$ corresponding to positive values of $\Omega \equiv 3\bar{C}E\omega(\tau)$ for $\tau > 0$. This is due to the positive sign of $\mathcal{Q}(H)$ for all positive $\tau$ in the case at hand. The width of the instability range expands with time $\tau$ owing to the fast decrease of the strength of the stabilizing capileary term, $\sim H^3$. The growth rate of the disturbance increases with time $\tau$ as well.

In the case of $K = 1$, $Q = 1$, $\mathcal{E} = 1$, and $\mathcal{M} = 2/3$, shown in Fig. 1.4(b), the term $\mathcal{Q}(H)$ is initially negative, which suggests stability of the corresponding basic state equation (3.3) for short times $\tau$. However, as time $\tau$ goes by, the thickness of the base state H decreases

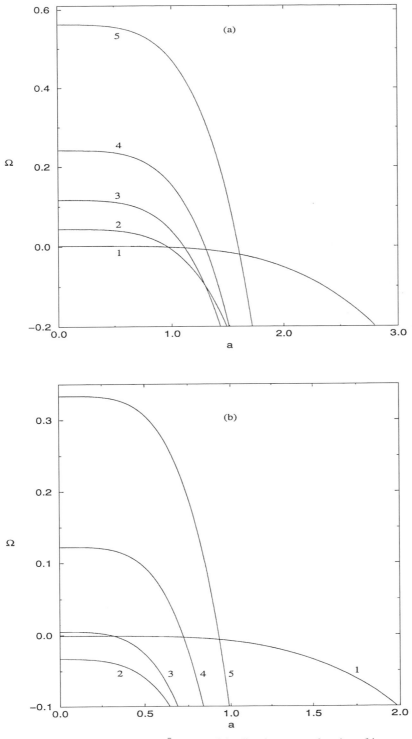

Figure 1.4. The growth rate $\Omega \equiv 3\bar{C}E\omega(\tau)$ of the disturbance as a function of its wave number $a$ for various times $\tau$: (a) $K = 0$, $Q = 1$, $\mathcal{E} = 1$, $\mathcal{M} = 0$; curve 1, $\tau = 0.005$; curve 2, $\tau = 0.1$; curve 3, $\tau = 0.2$; curve 4, $\tau = 0.3$; curve 5, $\tau = 0.4$. The film disappearance time $\tau_d \equiv ET_d \approx 0.4055$. (b) $K = 1$, $Q = 1$, $\mathcal{E} = 1$, $\mathcal{M} = 2/3$; curve 1, $\tau = 0.005$; curve 2, $\tau = 0.36$; curve 3, $\tau = 0.7$; curve 4, $\tau = 1.0$; curve 5, $\tau = \tau_d \approx 1.2759$.

11

because of evaporation, and $Q(H)$ becomes positive when $\tau = \tilde{\tau} \approx 0.36$. The critical thickness $\tilde{H} = \sqrt{K^2 + (2/Q)} - K$ corresponding to $\tau = \tilde{\tau}$ represents the most stable state of the film [see curve 2 in Fig. 1.4.(b)], so that the growth rate of the disturbance is negative and minimal. Some time later, when $\tau$ is slightly below 0.7 [see curve 3 in Fig. 1.4(b)], the basic state becomes unstable. This behavior can be understood when the following arguments are made. In the case at hand, the mass loss due to evaporation is nonmonotonic, i.e., decreases with H when $0 < H < \tilde{H}$ and increases with H when $H > \tilde{H}$. The critical thickness $\tilde{H}$ is smaller than 1 (the initial thickness of the film) if

$$Q > Q_c \equiv \frac{2}{1 + 2K}. \tag{3.11}$$

Therefore, if Eq. (3.11) is satisfied, the amplitude of the interfacial disturbance first decreases and then increases because of the corresponding variation of the evaporative mass flux. In the former stage the basic state is stable, while in the latter one it is unstable. As in the previous case, the unstable range $0 < a < a_0$ expands with time $\tau$. It is interesting to note that the critical value $Q_c$ coincides with that given by Oron and Peles (1998) for unconditional stability of a nonevaporating film subject to the Marangoni effect only, when $K$ represents the inverse Biot number. On the other hand, if $Q < Q_c$, then $\tilde{H} \geq 1$, and the evaporative mass flux monotonically decreases with H during the evolution of the interface. This leads to instability of the basic state equation (3.3) in the range $Q \leq Q_c$.

## 1.4. Nonlinear Behavior of the Film

The two-dimensional version of Eq. (2.9) in terms of $H = H(X, T)$ that is given by

$$\partial_T H + \frac{E}{H + K}\left(\Theta_\omega + \frac{Q}{2}H^2\right) + \frac{1}{3}\bar{C}^{-1}\partial_X\left(H^3\partial_X^3 H\right)$$
$$+ \partial_X\left[KMH^2\frac{2\Theta_w - Q(H^2 + 2KH)}{4(H + K)^2}\partial_X H\right] = 0 \tag{4.1}$$

is numerically solved with periodic boundary conditions in the interval $0 \leq X \leq \Delta$ and the initial condition $H(X, T = 0) = 1 + 0.05 \cos(2\pi X/\Delta)$. A few typical examples directly connected to what was discussed in Section 1.3 are now considered. Figure 1.5 displays the spatiotemporal nonlinear behaviors of the film interface in the two cases presented in Fig. 1.4, as governed by Eq. (4.1). In both cases the film ruptures ($H = 0$) before the disappearance time $\tau_d$ predicted by the analysis of the corresponding basic states. In the case of $K = 0$, corresponding to the case shown in Fig. 1.4(a) and presented now in Fig. 1.5 by the solid curves, the formation of a sharp (but not cuspy) trough occurs at rupture. This is due to a large local mass loss owing to evaporation, when the local film thickness is small and $K = 0$. This behavior is in contrast with the other case, $K \neq 0$, corresponding to the case shown in Fig. 1.4(b) and presented now in Fig. 1.5 by the dashed curves, when the wide depression is formed before rupture by simultaneous action of evaporation and the Marangoni effect.

Figure 1.6 presents the time evolution of the amplitude $A$ of the interfacial disturbance calculated as the difference between the maximal and the minimal local thicknesses of the film in the two cases shown in Fig. 1.5. As predicted by the analysis of the basic states in Section 1.3 and shown in Fig. 1.4, the amplitude $A$ grows monotonically, driven by the instability of the basic state in the case given in Fig. 1.4(a) [see curve (a) in Fig. 1.6] and decays first and grows thereafter

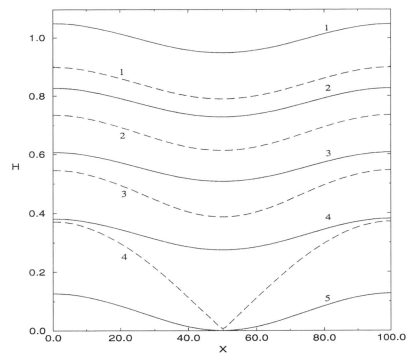

Figure 1.5. Temporal evolution of the film interface, as described by Eq. (4.1), in the case shown in Fig. 1.4(a), solid curves: curve 1, $T = 0$; curve 2, $T = 0.3$; curve 3, $T = 0.6$; curve 4, $T = 0.9$; curve 5, $T = 1.2085$; and in Fig. 1.4(b), dashed curves: curve 1, $T = 0.1$; curve 2, $T = 0.2$; curve 3, $T = 0.3$; curve 4, $T = 0.3724$. The initial condition in both cases is the same, as given by solid curve 1. $\bar{C}^{-1} = 3$ and $\Delta = 100$.

in the case given in Fig. 1.4(b) [see curve (b) in Fig. 1.6]. In the latter case the minimal value of $A$ occurs at $\tau \approx \tilde{\tau}$, as predicted by the linear theory [see curve 2 in Fig. 1.4(b)]. Also, the value of $A$ returns to its initial magnitude at $\tau$ slightly below 0.7, which compares well with the linear case [see curve 3 in Fig. 1.4(b)].

Finally, Eq. (4.1) is solved in the case of $\Theta_\omega = 0$. The evolution of the interface in this case exhibits a decrease of the amplitude of the interfacial disturbance, and the local thickness of the film does not vanish for an indefinitely long time. This is in agreement with the results obtained in Subsection 1.3.1.

## 1.5. Summary

The nonlinear behavior of a thin volatile liquid film with volumetric heat sources has been considered subject to the action of the capillary and thermocapillary forces. Space-uniform basic states are found in the presence and the absence of a temperature gradient across the film. It is shown that in the absence of such a temperature gradient the corresponding basic state is linearly stable and has an infinite lifetime span. However, when a transverse temperature gradient is applied, the corresponding basic state disappears in a finite time, and the presence of heat sources accelerates the dryout of the film. It is shown that the basic state in this case may be either linearly unstable for all times or linearly stable for short times and linearly unstable thereafter. Linear instability of the basic state is always associated with a finite bandwidth

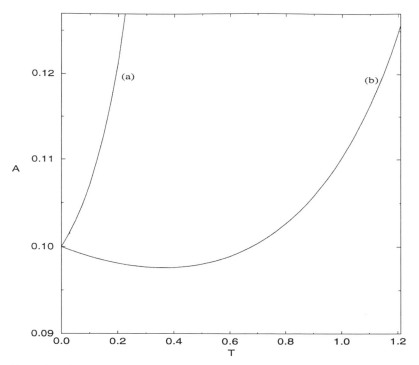

Figure 1.6. Temporal evolution of the amplitude $A$ of the interfacial disturbance, as found from Eq. (4.1) in the case presented in Fig. 1.5. Curves (a) and (b) correspond to the cases shown in Figs. 1.4(a) and 1.4(b), respectively.

of unstable disturbances whose wave numbers range from zero to the value growing with time. The numerical study of the two-dimensional version of the governing evolution equation qualitatively confirms the results of the linear stability analysis and suggests that the rupture time of the film is shorter than the disappearance time of the corresponding basic state.

## Acknowledgments

The research was partially supported by the Binational Israel–U.S. Science Foundation contract 96-00395 and the Fund for the Promotion of Research at the Technion.

## References

Burelbach, J. P., Bankoff, S. G., and Davis, S. H. 1988. Nonlinear stability of evaporating/condensing liquid films. *J. Fluid Mech.* **195**, 463–494.

Char, M.-I. and Chiang, K.-T. 1994. Stability analysis of Benard–Marangoni convection in fluids with internal heat generation. *J. Phys. D: Appl. Phys.* **27**, 748–755.

Oron, A., Bankoff, S. G., and Davis, S. H. 1996. Thermal singularities in film rupture. *Phys. Fluids A* **8**, 3433–3435.

Oron, A., Davis, S. H., and Bankoff, S. G. 1997. Long-scale evolution of thin liquid films. *Rev. Mod. Phys.* **69**, 931–980.

Oron, A. and Peles, Y. 1998. Stabilization of thin liquid films by internal heat generation. *Phys. Fluids A* **10**, 537–539.

Plesset, M. S. and Prosperetti, A. 1976. Flow of vapor in a liquid enclosure. *J. Fluid Mech.* **78**, 433–444.

Wilson, S. K. 1997. The effect of uniform internal heat generation on the onset of steady Marangoni convection in a horizontal layer of fluid. *Acta Mech.* **124**, 63–78.

# 2

# The Effect of Air Height on the Pattern Formation in Liquid–Air Bilayer Convection

D. JOHNSON, R. NARAYANAN, AND P. C. DAUBY

## 2.1. Introduction

This chapter presents a study of the effect of air height on the pattern formation in liquid–air bilayer convection. Bilayer convection results from fluid interaction in two immiscible, superposed fluid layers, and it is a transport problem that is affected by the ratios of various thermophysical properties and geometries. It is also rich in the physics of nonlinear dynamics. With both calculations and experiments, an interesting mechanism for bilayer convection, resulting from air interaction, is revealed in this study. It is shown that large air heights cause a predominance of convection in the gas layer, simultaneously generating weak surface-driven motion in the lower liquid layer through thermal coupling of a special kind. This coupling is contrasted with traditional studies in which the air is either treated as passive or is driven by being coupled mechanically with the lower liquid layer in which the convection is dominant.

When a temperature difference is applied across a pure liquid layer with a free surface, convection can occur by two mechanisms: Marangoni (interfacial tension) and Rayleigh (buoyancy) convection. Interfacial-tension-gradient-driven convection is caused by the variation of interfacial tension with temperature and may occur whether the liquid is being heated from above or from below. Buoyancy-driven convection occurs only when the fluid, with a negative thermal expansion coefficient, is heated from below. Kinematic viscosity and thermal diffusivity act in a manner to dissipate the convective flow. The imposed temperature gradient is presumed to act perpendicularly to an initially quiescent flat surface in a fluid that is conducting heat in the base state.

Many researchers have studied the problem of Rayleigh–Marangoni convection in a single liquid layer with an overlying passive gas. A compilation of recent results is given in Koschmieder's book (1993). Some work has been done on convection in bounded containers (Rosenblat et al., 1982; Dijkstra, 1992, 1995A, 1995B, 1995C; Dauby and Lebon, 1996; Zaman and Narayanan, 1996). In addition, some experimental work on the effect of narrow containers has been done (Koschmieder and Prahl, 1990; Ondarçuhu et al., 1993; Johnson and Narayanan, 1996), but none has been done on the effect of layer depths on the onset conditions and associated patterns.

In the experiments performed by Koschmieder and Prahl (1990), a layer of silicone oil, with a very thin overlying layer of air, was placed into a cylindrical container with nearly insulating walls. The oil–air bilayer was then heated from below. In each experiment, the liquid's aspect ratio was increased when the liquid layer depth was decreased. In their study, Koschmieder and Prahl were able to show that the number of convective flow cells increased monotonically as the aspect ratio increased.

Experiments performed by Johnson and Narayanan (1996) for cylindrical containers were able to illustrate an interesting phenomenon at aspect ratios at which the onset of convection occurs with two different flow patterns becoming simultaneously unstable. At these aspect ratios, called codimension two points, the fluid can oscillate between two or more flow patterns. Calculations performed by Zaman and Narayanan (1996) and also by Dauby and co-workers (1996, 1997) generally agree qualitatively with Koschmieder's experiments. In each of these works, the air layer was considered to be at rest and only conducting heat. In a bilayer system, the interactions of convection are far more complex (Johnson and Narayanan, 1997) and depend on the phase that initiates convection. Throughout this chapter phrases such as "convection initiates or begins in one phase" are used. Clearly in a mathematical sense, there is only a single condition for the onset of convection, and this onset must occur simultaneously in both layers. The notion of convection initiating in one layer or another is ultimately a physical one and is perhaps best explained qualitatively as one in which convection is more vigorous in the initiating layer. Moreover, in the layer in which convection is said to initiate, the magnitude of the velocities is significantly higher than in the adjacent layer, which is the respondent layer.

One of the important parameters that determines the layer that is the initiator is the depth of each fluid. For the sake of simplicity, assume temporarily that each fluid has comparable thermophysical properties and that the Marangoni effect is absent. When the depth of the lower fluid is much greater than the depth of the upper fluid, buoyancy-driven convection will occur primarily in the lower layer while the upper fluid will respond by being dragged by the lower fluid. If the depths are approximately equal, then buoyancy effects are also approximately equal and both layers become simultaneously unstable. In fact, when both fluid layers become simultaneously unstable, the convection can couple in two different ways (Fig. 2.1): viscously or thermally. In viscous coupling the convection rolls rotate in opposite directions at the interface. In thermal coupling, the convection rolls rotate in the same direction. Thermal coupling is caused by rising hot plumes from the lower layer, creating hot spots at the interface, and the no-slip condition at the interface is satisfied by vanishing tangential components of velocity. When the thermophysical properties of the two layers are unequal, the type of coupling depends not only on the relative thickness of the layers but also on these properties. Further details are available in the paper by Johnson and Narayanan (1997) and in the references therein.

Viscous Coupling                          Thermal Coupling

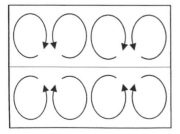
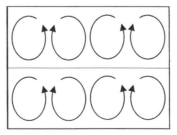

Figure 2.1. Schematic of thermal and viscous coupling. Counterrotating rolls, in which up-flow in the bottom layer meets down-flow in the upper layer, characterize viscous coupling. Corotating rolls, in which up-flow in the lower layer meets up-flow in the upper layer, characterize thermal coupling.

Consider now the special case in which the upper fluid is a gas and assume that the Marangoni effect is included. Large upper gas depths will promote buoyancy-driven convection in the gas. Because the viscosity of the gas is small, the convecting gas is incapable of shearing the more viscous liquid below it. Therefore, when the onset of convection is seen in the lower liquid layer, while it is known to be dominant in the upper gas layer, then some mechanism other than mechanical coupling must be involved in order to drive convection in the liquid. It is our view that this mechanism is a new form of thermal coupling unlike the one explained above. Instead it is proposed that buoyancy-driven convection is initiated in the upper layer and the convecting gas takes on a particular flow pattern, which depends on its aspect ratio. This particular pattern simultaneously imposes a temperature profile at the gas–liquid interface, immediately generating surface-driven flow, even if it is only weak, by both surface-tension and horizontal density gradients in the lower liquid near the interface. Further understanding of the coupling mechanism in the liquid–gas bilayer problem requires both calculations and careful experiments. It is to these tasks that we now turn.

## 2.2. Models and Experiments

The experimental system that was investigated was a bilayer of air superposing a layer of viscous silicone oil. The convection was experimentally observed by means of an infrared camera that measures the temperature profile at the gas–liquid interface. Calculations were made assuming the thermophysical properties of the fluids that were used in the experiment. The results of these calculations are in good qualitative agreement with the experiments. The calculations involved the linearized Boussinesq equations, which were solved with the Chebyshev spectral tau method and provided the value of the critical temperature difference needed to drive convection, as well as the associated patterns.

Two different linear calculations were performed. The first set of calculations assumed laterally unbounded layers of two fluids. The second set of calculations assumed a single, bounded liquid layer with a passive overlying gas in a rigid, circular cylinder with no-slip lower and radial walls. By a passive gas, we mean that the gas is completely quiescent and the temperature is not perturbed by the instability mechanism. The first set of calculations was done to analyze the different types of convection coupling (for example, thermal coupling) as the silicone oil and the air height were changed. It may be noted that calculations for which laterally unbounded geometries are assumed can be easily extended to reveal qualitative features of convection in laterally bounded circular cylinders with unrealistic boundary conditions that correspond to vanishing tangential and vertical components of vorticity along the radial walls. Johnson and Narayanan (1997) have exploited this feature of the unbounded geometry calculations more fully. Since calculations in which bounded geometries with realistic sidewalls are assumed are complicated, they were simplified by the assumption that a passive gas layer superposed the liquid. Even with this assumption, the calculations were able to explain qualitatively many observations of the small aspect ratio experiments when the air height was small compared with the liquid height.

**Model Equations for the Unbounded Bilayer.** The equations model convection in two, superposed immiscible fluid layers. Ferm and Wollkind (1982) give complete derivations of the model for convection in two unbounded fluid layers. It is assumed that the two fluid layers are bounded vertically by two rigid, semi-infinite plates. The plates are held at constant temperatures and located at $z = -d_1$ and $z = d_2$. The interface separating the two fluids is assumed to have negligible mass and interfacial viscosity. The interface is located at $z = \eta(x, t)$ and has a mean position of $z = 0$. The density of each fluid and the interfacial tension between the fluids are assumed to vary linearly with temperature:

$$
\begin{aligned}
\sigma &= \bar{\sigma} + \sigma_T(T_1 - T_{\text{int}}), \\
\rho_1 &= \bar{\rho}_1[1 - \alpha_1(T_1 - T_b)], \\
\rho_2 &= \bar{\rho}_2[1 - \alpha_2(T_2 - T_{\text{int}})].
\end{aligned}
\tag{2.1}
$$

Here $\sigma$ is the surface tension, $\rho$ is the density, $T_b$ is the temperature at the bottom plate, $T_{\text{int}}$ is the temperature at the interface, $\sigma_T = \partial\sigma/\partial T$, and $\alpha_i = -\frac{1}{\rho_i}\frac{\partial\rho_i}{\partial T}$ are evaluated at the reference states. The subscript 1 denotes the lower fluid, and the subscript 2 denotes the upper fluid. The overbar represents the reference state. The length, velocity, time, and pressure are scaled with $d_1, \kappa_1/d_1, d_1^2/\kappa_1$, and $\mu\kappa_1/d_1^2$, respectively. The temperature is scaled with respect to the temperature difference across the lower liquid layer in the linear conductive state. In this chapter the following symbols are used for the thermophysical properties: $\alpha_i$ are the thermal expansivities, $\kappa_i$ are the thermal diffusivities, $\mu_i$ are the viscosities, $\rho_i$ are the densities, $k_i$ are the thermal conductivities, and $\nu_i = \mu_i/\bar{\rho}_i$ are the kinematic viscosities. The following symbols are used for the ratio of thermophysical properties: $\alpha = \alpha_2/\alpha_1$, $l = d_2/d_1$, $k = k_2/k_1$, $\kappa = \kappa_2/\kappa_1$, $\rho = \bar{\rho}_2/\bar{\rho}_1$, and $\mu = \mu_2/\mu_1$. In what follows, $x$ indicates the horizontal spatial variable. The extension of the analysis to two horizontal dimensions is trivial and not performed here.

A linear stability analysis is subsequently performed on the model that uses the familiar Boussinesq approximation. Only stationary convection was investigated and it was seen in a separate series of calculations, which are not discussed here, that the onset of convection was not oscillatory for all cases of experimental interest. The domain equations in the lower phase are

$$
\begin{aligned}
DW_1 + i\omega U_1 &= 0, \\
(D^2 - \omega^2)U_1 - i\omega\Pi_1 &= 0, \\
(D^2 - \omega^2)W_1 - D\Pi_1 + \text{Ra}\Theta_1 &= 0, \\
(D^2 - \omega^2)\Theta_1 + W_1 &= 0.
\end{aligned}
\tag{2.2}
$$

Here $\omega$ is the wave number of the planform at the onset of convection. The first of these equations is obtained from continuity, while the next two equations are a result of linearizing the momentum equation, and the last equation is obtained from the energy equation. Likewise, the domain equations in the upper phase are

$$
\begin{aligned}
DW_2 + i\omega U_2 &= 0, \\
\frac{\mu}{\rho}(D^2 - \omega^2)U_2 - \frac{i\omega}{\rho}\Pi_2 &= 0, \\
\frac{\mu}{\rho}(D^2 - \omega^2)W_2 - \frac{1}{\rho}D\Pi_2 + \alpha\,\text{Ra}\Theta_2 &= 0, \\
\kappa(D^2 - \omega^2)\Theta_2 + \frac{1}{k}W_2 &= 0.
\end{aligned}
\tag{2.3}
$$

The thirteen boundary conditions are

$$W_2 = W_1 = 0 \quad \text{at } z = 0,$$

$$DW_2 = DW_1 \quad \text{at } z = 0,$$

$$\Pi_2 - \Pi_1 + \left(\frac{G + \omega^2}{C}\right)\eta_1 + 2(DW_1 - \mu DW_2) = 0 \quad \text{at } z = 0,$$

$$(D^2 + \omega^2)W_1 - \mu(D^2 + \omega^2)W_2 = \omega^2 \, \text{Ma}(\eta_1 - \Theta_1) \quad \text{at } z = 0, \tag{2.4}$$

$$kD\Theta_2 = D\Theta_1 \quad \text{at } z = 0,$$

$$\Theta_1 = \Theta_2 + \eta_1\left(1 - \frac{1}{k}\right) \quad \text{at } z = 0,$$

$$DW_1 = W_1 = \Theta_1 = 0 \quad \text{at } z = -1,$$

$$DW_2 = W_2 = \Theta_2 = 0 \quad \text{at } z = l,$$

where $D \equiv (d/dz)$.

The dependent variables in each phase are $W_i$ for the vertical component of velocity, $\Theta_i$ for the temperature, $\Pi_i$ for the pressure, $U_i$ for the tangential component of velocity, and $\eta_1$ for the surface-deflection term. In further calculations, substituting the equation of continuity eliminates the horizontal components of velocity in both fluid layers. Ma and Ra given in Eqs. (2.2)–(2.4) are the Marangoni and the Rayleigh numbers, respectively, where $\text{Ra} = \alpha_1 g \Delta T_1 d_1^3 / \kappa_1 \nu_1$ and $\text{Ma} = \sigma_T \Delta T_1 d_1 / \kappa_1 \mu_1$. $G$ is the Weber number, where $G = \{[(\bar{\rho}_1 - \bar{\rho}_2)gd_1^2]/(\sigma_0)\}$ and $C$ is the Crispation number, where $C = [(\mu_1\kappa_1)/(\sigma_0 d_1)]$. The Rayleigh number of the upper gas phase will be important and will play a role in the explanation of the physics. It is defined in a manner similar to the Rayleigh number of the lower fluid instead by use of all of the thermophysical properties of the upper gas as well as the temperature difference of the gas phase.

It is instructive to see how the above model relates to some of the earlier models in which the gas phase was assumed to be at rest. The modeling of the perturbation of the heat transfer in the gas can take three different forms: the gas can be considered as mechanically and thermally passive, as only mechanically passive, or as both mechanically and thermally active. Saying that the gas is mechanically and thermally passive means that this gas always remains at rest and that the temperature gradient within this phase remains equal to its value in the purely conductive state. In this situation, we set the perturbed velocity and temperature of the upper fluid to zero and replace the temperature boundary conditions at the interface with Newton's law of cooling:

$$D\Theta_1 + \text{Bi}\Theta_1 = 0, \tag{2.5}$$

where Bi is the Biot number given by

$$\text{Bi} = \frac{k_{\text{air}} \, d_{\text{oil}}}{k_{\text{oil}} \, d_{\text{air}}}. \tag{2.6}$$

In this expression, $k_{\text{air}}$ is the thermal conductivity of the air, $k_{\text{oil}}$ is the thermal conductivity of the silicone oil, and $d_{\text{air}}$ and $d_{\text{oil}}$ are the depths of the air and the silicone oil, respectively.

Considering the gas to be mechanically passive but thermally active is tantamount to assuming that only the temperature perturbations with respect to the purely conductive reference state are possible in the gas. In this case, a Biot number also appears as in Eq. (2.5) but its expression depends on the wave number (Normand et al., 1977):

$$\text{Bi}(\omega) = \frac{k_{\text{air}}}{k_{\text{oil}}} \omega \coth\left(\omega \frac{d_{\text{air}}}{d_{\text{oil}}}\right). \tag{2.7}$$

In the third case of a mechanically and thermally active gas, velocity and temperature perturbations are taken into account in both the liquid and the gas and a full bilayer model describes the system.

Note that the constant Biot number given in Eq. (2.6), which was used by several earlier workers (Nield, 1964; Koschmieder, 1993), can be obtained from Eq. (2.7) by taking the limit as the wave number goes to zero (the long-wavelength assumption). As the upper gas height becomes smaller, the Biot number from Eq. (2.6) as well as from Eq. (2.7) approaches infinity and consequently the Marangoni effect is neglected in this limit. However, as the upper gas depth approaches infinity, the Biot number given by Eq. (2.6) reaches zero while the Biot number given by Eq. (2.7) approaches a value that is proportional to the wave number. If the wave number is taken to be 2 (a typical value for a number of situations in which the gas is assumed to be at rest), this asymptotic value of the Biot number is ∼0.3 for the silicone oil–air system. Calculations for which the buoyancy effect is active in both phases are presented and compared with the simplified cases in which the Biot number is given by Eq. (2.6) or by Eq. (2.7).

A Chebyshev spectral tau method (Canuto et al., 1988; Johnson, 1996) was used to solve the eigenproblem given by Eqs. (2.2)–(2.4). This method easily incorporates the complicated boundary conditions and provides the accuracy needed by using only a few numbers of terms.

Note that the Marangoni number Ma and the Rayleigh number Ra are not independent of each other for a given experiment. The ratio Ma/Ra = Γ is a constant, which depends on the thermophysical properties of the fluid and the height of the lower layer. Johnson and Narayanan (1997) give the properties of silicone oil and air that were used in the calculations in the paper. The equation Ma = ΓRa is substituted into Eqs. (2.4) and makes the Rayleigh number the eigenvalue parameter. The purpose of the linear calculations is to find the vertical components of velocity, $W_1$ and $W_2$, and the temperature perturbations, $\Theta_1$ and $\Theta_2$, and plot them against the height of the liquid layers. The results are then used to explain the type of coupling, depending on the orientation of the maximums of the velocities and temperature perturbations.

**Model for the No-Slip, Bounded Calculations.** Details for the derivations and calculations of the no-slip, bounded model are given in the papers by Dauby and Lebon (1996) for rectangular boxes and by Dauby et al. (1997) for cylindrical containers. In both papers, a conductive, passive gas superposed the liquid layer and the surface of the liquid was assumed to be flat. All other sides of the cylindrical container had no-slip boundaries. A fixed value of the Biot number (0.30) was used to model the effective heat transfer from the free liquid surface. The lower vertical boundary was a rigid, no-slip, conducting plate. The lateral sidewalls were assumed to be either perfect conductors or perfect insulators. The calculations were plotted and gave the critical temperature difference (or Rayleigh number of Marangoni number) for different azimuthal modes over a range of the cylinder's aspect ratio.

**Experimental Apparatus and Procedure.** The objectives of the experiments were to study the effect of the air height on the flow pattern of the lower liquid. The experimental apparatus

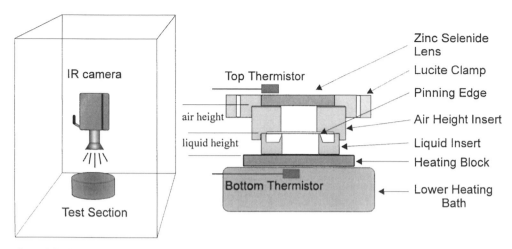

Figure 2.2. Schematic of the experiment and cross-sectional view of the test section.

was designed so that various air heights, as well as various liquid aspect ratios, could be used. A schematic of the test section is given in Fig. 2.2.

The lower heating element consisted of a heating plate under a hollowed cylinder made of Lucite, 3/4 in. thick, called the lower bath. The top and the bottom of the cylinder were capped with thin copper plates, 1/4 in. thick, and the interior of the cylinder was filled with water. A magnetic stir bar was placed in the water, and the entire lower bath sat on top of a magnetic stirrer. The stirred water helped to stabilize any temperature fluctuations from the heating block, ensuring a uniform temperature within $\pm 0.05\,°C$.

The test section itself consisted of four pieces: the liquid insert and the air insert, the clamp, and a zinc selenide window. The thermal conductivities (in units of $erg^{-1}\ cm^{-1}\ s^{-1}\ K^{-1}$) of copper and zinc selenide were $4.1 \times 10^6$ and $1.8 \times 10^6$ respectively, while Lucite had a thermal conductivity of $1.7 \times 10^4$. The liquid insert, the air insert, and the clamp were made of Lucite. To ensure that the liquid–gas interface was flat, a pinning edge was used in the liquid insert. The liquid insert could be made of different radii and heights to achieve the desired liquid aspect ratio. The inner radius of the air insert was always the same as the liquid insert's inner radius. However, different air heights could be used for the same liquid insert. The Lucite clamp fastened the liquid and air inserts down, preventing silicone oil from leaking. The outer radius of the liquid inserts was 1 13/16 in. The outer radius of the air inserts was 2 1/4 in. The zinc selenide window had a diameter of 2 in. and was 5 mm thick. The accuracy of each piece was machined to within 0.1 mm.

An Inframetrics, model 760, infrared camera was used to visualize the flow patterns. This particular model of infrared camera is capable of measuring in the 3–5-$\mu$m range or the 8–12-$\mu$m range. However, only the 8–12-$\mu$m range was used. The infrared camera was placed directly above the test section and measured the infrared radiation being emitted by the silicone oil. As silicone oil readily absorbs infrared radiation, only the radiation from the silicone oil interface could be detected. An effective emissivity could be calibrated and programmed into the infrared camera to find the temperature of the interface. Zinc selenide is 60% transparent to infrared radiation in the 8–12-$\mu$m range. Additionally, an antireflective infrared polymer was coated on the zinc selenide window by II-VI Inc. This coating was necessary to eliminate any false images generated by reflected, ambient infrared radiation.

To control the temperature at the top of the test section, an infrared transparent medium was needed. The medium of choice was simply air. A Lucite box enclosed the test section, the lower bath, the magnetic stirrer, the infrared camera, and the heating control elements. A heater, a fan, and a radiator were used to control the temperature of the ambient air in the Lucite box. To increase the ambient temperature, the heater would turn on. Cool water, pumped through the radiator, continuously removed heat from the air. The blowing fan ensured proper mixing. The overall temperature control across the top of the zinc selenide and the bottom of the top copper plate of the lower bath never deviated by more than 2.4% from the setpoint. When the smoothing of the temperature fluctuations, due to the highly conductive plates (copper and zinc selenide), was taken into consideration, the actual control across the two fluid layers was probably better. The temperature of the top and the bottom plates and the temperature of the cooling water were measured. All of the temperatures were recorded and controlled with a computerized data-acquisition system. When the control algorithm was rewritten, the temperature control was later improved to $\pm 0.05\,°C$, with no noticeable effect on the experimental results.

Each experiment was conducted in the same systematic manner. First the silicone oil was placed into the liquid insert. The corresponding air insert was chosen and placed on top of the liquid insert. The Lucite clamp was placed on top of the air insert and screwed into the lower bath. Both a standard bubble level and the infrared camera could check the level of the oil–air interface. As it turns out, the oil–air interface acts as a lens to infrared radiation, concentrating infrared radiation if the surface is depressed in the center or diffusing infrared radiation if the interface is elevated in the center. If there is an insufficient amount of silicone oil in the insert, the liquid–air interface will be depressed and the temperature will appear to be higher in the center, even though the temperature is uniform. If there is too much oil in the insert, the liquid–air interface will be elevated in the center and the temperature will appear to be lower in the center. Checking the oil level with the infrared camera proved to be more accurate than checking with the bubble level.

Once the test section was secured, a temperature difference was applied across the liquid–gas bilayer. The initial temperature difference was such that the temperature differences in each layer were less than the critical temperature difference necessary to initiate convection in either layer. When a temperature difference was applied, it was held constant for several time constants. The longest time constant in these experiments was the horizontal thermal diffusion time. Here the thermal diffusivity of silicone oil was $\kappa = 1.1 \times 10^{-3}$ cm$^2$/s and the typical diameter was $\sim 25$ mm to give a horizontal time constant of $d^2/\kappa = 1.6$ h. The temperature difference for these experiments was held constant for 4 h. However, steady state was usually reached well within the 4-h period. After the temperature difference was held constant for 4 h, the temperature difference was increased to a new setpoint and held constant for another 4 h. This procedure was repeated until the fluid began to flow. Once the fluid began to flow, the temperature difference and the flow pattern were noted.

### 2.3. Discussion

**Observations from Calculations.** Calculations were performed to determine both the critical temperature difference and the flow pattern at the onset of convection. These computations were based on a linearized instability analysis for both the laterally unbounded as well as the bounded geometries. The calculations in which two laterally unbounded layers were assumed were done to obtain qualitative features of the physics of bilayer convection. The corresponding results are presented in Table 2.1. Two features in particular were investigated. First, the effect

Table 2.1. *Critical Rayleigh Numbers and Wave Numbers Calculated with Four Different Conditions: a Single Layer with Eq. (2.6) as the Biot Number, a Single Layer with Eq. (2.7) as the Biot Number, and the Bilayer Calculations*

| Air Height (mm) | Results with Nield's Model | Results with the Biot Number from Eq. (2.6) | Results with the Biot Number from Eq. (2.7) | Full Bilayer Calculation: Rayleigh Number for Silicone Oil | Full Bilayer Calculation: Rayleigh Number for Air |
|---|---|---|---|---|---|
| 0.1 | 513.3 ($\omega = 2.55$) | 514.2 ($\omega = 2.55$) | 514.5 ($\omega = 2.55$) | 518.3 ($\omega = 2.56$) | $1.350 \times 10^{-4}$ |
| 1 | 237.6 ($\omega = 2.18$) | 237.8 ($\omega = 2.18$) | 241.7 ($\omega = 2.16$) | 241.9 ($\omega = 2.16$) | 0.6301 |
| 3 | 205.4 ($\omega = 2.07$) | 205.5 ($\omega = 2.05$) | 216.2 ($\omega = 2.05$) | 216.5 ($\omega = 2.03$) | 45.67 |
| 5 | 198.6 ($\omega = 2.04$) | 198.6 ($\omega = 2.05$) | 213.7 ($\omega = 2.00$) | 212.5 ($\omega = 1.99$) | 345.9 |
| 7 | 195.7 ($\omega = 2.03$) | 195.7 ($\omega = 2.04$) | 213.3 ($\omega = 2.00$) | 200.2 ($\omega = 1.85$) | 1252 |
| 9 | 193.9 ($\omega = 2.02$) | 194.0 ($\omega = 2.00$) | 213.3 ($\omega = 2.00$) | 95.32 ($\omega = 1.42$) | 1629 |
| 14 | 191.8 ($\omega = 2.01$) | 191.9 ($\omega = 2.00$) | 213.2 ($\omega = 2.00$) | 16.60 ($\omega = 0.92$) | 1660 |
| 20 | 188.2 ($\omega = 2.00$) | 188.3 ($\omega = 2.00$) | 213.2 ($\omega = 1.99$) | 4.009 ($\omega = 0.65$) | 1670 |
| 100 | 188.0 ($\omega = 2.00$) | 188.1 ($\omega = 2.00$) | 213.2 ($\omega = 1.99$) | 0.0065 ($\omega = 0.13$) | 1699 |

*Note:* The Rayleigh numbers of silicone oil and air are defined with respect to their own thermophysical properties. In each calculation, a depth of 4.2 mm of 100-cS silicone oil was assumed. The wave numbers of the active air calculations are the same as those of the silicone oil.

of the height of the upper phase on the convective threshold was studied. This was done by assuming that the upper phase was either strictly passive, one that allowed thermal perturbations, or one that was both mechanically and thermally active. The second feature that was examined was the effect of the air height on the type of convective coupling.

In the first feature of the unbounded layer calculations, the three different cases of the amount of gas activity were considered by use of Eq. (2.5) in conjunction with either Eq. (2.6) or Eq. (2.7) and the bilayer model.

The calculations for which the constant Biot number was used, shown in the second column of Table 2.1, gave results that were very close to those found by Nield (1964), which are shown in the second column of Table 2.1, even though the current study assumed a deflecting interface. The reason for this is that the surface tension between silicone oil and air is quite large and the surface deflections do not substantially change the critical Rayleigh or Marangoni numbers for the liquid depths studied here. A comparison of the computed critical Rayleigh number and the critical wave number for the various cases of gas activity provides some insight into the physics of the problem. Several observations can be made from Table 2.1, which gives the Rayleigh number for various air depths. First, the critical Rayleigh number, when the Biot number is given by Eq. (2.7), is always greater than the critical Rayleigh number for the long-wavelength Biot number given by Eq. (2.6). This is understandable as the Biot number given in Eq. (2.7) is always greater than the Biot number given in Eq. (2.6) and a larger Biot number corresponds to a more conductive air layer, which more easily dampens the perturbations. The critical wave number, however, differs very little between these two cases. Observe that for small air heights, (less than 5 mm) the critical Rayleigh number for the active bilayer calculations (given in the fifth column of Table 2.1) is greater than the corresponding critical Rayleigh numbers for cases that assume either of the other two Biot number conditions. The increase in the critical Rayleigh number in the fifth column can be attributed to allowing fluid motion in the air layer, therefore more efficiently removing heat from the liquid and stabilizing the system. Also observe that the liquid Rayleigh number approaches the value obtained when the gas phase is assumed to be either purely passive or only thermally active, as the air height becomes smaller. This not only provides a check on the active bilayer calculations but also shows conclusively that as the air height becomes smaller, it really does become passive, both thermally and mechanically.

The second important point that can be made from Table 2.1 is when the air height becomes large. When the Biot numbers in Eqs. (2.6) and (2.7) are used, the critical Rayleigh number and the critical wave number reach an asymptotic value as the air height increases. The active air layer calculations, on the other hand, show a dramatic decrease in both the critical Rayleigh number and the critical wave number of the liquid. This can be explained by dominant convection in the air layer. Indeed, the magnitude of the temperature drop in each layer, in the conductive state, depends on the height and the conductivity of each layer. Now, as the height of the air layer increases, the temperature difference across it will increase relative to the temperature difference across the lower liquid for a fixed overall temperature drop. Consequently, under critical conditions, the Rayleigh number of the lower liquid is small and gets even smaller as the air height increases. By contrast, the Rayleigh number of the air becomes larger as its height is increased. It may also be observed that the Rayleigh number of the air increases and tends to the asymptotic value (Chandrasekhar, 1961) of 1708. This value is the critical Rayleigh number for pure buoyancy convection assuming rigid, perfectly conducting plates; indeed, note that the air acts like a fluid bounded by two rigid conducting walls. Because the convection is dominant in the air layer, the liquid layer simply responds to convection in the upper gas. Moreover, because convection in both layers is simultaneous, it is clear that the convection of air immediately sets up transverse temperature gradients in the interface, generating weak surface-tension-driven and buoyancy-driven convection in the liquid. The decrease in the crital wave number must therefore be a signature of the pattern because of dominant convection in the air layer.

Turning now to the second feature of the unbounded calculations, the vertical components of velocity, $W_1(z)$ and $W_2(z)$, for various air heights were calculated and are shown in Fig. 2.3. Each graph represents calculations assuming 5 mm of 100-cS silicone oil. The vertical component of velocity is displayed at the critical wave number in each graph. For a small air height, one would expect that the Rayleigh number in the oil, at onset, would be much greater than the Rayleigh number in the air and we would say that the oil convects first. Here motion in the air is caused by the silicone oil's dragging it. As the air height increases, the Rayleigh numbers in each layer become comparable and so does the convection. If the flow in the upper layer is in the same direction as the flow in the lower layer (corotating), then the convection is considered to be thermally coupled. If the flow in the upper layer is in the opposite direction as the flow in

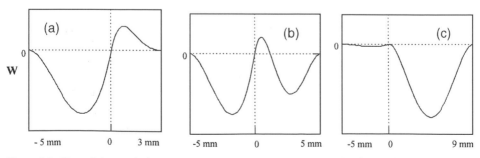

Figure 2.3. Plot of the vertical component of velocity versus fluid depths for (a) 3-mm, (b) 5-mm, (c) 9-mm air heights. The liquid–gas interface is represented by a vertical dotted line. For 3 mm, the flowing silicone oil is dragging air. For 5 mm, air is convecting because of thermal coupling. For 9 mm, most of the convection occurs in the air layer. Each calculation assumed 5 mm of 100-cS silicone oil.

the lower layer (counterrotating), then the convection is considered to be mechanically coupled. For the calculations given in Fig. 2.3(b), convection is slightly more dominant in the oil and the air would be termed thermally coupled had it not been for a small roll developed near the interface in the air layer. In other words, the silicone oil dragging the air causes a secondary roll near the interface. For a larger air height of 9 mm [Fig. 2.3(c)] the convection is almost entirely in the air layer, while the liquid appears mostly passive. The onset of the strong motion in the air simultaneously causes tangential gradients of temperature at the interface, inducing a weak motion in the oil. It is our view that the moving oil causes a small counter roll in the upper gas layer near the interface because of the condition of continuity of tangential components of velocity – a condition imposed by the model. Evidence of this small counter roll can be seen with difficulty in Fig. 2.3(c) but is clear from an inspection of the numerical results. The nature of the thermal coupling is unlike the explanation given above in the when discussion of Fig. 2.1 and is induced by convection in the gas while the liquid remains mostly at rest. Of course, the presence of even a very small counter roll does mean that there is some inevitable mechanical coupling.

We now turn to the calculations that assume bounded rigid radial walls in a cylinder. There are two extreme cases that interest us and for which calculations are readily available. The first case is concerned with thin gas layers for which the gas acts as if it were passive. This calculation was done by Dauby et al. and is reproduced here by Dauby. We compare the experiments for thin air layers with these calculations. The second case is one in which the gas depth is very large and it acts as if it were contained by rigid, conducting, horizontal and vertical walls. This case was computed quite thoroughly by Hardin et al. (1990), and we refer to their calculations to help interpret the experiments with large air heights. The intermediate case in which both layers might be active requires more intensive computations and is not investigated in this study.

Returning to the Marangoni–Rayleigh convection problem in a bounded cylinder, we consider Fig. 2.4, which is a depiction of the calculated critical Marangoni number for a silicone

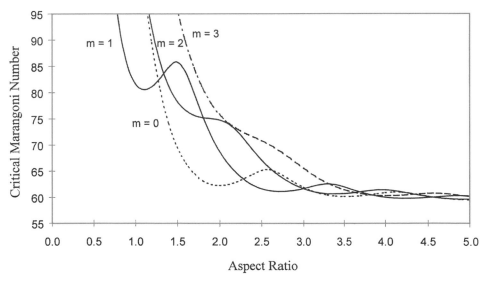

Figure 2.4. Plot of the bounded, linear calculations made with an insulating, no-slip radial wall. In the calculations, a depth of 5 mm of silicone oil is assumed. The Biot number is assumed to be 0.3. Calculations (not shown) conclude that an increase in the Biot number causes the curves to shift to smaller aspect ratios.

oil–air system assuming a Biot number of 0.3. As mentioned above, the calculations used properties pertaining to the silicone oil–air system, as these were the fluids that were used in the experiments. Three observations can be made from Fig. 2.4. First, the critical Marangoni number is not a monotonic function of the aspect ratio. At small aspect ratios, the minimum value of each mode is much greater than the asymptotic minimum reached at aspect ratios greater than 4.0. Second, the pattern changes as the aspect ratio changes. Third, at specific aspect ratios, two patterns can coexist, resulting in a codimension two point. These observations are recalled as the results from experiments are discussed in the next subsection.

**Observations from Experiments.** A series of careful experiments was carried out to investigate the effect of the air height on the pattern formation at the onset of convection and also to verify whether convection in the upper air layer could drive convection in the lower layer. Two different samples of Dow Corning silicone oil were used. From earlier experience (Zhao et al., 1995) most of the thermophysical properties of the oil, with the exception of the dynamic viscosity, can be assumed to be constant within the temperature range studied. Dow Corning silicone oils are a blend of poly-methylsiloxanes, and the viscosities are strong functions of temperature. The viscosities of two separate samples of silicone oil were measured with a Paar, cone, and plate viscometer over a temperature range. The functional relationship of viscosity with temperature for each sample is

$$
\begin{aligned}
\mu_a &= -1.5T + 560, \\
\mu_b &= -2.9T + 1050,
\end{aligned}
\tag{3.1}
$$

where the subscripts $a$ and $b$ refer to the two samples, the temperature $T$ is in degrees Kelvin and is valid between the temperatures of 25 and 50 °C. The viscosity is given in centipoise (cP). Sample $a$ had a lower nominal viscosity than sample $b$. Now, viscosity conveniently scales out of the problem if the fluid surface is flat and the gas is passive, and so the bounded layer model results given in Fig. 2.4 are independent of the value of viscosity.

For each of the experiments, a liquid layer insert, 5.0 mm deep and 20 mm in diameter (2.0 liquid aspect ratio) was used. Three different air heights of 3, 14, and 20 mm with silicone oils

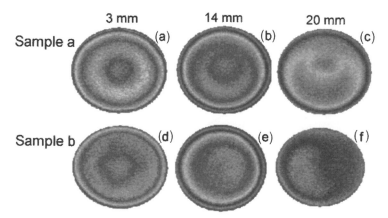

Figure 2.5. Infrared images of the flow pattern for different air heights and different viscosities of silicone oil. (a)–(c) used sample $a$'s silicone oil. (d)–(f) used sample $b$'s silicone oil. (a) and (d) had a 3-mm air height, (b) and (e) had a 14-mm air height, and (c) and (f) had a 20-mm at height.

Table 2.2. *Temperature Differences across the Silicone Oil and Overall Temperature Differences for the Six Experiments Shown in Fig. 2.5*

| | Sample *a* | | Sample *b* | |
| --- | --- | --- | --- | --- |
| Air Height (mm) | *Temperature Difference Across Silicone Oil (°C)* | *Overall Temperature Difference (°C)* | *Temperature Difference Across Silicone Oil (°C)* | *Overall Temperature Difference (°C)* |
| 3 | 0.9 | 8.4 | 1.5 | 14.5 |
| 14 | 0.9 | 16.2 | 0.8 | 14.4 |
| 20 | 0.35 | 8.7 | 0.4 | 9.7 |

from samples *a* and *b* were used. These experiments were done to study how convection, which initiates in the air, can affect the flow pattern in the silicone oil.

Figure 2.5 displays the different flow patterns observed at the onset of convection for the 2.0 liquid aspect ratio. The critical temperature differences for each experiment are given in Table 2.2. The critical temperature difference here refers to the temperature difference at which the flow was first observed. First the 3-mm air and sample *a* silicone oil were used. An $m = 0$, double-toroid pattern was observed [Fig. 2.5(a)]. This flow pattern agrees with the no-slip, three-dimensional, linear calculations in a bounded cylinder given in Fig. 2.4. Next, a 14-mm was used and then a 20-mm air insert was used, again with sample *a* silicone oil. This time the flow patterns seen at the onset of convection were much different than those predicted by the bounded linear calculations. For the 14-mm air height [Fig. 2.5(b)], the double-toroidal pattern became skewed. For the 20-mm air height [Fig. 2.5(c)], the pattern completely changed to a unicellular, $m = 1$ pattern. The flow pattern for sample *a* silicone oil with a 20-mm air height does not agree with the bounded linear calculations, that assume a passive air layer, even after the change in the Biot number is considered. This calls for an explanation, which is set forth in the next few paragraphs.

The skewed double-toroidal and the unicellular flow patterns for the 14- and the 20-mm air heights can be explained by dominant convection in the air. Assume for a moment that a buoyantly convecting fluid bounded by both rigid radial walls as well as rigid top and bottom plates can model the onset of convection in the air. This assumption is justified because, before convection, the oil, which is much more viscous than the gas, acts much like a rigid plate. Further, the heat transfer from the air to the liquid is given by the inverse of the Biot number defined by Eq. (2.6). The corresponding values of the reciprocal of the Biot number for the 14- and the 20-mm air heights are 17 and 20, respectively. At these values, the liquid acts effectively like a thermal conductor. This assumption is also substantiated by calculations given by Sparrow et al. (1963). As the thermal conductivity of the radial Lucite walls is much larger than the air's thermal conductivity, we can assume that the air is bounded radially by purely conductive walls. Indeed, finite-volume calculations that take sidewall properties into account showed that the isotherms were horizontal in the conduction problem. Calculations for a buoyantly convecting fluid in a rigid cylinder with conducting radial walls were performed by Hardin et al. (1990). Therefore, for large air heights, the temperature difference across the air, at the onset of convection, can be compared with the calculations given by Hardin et al. In this regard three observations may be made.

First, the critical temperature across the air for the 14- and the 20-mm air heights were 15.3 and 8.35 °C, respectively (Table 2.3). These can be compared with the theoretical temperature

Table 2.3. *Comparison of the Critical Temperature Differences in the Experiments with the Critical Temperature Difference Calculated from the Paper of Hardin et al.*

| Air height (mm) | Experiments (°C) | Hardin's Conducting Sidewalls (°C) | Hardin's Insulating Sidewalls (°C) |
|---|---|---|---|
| 14 | 15.3 | 15.3 | 9.1 |
| 20 | 8.4 | 8.8 | 4.2 |

*Note:* The experimental values reported are for sample $a$'s silicone oil. The thermophysical properties of air, used to calculate the critical temperature difference from the paper of Hardin et al., are those that were used for the calculations reported in Table 2.1.

differences by Hardin et al. of 15.3 and 8.8 °C, which show excellent agreement. The second observation is related to the fact that as the depth of the air layer increases, the Biot number at the liquid–gas interface decreases. From the laterally unbounded layer calculations, it follows that the critical wave number decreases as the Biot number decreases. In the bounded layer calculations, the increase of the Biot number shifts the curves to the right, causing a change in the observed pattern for a fixed aspect ratio. However, all of this is relevant only if the convection were to initiate in the lower liquid. To determine whether this occurred in the experiments for various air heights, each experiment was performed again with sample $b$'s silicone oil, whose viscosity was higher than that of sample $a$ by $\sim$60%. The results were very interesting. As the Rayleigh and the Marangoni numbers are inversely proportional to the viscosity, increasing the viscosity of the silicone oil from sample $a$ to sample $b$ should have proportionally increased the observed critical temperature difference, provided that the observed flow pattern is indeed the result of the onset of convection in the lower layer. Figure 2.5(d) shows the double-toroidal flow pattern observed at the onset of convection for the 2.0 liquid aspect ratio, 3-mm air insert, and sample $b$'s silicone oil. The critical temperature difference this time was 1.5 °C (14.5 °C overall); 67% more than the same experiment with sample $a$'s silicone oil, confirming the hypothesis that convection initiated in the lower layer. For the 14- and the 20-mm air heights, the change in viscosity changed the measured temperature difference (Table 2.2) across the lower liquid only slightly, indicating that convection initiated in the air. The third observation concerns the pattern at the onset of convection for the 14- and the 20-mm air heights. For the 20-mm air insert corresponding to a 0.5 aspect ratio of air, the observed pattern for both fluid viscosities was a unicellular, $m = 1$ type. This pattern was predicted by Hardin et al. for that aspect ratio, and once again the experiments and theory are in close agreement. For the 14-mm air height corresponding to a 0.71 aspect ratio of air, the observed pattern changed slightly between the experiments of the two different silicone oils. For sample $a$, a lopsided double toroid was seen and with sample $b$ a mixture of $m = 0$ and $m = 1$ was seen. The calculations of Hardin et al. for a 0.7 aspect ratio predict a codimension two point of $m = 0$ and $m = 1$. Therefore the patterns seen in the experiments conducted with the sample $b$ fluid are beautifully explained by the results of Hardin et al. The experiment with the lower-viscosity sample $a$ oil showed that the very weak convection in the oil still has some influence on the flow pattern.

Two additional experiments were conducted with smaller air heights of 2 mm and a depth of 5 mm of sample $a$ silicone oil for aspect ratios of 2.0 and 2.5. The critical temperature differences across the liquid for the 2.0 and the 2.5 aspect ratio experiments were 2.5 °C (8.5 °C overall) and 2.4 °C (8.3 °C overall), respectively. When Eq. (3.1) was used, the nominal dynamic viscosity of the silicone oil was 91.7 cP, resulting in critical Marangoni numbers of 62 and 59. The theoretical values of the Marangoni numbers are 64.3 and 63.5 for insulating sidewalls and 71 and 67 for conducting sidewalls for the two aspect ratios. These results are within 12% of the experiments, which is within the error of the thermophysical properties. Some of the crucial thermophysical properties of the Marangoni number such as $\sigma_1$ and $\kappa$ are known with poor accuracy. Therefore it is better to determine the ratios of the experimental critical Marangoni numbers and compare these with the theoretical ratios. This eliminates some, but not all, of the uncertainties in the values of the thermophysical properties. The ratios of the experimentally determined critical Marangoni numbers for aspect ratios of 2.0 and 2.5 are within 2% of the corresponding theoretical values for conducting sidewalls and within 5% of the calculated values for insulating sidewalls. However, there is a large difference between the results of the 2-mm and the results of the 3-mm air height experiments that used a 2.0 aspect ratio. The smaller the air height, the closer the results are to the theoretical predictions. This discrepancy gives even more credence to the fact that convection in the air plays an important role in bilayer convection.

In summary, we observe that results from the thin air layer experiments agree well with Fig. 2.4, while the large air layer experimental results agree very well with the calculations of Hardin et al. These are the two extreme cases. For moderate air heights, a full bounded bilayer calculation is desirable but not yet available.

## 2.4. Conclusion

By use of both calculations and experiments, a new convection coupling mechanism was revealed. This mechanism requires that convection be predominant in the upper fluid first. The convection in the upper fluid layer then simultaneously generates transverse thermal gradients along the fluid–fluid interface, immediately causing surface-tension-gradient-driven and buoyancy-driven convection in the lower fluid. In both the calculations and the experiment, silicone oil and air were used as the two fluids. As the viscosity of air is quite low compared with that of the silicone oil, convection in the air has no mechanical influence on the silicone oil, and it was this property that was advantageous in isolating the new convection mechanism.

## Acknowledgments

We thank the National Science Foundation for the funding and support through grants CTS 95-00393 and CTS 93-07819 as well as NASA through grant NGT 3-52320. We also thank A. A. Zaman for his help in the viscosity measurements. P. C. Dauby is pleased to acknowledge NATO for a research fellowship that allowed this collaboration.

## References

Canuto, C., Hussaini, M. Y., Quarteroni, A., and Zhang, T. A. 1988. *Spectral Methods in Fluid Dynamics*, Springer-Verlag, Berlin.

Chandrasekhar, S. 1961. *Hydrodynamic and Hydromagnetic Stability*, Oxford University Press, Oxford.

Dauby, P. C. and Lebon, G. 1996. J. Fluid Mech. **329**, 25.

Dauby, P. C., Lebon, G., and Bouhy, E. 1997. Phys. Rev. E **56**, 520.

Dijkstra, H. A. 1992. J. Fluid Mech. **243**, 73.

Dijkstra, H. A. 1995A. Microgravity Sci. Technol. VII/4, 307.

Dijkstra, H. A. 1995B. Microgravity Sci. Technol. VIII/2, 70.

Dijkstra, H. A. 1995C. Microgravity Sci. Technol. VIII/3, 155.

Ferm, E. N. and Wollkind, D. J. 1982. J. Non-Equilib. Thermodyn. **7**, 169.

Hardin, G. R., Sani, R. L., Henry, D., and Roux, B. 1990. Int. J. Num. Meth. Fluids **10**, 79.

Johnson, D. 1996. NASA Contractor Report 198451.

Johnson, D. and Narayanan, R. 1996. Phys. Rev. E **54**, R3102.

Johnson, D. and Narayanan, R. 1997. Phys. Rev. E **56**, 5462.

Koschmieder, E. L. 1993. *Bénard Cells and Taylor Vortices*, Cambridge University Press, Cambridge.

Koschmieder, E. L. and Prahl, S. A. 1990. J. Fluid Mech. **215**, 571.

Nield, D. A. 1964. J. Fluid Mech. **19**, 1635.

Normand, C., Pomeau, Y., and Velarde, M. G. 1977. Rev. Mord. Phys. **49**, 581.

Ondarçuhu, T., Mindlin, G. B., Mancini, H. L., Garcimartin, A., and Pérez-Garcia, C. 1993. Phys. Rev. Lett. **70**, 3892.

Rosenblat, S., Davis, S. H., and Homsy, G. M. 1982. J. Fluid Mech. **120**, 91.

Sparrow, E. M., Goldstein, R. J., and Jonsson, V. K. 1963. J. Fluid Mech. **18**, 33.

Zaman, A. and Narayanan, R. 1996. J. Colloid Interface Sci. **179**, 151.

Zhao, A. X., Moates, F. C., and Narayanan, R. 1995. Phys. Fluids **86**, 1576.

# 3

# The Third Type of Benard Convection Induced by Evaporation

WEN-JEI YANG

## Nomenclature

| | |
|---|---|
| $A$ | local heat transfer coefficient at the liquid–vapor interface defined by Eq. (6) [W/(m$^2$ °C)] |
| $a$ | thermal diffusivity (m$^2$/s) |
| $\hat{a}$ | accommodation coefficient |
| $B$ | Marangoni number defined by Eq. (22) |
| $D$ | derivative $\equiv$ d/d$\eta$ |
| $g$ | gravitational acceleration (m/s$^2$) |
| $h_{fg}$ | latent heat of evaporation (kJ/kg) |
| $k$ | thermal conductivity of the liquid [W/(m °C)] |
| $L$ | Biot number defined by Eq. (23) |
| $l$ | liquid layer thickness (m) |
| $M$ | molecular weight (kg/kmol) |
| Ma | Marangoni number |
| $Q$ | rate of heat loss per unit area [W/(m$^2$ °C)] |
| $q$ | change of heat flux with respect to temperature at the liquid surface, $= (\partial Q/\partial T)$ [W/(m$^2$ °C)] |
| $P$ | pressure (Pa) |
| Pr | Prandtl number, $\nu/a$ |
| $R$ | universal gas constant [kJ/(kg mol °C)] |
| $S$ | surface tension of the liquid (N/m) |
| $T$ | temperature (°C) |
| $v'$ | velocity in the $y$-direction (m/s) |

## Greek Letters

| | |
|---|---|
| $\alpha$ | nondimensional wave number |
| $\beta$ | temperature gradient in liquid layer (°C/m) |
| $\Delta T$ | temperature difference, $T_e - T_g$ (°C) |
| $\mu$ | absolute viscosity (Pa s) |
| $\nu$ | kinematics viscosity (m$^2$/s) |
| $\rho$ | density (kg/m$^3$) |
| $\sigma$ | thermocapillarity $= -(\partial S/\partial T)$ [N/(m °C)] |

## Superscripts

| | |
|---|---|
| $c$ | bottom surface |
| $e$ | free surface |

| $g$ | atmospheric environment |
|---|---|
| $l$ | liquid phase |
| $o$ | unperturbed system |
| 1 | upper liquid region |
| 2 | lower liquid region |

## 3.1. Introduction

It is well known that the Rayleigh–Benard-type convection induced by the buoyancy force (Rayleigh 1916), the Marangoni–Benard-type convection caused by the surface-tension force (Pearson 1958), or a type of convection caused by a combined action of the two forces (Nield 1964) is characterized by a negative temperature gradient in a liquid layer heated from below or cooled from above. The physics of natural convection in a fluid layer was described in detail by Velarde and Normand (1980) and Batchelor and Freund (1993).

Berg et al. (1966) conducted a study on natural convection in pools of evaporating liquid in an open environment. Cellular convection occurred within only a certain range of pool depth, outside of which convection would cease. Cellular patterns varied with pool depth. The mechanism is unique in that convection is induced by liquid evaporation at the free surface. They attributed cellular convection to be a surface-tension-induced mechanism because surface evaporation results in a cooled-from-above situation. Block (1956) carried out an experiment on a liquid film of $\sim$0.8-mm thickness with a free surface that was cooled at its base. Nevertheless, Benard cells appeared in the film, which was considered a surface-tension-driven convection too.

Guo (1988) studied the thermocapillary cellular convection in a thin liquid layer in an atmospheric environment at room temperature (20 °C). Three organic liquids were tested: ethanol, methanol, and acetone. Liquid-layer thickness was varied, and evaporation rate was measured. It was disclosed that there were two convective patterns of a polygonal tessellated cell in the liquid layer of 1-mm thickness. When Ma is less than 2000, the wave number of cellular convection continues to decrease from the onset of vaporization to a final stationary state. When Ma is sufficiently large, say 5000, the stationary tessellated cell pattern becomes unstable and keeps moving in a chaotic way. When the layer thickness was increased to 2 mm, the flow pattern remained similar but the ribs and cell partitions moved more vigorously. In such a circumstance, both surface-tension and buoyancy mechanisms are important, and the induced convective flow was so strong that no stationary convection cell persisted in the liquid and the flow pattern transmuted in a completely chaotic way.

The evaporation on the upper interface has a twofold effect on the liquid-layer system. First, it always sets up an unstable thermal stratification for cellular convection in the liquid phase because of evaporating cooling. Whenever the stratification becomes strong enough (i.e., whenever the Marangoni number exceeds its critical value), convection cells will appear in the liquid. Second, the phase change itself may stimulate the interface to cause interfacial instability.

Recently Chai and Zhang (1996a) and Zhang and Chai (1996b) extended Guo's flow visualization study by means of the tracer method (aluminum powder) intrusive technique to observe cellular convection in an evaporating thin liquid layer heated, adiabatic, or cooled from below. Twelve liquids of low evaporation point were tested. It was concluded that cellular convection occurs when the liquid layer is cooled from below. The temperature profile in the layer is nonlinear. However, no attempt was made to explain the mechanism. Yang et al. (1997a, 1997b) have proposed a new model to explain the mechanism of cellular convection in a thin liquid

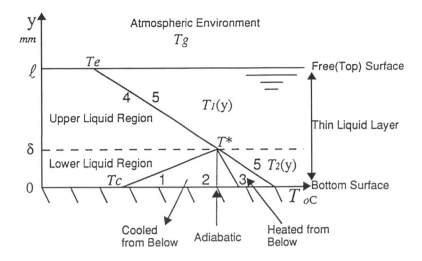

Line  1: positive temperature gradient due to cooling from below

2: zero temperature gradient due to no heat flow from below

3: negative temperature gradient due to heated from below

4: negative temperature gradient due to surface evaporation

5: Rayleigh-Benard and Marangoni-Benard type convection

Figure 3.1. A model for evaporation-induced natural convection in a horizontal, thin liquid layer with a thickness of 1 mm or less subjected to being cooled, insulated, and heated from below (lines 1, 2, and 3) compared with Rayleigh–Benard and Marangoni–Benard convection (line 5).

layer induced by surface evaporation. The mechanism is distinctly different from those of the Rayleigh–Benard- and the Marangoni–Benard-type cellular convections. The model shows that evaporation at the free surface of the thin liquid layer results in a negative temperature gradient in the upper stratum in which cellular convection may occur, irrespective of a positive, zero, or negative temperature gradient's prevailing in the remaining lower stratum, as shown in Fig. 3.1. That is to say, the model suggests the formation of two regions in the liquid layer with the occurrence of cellular flow patterns in the top region where the temperature gradient is negative, irrespective of the bottom region's being cooled, adiabatic, or heated.

This study develops a theoretical model for this new type of natural convection based on temperature measurements that indicate that the liquid layer consists of two strata with Benard cells appearing in the upper stratum and heat conduction taking place in the lower one. A stability analysis is performed by means of the separation-of-variables method. The governing dimensionless parameters are identified, and stability criteria are determined.

## 3.2. Analysis

Consider a liquid layer consisting of two strata with Benard cells appearing in the upper stratum and heat conduction taking place in the lower one, as shown in Fig. 3.1. The normal coordinate $y$ measures the distance from the solid surface. Line 4 is the temperature profile induced by evaporation cooling at the free surface, $y = l$, which is at a temperature of $T_e$. Line 1 is due to the layer's being cooled from below with the surface at a temperature $T_c$, which is lower than $T^*$. $T^*$ denotes the temperature at a location $y = \delta$ that separates the upper from the lower liquid regions. Line 2 corresponds to an adiabatic surface corresponding to an insulated case with $T_c = T^*$. Line 3 occurs when the liquid layer is heated from below with $T_c > T^*$. Superimposed in the figure is line 5 for both the Rayleigh–Benard-type and the Marangoni–Benard-type conduction for reference. In contrast to a monotonic temperature distribution of line 5, the temperature profiles in an evaporating liquid layer are, in general, nonlinear as monitored by thermocouples (Chai and Zhang, 1996a; Zhang and Chai, 1996b), irrespective of the layer's being cooled or heated from below or adiabatic.

The present analysis follows that of Pearson (1958) for a liquid layer being heated from below, except that evaporation takes place at the liquid surface. It is postulated that (i) the flow field is treated by a one-dimensional quasi-steady state and constant physical properties; (ii) the liquid layer thickness $l$ has a constant value; (iii) a saturated vapor phase exists over the liquid layer, and the liquid–vapor interface is a flat and nondeformable surface; (iv) the only physical quantities that are assumed to vary within the liquid are the temperature, the surface tension, which is regarded as a function of temperature only, and the rate of heat loss of the surface, also a function of temperature only.

Under the above assumptions, the rate of heat loss per unit area from the upper free surface in the unperturbed upper liquid region can be expressed as

$$Q_o = \dot{m}'' h_{fg} = -k\beta_1 - A\Delta T. \tag{1}$$

Here, $\dot{m}''$ denotes the mass flux of evaporation at the liquid surface, $h_{fg}$ is the latent heat of evaporation, $k$ is the thermal conductivity of the liquid, $\beta_1$ is the temperature gradient of the upper liquid region where the temperature gradient is negative, $A$ is the local heat transfer coefficient at the liquid–vapor interface, and $\Delta T$ is the temperature difference between the free surface and the atmospheric environment $T_e - T_g$. Note that positive $\dot{m}''$ implies condensation and that negative implies evaporation. Let $T_1(y)$ and $T_2(y)$ be the liquid temperatures in the upper and the lower regions, respectively. $\beta_1$ can be expressed as

$$\beta_1 = \frac{\partial T_1}{\partial y} = \frac{T_e - T^*}{l - \delta}. \tag{2}$$

For the lower liquid region, the temperature gradient $\beta_2$ can be expressed as

$$\beta_2 = \frac{\partial T_2}{\partial y} = \frac{T^* - T_c}{\delta}. \tag{3}$$

Thus the temperature profiles in the upper and the lower liquid regions shown in Fig. 3.1 can be written as

$$T_1 = T_e + \beta_1(y - l), \tag{4}$$

$$T_2 = T_c + \beta_2 y, \tag{5}$$

respectively. The local heat transfer coefficient $A$ at the liquid–vapor interface is obtained from the works of Carrey (1992) and Faghri (1995) as

$$A = \left(\frac{2\hat{a}}{2 - \hat{a}}\right) \sqrt{\frac{M}{2\pi R T_g}} \left(\frac{h_{fg}^2}{T_g v_{lg}}\right) \left(1 - \frac{P_g v_{lg}}{2h_{fg}}\right). \tag{6}$$

Here, $\hat{a}$ is an accommodation coefficient and its value is given by Faghri (1995), $M$ is the molecular weight, $R$ is the universal gas constant, $h_{fg}$ is the latent heat of evaporation, $T_g$ is the temperature in the vapor phase, $v_{lg}$ is the change in specific volume from liquid to vapor, and $P_g$ is the vapor pressure.

The equations of motion and heat conduction for the upper liquid region become

$$\left(\frac{\partial}{\partial t} - \nu \nabla^2\right) \nabla^2 v' = 0, \tag{7}$$

$$\left(\frac{\partial}{\partial t} - a \nabla^2\right) T' = \beta_1 v', \tag{8}$$

respectively. Here, $\nu$ is the kinematics viscosity, $v'$ is the velocity in the $y$ direction, $a$ is the thermal diffusivity of the liquid, $T'$ is the temperature disturbance, and $t$ is the time. Note that the equation for $v'$ contains no buoyancy term and is the same as that of Pearson (1958).

The surface-tension coefficient $S$ and the heat flux $Q$ are taken to depend linearly on $T'$ because we are considering an infinitesimal disturbance theory. The first two terms in a Taylor expansion give

$$S = S_o - \sigma T_e', \tag{9}$$

$$Q = Q_o + q T_e'. \tag{10}$$

Here $\sigma = -(\partial S/\partial T)_{T=T_e}$ represents the rate of change of surface tension with temperature; $q = (\partial Q/\partial T)_{T=T_e}$ denotes the change of heat flux with respect to temperature at the liquid surface. $\sigma$, $q$, $S_o$, and $Q_o$ are all evaluated at the undisturbed liquid surface temperature.

Since the lower liquid region is quiescent, four hydrodynamic conditions for the upper liquid can be described as follows:
At $y = \delta$,

$$v' = \frac{\partial^2 v'}{\partial y^2} = 0. \tag{11}$$

At $y = l$,

$$v' = 0, \qquad \rho \nu \frac{\partial^2 v'}{\partial y^2} = \sigma \nabla_1^2 T'. \tag{12}$$

Here,

$$\nabla_1^2 = \left(\frac{\partial^2}{\partial x^2} + \frac{\partial^2}{\partial z^2}\right),$$

and $\rho$ denotes the liquid density.

Thermal boundary conditions are

$$T' = Y \frac{\partial T'}{\partial y} \tag{13}$$

at $y = \delta$, and

$$-k \frac{\partial T'}{\partial y} - AT' = qT' \tag{14}$$

at $y = l$, where $y$ is a constant. Two cases of thermal boundary condition (13) are considered: One is the supply or removal of heat at the bottom surface of the upper liquid region by the lower liquid region under the conducting condition $T' = 0$ (i.e., $Y = 0$). It corresponds to lines 1 and 3 in Fig. 3.1. The other corresponds to an adiabatic or insulating boundary condition, corresponding to line 2 in Fig. 3.1, for which $\partial T'/\partial y = 0$ (i.e., $Y^{-1} = 0$).

Let dimensionless variables be defined as

$$(\xi, \eta, \varsigma) = \left( \frac{x}{l - \delta}, \frac{y'}{l - \delta}, \frac{z}{l - \delta} \right), \qquad \tau = \frac{ta}{(l - \delta)^2}. \tag{15}$$

Here, $y' = y - \delta$. $T'$ and $v'$ can be written in the forms

$$T' = \beta_1 (l - \delta) F(\xi, \varsigma) g(\eta) e^{p\tau}, \tag{16}$$

$$v' = -\frac{a}{(l - \delta)} F(\xi, \varsigma) f(\eta) e^{p\tau}, \tag{17}$$

which yield

$$\frac{\partial^2 F(\xi, \varsigma)}{\partial \xi^2} + \frac{\partial^2 F(\xi, \varsigma)}{\partial \varsigma^2} + \alpha^2 F(\xi, \varsigma) = 0, \tag{18}$$

where $\alpha$ is a nondimensional wave number derived from the application of the separation-of-variables method and $p$ is a constant (which can be complex). One also obtains

$$[p - \mathrm{Pr}(D^2 - \alpha^2)](D^2 - \alpha^2) f = 0, \tag{19}$$

$$[p - (D^2 - \alpha^2)]g = -f, \tag{20}$$

where $D \equiv d/d\eta$, and $\mathrm{Pr} = v/a$, which is called the Prandtl number.

Boundary conditions (11), (12), and (14) become

$$f(0) = f''(0) = 0, \qquad f(1) = 0, \qquad f''(1) = \alpha^2 B g(1), \qquad g'(1) = -L g(1), \tag{21}$$

where

$$B = \frac{\sigma \beta_1 (l - \delta)^2}{\rho v a} \tag{22}$$

and

$$L = \frac{(q + A)(l - \delta)}{k} \tag{23}$$

are the dimensionless parameters, called the Marangoni and the Biot numbers, respectively.

The thermal boundary condition for the conducting case ($Y = 0$) yields

$$g(0) = 0, \tag{24}$$

and that for the insulating case ($Y^{-1} = 0$) gives

$$g'(0) = 0. \tag{25}$$

For neutral stability, Eqs. (19) and (20) for $p = 0$ are reduced to

$$(D^2 - \alpha^2)(D^2 - \alpha^2)f = 0 \tag{26}$$

and

$$(D^2 - \alpha^2)g = f, \tag{27}$$

respectively. Equation (26), subject to the first three equations of Eqs. (21), is solved to yield

$$f(\eta) = a_f(\sinh \alpha \cdot \eta \cdot \cosh \alpha - \cosh \alpha \cdot \sinh \alpha\eta) \tag{28}$$

where

$$a_f = \frac{\alpha B g(1)}{2 \sinh^2 \alpha}.$$

Similarly, Eq. (27), subject to the last condition of Eqs. (21) and (24) or (25), gives

$$g(\eta) = a_f \cdot \frac{1}{4\alpha^2}\left[ \alpha \cdot \sinh \alpha \cdot \eta^2 \cdot \sinh \alpha\eta - \sinh \alpha \cdot \eta \cdot \cosh \alpha\eta - 2\alpha \cdot \cosh \alpha \cdot \eta \cdot \cosh \alpha\eta - \sinh \alpha \right.$$
$$\left. \cdot \frac{\alpha \sinh^2 \alpha - (1+\alpha^2) \sinh \alpha \cdot \cosh \alpha - 2\alpha \cosh^2 \alpha + L(\alpha \sinh^2 \alpha - \sinh \alpha \cdot \cosh \alpha - 2\alpha \cosh^2 \alpha)}{\alpha \cdot \cosh \alpha + L \sinh \alpha} \right] \tag{29}$$

for the conducting case and

$$g(\eta) = a_f \cdot \frac{1}{4\alpha^2}\left( \alpha \cdot \sinh \alpha \cdot \eta^2 \cdot \sinh \alpha\eta - \sinh \alpha \cdot \eta \cdot \cosh \alpha\eta - 2\alpha \cdot \cosh \alpha \cdot \eta \cdot \cosh \alpha\eta \right.$$
$$+ \left( \frac{\sinh \alpha}{\alpha} + 2 \cosh \alpha \right)(\cosh \alpha\eta + \sinh \alpha\eta) + \left\{ \frac{-\alpha(1+\alpha) \sinh^2 \alpha + \alpha^2(\alpha - 2) \sinh \alpha \cdot \cosh \alpha}{\alpha(\alpha \sinh \alpha + L \cosh \alpha)} \right.$$
$$\left. + \frac{L[2\alpha^2 \cosh^2 \alpha - (1+\alpha) \sinh \alpha \cdot \cosh \alpha - (1+\alpha^2) \sinh^2 \alpha - 2\alpha \cosh^2 \alpha]}{\alpha(\alpha \sinh \alpha + L \cosh \alpha)} \right\} \cosh \alpha\eta \right) \tag{30}$$

for the insulating case. A substitution of Eqs. (29) and (30) into the fourth boundary condition of Eqs. (21) results in

$$B = \frac{8\alpha \sinh^2 \alpha (\alpha \cosh \alpha + L \sinh \alpha)}{-2\alpha^2 \cosh \alpha + \sinh^2 \alpha \cdot \cosh \alpha + \alpha \sinh \alpha} \tag{31}$$

for the conducting case and

$$B = \frac{8\alpha \sinh \alpha (\alpha \sinh \alpha + L \cosh \alpha)}{\sinh^2 \alpha - \alpha^2} \tag{32}$$

for the insulating case.

### 3.3. Results and Discussions

Equations (31) and (32) are graphically illustrated for $B$ versus $\alpha$ in Figs. 3.2 and 3.3, respectively, with $L$ as the parameter. The obtained curves are marginal (neutral) stability curves separating the regions of growing disturbances from those of damping disturbances. All the curves are asymptotic to $B = 8\alpha^2$ at large values of $\alpha$. The values of the critical Marangoni number and of the critical wave number are, for the conducting case and $L = 0$, $B_{\text{crit}} = 51$ and $\alpha_{\text{crit}} = 1.75$, respectively. For reference, the case $L = 0$ for the insulating boundary condition is special with $\alpha_{\text{crit}} = 0$ and $B_{\text{crit}} = 24$. In general, larger positive values of $L$ lead to greater stability.

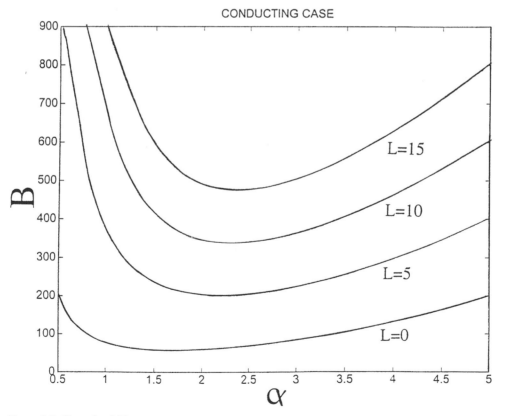

Figure 3.2. Neutral stability curves, conducting case ($L = 7.45$, $B_{\text{crit}} = 330.9$) (from Yang et al., 1997b).

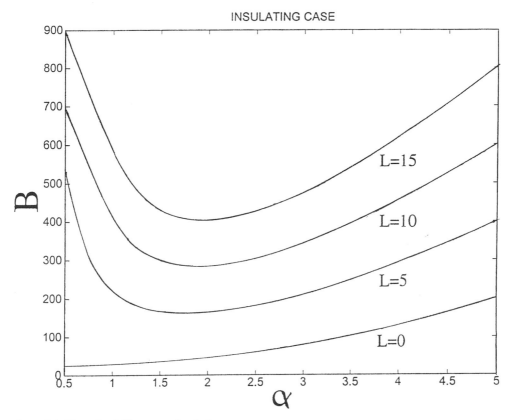

Figure 3.3. Neutral stability curves, insulating case.

### 3.3.1. *Comparison with Surface-Tension Mechanism (Pearson, 1958).*

In order to compare with the surface-tension mechanism, the marginal stability curves of Pearson (1958) are reproduced in Figs. 3.4 and 3.5 for the conducting and the insulating cases, respectively. These curves are all asymptotic to $B = 8\alpha^2$ for large values of $\alpha$. Of particular interest is the case $L = 0$ for which the insulating boundary condition is that the critical values are $\alpha = 0$ and $B = 48$, which is twice $B_{\text{crit}}$ obtained in the present study. For the conducting case and $L = 0$, Fig. 3.4 yields $B_{\text{crit}} = 79.607$ and $\alpha_{\text{crit}} = 1.993$, both of which are higher than their counterparts in the present study. It is concluded from the comparison that with the same temperature difference in the liquid phase, $T_c - T_e$, the layer thickness required for inducing natural convection would be smaller for the evaporation mechanism (in the present study) than for the surface-tension mechanism (Pearson, 1958). This implies that cellular patterns can be induced by the evaporation mechanism in a very thin liquid layer. Accordingly, the critical Marangoni number for the onset of cellular convection would be lower than that for the surface-tension mechanism. The latter corresponds to the case in which the hydrodynamic conditions at $y = \delta$, Eq. (11), are replaced with

$$v' = \frac{\partial v'}{\partial y} = 0 \quad \text{(for a solid wall).} \tag{33}$$

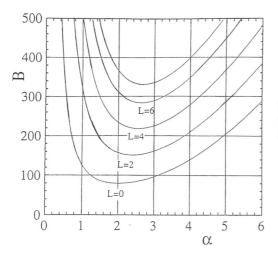

Figure 3.4. Neutral stability curves, conducting case under the surface-tension mechanism from Pearson (1958).

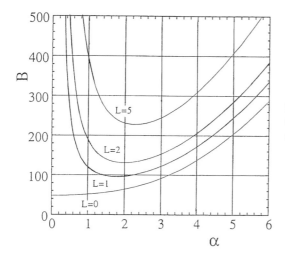

Figure 3.5. Neutral stability curves, insulating case under the surface-tension mechanism from Pearson (1958).

Pearson (1958) compared the modified Biot number, defined by Eq. (23), with $\delta = 0$, which is relevant for the surface-tension mechanism, and with the Rayleigh number, which is relevant for the density-dependent mechanism. From the $L = 0$ case, he found that for most liquids with a thickness of less than the order of 1 cm, surface-tension forces become more effective than buoyancy forces in producing instability and that for a liquid-layer thickness as small as 1 mm, the onset of cellular motion could be attributed to surface tension rather than to buoyancy. His observations were supported by Nield (1964) and Berg et al. (1966). However, the present study deals with the evaporation mechanism for instability in liquid layers of less than 1-mm thickness, as indicated in Subsection 3.3.2.

### 3.3.2. *Comparison with Experiment*

For thermocapillary cellular convection in a thin liquid layer (with a thickness of 1 mm or less) evaporating in an atmospheric environment at room temperature (20 °C), Yang et al. (1997a, 1997b) have monitored the temperature–time history in a thin liquid layer for ethyl alcohol and Freon-113, heated, insulated, and cooled from below. Even though the temperature profile was strongly nonlinear, the temperature difference in the evaporating upper layer was found to range from 0.4 to 0.5 °C in ethyl alcohol and from 0.8 to 1.0 °C in Freon-113, irrespective of the lower liquid layer's being heated, cooled, or insulated below. Hence the temperature gradient in the thin liquid layer of Freon-113 is higher than that of ethyl alcohol. At a temperature difference of 0.5 °C in the evaporating upper liquid layer, its thickness $l - \delta$, obtained from Eq. (22) for the conducting case and $L = 0$, is 0.236 mm for ethyl alcohol and 0.045 mm for Freon-113. In other words, the temperature gradient in the Freon-113 layer is larger than that of the ethyl alcohol layer, in agreement with experimental data. When the temperature difference in a thin liquid layer is fixed, the thickness $l - \delta$ increases with an increase in the modified Biot number $L$ as well as the critical Marangoni number $B_{\text{crit}}$. When the Biot number $L$ is fixed at 7 and 35 for ethyl alcohol and Freon-113, respectively, their $l - \delta$ values reach 0.79 and 0.69 mm, respectively, both being less than 1 mm as in the experiment.

In general, the present analysis shows that the thickness $l - \delta$ for Freon-113 is less than that of ethyl alcohol when both the critical Marangoni number $B_{\text{crit}}$ and the temperature difference in the layer are maintained constant. It results in the temperature gradient in the Freon-113 layer being larger than the ethyl alcohol layer, as experimentally observed.

It is difficult to determine the Biot number accurately because heat transfer mechanisms involve convection and radiation, as well as liquid–vapor interfacial phenomena. However, the magnitudes of $q + A$ are estimated to be of the order of $2.7 \times 10^3$ W/(m² K) for ethyl alcohol and of $5.0 \times 10^3$ W/(m² K) for Freon-113 by use of the definition of Eq. (6) and the physical properties for both ethyl alcohol and Freon-113. On the other hand, the magnitudes of $q + A$ obtained by Eq. (23) are of the order of $2.5 \times 10^3$ W/(m² K) for ethyl alcohol and $4.9 \times 10^3$ W/(m² K) for Freon-113 when the Biot numbers $L$ are 7 and 35 for ethyl alcohol and Freon-113, respectively.

The fact that estimations made with Eqs. (6) and (23) are quite close implies that the theory is supported by experiments.

## 3.4. Conclusions

Temperature measurements revealed a thin liquid layer consisting of two strata, with Benard cells appearing in the evaporating upper stratum and heat conduction taking place in the lower one. A theoretical model has been developed to describe the new type of natural convection in the present study. The following conclusions have been derived:

1. Cellular convection in a thin liquid layer induced by surface evaporation can be explained by the present stability analysis.
2. The governing dimensionless parameters, Marangoni and Biot numbers, are derived by means of the separation-of-variables method.
3. The analytical results are verified by the experimental data.

**Acknowledgment**

The author thanks M. Sakamoto for assistance in the initial stage of the study during his stay as Visiting Scholar with the Thermal-Fluid Laboratory.

# References

Batchelor, G. K. and Freund, L. B. 1993. *Benard Cells and Taylor Vorticies*. Cambridge University, Cambridge.

Berg, J. C., Boudart, M., and Acrivos, A. 1966. Natural convection in pools of evaporating liquids. J. Fluid Mech. **24**, 721–735.

Block, M. J. 1956. Surface tension as the cause of Benard cells and surface deformation in liquid film. Nature (London) **178**, 650–651.

Carrey, V. P. 1992. *Liquid–Vapor Phase Change Phenomena: An Introduction to the Thermophysics of Vaporization and Condensation Process in Heat Transfer Equipment*, pp. 112–120. Hemisphere, New York.

Chai, A. T. and Zhang, N. 1996a. Marangoni-Benard convection in an evaporating liquid thin layer, presented at the Fourth International Symposium on Heat Transfer, 7–11 October 1996, Beijing.

Faghri, A. 1995. *Heat pipe science and technology*, pp. 72–75. Taylor & Francis, New York.

Guo, K. H. 1988. Cellular thermocapillary convection under evaporation. Ph.D. Dissertation, Department of Mechanical Engineering and Applied Mechanics, University of Michigan, Ann Arbor, Michigan.

Nield, D. A. 1964. Surface tension and buoyancy effects in cellular convection. J. Fluid Mech. **19**, 341–352.

Pearson, J. R. A. 1958. On convective cells induced by surface tension. J. Fluid Mech. **4**, 489–500.

Rayleigh, L. 1916. On convection currents in a horizontal layer of fluid when the high temperature is on the under side. Philos. Mag. **32**(192), 529–546.

Velarde, M. G. and Normand, C. 1980. Convection. Sci. Am. **243**, 92–108.

Yang, W. J., Guo, K. H., and Sakamoto, M. 1997a. Evaporation-induced cellular convection in the thin liquid layers. Exp. Heat Transfer **10**.

Yang W. J., Zhang, N., Chai, A. T., Guo, K. H., and Sakamoto, M. 1997b. Evaporation-induced Benard convection – A new type and its mechanism, in *ASME Proceedings of the 32nd National Heat Transfer Conference*, Vol. 11, HTD349, pp. 37–50.

Zhang, N. and Chai, A. T. 1996b. Effect of evaporation on thermocapillary flows in a liquid thin layer, presented at the Fourth International Symposium on Heat Transfer. 7–11 October 1996, Beijing.

# 4

# Waves Generated by Surface-Tension
# Gradients and Instability

M. G. VELARDE, A. YE. REDNIKOV, AND H. LINDE

## 4.1. Introduction

When there is an open surface or an interface exists between two liquids, the interfacial tension $\sigma$ accounts for the jump in normal stresses proportional to the surface curvature across the interface; hence this Laplace force affects its shape and stability. Gravity competes with it in accomodating equipotential levels with curvature. Their balance defines, for instance, the stable equilibrium of spherical drops or bubbles. If the surface tension varies with temperature or composition and, eventually, with position along an interface, its change takes care of the jump in the tangential stresses. Hence its gradient acts as a shear stress applied by the interface on the adjoining bulk liquid (Marangoni stress) and thereby generates flow or alters an existing one (Marangoni effect). Surface-tension-gradient-driven flows are known to affect the evolution of growing fronts and measurements of transport phenomena. The variation of surface tension along an interface may be due to the existence of a thermal gradient along the interface or perpendicular to it. In the former case we have instantaneous convection, while in the latter flow occurs past an instability threshold (Levich, 1965; Ostrach, 1982; Davis, 1987; Velarde, 1998). The Marangoni effect is the engine transforming physicochemical energy into flow whose form and time dependence for standard liquids rests on the sign of the thermal gradient and the ratios of viscosities and diffusivities of adjacent fluids (Sternling and Scriven, 1959; Chu and Velarde, 1989; Edwards et al., 1991).

## 4.2. Overstability and Waves in Bénard Layers (Transverse and Longitudinal Waves, Surface and Internal Waves, and Their Mode Mixing)

Let us consider a liquid layer of depth $d$ heated from the air side or open to suitable mass adsorption of a light component from a vapor phase above, with subsequent absorption in the bulk, hence creating a (stabilizing) thermal gradient inside the liquid layer. Contrary to the case of a layer heated from the liquid side, here we consider that, as the layer is stably stratified, the problem refers to oscillatory motions, waves, and not to Bénard cells (Koschmieder, 1993; Normand et al., 1977; Velarde and Normand, 1980; Bragard and Velarde, 1997, 1998).

Generally the Bénard problem with Marangoni stresses, gravity, and buoyancy involves several time scales. On the one hand we have the viscous and the thermal scales, $t_{\mathrm{vis}} = (d^2/\nu)$ and $t_{\mathrm{th}}(d^2/\chi)$, respectively. There are also two time scales associated with gravity and surface tension (Laplace force) that tend to suppress surface deformation, $t_{\mathrm{gr}} = (d/g)^{1/2}$ and $t_{\mathrm{cap}} = [(\rho d^3)/\sigma]^{1/2} \cdot g$ is gravity acceleration and $\rho$ is density or density contrast; $\upsilon$ and $\chi$ denote kinematic viscosity and thermal diffusivity, respectively. The time scale related to the

Marangoni effect is

$$t_{mar} = \left( \frac{\rho d^2}{|\sigma_T \beta|} \right)^{1/2}.$$

$\sigma_T = d\sigma/dT < 0$; $\beta = \Delta T/d$; $\beta > 0$ when heating from below, $T$ denotes temperature. There is also another time scale related to buoyancy due to the stratification imposed by the temperature gradient

$$t_{st} = \left( \frac{1}{|\alpha \beta| g} \right)^{1/2}.$$

$\alpha$ is the thermal-expansion coefficient, positive in the standard case. The ratios of the time scales give rise to the dimensionless groups

$$\Pr = \frac{t_{th}}{t_{vis}} = \frac{\upsilon}{\chi}, \qquad M = \sigma_T \frac{t_{th} t_{vis}}{t_{st}^2} = -\frac{\sigma_T \beta d^2}{\eta \chi}, \qquad R = \frac{t_{th} t_{vis}}{t_{st}^2} = \frac{\alpha \beta g d^1}{\upsilon \chi},$$

$$G = \frac{t_{th} t_{vis}}{t_{gr}^2} = \frac{g d^3}{\upsilon \chi}, \qquad B = \frac{t_{cap}^2}{t_{gr}^2} = \frac{\rho g d^2}{\sigma},$$

which are the Prandtl, Marangoni, Rayleigh, Galileo, and (static) Bond numbers, respectively.

These time scales are not always of the same quantitative order. For example, for the Pearson problem (Pearson 1958), dealing with the onset of steady monotonic instability in a liquid layer with undeformable surface, $M \approx 1$, but $G \gg 1$ and $R \ll, M$. Indeed, although Pearson neglects gravity, his assumption was equivalent to gravity had he been able to keep the surface level, whatever flows and thermal inhomogeneities existed. He also neglected buoyancy in the bulk. The characteristic time scale of the problem is $t_{th} \approx t_{vis} \approx t_{mar} (at \Pr \approx 1)$. For monotonic instability, hence the case leading to Bénard cells when heating the liquid layer from the liquid side, there is a finite limit of the critical Marangoni number as $G \to \infty$. As oscillatory instability does not appear in the one-layer problem with undeformable surface, we expect that if such instability is possible the critical Marangoni number tends to infinity with $G \to \infty$. Thus the critical Marangoni number should be scaled with $G$, as $G$ becomes very large ($G \to \infty$), in agreement with earlier studies (Chu and Velarde, 1988; García-Ybarra and Velarde, 1987).

It is indeed known that for high enough values of $G$ an oscillatory mode is the capillary-gravity wave. The time scales $t_{gr}$ and $t_{cap}$ associated with this twofold wave are much smaller than the viscous and the thermal time scales (at least for $\Pr \approx 1$, $B \approx 1$). Then dissipative effects are relatively weak and the dispersion relation is $\omega^2 = G \Pr k[1 + (k^2/B)] \tanh(k)$ (to nondimensionalize $\omega$, the thermal time scale is used hereafter; $k$ is the dimensionless wave number in units of $d^{-1}$). Clearly, the higher $G$ is (and the wave frequency), the stronger should be the work of the Marangoni stresses (i.e., the higher the critical Marangoni number) to excite and sustain capillary-gravity waves, in agreement with the above discussion. For a standard liquid layer, $\sigma_T < 0$, this instability appears when the liquid layer is heated from the air side ($M < 0$), as expected.

The capillary-gravity wave is not the only possible wave motion in the liquid layer. When the Marangoni number is high enough (and negative), there is also another high-frequency oscillatory mode. Indeed, when a liquid element rises to the surface, it creates a cold spot there. Then the surface-tension gradient acts toward this spot, pushing the element back to the bulk, hence overstability. High values of $M$ ensure that the oscillations exist. Let their characteristic time scale also be $t_{mar}$. The corresponding wave is called longitudinal or dilational, viscous, as it is due to the Marangoni stresses along the surface in contrast to capillary-gravity waves with essentially transverse motion of the surface (Lamb, 1945; Lucassen, 1968). Calculation yields the following expression for the frequency of the longitudinal wave (in the limit $M \to -\infty$):

$$\omega^2 = -M \frac{Pr}{Pr^{1/2} + 1} k^2.$$

Although this dilational wave has a genuinely dissipative nature, the damping rate is asymptotically smaller, $O(|M|^{1/4})$, than its frequency. Up to some extent the flow field accompanying the dilational wave is qualitatively similar to that of the capillary-gravity wave. Potential flow can be assumed in the bulk of the layer, while vorticity is present only in boundary layers at the bottom rigid plate and at the upper open surface. The boundary-layer thickness is of the order of $O(|M|^{-1/4})[O(G^{-1/4})]$. For the longitudinal wave, the horizontal velocity field in the surface boundary layer is stronger than the potential flow in the bulk [by $O(|M|^{1/4})$] at variance with the capillary-gravity wave. Thus the longitudinal motion is really concentrated near the surface. Furthermore, it appears that with an undeformable surface ($1 \ll |M| \ll G$), the longitudinal mode is always damped. Indeed, as stated above, oscillatory instability does not appear in the one-layer Marangoni problem without surface deformability. However, if the longitudinal wave is accompanied by a nonnegligible surface deformation ($|M| \geq G$), it can be amplified, a striking result.

Thus, at $G \gg 1$, two thresholds for oscillatory Marangoni instability are expected with corresponding two (high-frequency) wave modes, capillary-gravity and longitudinal/viscous/dilational wave motions. As stated above, to sustain the longitudinal wave one needs surface deformability. Alternatively, to sustain a capillary-gravity wave one needs the Marangoni effect. This is the tight coupling between capillary-gravity and longitudinal waves if they are to appear and be sustained past an instability threshold. The most dramatic manifestation of this coupling occurs at resonance, when frequencies (1) and (2) are equal to each other. Near resonance there is mode mixing, namely, the capillary-gravity mode in the parameter half-space from one side of the resonance manifold is swiftly converted into the longitudinal one when crossing the manifold, and vice versa. Another feature of resonance is that the damping/amplification rates are drastically enhanced here, namely, $O(G^{3/5})$ versus $O(G^{1/4})$ far from resonance.

If the liquid layer is deep enough and has an undeformable surface, the possibility also exists of coupling surface longitudinal waves to internal (negative-buoyancy-driven) waves with $|R| \ll G$. This may be called the Rayleigh–Marangoni problem. Indeed, the role of the capillary-gravity wave is now played by the Brunt–Väisälä frequency

$$\omega^2 = -R \, Pr \frac{k^2}{k^2 + \pi^2 n^2} \quad (n = 1, 2, \ldots)$$

for a stably stratified layer when it is heated from above. The wave–wave coupling is now between internal and surface longitudinal waves. In the absence of the Marangoni effect, no oscillatory motion by means of instability is possible, which again stresses the crucial role played by the coupling of the two wave disturbances.

Although general features of mode coupling are similar in the two problems, there are differences. In the Rayleigh–Marangoni case we have a countable number ($n = 1, 2, \ldots$) of internal wave modes, and the surface longitudinal wave can be coupled to each of them; hence there are a countable number of marginal stability conditions. The form of the marginal curves is qualitatively different. Furthermore, there is the minimally possible Rayleigh number (in absolute value), below which there is no oscillatory instability. No such bound was found for the Galileo number in the first problem (at least in the region where $G$ remains high).

## 4.3. Nonlinear Theory for Long-Wavelength Motions in Bénard Layers

The nonlinear evolution past threshold of either capillary-gravity or longitudinal waves poses formidable tasks. Let us then concentrate, for one of two possible waves, on a simplified analysis that is amenable to an experimental test. Consider a horizontal liquid layer open to air and heated from above in which long (a term that is made precise below) capillary-gravity waves can be excited. The liquid layer is placed on a flat rigid support, and the air layer is bounded from above by a flat rigid top that fits well with experimental setups. For simplicity, we assume that the layers are of infinite horizontal extent and treat the problem in two-dimensional (2D) geometry. At rest, there is a linear vertical temperature distribution.

Let $h_j$ denote vertical depth, $\rho_j$ density, $\upsilon_j$ kinematic viscosity, $\chi_j$ thermal diffusivity, and $\kappa_j$ thermal conductivity, where the subscripts $j = 1, 2$ refer to the liquid and the gas layers, respectively. The corresponding symbols without subscripts denote ratios: $h \equiv h_2/h_1$, $\rho \equiv \rho_2/\rho_1$, $\upsilon \equiv \upsilon_2/\upsilon_1$, $\chi \equiv \chi_2/\chi_1$, and $\kappa \equiv \kappa_2/\kappa_1$. We assume that $h$ is of the order of unity, while $\upsilon$ and $\chi$ are large enough, and $\rho$ and $\kappa$ are smaller than unity, in accordance with standard gas and liquid properties. The ratio of the dynamic viscosities, $\rho\upsilon$, is also small enough. The Prandtl numbers for both the liquid, $\mathrm{Pr} \equiv \upsilon_1/\chi_1$, and the gas, $P \equiv \upsilon_2/\chi_2$, are also taken to be of the order of unity.

As we consider only long waves, let us define a smallness parameter $\varepsilon$ as the ratio of the depth to a characteristic wavelength. Then there are two time scales in our problem. One of them is defined by heat diffusion, $t_{\mathrm{th}} = h_1^2/\chi_1$ (as Pr is assumed to be of the order of unity, the viscous time scale $t_{\mathrm{vis}} = h_1^2/\upsilon_1$ is of the order of the thermal one). The other time scale, $t_{\mathrm{gr}} = \varepsilon(h_1/g^{1/2})$, where $g$ is the gravity acceleration, is associated with long gravity waves. When $t_{\mathrm{th}} \ll t_{\mathrm{gr}}$, the heat and the viscous effects are predominant, which, in practice, corresponds to very shallow liquid layers or microgravity conditions. In the opposite situation, $t_{\mathrm{th}} \gg t_{\mathrm{gr}}$, the dissipation is limited to the boundary layers at the bottom and, if the Marangoni effect is significant, at the upper surface. In terms of the Galileo number, the first case corresponds to $G\varepsilon^2 \ll 1$, while the second corresponds to $G\varepsilon^2 \gg 1$. We shall consider $G \gg 1$.

The thickness $d$ of the boundary layers can be estimated as follows. We have $t_{\mathrm{gr}} \approx t_{\mathrm{th}}$, where $t_{\mathrm{gr}}$ is as defined above while for $t_{\mathrm{th}}$ we now have $t_{\mathrm{th}} = d^2/\chi_1$. Then we get

$$\frac{d}{h_1} \approx \frac{1}{\varepsilon^{1/2}G^{1/4}}. \tag{1}$$

We define dimensionless quantities by using suitable scales: $h_1$ for length, $(gh_1)^{1/2}$ for velocity, $(h_1/g)^{1/2}$ for time, $\rho_1 g h_1$ for pressure in the liquid, $\rho_2 g h_1$ for pressure in the air, $\beta h_1$ for temperature in the liquid, and $\kappa^{-1}\beta h_1$ for temperature in the air, where $\beta$ is the vertical temperature gradient in the liquid layer (note the convention taken here: $\beta > 0$ corresponds to heating from above, i.e., from the air side and $\beta < 0$ if heating is from below, i.e., from the liquid side). By pressure and temperature we denote deviations of the corresponding quantities from their stationary distribution, linear with the vertical coordinate.

Let $x$ and $z$ be the horizontal and the vertical coordinates, respectively. The bottom of the layer is taken at $z = -1$, the free surface at $z = \eta(x, t)$, and the top of the air layer at $z = h$, where

$t$ is time and $\eta(x, t)$ describes the liquid surface deformation. Thus, as already mentioned, we restrict consideration to 2D flow motions. To search for only long traveling-wave motions in a shallow layer, we redefine the horizontal variable $\xi = \varepsilon(x - Ct)$, where $C$ is a phase velocity to be determined. In addition, we scale horizontal velocity, pressure, and deformation of the surface $\eta$ with $\varepsilon^2$ and vertical velocity with $\varepsilon^3$ and introduce the slow time scale $\tau = \varepsilon^3 t$. The scale for temperature is determined by the leading convective contribution to the temperature field, which is of the order of $\varepsilon^2$. Accordingly, the equations governing long-wave disturbances are

$$u_\xi + w_z = 0, \tag{2}$$

$$\varepsilon^2 u_\tau - Cu_\xi + \varepsilon^2 wu_z = -P_\xi + \varepsilon^{-1}\left(\frac{\text{Pr}}{G}\right)^{1/2}(\varepsilon^2 u_{\xi\xi} + u_{zz}), \tag{3}$$

$$\varepsilon^4 w_0 - \varepsilon^2 Cw_\xi + \varepsilon^4 uw_\xi + \varepsilon^4 ww_z = -P_z + \varepsilon\left(\frac{\text{Pr}}{G}\right)^{1/2}(\varepsilon^2 w_{\xi\xi} + w_{zz}), \tag{4}$$

$$\varepsilon^2 T_\tau - CT_\xi + \varepsilon^2 uT_\xi + \varepsilon^2 wT_z + w = \varepsilon^{-1}\frac{1}{(\text{Pr }G)^{1/2}}(\varepsilon^2 T_{\xi\xi} + T_{zz}), \tag{5}$$

$$U_\xi + W_z = 0, \tag{6}$$

$$\varepsilon^2 U_\tau - CU_\xi + \varepsilon^2 UU_\xi + \varepsilon^2 WU_z = -\prod{}_\xi + \varepsilon^{-1}\left(\frac{\text{Pr}}{G}\right)^{1/2}\upsilon(\varepsilon^2 U_{\xi\xi} + U_{zz}), \tag{7}$$

$$\varepsilon^4 W_\tau - \varepsilon^2 CW_\xi + \varepsilon^4 UW_\xi + \varepsilon^4 WW_z = -\prod{}_z + \varepsilon\left(\frac{\text{Pr}}{G}\right)^{1/2}\upsilon(\varepsilon^2 W_{\xi\xi} + W_{zz}), \tag{8}$$

$$\varepsilon^2 \theta_\tau - C\theta_\xi + \varepsilon^2 U\theta_\xi + \varepsilon^2 W\theta_z + W = \varepsilon^{-1}\left(\frac{\text{Pr}}{G}\right)^{1/2}\frac{\upsilon}{P}(\varepsilon^2 \theta_{\xi\xi} + \theta_{zz}), \tag{9}$$

with the following boundary conditions (BCs):
at $z = -1$,

$$u = w = T = 0; \tag{10}$$

at $z = h$,

$$U = W = \theta = 0; \tag{11}$$

and at $z = \eta(\xi, \tau)$,

$$w = \varepsilon^2 \eta_\tau - C\eta_\xi + \varepsilon^2 u\eta_\xi = W, \tag{12}$$

$$u = U \tag{13}$$

$$p = \eta - \varepsilon^2\left[\frac{1}{B} - \varepsilon^2\frac{M}{G}(T + \eta)\right]\frac{\eta_{\xi\xi}}{N^3} + \frac{2}{N^2}\varepsilon\left(\frac{\text{Pr}}{G}\right)^{1/2}\left[w_z - \varepsilon^2\eta_\xi(u_z + \varepsilon^2 w_\xi) + \varepsilon^6 u_\xi\eta_\xi^2\right], \tag{14}$$

$$(u_z + \varepsilon^2 w_\xi)(1 - \varepsilon^6\eta_\xi^2) + 2\varepsilon^4\eta_\xi(w_z - u_\xi) + \frac{MN\varepsilon}{(\text{Pr }G)^{1/2}}(\eta_\xi + T_\xi + \varepsilon^2\eta_\xi T_z) = 0, \tag{15}$$

$$\eta + \theta = 0, \tag{16}$$

$$T_z - \varepsilon^4\eta_\xi T_\xi = \theta_z - \varepsilon^4\eta_\xi\theta_\xi, \tag{17}$$

with

$$M \equiv -\frac{d\sigma}{dT}\frac{\beta h_1^2}{\rho_1 \upsilon_1 \chi_1}; \; B \equiv \frac{\rho_1 g h_1^2}{\sigma}; \; N \equiv \left(1 + \varepsilon^6 \eta_\xi^2\right)^{1/2}.$$

Here $u$, $w$, $p$, and $T$ denote the horizontal and the vertical velocity components, pressure, and temperature fields in the liquid layer, respectively. $U$, $W$, $\Pi$, and $\theta$ denote the corresponding fields in the air layer. $M$ and $B$ correspond to the above-defined (static) Bond numbers. $B$ is assumed to be of the order of unity, while $M$ is taken to be large enough as, indeed, the Marangoni effect is the agent leading to instability past a (high enough) threshold. The scaling of $M$ with $\varepsilon$ is provided when the problem is solved.

Note that the dynamic properties of air are neglected in the normal [Eq. (14)] and the tangential [Eq. (15)] stress balances. The smallness of $\kappa$ permits us to write the BC representing continuity of temperature across the surface, $T + \eta = \kappa^{-1}(\theta + \eta)$ in the form of Eq. (16). Thus, $\kappa$ as well as $\rho$ dissappear from the equations.

Now let us discuss what should be the relation between the smallness parameters $\varepsilon$ and $G^{-1}$. As long as we limit ourselves to the case in which $\varepsilon^{-1/2} G^{-1/4} \ll 1$, i.e., to the case in which the liquid layer can be subdivided into the bulk for which the flow is potential, and the boundary layers, the most interesting asymptotics, correspond to the case in which the boundary-layer thickness and the deformation of the surface are of the same order. From Eq. (1), it follows that $\varepsilon^{-1/2} G^{-1/4} \approx \varepsilon^2$, i.e.,

$$\varepsilon = G^{-1/10}. \tag{18}$$

(We write $=$ to define $\varepsilon$ in terms of $G$). Then the effects of energy output (due to heat and viscous dissipation) and input (due to the Marangoni effect) will be of the same order as that of nonlinearity and dispersion. The latter two are in appropriate balance for the ideal liquid Korteweg–de Vries (KdV) equation for long waves in the shallow layer.

Turning to the equations in the air layer, we assume that, because of its relatively large kinematic viscosity ($\upsilon \gg 1$) and thermal diffusivity ($\chi \gg 1$), inertia in the air is no longer dominating over dissipative effects. Then, in the most general case,

$$a^2 \equiv \varepsilon^{-1} \mathrm{Pr}^{1/2} G^{-1/2} \upsilon \approx 1.$$

The coefficients of the Laplacians in Eqs. (7) and (9) are $a^2$ and $a^2/P$, respectively. From Eq. (18), in the liquid layer, Eqs. (3) and (5), it follows that they are $\varepsilon^4$ and $\varepsilon^4/\mathrm{Pr}$, respectively.

To solve problems (2)–(17), all components of $f = (u, w, p, T, U, W, \Pi, \theta)$ are suitably expanded with $\varepsilon$ ($\eta$ is not yet expanded here). In the bulk of the liquid layer and in the air layer we set

$$f(\xi, z, \tau) = f_0 + \varepsilon^2 f_1 + \cdots, \tag{19}$$

while in the boundary layer near the open surface

$$f(\xi, \bar{z}, \tau) = \bar{f}_0 + \varepsilon^2 \bar{f}_1 + \cdots, \tag{20}$$

and in the bottom boundary layer

$$f(\xi, \tilde{z}, \tau) = \tilde{f}_0 + \varepsilon^2 \tilde{f}_1 + \cdots, \tag{21}$$

where new vertical coordinates have been introduced,

$$\bar{z} \equiv -\frac{z}{\varepsilon^2}, \qquad \tilde{z} \equiv \frac{z+1}{\varepsilon^2},$$

in both the surface and the bottom boundary layers.

After substituting Eqs. (19)–(21) into Eqs. (2)–(17), we get a hierarchy of linear problems corresponding to $\varepsilon^n (n = 0, 2 \ldots)$. At each step a solvability condition must be satisfied that eventually yields an evolution equation for $\eta$. Without loss of generality, we restrict consideration to solutions with a zero mean value of $\eta$.

In the boundary layers, Eqs. (2) and (4) yield $\bar{p}_{0\bar{z}} = \bar{w}_{0\bar{z}} = \tilde{p}_{0\tilde{z}} = \tilde{w}_{0\tilde{z}} = 0$; hence these functions are constants. Thus BCs (10), (12), and (14) can be directly applied to $p_0$ and $w_0$. Finally, the leading-order solution in the bulk coincides with that in the inviscid liquid case:

$$P_0 = \eta, \quad u_0 = \eta, \quad w_0 = -\eta_\xi (1 + z)$$

with $C = 1$. For simplicity here we consider only the right propagating wave ($C > 0$). The results for the left propagating wave ($C = -1$) can then be deduced by symmetry.

At the same time we can write

$$\bar{p}_0 = \tilde{p}_0 = \eta, \quad \bar{w}_0 = -\eta_\xi, \quad \tilde{w}_0 = 0. \tag{22}$$

For the horizontal velocity in the bottom boundary layer, Eqs. (3) and (22) yield

$$-\tilde{u}_{0\xi} = -\eta_\xi + \tilde{u}_{0\tilde{z}\tilde{z}}.$$

When BC (10) is taken into account, the solution is

$$\tilde{u}_0 = \eta - \frac{\tilde{z}}{2\pi^{1/2}} \int_\xi^\infty \frac{\eta(\xi')}{(\xi' - \xi)^{3/2}} \exp\left[-\frac{\tilde{z}^2}{4(\xi' - \xi)}\right] d\xi'.$$

Then, when Eq. (2) and BC (10) are used, $\tilde{w}_1$ is

$$\tilde{w}_1 = -\int_0^{\tilde{z}} \tilde{u}_{0\xi}(y') \, d\tilde{z}' = -\eta_\xi \tilde{z} + \frac{1}{\pi^{1/2}} \frac{d}{d\xi} \int_\xi^\infty \frac{\eta(\xi')}{(\xi' - \xi)^{1/2}} \left[1 - \exp\frac{\tilde{z}^2}{4(\xi' - \xi)}\right] d\xi'.$$

The matching condition between $\tilde{w}_0 + \varepsilon^2 \tilde{w}_1$ and $w_0 + \varepsilon^2 w_1$ yields the BC for the function $w_1$ in the bulk at $z = -1$:

$$w_1 = \frac{1}{\pi^{1/2}} \frac{d}{d\xi} \int_\xi^\infty \frac{\eta(\xi')}{(\xi' - \xi)^{1/2}} \, d\xi. \tag{23}$$

In the surface boundary layer, Eqs. (3) and (22) yield

$$-\bar{u}_{0\xi} + \eta_\xi \bar{u}_{0\bar{z}} = -\eta_\xi + \bar{u}_{0\bar{z}\bar{z}}.$$

This equation poses a difficult task because of the appearance of the variable coefficient, which moreover is the unknown function $\eta_\xi$. However, redefining the vertical coordinate (Prandtl transformation) as

$$\phi = \bar{z} + \eta(\xi)$$

yields

$$-\bar{u}_{0\xi} = -\eta_\xi + \bar{u}_{0\phi\phi},$$

thus greatly simplifying the problem. Then

$$\bar{u}_0 = \eta + V(\xi, \phi) \tag{24}$$

with

$$V = \frac{\phi}{2\pi^{1/2}} \int_\xi^\infty \frac{c(\xi')}{(\xi' - \xi)^{3/2}} \exp\left[ -\frac{\phi^2}{4(\xi' - \xi)} \right] d\xi',$$

where $c(\xi)$ is yet to be found.

Continuity equation (2) becomes

$$\bar{u}_{0\xi} + \bar{u}_{0\phi}\eta_\xi - \bar{w}_{1\phi} = 0.$$

Taking BC (12) into account, which is now at $\phi = 0$, we get

$$\bar{w}_1 = \eta_\xi \phi + V\eta_\xi + \eta_\tau + \eta\eta_\xi + \frac{1}{\pi^{1/2}} \frac{d}{d\xi} \int_\xi^\infty \frac{c(\xi')}{(\xi' - \xi)^{1/2}} \left\{ 1 - \exp\left[ -\frac{\phi^2}{4(\xi' - \xi)} \right] \right\} d\xi' \tag{25}$$

Finding $\bar{p}_1$ with the help of Eq. (4) and BC (14), and using the matching procedure, we get, at $z = 0$,

$$p_1 = -\eta_{\xi\xi}/B. \tag{26}$$

BCs (23) and (26) are already sufficient to find the potential solution in the main bulk. Equations (2)–(4) become

$$u_{1\xi} + w_{1z} = 0, \tag{27}$$

$$-u_{1\xi} + \eta_\tau + \eta\eta_\xi = -p_{1\xi}, \tag{28}$$

$$\eta_{\xi\xi} = -p_{1z}. \tag{29}$$

The solution of problem (23), Eqs. (26)–(29), is

$$w_1 = -\eta_\tau z - \eta\eta_\xi z - \left( \frac{1}{2} - \frac{1}{B} \right)\eta_{\xi\xi\xi} + \eta_{\xi\xi\xi} + \frac{1}{\pi^{1/2}} \frac{d}{d\xi} \int_\xi^\infty \frac{\eta(\xi')}{(\xi' - \xi)^{1/2}} d\xi' \tag{30}$$

($p_1$ and $u_1$ are not needed later on). Compared with the inviscid liquid case, an additional integral term appears in Eq. (30), which is due to a delay of the flow in the bottom boundary layer.

The matching condition between $w_0 + \varepsilon w_1$ and $\bar{w}_0 + \varepsilon \bar{w}_1$ yields an implicit representation for $c(\xi)$:

$$\frac{1}{\pi^{1/2}} \frac{d}{d\xi} \int_\xi^\infty \frac{c(\xi')}{(\xi' - \xi)^{1/2}} d\xi' = \left[ 2\eta_\tau + 3\eta\eta_\xi + \left( \frac{1}{3} - \frac{1}{B} \right)\eta_{\xi\xi\xi} \right]$$

$$+ \frac{1}{\pi^{1/2}} \frac{d}{d\xi} \int_\xi^\infty \frac{\eta(\xi')}{(\xi' - \xi)^{1/2}} d\xi'. \tag{31}$$

Let us now consider the temperature field. In the surface boundary layer, Eqs. (5) and (22) and BC (17), here reduced to $\bar{T}_{0\bar{z}} = 0$, yield

$$\bar{T}_0 = -\eta$$

Thus the contribution to the thermocapillay term in BC (15) is zero in this order.

For the first correction in the surface boundary layer, $\bar{T}_1$, Eq. (5) and BC (17) yield

$$\bar{T}_{0\tau} - \bar{T}_{1\xi} + \bar{u}_0 \bar{T}_{0\xi} - \bar{w}_0 \bar{T}_{1\bar{z}} + \bar{w}_1 = \frac{1}{\text{Pr}} \bar{T}_{1\bar{z}\bar{z}},$$

and, at $\bar{z} = -\eta$,

$$\bar{T}_{1\bar{z}} = -Q(\tau, \xi),$$

where $Q$ is obtained when the problem is solved in the air layer.

By introducing the known functions and using $\phi$ instead of $\bar{z}$, we get

$$-\bar{T}_{1\xi} + \eta_\xi \phi + \frac{1}{\pi^{1/2}} \frac{d}{d\xi} \int_\xi^\infty \frac{c(\xi')}{(\xi' - \xi)^{1/2}} \left\{ 1 - \exp\left[ -\frac{\phi^2}{4(\xi' - \xi)} \right] \right\} d\xi' = \frac{1}{\text{Pr}} \bar{T}_{1\phi\phi},$$

and, at $\phi = 0$,

$$\bar{T}_{1\phi} = -Q(\tau, \xi), \tag{32}$$

whose solution is

$$\begin{aligned}
\bar{T}_{1\xi} &= \eta\phi + \frac{1}{\pi^{1/2}} \int_\xi^\infty \frac{c(\xi')}{(\xi' - \xi)^{1/2}} d\xi' + \frac{1}{\pi^{1/2}} \frac{\text{Pr}}{1 - \text{Pr}} \int_\xi^\infty \frac{c(\xi')}{(\xi' - \xi)^{1/2}} \exp\left[ -\frac{\phi^2}{4(\xi' - \xi)} \right] d\xi' \\
&\quad - \frac{1}{\pi^{1/2}} \frac{\text{Pr}^{1/2}}{1 - \text{Pr}} \int_\xi^\infty \frac{c(\xi')}{(\xi' - \xi)^{1/2}} \exp\left[ -\frac{\text{Pr}\,\phi^2}{4(\xi' - \xi)} \right] d\xi' \\
&\quad + \frac{1}{\pi^{1/2}} \frac{1}{\text{Pr}^{1/2}} \int_\xi^\infty \frac{\eta(\xi') + Q(\xi')}{(\xi' - \xi)^{1/2}} \exp\left[ -\frac{\text{Pr}\,\phi^2}{4(\xi' - \xi)} \right] d\xi'.
\end{aligned} \tag{33}$$

Note that as we do not need an explicit expression for the temperature field in the bottom boundary layer, the results do not depend on the type of heat exchange at the solid support.

Finally, let us use BC (15) for the tangential stress balance. It becomes, at $\phi = 0$,

$$-\bar{u}_{0\phi} + m\bar{T}_{1\xi} = 0, \tag{34}$$

where for convenience we have redefined the Marangoni number,

$$m \equiv \frac{M\varepsilon^{10}}{\text{Pr}} = \frac{M}{G}. \tag{35}$$

This modified Marangoni number, $m$, of the order of unity, corresponds to the most general case [viscous and thermocapillary stresses are of the same order as that in Eq. (34)]. This also means that $M$ is of the order of $G$. It is an inverse (dynamic) Bond number.

Substituting Eqs. (24) and (33) into BC (34) and using Eq. (31), we get

$$
2\left(1 - \frac{m}{\mathrm{Pr}^{1/2} + 1}\right)\left[\eta_\tau + \frac{3}{2}\eta\eta_\xi + \left(\frac{1}{6} - \frac{1}{2B}\right)\eta_{\xi\xi\xi}\right] + \left(m\frac{2\mathrm{Pr}^{1/2} + 1}{\mathrm{Pr} + \mathrm{Pr}^{1/2}} - 1\right)
$$
$$
\times \frac{1}{\pi^{1/2}}\frac{d}{d\xi}\int_\xi^\infty \frac{\eta(\xi')}{(\xi' - \xi)^{1/2}}\,d\xi' + \frac{m}{\mathrm{Pr}^{1/2}}\frac{1}{\pi^{1/2}}\frac{d}{d\xi}\int_\xi^\infty \frac{Q(\xi')}{(\xi' - \xi)^{1/2}}\,d\xi' = 0.
\tag{36}
$$

Expression (36) is a dissipation-modified KdV reinstate with $Q$ from Eq. (32) yet to be determined.

The leading-order problems (6)–(9), (11)–(13), and (16) in the air layer are linear. Their solution is found in terms of the Fourier components

$$
f_k = \int_{-\infty}^{+\infty} f(\xi)\exp(-ik\xi)\,d\xi, \qquad f(\xi) = \frac{1}{2\pi}\int_{-\infty}^{+\infty} f_k\exp(ik\xi)\,dk.
\tag{37}
$$

Thus (36) becomes

$$
2\left(1 - \frac{m}{\mathrm{Pr}^{1/2} + 1}\right)\left[\eta_\tau + \frac{3}{2}\eta\eta_\xi + \left(\frac{1}{6} - \frac{1}{2B}\right)\eta_{\xi\xi\xi}\right]
$$
$$
- \frac{2m}{\mathrm{Pr}^{1/2}}\frac{1}{2\pi}\int_{-\infty}^{+\infty} Q_2(\xi - \xi')\left[\eta_\tau + \frac{3}{2}\eta\eta_{\xi'} + \left(\frac{1}{6} - \frac{1}{2B}\right)\eta_{\xi'\xi'\xi'}\right]d\xi'
$$
$$
+ \left(m\frac{2\mathrm{Pr}^{1/2} + 1}{\mathrm{Pr} + \mathrm{Pr}^{1/2}} - 1\right)\frac{1}{\pi^{1/2}}\frac{d}{d\xi}\int_\xi^\infty \frac{\eta(\xi')}{(\xi' - \xi)^{1/2}}\,d\xi'
$$
$$
- \frac{m}{\mathrm{Pr}^{1/2}}\frac{1}{2\pi}\int_{-\infty}^{+\infty} Q_1'(\xi - \xi')\eta(\xi')\,d\xi' = 0,
\tag{38}
$$

where $Q_1'$ and $Q_2$ are the originals of $Q_{1k}\sqrt{-ik}$ and $Q_{2k}$, respectively, and

$$
Q_{1k} = -c_1 - \frac{P}{1 - P}c_1C_1 - \frac{P}{1 - P}c_2S_1 - \frac{P^{1/2}}{1 - P}c_2\frac{1}{S_2} + \frac{\sqrt{-ik}}{a}hP^{1/2}c_1\frac{C_2}{S_2}
$$
$$
+ \frac{P^{3/2}}{1 - P}c_1S_1\frac{C_2}{S_2} + \frac{P^{3/2}}{1 - P}c_2C_1\frac{C_2}{S_2} + c_2P^{1/2}\frac{C_2}{S_2} + \frac{\sqrt{-ik}}{a}P^{1/2}\frac{C_2}{S_2},
$$

$$
Q_{2k} = -s_1 - \frac{P}{1 - P}s_1C_1 - \frac{P}{1 - P}s_2S_1 - \frac{P^{1/2}}{1 - P}s_2\frac{1}{S_2} + \frac{\sqrt{-ik}}{a}hP^{1/2}s_1\frac{C_2}{S_2}
$$
$$
+ \frac{P^{3/2}}{1 - P}s_1S_1\frac{C_2}{S_2} + \frac{P^{3/2}}{1 - P}s_2C_1\frac{C_2}{S_2} + s_2P^{1/2}\frac{C_2}{S_2},
$$

$$
c_1 = \frac{2\sqrt{-ik}a(C_1 - 1) - ikS_1}{2\sqrt{-ik}a(C_1 - 1) + ikhS_1}, \qquad s_1 = \sqrt{-ik}a\frac{C_1 - 1}{2\sqrt{-ik}a(C_1 - 1) + ikhS_1},
$$

$$
c_2 = c_1\frac{1 - C_1}{S_1} - \frac{2}{S_1}, \qquad s_2 = s_1\frac{1 - C_1}{S_1} - \frac{1}{S_1},
$$

$$S_1 = \sinh\left(\frac{\sqrt{-ik}}{a}h\right), \qquad C_1 = \cosh\left(\frac{\sqrt{-ik}}{a}h\right), \qquad S_2 = \sinh\left(\frac{\sqrt{-ik}}{a}P^{1/2}h\right),$$

$$C_2 = \cosh\left(\frac{\sqrt{-ik}}{a}P^{1/2}h\right).$$

Equation (38) is the most general dissipation-modified KdV equation describing long surface-tension-gradient-driven waves in a Bénard layer. They significantly generalize earlier work (Chu and Velarde, 1991; Garazo and Velarde, 1991; Velarde et al., 1991; Nepomnyashchy and Velarde, 1994, Rednikov et al., 1995). On the other hand, another earlier result with no Marangoni effect (Miles, 1976) follows from Eq. (36) [or Eqs. (38)] by setting $m = 0$, as expected.

Equation (38) is easier solved in Fourier space:

$$2\left(1 - \frac{m}{\mathrm{Pr}^{1/2}+1} - \frac{m}{\mathrm{Pr}^{1/2}}Q_{2k}\right)\left[\eta_{k\tau} + \frac{3}{4}ik\int_{-\infty}^{+\infty}\eta_{k-k'}\eta_{k'}\mathrm{d}k' - ik^3\left(\frac{1}{6} - \frac{1}{2B}\right)\eta_k\right]$$

$$-\left(m\frac{2\mathrm{Pr}^{1/2}+1}{\mathrm{Pr}+\mathrm{Pr}^{1/2}} + \frac{m}{\mathrm{Pr}^{1/2}}Q_{1k} - 1\right)\sqrt{-ik}\eta_k = 0. \tag{39}$$

The dissipation-modified KdV equations (38) or (39) possess the necessary ingredients to have solutions in the form of stationary propagating waves: there is an unstable wave-number interval, in which the energy is brought by the Marangoni effect and dissipation occurs on the wave numbers belonging to the stability interval. The convective nonlinearity does not participate in the input/output of energy. However, it redistributes the energy from long to short waves, making possible the dynamic equilibrium and the appropriate energy balance for the dissipative wave. We may expect solutions such as sustained dissipative periodic wave trains and solitary waves or dissipative solitons as in the above-mentioned drastically simplified dissipation-modified KdV equations (Christov and Velarde, 1995; Nekorkin and Velarde, 1994; Nepomnyashchy and Velarde, 1994; Rednikov et al., 1995; Velarde et al., 1995).

Let us consider the case of a thin air gap above the liquid layer ($h \ll 1$; however, $h$ should remain larger than the surface deformation). In this case, the air motion is dissipation dominated. Then $\theta = \eta(z - h)/h$ and $Q = \eta/h$. By using this in Eq. (36), we get

$$2\left(1 - \frac{m}{\mathrm{Pr}^{1/2}+1}\right)\left[\eta_\tau + \frac{3}{2}\eta\eta_\xi\left(\frac{1}{6} - \frac{1}{2B}\right) + \eta_{\xi\xi\xi}\right]$$

$$+\left(m\frac{2\mathrm{Pr}^{1/2}+1}{\mathrm{Pr}+\mathrm{Pr}^{1/2}} + m\frac{1}{h\,\mathrm{Pr}^{1/2}} - 1\right)\frac{1}{\pi^{1/2}}\frac{d}{d\xi}\int_\xi^\infty\frac{\eta(\xi')}{(\xi'-\xi)^{1/2}}\,\mathrm{d}\xi' = 0. \tag{40}$$

The critical modified Marangoni number is now

$$m_b = \left(\frac{2\mathrm{Pr}^{1/2}+1}{\mathrm{Pr}+\mathrm{Pr}^{1/2}} + \frac{1}{h\,\mathrm{Pr}^{1/2}}\right)^{-1}. \tag{41}$$

An important fact is that $m_b$ considerably decreases with decreasing $h$. Thus, for observing sustained gravity-capillary waves, the thinner the air gap the better. Take, for example, a water like liquid. For illustration we take $h_1 = 0.1$, $h_2 = 0.01 (h = 0.1)$, $d\sigma/dT = -0.15$, $g = 10^3$, and $Pr = 6$ (for dimensional quantities, the cgs system is used). Then, according to Eq. (41), the temperature difference applied to the liquid layer needed to excite and sustain capillary-gravity waves is 15 K or 150 K/cm gradient.

However, in the supercritical case, with expression (40), all wave numbers are unstable. This is related to the fact that, from Eq. (39), in the limit of high dissipation in the air layer ($h \ll 1$ or $a \ll 1$), the band of unstable wave numbers shifts to higher and higher $k$, where our long-wave approximation ceases to be valid. In this case, to obtain a suitable energy balance to maintain the waves, we must proceed to a higher-order approximation [$\varepsilon^2$ in expression (40); hence $m - m_b \approx \varepsilon^2$].

Finally, let us briefly mention that the vanishing of the coefficient of the ideal KdV terms in expression (40), at $m = Pr^{1/2} + 1$, corresponds to the resonance between capillary-gravity and longitudinal waves. Our approach is not valid in the vicinity of this resonance point. However, note that the resonance value of $m$ is always higher than $m_b$.

### 4.4. A Few Experiments

Both mass absorption and desorption and heat transfer experiments have been carried out with Bénard layers (Linde et al., 1993a, 1993b, 1997; Wierschem et al., 1999). For the case of heat transfer (liquid depths from 0.3 to 0.8 cm), liquid octane is poured in a square or a cylindrical vessel and in an annular channel (1.5- and 2.0-cm inner and outer radii, respectively). The bottom is cooled by air or water at 20 °C and the cover made of quartz placed at 0.3 cm above the liquid is heated, hence establishing in the liquid layer a temperature gradient. For values of this gradient ranging from 10 to 200 K/cm, solitary waves and periodic wave trains have been observed that show properties similar to those of the waves also observed in mass transfer experiments.

For mass transfer the following setup was devised. In vessel A, either a cylindrical container or an annular channel is filled with liquid (liquid depth 1.8 cm). Two vessels, $B_1$ and $B_2$, are also filled with another liquid. With pentane in $B_1$ and $B_2$, either xylene, nonane, trichloroethylene or benzene are used as absorbing liquid in A, while with toluene as absorbing liquid in A either hexane, pentane, acetone or diethylether in $B_1$ and $B_2$ are used. In all cases the results are qualitatively the same. Glass cover C is placed on top of vessels $B_1$ and then when the vessel C is full of hexane vapor, say, it is placed on top of A, thus allowing the absorption of hexane by the toluene liquid in A. The adsorption and the subsequent absorption processes are rather strong, hence creating Marangoni stresses high enough to trigger and sustain instability. During the whole duration of the experiment, hexane vapor is also allowed to diffuse from the two vessels, $B_2$ to A.

Observation and recording with a CCD camera is made by shadowgraph from the top with a pointlike illumination from the bottom up. For instance, with cylindrical or annular cylindrical containers, at first rather violent chaotic motions occur along the surface in A, with waves moving in practically all directions, but finally, after $\sim$1 min, when most of the vapor in C has been adsorbed, a dramatic self-organization leads to strikingly regular wave motion. Long-time-lasting, synchronically colliding counterrotating periodic wave trains have been observed in an annular channel for $\sim$50–200 s, while a single (periodic) wave train with either clockwise or counterclockwise rotation remained up to 450 s. When waves or counterrotating wave trains collide, typical mean wave velocities at the outer wall of the annular channel before and after collision are, respectively, 2.7 and 1.7 cm/s (corresponding to angular speeds of 71.4 and 45.7 deg/s, respectively). Thus the mean wave velocity right after collision is $\sim$64% (with less than 2% error), the mean wave speed measured before collision. Approximately 0.2 s after collision the original wave speed is recovered. Postcollision trajectories of solitary waves or wave crests experience phase shifts as in the case of solitons (Zabusky and Krustal, 1965). Reflections at walls also illustrate the solitonic or shock behavior of the waves that occur with and without the formation of a (phase-locked, third wave) Mach stem according to the angle of incidence (Courant and Friedrichs, 1948; Wu, 1987; Krehl and van der Geest, 1991; Russell, 1885). The phenomena observed are complex, and only recently has there been a clear-cut distinction between mostly surface waves and (mostly) internal waves (Rayleigh–Marangoni problem), all of them triggered and sustained by the Marangoni effect. Further details about the experiments can be found in recent literature (Linde et al., 1997; Wierschem et al., 1999; Santiago-Rosanne et al., 1997).

## 4.5. Acknowledgments

The authors acknowledge fruitful discussions with Michèle Adler, J. Bragard, C. I. Christov, X. L. Chu, K. Loeschcke, A. A. Nepomnyashchy, M. Santiago-Rosanne, W. Waldhelm, and A. Wierschem. The research was supported by various grants: EU Network ERBCHRX-CT96-0107, EU Network ERBFMRX-CT96-0010 and DGICYT PB 96-599.

## References

Bragard, J. and Velarde, M. G., J. Non-Equilib. Thermodyn. **22**, 95 (1997).

Bragard, J. and Velarde, M. G., J. Fluid Mech. **368**, 165 (1998).

Courant, R. and Friedrichs, K. O., *Supersonic Flow and Shock Waves* (Interscience, New York, 1948).

Christov, C. I. and Velarde, M. G., Physica D **86**, 323 (1995).

Chu, X.-L. and Velarde, M. G., Phys. Chem. Hydrodyn. **10**, 727 (1988).

Chu, X.-L. and Velarde, M. G., J. Colloid Interface Sci., **131**, 471 (1989).

Chu, X.-L. and Velarde, M. G., Phys. Rev. A **43**, 1094 (1991).

Davis, S. H., Ann. Rev. Fluid Mech. **19**, 403 (1987).

Edwards, D. A., Brenner, H., and Wasan, D. T., *Interfacial Transport Processes and Rheology* (Butterworth-Heineman, Boston, 1991).

Garazo, A. N. and Velarde, M. G., Phys. Fluids A **3**, 2295 (1991).

García-Ybarra, P. L. and Velarde, M. G., Phys. Fluids **30**, 1649 (1987).

Koschmieder, E. L., *Bénard Cells and Taylor Vortices* (Cambridge U. Press, Cambridge, 1993).

Krehl, P. and van der Geest, M., Shock Waves **1**, 3 (1991).

Lamb, H., *Hydrodynamics* (Dover, New York, 1945), p. 349.

Levich, B. G., *Physicochemical Hydrodynamics* (Prentice-Hall, Englewood Cliffs, NJ, 1965).

Linde, H., Chu X.-L., and Velarde, M. G., Phys. Fluids A **5**, 1068 (1993a).

Linde, H., Chu X.-L., Velarde, M. G., and Waldhelm, W., Phys. Fluids A **5**, 3162 (1993b).

Linde, H., Velarde, M. G., Wierschem, A., Waldhelm, W., Loeschcke, K., and Rednikov, A. Ye., J. Colloid Interface Sci. **188**, 16 (1997).

Lucassen, J., Trans. Faraday Soc. **64**, 2221 (1968).

Miles, J. M., Phys. Fluids **19**, 1063 (1976).

Nekorkin, V. I. and Velarde, M. G., Int. J. Bif. Chaos **4**, 1135 (1994).

Nepomnyashchy, A. A. and Velarde, M. G., Phys. Fluids **6**, 187 (1994).

Normand, Ch., Pomeau Y. and Velarde, M. G., Rev. Mod. Phys. **49**, 581 (1977) and references therein.

Ostrach, S., Ann. Rev. Fluid Mech. **14**, 313 (1982).

Pearson, J. R., J. Fluid Mech. **4**, 489 (1958).

Rednikov, A. Ye., Velarde, M. G, Ryazantsev, Yu. S., Nepomnyashchy A. A., and Kurdyumov, V. N., Acta Appl. Math. **39**, 457 (1995).

Russell, J. S., *The Wave of Translation in the Oceans of Water, Air and Ether* (Trübner, & London, 1885) (An Appendix reproduces Russell's celebrated *Report on Waves* made to the meetings of the British Association in 1842 and 1843).

Santiago-Rosanne, M., Vignès-Adler, M., and Velarde, M. G., J. Colloid. Interface Sci. **191**, 65–80 (1997).

Sternling, C. V. and Scriven, L. E., AIChE J. **5**, 514 (1959).

Velarde, M. G., Phil. Trans. R. Soc. London A **356**, 829 (1998).

Velarde, M. G., V. I. Nekorkin, V. I., and Maksimov, A., Int. J. Bif. Chaos **5**, 831 (1995).

Velarde, M. G., Chu, X.-L., and Garazo, A. N., Phys. Scripta T **35**, 71 (1991).

Velarde, M. G. and Normand, C., Sci. Am. **243**, 92 (1980).

Wierschem, A., Velarde, M. G., Linde, H., and Waldhelm, W., J. Colloid Interface Sci., **212**, 365 (1999).

Wu, T. Y., J. Fluid Mech., **184**, 75 (1987).

Zabusky, N. J. and Kruskal, M. D., Phys. Rev. Lett. **15**, 57 (1965).

# 5

# Thermocapillary-Coriolis Instabilities

ABDELFATTAH ZEBIB AND CÉDRIC LE CUNFF

## 5.1. Introduction

Understanding thermocapillary instabilities is important to successful material processing in the microgravity environment of space (Ostrach, 1982). Accordingly, there are a large number of studies, both theoretical and experimental, to predict these transitions (Smith and Davis, 1983; Xu and Davis, 1983; Neitzel, et al., 1993; Wanschura, et al., 1995; Levenstam and Amberg, 1995; Xu and Zebib, 1998). However, experiments conducted in space (Kamotani, et al., 1992) have not been in agreement with theoretical predictions of critical conditions for transition to oscillatory states. One reason proposed to explain the disagreement is that theoretical models assume nondeformable interfaces that are typically very small. However, it was shown (Mundrane and Zebib, 1995) that inclusion of free-surface deformation in one model does not influence transition at small capillary numbers. In this chapter we show that a candidate for explaining the discrepancy is that existing theoretical and laboratory models do not take into account the Coriolis force resulting from the orbiting motion. We demonstrate this here by assessing the influence of system rotation on the instabilities of the thermocapillary return flows in infinite plane layers (Zebib, 1996) and cylindrical liquid bridges.

## 5.2. Governing Equations

The infinite cylinder has a radius $a$ and is oriented along the $z$ axis. Perpendicular to the $z$ direction, the fixed $x$ and $y$ directions are defined, as well as the usual polar coordinates $(r, \theta)$. The flow field is represented by $u$, $v$, and $w$ in the $r, \theta$, and $z$ directions, respectively, and $P$ and $T$ are the pressure and the temperature fields, respectively. The equations are solved in polar coordinates that rotate about the fixed frame with an angular velocity $\vec{\Omega}$. Fluid properties are taken constant, except for the surface tension, which is taken as a linear function of the temperature (* indicates a dimensional quantity):

$$\sigma^* = \sigma_r^* - \gamma(T^* - T_r^*). \tag{1}$$

We define the temperature gradient $b = -\partial T^*/\partial z$, as well as the dynamic viscosity $\mu$, the kinematic viscosity $\nu$, the density $\rho$, the thermal diffusivity $\kappa$, and the vector Taylor number $\vec{\tau} = \vec{\Omega} a^2/\nu$ (with components $\tau_1$, $\tau_2$, and $\tau_3$, in Cartesian coordinates). The scales used, respectively, for the length, velocity, time, temperature, and pressure are $a$, $\gamma ba/\mu$, $\mu/\gamma b$, $ba$, and $\gamma b$.

The resulting equations are therefore

$$\nabla \cdot \vec{v} = 0, \tag{2}$$

$$\text{Re}\left[\frac{\partial \vec{v}}{\partial t} + \nabla \cdot (\vec{v}\vec{v})\right] = -\nabla P + \nabla^2 \vec{v} - 2\vec{\tau} \times \vec{v}, \tag{3}$$

$$\text{Ma}\left[\frac{\partial T}{\partial t} + \nabla \cdot (\vec{v}T)\right] = \nabla^2 T, \tag{4}$$

with two nondimensional quantities, the Reynolds number, $\text{Re} = \rho \gamma b a^2/\mu^2$, and the Marangoni number, $\text{Ma} = \text{RePr}$, where $\text{Pr} = \nu/\kappa$ is the Prandtl number.

The boundary conditions are applied at the free surface ($r = 1$). The assumption that the surface is nondeformable requires that

$$u = 0. \tag{5}$$

From the balance between viscous and thermocapillary stresses, we have

$$\frac{\partial w}{\partial r} + \frac{\partial T}{\partial z} = 0, \tag{6}$$

$$r \frac{\partial}{\partial r}\left(\frac{v}{r}\right) + \frac{1}{r}\frac{\partial T}{\partial \theta} = 0. \tag{7}$$

Finally, the heat transfer condition at the surface implies that

$$\frac{\partial T}{\partial r} + \text{Bi}(T - T_r) = 0, \tag{8}$$

with the Biot number defined as $\text{Bi} = ha/k$, where $h$ is the heat transfer coefficient and $k$ is the thermal conductivity. At $r = 0$, the solution has to be well behaved since this location is a singular point for the cylindrical coordinates system.

We also consider an infinite layer of depth $a$ oriented along the $x$ axis. The velocities $(u, v, w)$ correspond to the $(x, y, z)$ rotating coordinates. The equations of motion are again Eqs. (2)–(4), and the boundary conditions at the rigid bottom surface $z = 0$ are taken as

$$u = v = w = \frac{\partial T}{\partial z} = 0, \tag{9}$$

while on the free surface $z = 1$ we have

$$w = 0, \tag{10}$$

$$\frac{\partial u}{\partial z} + \frac{\partial T}{\partial x} = 0, \tag{11}$$

$$\frac{\partial v}{\partial z} + \frac{\partial T}{\partial y} = 0, \tag{12}$$

$$\frac{\partial T}{\partial z} + \text{Bi}(T - T_r) + Q = 0. \tag{13}$$

Here $Q$ is an imposed heat flux such that $T = T_r$ on $z = 1$ in the basic steady state.

## 5.3. Results and Discussion

### 5.3.1. *Plane Layer*

#### 5.3.1.1. *Base Flow*

The basic return flow of Smith and Davis (1983) is modified by rotation. The flow field and the temperature are given by

$$U = \bar{A} \sin(az) \sinh(az) + \bar{B} \sin(az) \cosh(az) + \bar{C} \cos(az) \sinh(az), \tag{14}$$

$$V = -\frac{\bar{G}}{2a^2} - \bar{A} \cos(az) \cosh(az) - \bar{B} \cos(az) \sinh(az) - \bar{C} \sin(az) \cosh(az), \tag{15}$$

$$T = -x + \frac{\bar{M}}{2a^2} \{\bar{A} [\cos(az) \cosh(az) - \cos(a) \cosh(a)] + \bar{B} [\cos(az) \sinh(az)$$
$$- \cos(a) \sinh(a)] - \bar{C} [\sin(az) \cosh(az) - \sin(a) \cosh(a)]\}, \tag{16}$$

where $a^2 = \tau_3$ is assumed greater than 0 without loss of generality (if $\tau_3 < 0$ then $a^2 = -\tau_3$ and $V$ change sign). A pressure gradient $\partial p / \partial x = -\bar{G} = 2a^2 \bar{A}$ develops such that $\int_0^1 U \, dz = 0$. The constants $\bar{A}$, $\bar{B}$, and $\bar{C}$ are determined by the boundary conditions to be

$$\bar{A} = \frac{\sin a \, \sinh a}{a(\sinh a \, \cosh a - \sin a \, \cos a)}, \tag{17}$$

$$\bar{B} = \bar{C} = \frac{(\cos a \, \sinh a - \sin a \, \cosh a)}{2a(\sinh a \, \cosh a - \sin a \, \cos a)}, \tag{18}$$

and $Q = 0$. It is easily shown that the return flow of Smith and Davis (1983) is recovered when $a = 0$. The components of rotation $\tau_1$ and $\tau_2$ do not contribute to the flow field and only modify the normal pressure gradient according to

$$\frac{\partial p}{\partial z} = 2(\tau_2 U - \tau_1 V). \tag{19}$$

Linear stability analysis is performed assuming perturbations of the form

$$f(z) \exp i(\alpha z + i\beta - \sigma t), \tag{20}$$

where $\alpha$ and $\beta$ are wave numbers, the phase speed of the disturbance is given by $\sigma_r / \sqrt{\alpha^2 + \beta^2}$, and the critical states correspond to $\sigma = 0$, where $\sigma = \sigma_r + i\sigma_i$. We find

$$L_{\text{Re}} u - \text{Re} \, U' w - 2(\tau_2 w - \tau_3 v) = i\alpha p, \tag{21}$$

$$L_{\text{Re}} v - \text{Re} \, V' w - 2(\tau_3 u - \tau_1 w) = i\beta p, \tag{22}$$

$$L_{\text{Ma}} \theta + \text{Ma} \, u - \text{Ma} \, T' w = 0, \tag{23}$$

$$i\alpha u + i\beta v + w' = 0, \tag{24}$$

where $z$ derivatives are denoted by primes and $L_{\text{Re}} = d^2/dz^2 - (\alpha^2 + \beta^2) - i(\alpha U + \beta V) \, \text{Re}$

$+ i\sigma$ Re. The boundary conditions become

$$u = v = w = \theta' = 0 \quad \text{on } z = 0, \tag{25}$$

$$w = u' + i\alpha\theta = v' + i\beta\theta = \theta' + \text{Bi}\,\theta = 0 \quad \text{on } z = 1. \tag{26}$$

For each Pr, $\tau_1$, $\tau_2$, $\tau_3$, and Bi we find the neutral point of a disturbance with wave numbers $\alpha$ and $\beta$ by using a Chebyshev pseudospectral method. This neutral point occurs at $\text{Ma}_n$ and an associated $\sigma_r$. In order that we may find the critical disturbance at the lowest $\text{Ma}_{\text{cr}}$ (i.e., the preferred mode of instability), we couple our eigenvalue solver with a minimization scheme (Press, et al., 1986). It is evident that the parameter space requires an extensive study. Here we include some key results only to emphasize the important influence of rotation with Bi = 0.

### 5.3.1.2. *Linear Stability of the Modified Base Flow*

Figure 5.1 shows the influence of $\tau_2 \neq 0$ with $\tau_1$ and $\tau_3$ zero. In this case Coriolis instability is well understood (Hart, 1971; Lezius and Johnston, 1976; Alfredsson and Persson, 1989) and,

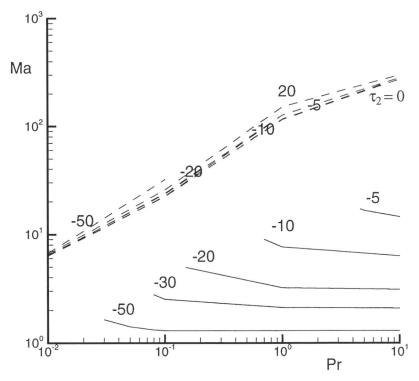

Figure 5.1. Influence of $\tau_2$ on the dependence of the critical Marangoni number Ma on the Prandtl number Pr with $\tau_3 = \tau_1 = 0$. Hydrothermal waves (dashed curves) are predicted with $\tau_2 = 0$, in complete agreement with Smith and Davis (1983). For $\tau_2 \leq -76$ stationary convection (solid curves) is preferred for Pr > 0.01 at much lower values of Ma. For $-50 < \tau_2 < 0$ there is a cutoff value of the Prandtl number $\text{Pr}_2$ that increases as $\tau_2$ increases toward zero. Rotation-stabilized traveling waves at Ma values very close to those with $\tau_2 = 0$ are preferred for Pr < $\text{Pr}_2$, while relatively highly destabilized stationary modes occur with Pr > $\text{Pr}_2$. Note that $\text{Pr}_2$ corresponding to $-5 < \tau_2 < 0$ is greater than 10, which is outside the range of the figure. Rotation-stabilized traveling waves are the preferred form of convection with $\tau_2 > 0$.

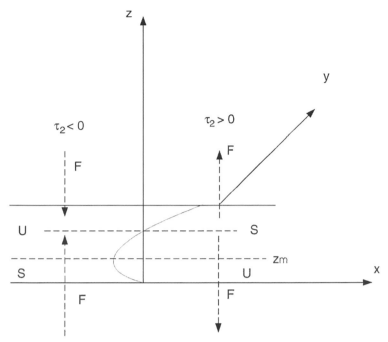

Figure 5.2. Regions of stable (S) and unstable (U) stratification of the Coriolis force (F) with $\tau_3 = \tau_1 = 0$. The upper dashed line is located where the basic return flow changes sign and hence F also changes sign. The lower dashed line at $z_m$ is located at maximum negative flow where |F| has a local maximum. $\tau_2 < 0$ is shown on the left where the upper region $z_m < z < 1$ is unstable and the lower region $0 < z < z_m$ is stable. Stratifications corresponding to $\tau_2 > 0$ are shown on the right.

when $\tau_2 < 0$, is due to the generation of an unstably stratified Coriolis force in the fluid region between the point of maximum negative $U$, say $z_m$, and the free surface at $z = 1$, as shown in Fig. 5.2. Note however that the region $0 < z < z_m$ is stabilized by rotation. For low values of $|\tau_2|$ a competition develops between the Marangoni and stationary Coriolis modes with the former prevailing at Pr less than a cutoff value $Pr_2$ and the latter at $Pr \geq Pr_2$. While the classical Coriolis–Poiseuille instabilities (Hart, 1971; Lezius and Johnston, 1976; Alfredsson and Persson, 1989) are stationary roll-cell modes, the new stationary Coriolis–Marangoni modes are not pure longitudinal rolls; however, they correspond to disturbances with $|\alpha|$ approximately an order of magnitude smaller than $|\beta|$. With $\tau_2 > 0$ the region of stable Coriolis force stratification is now $z_m < z < 1$, which includes the free-surface region, and the lower region is destabilized. The preferred mode of instability is rotation-stabilized traveling waves for $0.01 \leq Pr \leq 10$.

Figure 5.3 shows the influence of $\tau_1$ (taken $> 0$ because of invariance) with $\tau_2$ and $\tau_3$ zero. It is seen that a small $\tau_1$ stabilizes the hydrothermal waves at low values of Pr and that for each $\tau_1$ there is a critical $Pr_1$ above which preferred stationary solutions appear. With $\tau_1$ greater than $\sim$87, the stationary solutions are available for the range of Pr considered. The qualitative similarities shown in Figs. 5.1 (for $\tau_2 < 0$) and 5.3 suggest similar physics. The quantitative differences are due to the fact that in the case of Fig. 5.1 inviscid roll-cell disturbances are possible (Pedley, 1969). These issues as well as clustering of modes near Pr of 0.01 in Figs. 5.1

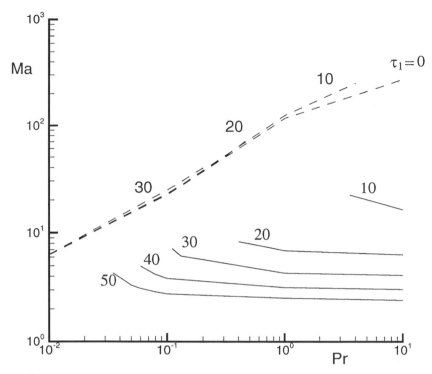

Figure 5.3. Influence of $\tau_1$ on the dependence of the critical Marangoni numbr Ma on the Prandtl number Pr with $\tau_2 = \tau_3 = 0$. For $\tau_1 \geq 87$, stationary convection (solid curves) is preferred for Pr $> 0.01$ at much lower values of Ma. For $0 < \tau_1 < 50$, there is a cutoff value of the Prandtl number $Pr_1$ that increases as $\tau_1$ decreases. Rotation-stabilized traveling waves (dashed curves), at Ma values very close to those with $\tau_1 = 0$, are preferred for Pr $< Pr_1$, while relatively highly destabilized stationary modes occur with Pr $> Pr_1$. Note that $Pr_1$ corresponding to $\tau_1 < 10$ is greater than 10, which is outside the range of the figure.

and 5.3 require further study and are best handled by asymptotic methods for small and large values of Pr and $\tau_1$.

Figure 5.4 shows the influence of $\tau_3$ with $\tau_1$ and $\tau_2$ zero. An important difference in this situation from those in Figs. 5.1 and 5.3 is that $V \neq 0$. It is seen that stationary modes at values of Pr lower than some value $Pr_3$, which increases with $\tau_3$, are destabilized with increasing $\tau_3$ until it reaches $\sim 8$. Increasing $\tau_3$ beyond 8 stabilizes the flow, and at a value of $\sim 50$, it is more stable compared with the hydrothermal waves at $\tau_3 = 0$. This initial stabilization and ultimate reversal with increasing $\tau_3$ are consistent with theoretical findings in the study of the rotating Poiseuille channel flow by Wollkind and DiPrima (1973). Further research is needed to investigate the spatial structure of instabilities and the influence of general rotation.

### 5.3.2. *Cylindrical Bridge*

Here we consider the linear stability of the rotation-modified base flow proposed by Xu and Davis (1983). The so-called return flow in a cylinder has been defined, so that the total mass flux

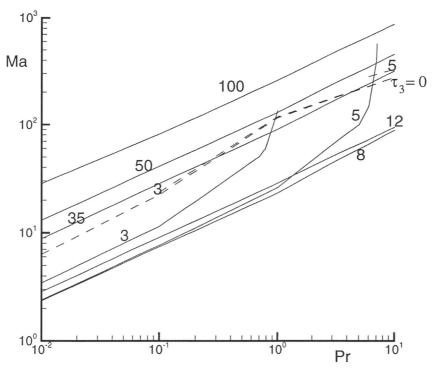

Figure 5.4. Influence of $\tau_3$ on the dependence of the critical Marangoni number Ma on the Prandtl number Pr with $\tau_1 = \tau_2 = 0$. For $0 < \tau_3 \leq 8$, there is a cutoff value of the Prandtl number $Pr_3$ that increases as $\tau_2$ increases toward 8. Rotation-stabilized traveling waves (dashed curves), at Ma values very close to those with $\tau_3 = 0$, are preferred for $Pr > Pr_3$, while relatively highly destabilized stationary modes (solid curves) occur with $Pr < Pr_3$. Note that $Pr_3$ corresponding to $\tau_3 > 5$ is greater than 10, which is outside the range of the figure. Increasing $\tau_3$ above $\sim 8$ is stabilizing with stationary modes preferred in this range of Pr.

at a section of the cylinder is 0. It is the particular solution of Eqs. (2)–(4) with zero rotation, where the velocity is function of $r$ only:

$$\bar{U} = 0, \tag{27}$$

$$\bar{V} = 0, \tag{28}$$

$$\bar{W} = \frac{1}{2}\left(r^2 - \frac{1}{2}\right), \tag{29}$$

$$\bar{T} = -z - \frac{\text{Ma}}{32}(1 - r^2)^2. \tag{30}$$

The base flow with no rotation is useful since it is also a solution of our system of equations with rotation along the $z$ axis. A perturbation is superimposed, of the form

$$f(r) \exp(i\alpha z + im\theta + st). \tag{31}$$

An eigenvalue problem results that provides the growth rate and the frequency of the preferred perturbation as well as its shape.

To study the effects of rotation about the $x$ (or the $y$) direction, we compute a correction to the base flow by assuming a small rotation $\bar{\epsilon}$ because the Coriolis force is two dimensional and no simple solution is obtainable for the resulting two-dimensional base flow. Asymptotic methods for $\bar{\epsilon} \to 0$ lead to a linear system that is solved to obtain the values of the new flow field. The new base flow (return flow + correction) was also studied by linear analysis to predict the influence of an $x$ rotation on the stability of the flow. The perturbations are now given by

$$\tilde{f}(r, \theta) \exp(i\alpha z + st). \tag{32}$$

### 5.3.2.1. *Linear Stability Results for z Rotation*

After introducing perturbation (31) into the equations, we have to solve the following system,

$$\left( D_*D - \alpha^2 - \frac{m^2 + 1}{r^2} - i\alpha \mathrm{Re}\,\bar{W} - s\mathrm{Re} \right) u - \frac{2im}{r^2} v = Dp - 2\tau_3 v, \tag{33}$$

$$\left( D_*D - \alpha^2 - \frac{m^2 + 1}{r^2} - i\alpha \mathrm{Re}\,\bar{W} - s\mathrm{Re} \right) v + \frac{2im}{r^2} u = \frac{im}{r} p + 2\tau_3 u, \tag{34}$$

$$\left( D_*D - \alpha^2 - \frac{m^2}{r^2} - i\alpha \mathrm{Re}\,\bar{W} - s\mathrm{Re} \right) w = i\alpha p + \mathrm{Re}\,\bar{W}_r u, \tag{35}$$

$$\left( D_*D - \alpha^2 - \frac{m^2}{r^2} - i\alpha \mathrm{Ma}\,\bar{W} - s\mathrm{Ma} \right) T = \mathrm{Ma}\bar{T}_r u + \mathrm{Ma}\bar{T}_z w, \tag{36}$$

and boundary conditions at $r = 1$,

$$u = Dw + i\alpha T = Dv - v + imT = DT + \mathrm{Bi}T = 0, \tag{37}$$

with $D = \partial_r$, $D^* = D + 1/r$, and $D^*D = D^2 + D/r$.

Pseudospectral solutions with 20 Chebychev polynomials reproduce very accurately those of Xu and Davis (1983) for zero rotation. We agreed with all their results except that with $m = 1$, for large Pr and both Bi $= 0$ and 1, we observed a more unstable mode. For instance, in Fig. 5.5(b), the mode $m = 1$, at Pr $= 100$ and Bi $= 0$, becomes unstable at $\mathrm{Ma}_c = 263$ and $\alpha_c = 1.802$ instead of 478.04 and 1.003, respectively. If we used their results at Pr $= 500$, Bi $= 0$, and the critical conditions they determined, $\alpha_c = 1.163$ and $\mathrm{Ma}_c = 1150.3$, we find the frequency $\omega_c = 1.023$, which is precisely their results, with zero growth rate since it is a neutral point. However, we also observed a more unstable mode with growth rate 0.0498 and frequency 0.1911. Furthermore, to verify our scheme, we used two types of conditions at $r = 0$, and they agree to machine precision. First, we solved the equations on $[\delta, 1]$, where $\delta$ goes to zero as the number of Chebyshev polynomials is increased. Then, we solved the equations on $[0, 1]$ with analytically determined boundary conditions at $r = 0$.

The influence of $\tau_3$ on the stability of the first mode is shown in Fig. 5.5. At low Pr, the rotation stabilizes the flow. Because of the symmetry of the equations, the same results are obtained for negative rotation. Therefore the steady state is observed for a larger range of parameters with rotation, but the transition is still characterized by the development of time-dependent waves. At large Pr, the opposite trend is observed. The rotation not only destabilizes the flow, but also displaces the critical wave number toward zero. These results might be affected by a finite cylinder since boundaries will restrict the range of the possible wavelengths.

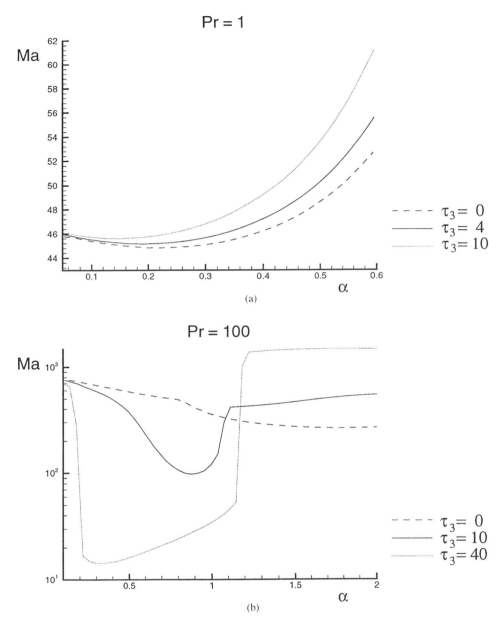

Figure 5.5. Neutral curves with $Bi = 0$ for the first and most unstable mode $m = 1$ at different $\tau_3$ at (a) $Pr = 1$, (b) $Pr = 100$. Rotation stabilizes the flow at $Pr = 1$. It has a destabilizing effect at $Pr = 100$, while shifting the critical wave number toward 0.

### 5.3.2.2. *Linear Correction to the Base Flow for an x Rotation*

The base flow is the sum of the basic profile with no rotation and $\mathcal{O}(\bar{\epsilon})$ profile due to $\tau_1 = \bar{\epsilon}$:

$$(U, W, P, T) = (0, \bar{W}, \bar{P}, \bar{T}) + \bar{\epsilon}(\hat{u}, \hat{w}, \hat{p}, \hat{T}) \sin\theta, \tag{38}$$

$$V = \bar{\epsilon}\hat{v} \cos\theta, \tag{39}$$

where $(\hat{u}, \hat{v}, \hat{w}, \hat{p}, \hat{T})$ satisfy

$$D_*\hat{u} - \frac{\hat{v}}{r} = 0, \tag{40}$$

$$\left(D_*D - \frac{2}{r^2}\right)\hat{u} + \frac{2\hat{v}}{r^2} - D\hat{p} = -2\bar{W}, \tag{41}$$

$$\left(D_*D - \frac{2}{r^2}\right)\hat{v} + \frac{2\hat{u}}{r^2} - \frac{\hat{p}}{r} = -2\bar{W}, \tag{42}$$

$$\left(D_*D - \frac{1}{r^2}\right)\hat{w} - \mathrm{Re}\,\bar{W}_r\hat{u} = 0, \tag{43}$$

$$\left(D_*D - \frac{1}{r^2}\right)\hat{T} - \mathrm{Ma}\bar{T}_r\hat{u} - \mathrm{Ma}\bar{T}_z\hat{w} = 0, \tag{44}$$

together with boundary conditions at $r = 1$,

$$\hat{u} = D\hat{w} = D\hat{v} - \hat{v} + \hat{T} = D\hat{T} + \mathrm{Bi}\hat{T} = 0. \tag{45}$$

An analytical solution can be obtained in the form

$$\hat{u} = (A + Br^2 + Cr^4), \tag{46}$$

$$\hat{v} = (A + 3Br^2 + 5Cr^4), \tag{47}$$

$$\hat{w} = \mathrm{Re}\left(Dr + \frac{A}{8}r^3 + \frac{B}{24}r^5 + \frac{C}{48}r^7\right), \tag{48}$$

$$\hat{p} = \left[\left(8B - \frac{1}{2}\right)r + \frac{1}{4}r^4\right], \tag{49}$$

$$\hat{T} = (T_1 r + T_3 r^3 + T_5 r^5 + T_7 r^7 + T_9 r^9), \tag{50}$$

where the coefficients have been determined by use of Maple.

The flow field in the $x, y$ plane is due to only the perturbation. Two cases are shown in Fig. 5.6 at $\mathrm{Pr} = 1$ and $\mathrm{Bi} = 0$: $\mathrm{Re} = 10$ [Fig. 5.6(a)] and $\mathrm{Re} = 50$ [Fig. 5.6(b)]. Two different patterns can be observed. At small Re, we have one vortex per semicircle. When Re increases, the norm of $v$ goes to zero and $v$ changes sign. At $\mathrm{Re} = 50$, Fig. 5.6(b) exhibits a new layer along the free surface with a recirculating flow. The cutoff Re number between the two patterns can be found if $v = 0$ is imposed or

$$\mathrm{Re} = \sqrt{\frac{7680}{13}\frac{1 + \mathrm{Bi}}{\mathrm{Pr} + \mathrm{Pr}^2}}, \tag{51}$$

which leads to $\mathrm{Re} = 24.3$ for the parameters used in Fig. 5.6.

### 5.3.2.3. *Linear Stability of the Modified Base Flow*

In this subsection, we use for the base flow the sum of the flow of Xu and Davis and the perturbations due to a small $x$ rotation given by Eqs. (38) and (39). The magnitude of the latter is a function of the nondimensional parameter whereas the former is constant. To provide a

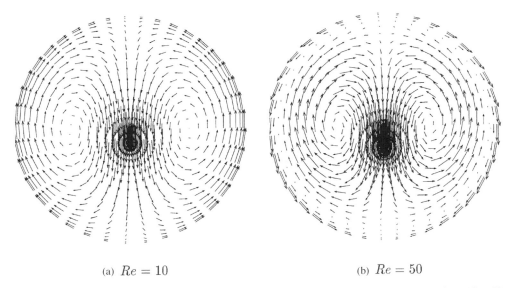

(a) $Re = 10$      (b) $Re = 50$

Figure 5.6. Flow field in the $x$, $y$ plane due to a small $x$ rotation, $Pr = 1$, $Bi = 0$. As Re increases from 10 to 50, the $v$ component of the velocity changes sign (at $Re = 24.3$ for this set of parameters), creating a recirculating region at the free surface.

measure of the difference between these two flows, a new norm $\epsilon$ is introduced for the rotation vector,

$$\epsilon = 4\bar{\epsilon}\hat{w}_m,\tag{52}$$

where $\hat{w}_m$ is the maximum correction in the $z$ direction, and we use the fact that the maximum velocity in the $z$ direction without rotation is 0.25. We then require that $\epsilon \ll 1$.

The perturbations given by expression (32) are inserted into the governing equations to arrive at continuity;

$$\tilde{u}_r + \frac{\tilde{u}}{r} + \frac{\tilde{v}_\theta}{r} + i\alpha\tilde{w} = 0;\tag{53}$$

$r$ momentum,

$$\mathrm{Re}\left(s\tilde{u} + \tilde{u}U_r + U\tilde{u}_r + \frac{\tilde{v}}{r}U_\theta + \frac{V}{r}\tilde{u}_\theta - 2\frac{V\tilde{v}}{r} + i\alpha W\tilde{u}\right)$$
$$= -\tilde{p}_r + \tilde{u}_{rr} + \frac{\tilde{u}_r}{r} - \frac{\tilde{u}}{r^2} + \frac{\tilde{u}_{\theta\theta}}{r^2} - \alpha^2\tilde{u} - 2\frac{\tilde{v}_\theta}{r^2} + 2\epsilon\sin\theta\,\tilde{w};\tag{54}$$

$\theta$ momentum,

$$\mathrm{Re}\left(s\tilde{v} + \tilde{u}V_r + U\tilde{v}_r + \frac{\tilde{v}}{r}V_\theta + \frac{V}{r}\tilde{v}_\theta + \frac{\tilde{u}V}{r} + \frac{U\tilde{v}}{r} + i\alpha W\tilde{v}\right)$$
$$= -\frac{\tilde{p}_\theta}{r} + \tilde{v}_{rr} + \frac{\tilde{v}_r}{r} - \frac{\tilde{v}}{r^2} + \frac{\tilde{v}_{\theta\theta}}{r^2} - \alpha^2\tilde{v} + 2\frac{\tilde{u}_\theta}{r^2} + 2\epsilon\cos\theta\,\tilde{w};\tag{55}$$

$z$ momentum,

$$\mathrm{Re}\left(s\tilde{w} + \tilde{u}W_r + U\tilde{w}_r + \frac{\tilde{v}}{r}W_\theta + \frac{V}{r}\tilde{w}_\theta + i\alpha W\tilde{w}\right)$$

$$= -i\alpha\tilde{p} + \tilde{w}_{rr} + \frac{\tilde{w}_r}{r} + \frac{\tilde{w}_{\theta\theta}}{r^2} - \alpha^2\tilde{w} - 2\epsilon\cos\theta\tilde{v} - 2\epsilon\sin\theta\tilde{u}; \qquad (56)$$

and energy,

$$\mathrm{Ma}\left(s\tilde{T} + \tilde{u}T_r + U\tilde{T}_r + \frac{\tilde{v}}{r}T_\theta + \frac{V}{r}\tilde{T}_\theta + i\alpha W\tilde{T} + \tilde{w}T_z\right)$$

$$= \tilde{T}_{rr} + \frac{\tilde{T}_r}{r} + \frac{\tilde{T}_{\theta\theta}}{r^2} - \alpha^2\tilde{T}. \qquad (57)$$

The Fourier decomposition

$$\tilde{u}(r,\theta) = \sum_{m=-M}^{M} \tilde{u}_m \exp(im\theta), \qquad (58)$$

with similar expressions for $\tilde{v}$, $\tilde{w}$, $\tilde{p}$, and $\tilde{T}$, is assumed. The resulting eigenvalue problem is solved both on $[\delta, 1]$ and by the imposition of boundary conditions at $r = 0$, obtained analytically, which provides a check on the results.

In Fig. 5.7, we study the influence of Pr on the stability of the flow. Rotation stabilizes the flow with Pr = 1 and Bi = 0, as shown in Fig. 5.7(a). At large Pr, a new mode appears and the flow becomes unsteady at much lower Ma. Such a trend is also observed for Bi = 1. Linear analysis therefore indicates that $x$ and $z$ rotations have a destabilizing effect at large Pr and a stabilizing one at small Pr.

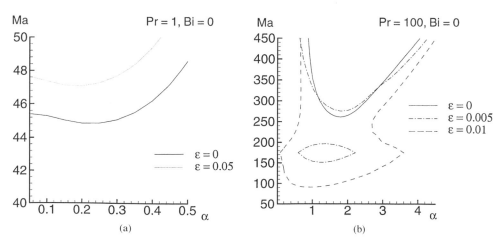

Figure 5.7. Neutral curves at Bi = 0 at different $x$ rotations; (a) Pr = 1, (b) Pr = 100. Rotation stabilizes the flow at Pr = 1. At Pr = 100, a new mode becomes excited. Rotation therefore has a destabilizing effect. Note that with $\epsilon = 0.005$, there are two distinct regions of instability.

**5.3.2.4.** *Conclusions*

We studied the effect of rotation on thermocapillary instabilities in both the infinite plane layer and the infinite cylinder. For the plane layer, rotation produces stationary modes that can become predominant. For the cylinder, rotations, both about the $x$ and the $z$ axes, are stabilizing at low Pr and destabilizing at large Pr. Traveling azimuthal waves are still the preferred mode of convection.

A typical space laboratory (Yuferev, et al., 1996), on the Shuttle or the future Space Station, orbits the Earth at approximately one revolution approximately every 2 h, which results in an angular velocity of $\sim 10^{-3}$ s$^{-1}$. Assuming a length scale in the range of 1–10 cm and a kinematic viscosity in the range from $10^{-2}$ to $10^{-4}$ cm$^2$/s results in Taylor numbers in the range $10^{-1}$–$10^3$. The results of this chapter suggest that the design of future crystal growth space experiments should take the influence of rotation into consideration.

## Acknowledgments

This research was partially supported by NASA through grant NAG3-1453. We also acknowledge the computer time provided by the Pittsburgh Supercomputer Center under grant CTS970014P.

## References

Alfredsson, P. H. and H. Persson, 1989, "Instabilities in channel flow with system rotation," J. Fluid Mech. **202**, 543–557.

Hart, J. E., 1971, "Instability and secondary motion in a rotating channel flow," J. Fluid Mech. **45**, 341–351.

Kamatoni, Y., J. H. Lee, S. Ostrach, and A. Pline, 1992, "An experimental study of oscillatory convection in cylindrical containers," Phys. Fluids A **4**, 955–962.

Levenstam, M. and G. Amberg, 1995, "Hydrodynamical instabilities of thermocapillary flow in a half-zone," J. Fluid Mech. **297**, 357–372.

Lezius, D. K. and J. P. Johnston, 1976, "Roll-cell instabilities in rotating laminar and turbulent channel flows," J. Fluid Mech. **77**, 153–175.

Mundrane, M. and A. Zebib, 1995, "Low Prandtl number Marangoni convection with a deformable interface," AIAA J. Thermophys. Heat Transfer **9**, 795–797.

Neitzel, G. P., K. T. Chang, D. F. Jankowski, and H. D. Mittleman, 1993, "Linear stability theory of thermocapillary convection in a model of the float-zone crystal growth process," Phys. Fluids A **5**, 108–114.

Ostrach, S., 1982, "Low-gravity fluid flows," Ann. Rev. Fluid Mech. **14**, 313–345.

Pedley, T. J., 1969, "On the instability of viscous flow in a rapidly rotating pipe," J. Fluid Mech. **35**, 97–115.

Press, W. H., B. P. Flannery, S. A. Teukolsky, and W. T. Vetterling, 1986, *Numerical Recipes* (Cambridge U. Press, Cambridge), pp. 289–293.

Smith, M. K. and S. H. Davis, 1983, "Instabilities of dynamic thermocapillary liquid layers Part 1. Convective instabilities," J. Fluid Mech. **132**, 119–144.

Wanschura, M., V. M. Shevtsova, H. C. Kuhlmann, and H. J. Rath, 1995, "Convective instability mechanisms in thermocapillary liquid bridges," Phys. Fluids **7**, 912–925.

Wollkind, R. and R. C. DiPrima, 1973, "Effect of a Coriolis force on the stability of plane Poiseuille flow," Phys. Fluids **16**, 2045–2051.

Xu, J.-J. and S. H. Davis, 1983, "Convective thermocapillary instabilities in liquid bridges," Phys. Fluids **27**, 1102–1107.

Xu, J., and A. Zebib, 1998, "Oscillatory two- and three-dimensional thermocapillary convection," J. Fluid Mech., in press.

Yuferev, V. S., E. N. Kolesnikova, Y. A. Polovko, and A. I. Zhmakin, 1996, "Effect of spacecraft rotation on fluid convection under microgravity," presented at the Third Microgravity Fluid Physics Conference, Cleveland, OH, June 13–15.

Zebib, A., 1996, "Thermocapillary instabilities with system rotation," Phys. Fluids **8**, 3209–3211.

PART TWO

# Shear and Pressure Driven Instabilities

# 6

# Control of Instability in a Liquid Film Flow

S. P. LIN AND J. N. CHEN

## 6.1. Introduction

The onset of instability of a liquid film flow down a stationary inclined plane was analyzed by Benjamin (1957) and Yih (1963). In particular, they showed that the film on the vertical plate is always unstable. The instability manifests itself as gravity-driven surface waves with a wavelength much longer than the film thickness. When the angle of inclination of the plate is smaller than 0.56′ but greater than zero the instability in a thick liquid layer may manifest itself as relatively short shear waves (Woods and Lin, 1996; Floryan et al., 1987; DeBruin, 1974). When the inclined plane has zero angle of inclination, the horizontal layer of liquid is stable. However, when the horizontal plate is made to oscillate in its own plane, unstable waves synchronous to the plate oscillation can be generated (Yih, 1968; Or, 1997). Control of instability by use of external forcing has attracted attention recently. Coward and Renardy (1995) studied a modulated two-layered problem, Or and Kelly (1996) studied the control of thermocapillary instabilities in a heated layer by external forcing, and Woods and Lin (1995) investigated the interaction of the shear waves, the surface waves, and the Faraday waves in a liquid film on the inclined plate that vibrates in a direction perpendicular to the plate. They found that the amplification rate of the unstable surface wave may be reduced by such an oscillation, but cannot be completely suppressed.

Bauer and Kerczek (1991) looked into the possibility of suppressing the unstable two-dimensional waves in a liquid film flow down an inclined plane by imparting an in-plane oscillation to the wall. Their results suggest that such a possibility may exist. However, as they pointed out in their concluding remarks, the results obtained with the long wave and the oscillation amplitude expansions did not have a sufficiently large parameter range of validity to allow them to reach a definitive statement. Lin et al. (1996) obtained the solution to the same problem for arbitrary wavelength and oscillation amplitude. They showed that unstable two-dimensional waves can be suppressed by use of appropriate amplitudes and frequencies of the plate oscillation. The windows of stability depend on the flow parameters. The question of whether these windows will be closed by the three dimensionality of the waves was answered by Lin and Chen (1997). For steady parallel flows, the Squire (1933) theorem states that three-dimensional infinitesimal disturbances are more stable than two-dimensional ones for all wave numbers. Lin and Chen showed that this is not true for the present problem. However, they showed with numerical results that the three-dimensional disturbances do not close the stability window opened by the two-dimensional disturbances, although they can be more dangerous than the two-dimensional disturbances. For the sake of easy comparison with the case of nonoscillatory film flow, their stability analysis is renormalized and summarized in Section 6.2. The renormalization also enables Lin and Chen (1998) to elucidate from an energy consideration the physical mechanism of the film-flow stabilization. The physical mechanism is further expounded in Section 6.3. Some definitive conclusions are drawn and some immediate questions needed to be answered are stated in the last section.

## 6.2. Stability Analysis

Consider the stability of a Newtonian liquid film flow down an inclined plane that makes an angle $\theta$ with the horizontal, as shown in Fig. 6.1. The rigid inclined plane oscillates sinusoidally with a constant frequency $\Omega$. The oscillation is parallel to the flow direction, and its amplitude is $\xi$. The liquid is assumed to be incompressible, and the effect of the ambient gas is neglected. The governing equations in a dimensionless form are

$$(\text{St}\,\partial_\tau + \mathbf{V} \cdot \nabla)\mathbf{V} = -\nabla P + \text{Re}^{-1}\nabla \cdot \boldsymbol{\sigma} + \mathbf{Fr}^{-1} - \mathbf{a}, \tag{1}$$

$$\nabla \cdot \mathbf{V} = 0, \tag{2}$$

where $\nabla$ is the gradient operator nondimensionalized with the half-film thickness $D$, $\mathbf{V}$ is the velocity normalized with the maximum velocity $U_m = 2g \sin\theta D^2/\nu$ in a nonoscillatory film flow, $g$ and $\nu$ being, respectively, the gravitational acceleration and the kinematic viscosity of the liquid, $\tau$ is the time normalized with $\Omega$, $\boldsymbol{\sigma}$ is the rate of strain tensor nondimensionlized with $U_m/D$, $p$ is the pressure normalized with $\rho U_m^2$, $\rho$ being the liquid density, $\text{Re} = U_m D/\nu$ is the Reynolds number, $\text{St} = \Omega D/U_m$ is the Strouhal number, and $\mathbf{Fr}^{-1}$ and $\mathbf{a}$ are, respectively, the dimensionless gravitational body force per unit mass and the D'Alembert apparent body force per unit mass, i.e.,

$$\mathbf{Fr}^{-1} = \left(F_r^{-1}\right)(\mathbf{i}\,\sin\theta - \mathbf{j}\,\cos\theta), \qquad F_r = U_m^2\big/gD, \qquad \mathbf{a} = \mathbf{i}d_x\,\text{St}^2\,\sin\tau,$$

in which $\mathbf{i}$ and $\mathbf{j}$ are, respectively, the unit vectors in the direction of and perpendicular to the inclined plane, and $d_x = \xi/D$. The corresponding boundary conditions relative to the Cartesian

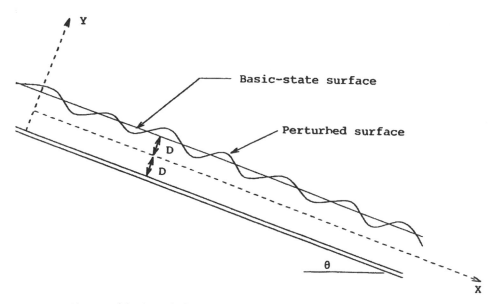

Figure 6.1. Diagram of the theoretical model.

coordinate system $(x, y, z)$ in Fig. 6.1 are the no-slip condition at the inclined plane,

$$\mathbf{v} = 0 \quad \text{at} \quad y = -1, \tag{3}$$

and the dynamic and the kinematic conditions at the free surface $y = h_1$, respectively given by

$$R_e^{-1}\sigma \cdot \mathbf{n} + \mathbf{n}\mathrm{We}^{-1}\nabla \cdot \mathbf{n} = 0, \tag{4}$$

$$\mathbf{V} \cdot \mathbf{j} = (\mathrm{St}\,\partial_\tau + \mathbf{V} \cdot \nabla)h_1, \tag{5}$$

where $\mathbf{n}$ is the unit normal vector at the free surface and We is the Weber number defined by $\mathrm{We} = U_m^2 D/S$, $S$ being the surface tension.

The basic flow velocity field $(U, V, W)$ in the Cartesian coordinator $(x, y, z)$ and the pressure field $P$ that satisfy Eqs. (1)–(5) exactly are given by

$$V = W = 0,$$

$$U(y, \tau) = \frac{3}{4}(3 + 2y - y^2) + \mathrm{St}\,d_x \cos\tau + U_0(y, \tau), \tag{6}$$

$$P(y, \tau) = P_0 - \mathrm{Fr}^{-1}\cos\theta \cdot (y - 1), \tag{7}$$

where

$$U_0(y, \tau) = \mathrm{St}\,d_x[f_1(\tau)G_1(y) + f_2(\tau)G_2(y)],$$

$$f_1(\tau) = -\cos\tau, \qquad f_2 = \sin\tau,$$

$$G_1(y) = \frac{ac + bd}{c^2 + d^2}, \qquad G_2(y) = \frac{bc - ad}{c^2 + d^2},$$

$$a = e^{\beta y}\cos\beta y + e^{\beta(2-y)}\cos[\beta(2 - y)], \qquad \beta = (\mathrm{Re}\,\mathrm{St}/2)^{1/2},$$

$$b = e^{\beta y}\sin\beta y + e^{\beta(2-y)}\sin[\beta(2 - y)],$$

$$c = a(-1), \qquad d = b(-1),$$

and $P_0$ is a reference pressure.

To investigate the onset of instability of this basic flow, we introduce the normal mode velocity $(\mathbf{v}')$ and pressure $(p')$ perturbations,

$$(\mathbf{v}', p') = [\hat{u}(y, \tau), \hat{v}(y, \tau), \hat{w}(y, \tau), \hat{p}(y, \tau)]\exp[i(\alpha x + \kappa z)], \tag{8}$$

into the Navier–Stokes equations linearized about the basic flow:

$$[\mathrm{Re}(\mathrm{St}\,\partial_\tau + U\partial_x) - \nabla^2](u', v', w') + \mathrm{Re}D\bar{u}(0, v', 0) = -\mathrm{Re}\nabla p', \tag{9}$$

$$\nabla \cdot \mathbf{v} = 0, \tag{10}$$

where $(\alpha, \kappa)$ are the wave numbers in the $(x, z)$ directions, $D = \mathrm{d}/\mathrm{d}y$, $\nabla$ is the gradient operator, and $\nabla^2 = \nabla \cdot \nabla$. Thus the disturbance $(\hat{u}, \hat{v}, \hat{w}, \hat{p})$ amplitudes must satisfy,

$$[\mathrm{Re}(\mathrm{St}\,\partial_\tau + i\alpha U) + (\alpha^2 + \kappa^2) - D^2](\hat{u}, \hat{v}, \hat{w}) + \mathrm{Re}DU(0, \hat{v}, 0) = -\mathrm{Re}(i\alpha, D, i\kappa)\hat{p}, \tag{11}$$

$$i(\alpha\hat{u} + \kappa\hat{w}) + D\hat{v} = 0. \tag{12}$$

The corresponding boundary conditions are the no-slip condition

$$\hat{u}(-1, \tau) = \hat{v}(-1, \tau) = \hat{w}(-1, \tau) = 0, \tag{13}$$

the linearized tangential components and the normal component of dynamic free-surface condition (4) at $y = 1 + \hat{\eta}(t)\exp[i(\alpha x + \kappa z)]$ respectively given by

$$\hat{\eta}D^2U + D\hat{u} + i\alpha\hat{v} = 0, \tag{14}$$

$$D\hat{w} + i\kappa\hat{v} = 0, \tag{15}$$

$$(\alpha^2 + \kappa^2)\text{We}^{-1}\hat{\eta} + (2/\text{Re})D\hat{v} - \hat{p} + \hat{\eta}\,\text{Fr}^{-1}\cos\theta = 0, \tag{16}$$

and the linearized free-surface kinematic boundary condition

$$\hat{v} = i\alpha U\hat{\eta} + \text{St}\,\partial_\tau\hat{\eta}. \tag{17}$$

All of the variables appearing in free-surface conditions (14)–(17) are to be evaluated at $y = 1$, since all of them are expanded around $y = 1$ in Taylor's series in the linearization.

By use of the Squire transformation extended to unsteady flows with a free surface,

$$\gamma\tilde{u} = \alpha\hat{u} + \kappa\hat{w}, \qquad \tilde{v} = \hat{v}, \qquad \gamma^2 = \alpha^2 + \kappa^2, \qquad \alpha\tilde{p} + \gamma\hat{p}, \qquad \alpha\tau = \gamma T,$$

$$\gamma\tilde{\eta} = \alpha\hat{\eta}, \qquad \gamma\,\tilde{\text{Re}} = \alpha\,\text{Re}, \qquad \gamma^2\,\text{We}^{-1} = \alpha^2\,\tilde{\text{We}}^{-1}, \qquad \gamma^2\,\text{Fr}^{-1} = \alpha^2\,\tilde{\text{Fr}}^{-1},$$

the three-dimensional differential system (11)–(17) can be reduced to

$$[\tilde{\text{Re}}(\text{St}\,\partial_T + i\gamma\tilde{u}) + \gamma^2 - D^2](\tilde{u}, \tilde{v}) + \tilde{\text{Re}}D\bar{u}(\tilde{v}, 0) = -\tilde{\text{Re}}(i\gamma, D)\tilde{p}, \tag{18}$$

$$i\gamma\tilde{u} + D\tilde{v} = 0, \tag{19}$$

$$\tilde{u}(-1, T) = \tilde{v}(-1, T) = 0, \tag{20}$$

$$\tilde{\eta}D^2U + D\tilde{u} + i\gamma\tilde{v} = 0, \tag{21}$$

$$\gamma^2\,\tilde{\text{We}}^{-1}\tilde{\eta} + (2/\tilde{\text{Re}})D\tilde{v} - \tilde{p} + \tilde{\eta}\,\tilde{\text{Fr}}^{-1}\cos\theta = 0, \tag{22}$$

$$\tilde{v} = i\gamma U\tilde{\eta} + \text{St}\,\partial_T\tilde{\eta}. \tag{23}$$

The system (18)–(23) is identical in form to the governing differential system solved by Lin et al. (1996) for the two-dimensional disturbances except $\tilde{\text{We}}$ and $\tilde{\text{Fr}}$ are inverses of their We and Fr because of the inverted definitions of these parameters. However, we cannot conclude that the Squire theorem applies here, i.e., the onset of instability due to three-dimensional disturbances will occur at a larger Reynolds $\text{Re} = (\tilde{\text{Re}}\,\gamma/\alpha)$ number than that for two-dimensional disturbances of all wave numbers because of this relation if the rest of flow parameters are held constant. This is because the Reynolds number not only appears explicitly in the differential system, it also appears implicitly in the basic flow. It is seen from Eq. (6) that the spatial variation of the unsteady part of the basic flow depends exponentially on $(\text{St}\,\text{Re}/2)^{1/2}$. Thus as one varies Re in the differential system, with the rest of the flow parameters held fixed, one refers to different basic flows. For the case of steady basic flow, the unsteady part in Eq. (6) drops out. The steady part of the basic flow remains the same for all Re. Thus the Squire theorem applies. The validity of Squire's theorem for steady parallel basic flows with a deformable free surface was actually pointed out earlier by Yih (1955).

The differential system (18)–(23) constitutes an eigenvalue problem. The eigenvalues for the unsteady two-dimensional problem have already been obtained by Lin and Chen (1997), who used Chebyshev series expansion with Floquet theory for various flow parameters. Their differential system for the time-dependent Chebyshev coefficients, i.e., their equations (27)–(32), remain valid except Re, Fr, and We in these equations must be replaced respectively by St Re, $(\text{St}^2 \text{Fr})^{-1}$ and $(\text{St}^2 \text{We})^{-1}$. This is because while the velocity was normalized with $\Omega D$ in the previous work of Lin and Chen it is normalized here with $U_m$. Then $\alpha$, Re, Fr, We, and $h$ in the resulting equations must be identified as $\gamma$, $\tilde{\text{Re}}$, $\tilde{\text{Fr}}$, $\tilde{\text{We}}$, and $\tilde{\eta}$ respectively. For a given set $(\alpha, \kappa, \text{We}, \text{Fr}, \text{Re})$ the values of $(\gamma, \tilde{\text{We}}, \tilde{\text{Fr}}, \tilde{\text{Re}})$ are calculated, and then the corresponding characteristic exponents are calculated by use of Floquet theory. According to the Floquet theory, the disturbance amplitudes can be written as the product of periodic functions of the same period as the forcing function and an exponential function $\exp(ft)$, where $f$ is the complex Flouqet exponent. The flow is stable or unstable depending on whether the real part of the characteristic exponent is negative or positive. The stabilization of the liquid film flow is discussed in Section 6.4 by use of the obtained numerical results.

## 6.3. Mechanism of Forced Stabilization

In practice, stabilization of a liquid film flow only in the regions of parameter space where both the two- and the three-dimensional disturbances can be suppressed is useful. In these regions both the two- and the three-dimensional disturbances are stable, and therefore only the mechanical energy budget of the two-dimensional disturbance will be obtained. By forming the dot product of Eq. (9) with $(u, v)$, integrating over the control volume covered by the perturbed free surface of length $\lambda = 2\pi/\alpha$, transforming some of the volume integrals to surface integrals by use of the Gauss theorem with the aid of continuity equation (10) and the dynamic free-surface boundary conditions, and averaging over one wavelength $\lambda$ of any spatially periodic disturbance as well as over one period of the plate oscillation, we obtain the average rate of the disturbance kinetic energy change per unit width of the film covered by one wavelength (Lin and Chen 1998):

$$\text{RKINE} = \text{SHEST} + \text{REYNS} + \text{DISSI} + \text{SURTE} + \text{HYDPR}, \tag{24}$$

where

$$\text{RKINE} = \frac{\text{St}}{4\pi\lambda} \int_0^{2\pi} \int_{-1}^{1} \int_0^{\lambda} (u^2 + v^2)_{/\tau} \, \mathrm{d}x \, \mathrm{d}y \, \mathrm{d}\tau,$$

$$\text{SHEST} = \frac{1}{2\pi\lambda\text{Re}} \int_0^{2\pi} \int_0^{\lambda} u(u_{/y} + v_{/x})_{y=1} \, \mathrm{d}x \, \mathrm{d}\tau,$$

$$\text{REYNS} = -\frac{1}{2\pi\lambda} \int_0^{2\pi} \int_{-1}^{1} \int_0^{\lambda} uvU_{/y} \, \mathrm{d}x \, \mathrm{d}y \, \mathrm{d}\tau,$$

$$\text{DISSI} = -\frac{1}{2\pi\lambda\text{Re}} \int_0^{2\pi} \int_{-1}^{1} \int_0^{\lambda} \left[ 2\left(u_{/x}^2 + v_{/y}^2\right) + (u_{/y} + v_{/x})^2 \right] \mathrm{d}x \, \mathrm{d}y \, \mathrm{d}\tau,$$

$$\text{SURTE} = \frac{1}{2\pi\lambda\text{We}} \int_0^{2\pi} \int_0^{\lambda} (v\eta_{/xx})_{y=1} \, \mathrm{d}x \, \mathrm{d}\tau,$$

$$\text{HYDPR} = \frac{1}{2\pi\lambda} \int_0^{2\pi} \int_0^{\lambda} (v\eta P_{/y})_{y=1} \, \mathrm{d}x \, \mathrm{d}\tau.$$

The term on the left-hand side of Eq. (24), i.e., RKINE, represents the time rate of change of the disturbance kinetic energy averaged over one wavelength and one period of the plate oscillation. The terms on the right-hand side of Eq. (24) represent distinctive physical factors that affect RKINE. Obviously these terms are also averaged quantities in the same way RKINE is. If the sum of those terms is positive, then the film is unstable since RKINE grows with time. If these terms add to be equal or less than zero, then the film is neutral or stable. SHEST represents the average rate of work done by the disturbance shear stress at the free surface on the bulk liquid. Note that $(u_{/y} + v_{/x})$ is not zero because of Eq. (21). That is to say, work must be done to satisfy boundary condition (21). REYNS represents the average time rate of work done by the Reynolds stress in the environment of highly unsteady basic flow. DISSI represents the average time rate of energy dissipation through viscosity. SURTE represents the average rate of work done by the surface tension. HYDPR represents the time rate of work done against the basic flow pressure gradient at the free surface, which is zero for the case of a vertical film. The sign of DISSI is always negative. The sign of the rest of terms on the right-hand side of Eq. (24) depends on the flow parameters. If it is positive, then it signifies the rate of work done on the disturbance. If it is negative then the work is done by the disturbance at the expense of its own kinetic energy. Note that SURTE represents the rate of work done by the surface tension on the bulk fluid. Therefore $-$SURTE is the rate of work done by the fluid on the surface. Hence if SURTE in Eq. (24) is moved to the left-hand side of the equation, $-$SURTE will represent the rate of surface energy change due to the work done on the surface. This view was taken by Kelly et al. (1989). These two views are equivalent if there is no surface viscosity.

For a reason that will become apparent, we also evaluate the energy dissipations per unit thickness at the free surface and at the wall, respectively given by

$$\text{DISF} = -\frac{1}{2\pi\lambda\text{Re}} \int_0^{2\pi} \int_0^{\lambda} \left[ 2\left(u_{/x}^2 + v_{/y}^2\right) + (u_{/y} + v_{/x})^2 \right]_{y=1} dx\, d\tau,$$

$$\text{DISW} = -\frac{1}{2\pi\lambda\text{Re}} \int_0^{2\pi} \int_0^{\lambda} \left[ 2\left(u_{/x}^2 + v_{/y}^2\right) + (u_{/y} + v_{/x})^2 \right]_{y=-1} dx\, d\tau.$$

The eigenvectors required for the evaluation of the integrals in Eq. (24) have been obtained by Lin and Chen (1998) for various flow parameter ranges. The results of the energy budget calculated by them are discussed below, after the results of stability analysis are discussed.

### 6.4. Results and Discussions

Figure 6.2 gives three sets of neutral curves on which the real part of the characteristic exponent vanishes in the $\alpha$-Re space for the given three sets of $\kappa$ and the flow parameters specified in the caption. The flow is unstable inside the neutral curves and stable outside the curves. As the values of $\kappa$ increase from $\kappa = 0$, the regions of instability shrink. Had the results been plotted in the $\alpha$-$\kappa$-Re space, the unstable regions would be inside the cones that stick out obliquely from the paper. The apex of each cone would be located at different values of $\kappa$. None of these cones intersect with each other. Thus the windows of elimination of two-dimensional waves remain open for three-dimensional disturbances, that is, there are regions in the parameter space in which both two- and three-dimensional disturbances are suppressed. However, the onset

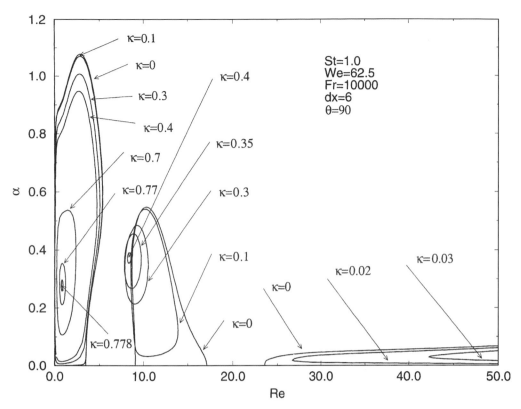

Figure 6.2. Stabilization of three-dimensional surface waves.

of instability with respect to three-dimensional disturbances may occur at smaller or larger Reynolds numbers than those for the two-dimensional disturbances, depending on the location of the neutral curve on the planes of $\kappa = $ constant. In fact, the three-dimensional disturbances of $\kappa = 0.3$, e.g., are more dangerous than the two-dimensional ones between $\alpha = 0.2$ and $0.5$ in the middle region of instability shown in Fig. 6.2. Similar results have been obtained for various flow parameters. This renders the idea of suppressing the waves in a film flow by use of plate oscillation more practically meaningful.

The neutral curve for a vertical film flow over a stationary plate is given by the dotted line in Fig. 6.3 for the case of We = 0.1. The flow is unstable in the region below this line, where the real part of the Floquet exponent is greater than zero, i.e., $f_r > 0$. This line passes through the origin in the $\alpha$-Re space. Hence the film flow is inherently unstable, since the onset of instability occurs at Re = 0, as was shown by Benjamin and Yih. A set of neutral curves for We = 0.1, $d_x = 10$, and various values of St is also given in Fig. 6.3. The film flow is unstable in the regions bounded by this set of curves and $\alpha = 0$. In these regions, $f_r > 0$. Outside of these regions, the flow is stable ($f_r < 0$). There are two branches of neutral curves for each given St, except the case of St = 2. The flow on an oscillating plate is stable between these two branches. Hence the inherently unstable steady vertical film flow is stabilized in the parameter space where the unstable region of the steady film flow overlaps with the stable region of the film flow with oscillation. Such an overlap region does not exist for the case of St = 2. It is necessary for the plate to oscillate at a sufficiently large dimensionless frequency (St > 2) in order to suppress the surface waves with $d_x = 10$. However, the existence of an overlapping region is not sufficient for the stabilization. While the second branch of the neutral curves with larger Re lies below the neutral curve for the steady case, part of the first branch of neutral curves with smaller Re lies above the neutral curve for the steady film flow in the region $\alpha \geq 0.21$. The relatively short waves with $\alpha \geq 0.21$ that are stable in a steady film flow are now unstable because of the plate oscillation. Thus the region of stabilization for $\alpha < 0.21$ described above may not be extended to the region of shorter wavelengths if the maximum Re on the first branch is greater than the minimum Re on the second branch for a given St. Fortunately, as St is increased beyond 2, the first and the second branches both shift toward the origin, and these two Reynolds numbers do not overlap. Thus the window of stabilization remains wide open as St is increased from 2, although the stabilization must be effected in a window with smaller Re. The minimum value of St and the width of the window for stabilization depend on other flow parameters. For the flow parameters specified in Fig. 6.3, $(St)_{min}$ is 2.5 and the window width $\Delta$Re maintains the same value of approximately 0.25 as St is increased from $(St)_{min}$.

In practice it is easier to search for the window of stabilization by fixing Re and $d_x$ and varying St. The two branches of neutral curves in the $\alpha$-St space are given in Fig. 6.4 for the case of Re $= 3.38$, We $= 0.00324$, and $d_x = 30$. These parameter values correspond to a water film of 0.12 mm at room temperature on a plate oscillating with an amplitude of 1.8 mm. A film of this thickness is several orders of magnitude thicker than that encountered in various surface coatings, including photographic film coatings. The vertical dotted line at St $= 2.0$ in

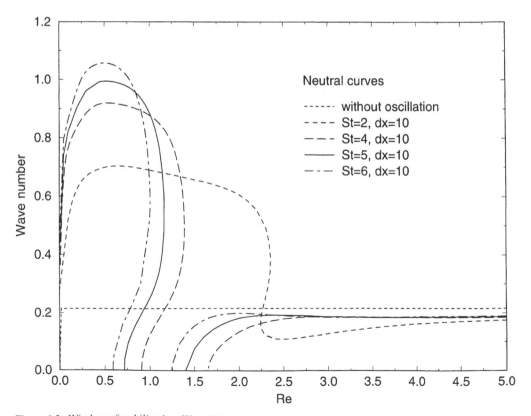

Figure 6.3. Window of stabilization, We $= 0.1$.

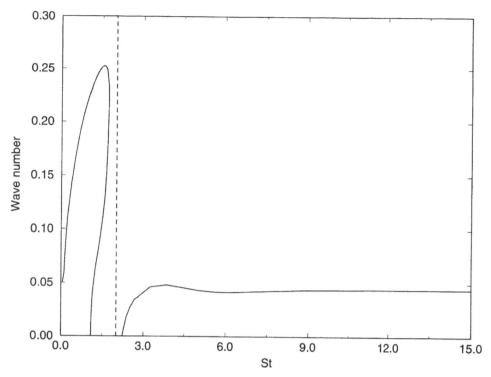

Figure 6.4. Neutral curves, ———, $Re = 3.38$, $We = 0.00324$, $d_x = 30$.

Fig. 6.4 indicates that the film is stable for $\alpha$ ranging from 0 to 0.3 and beyond, although the stable region for $\alpha > 0.3$ is not shown in this figure.

To see how the film flow on a stationary plate is stabilized by the plate oscillation, we obtain the energy budgets for the film on a stationary plate and on an oscillating plate for various values of $\alpha$ along the dotted line in Fig. 6.4: The results are given, respectively, in Tables 6.1 and 6.2. It should be pointed out that all energy items are renormalized with RKINE of the most amplified disturbance in a steady film flow. It is seen that while RKINE is positive for $\alpha \leq 0.050$ in Table 6.1, it is negative in Table 6.2. Thus it is the disturbances in this wave-number range that are stabilized. SURTE is negative for $\alpha \leq 0.050$ in Table 6.1 but it is positive in Table 6.2. Hence while work is done on the surface of the stationary film, the surface tension exerts work on the disturbance in the oscillating film. Therefore the stabilization by oscillation cannot be due to the surface tension. Comparisons of these two tables also show that while the absolute values of DISSI are greater than those of SHEST for the case of an oscillating film, the reverse is true for the case of a stationary plate. Moreover, REYNS either becomes negative or more negative from Table 6.1 to Table 6.2. Thus the Reynolds stress does participate in the stabilization. However, the stabilization effect of REYNS is relatively small compared with that of the energy dissipation. Although the plate oscillation increases the main energy production term SHEST, the corresponding increase in dissipation is even greater. We may conclude therefore that the stabilization is achieved for the water film by the increased dissipation rate promoted by the oscillation of the plate with appropriate frequencies and amplitudes (i.e., St and $d_x$). Mechanical energy dissipation was also found by Kelly and Hu (1993) to be the resonon for stabilization for the Rayleigh–Benard convection by means

Table 6.1. *Energy Budget for a Steady Vertical Film Flow (Re = 3.38, We = 0.00324)*

| α | RKINE | SURTE | REYNS | DISSI | SHEST | DISR |
|---|---|---|---|---|---|---|
| 0.001 | 0.611406E − 04 | −0.283427E − 07 | −0.105764E − 05 | −0.188269E + 01 | 0.188275E + 01 | 0.1000E + 01 |
| 0.010 | 0.569967E − 01 | −0.267836E − 02 | −0.981765E − 03 | −0.187075E + 02 | 0.187681E + 02 | 0.1018E + 01 |
| 0.034 | 0.100000E + 01 | −0.588389E + 00 | −0.110598E − 01 | −0.611113E + 02 | 0.627107E + 02 | 0.1154E + 01 |
| 0.045 | 0.743933E + 00 | −0.754417E + 00 | 0.184640E − 01 | −0.814898E + 02 | 0.829697E + 02 | 0.1142E + 01 |
| 0.050 | 0.398212E − 02 | −0.478899E − 02 | 0.552571E − 01 | −0.925977E + 02 | 0.925512E + 02 | 0.1079E + 01 |
| 0.051 | −0.218133E + 00 | 0.269858E + 00 | 0.651179E − 01 | −0.950693E + 02 | 0.945162E + 02 | 0.1061E + 01 |
| 0.055 | −0.143248E + 01 | 0.194603E + 01 | 0.114529E + 00 | −0.106087E + 03 | 0.102594E + 03 | 0.9728E + 00 |
| 0.100 | −0.112877E + 03 | 0.139780E + 03 | 0.316730E + 01 | −0.497838E + 03 | 0.242013E + 03 | 0.2254E + 00 |
| 0.110 | −0.172890E + 03 | 0.205817E + 03 | 0.523802E + 01 | −0.667231E + 03 | 0.283286E + 03 | 0.1908E + 00 |
| 0.121 | −0.257154E + 03 | 0.297247E + 03 | 0.869762E + 01 | −0.893688E + 03 | 0.330589E + 03 | 0.1668E + 00 |
| 0.191 | −0.143529E + 04 | 0.155774E + 04 | 0.885215E + 02 | −0.370952E + 04 | 0.627968E + 03 | 0.1060E + 00 |
| 0.200 | −0.169331E + 04 | 0.183283E + 04 | 0.109039E + 03 | −0.429656E + 04 | 0.661384E + 03 | 0.1013E + 00 |
| 0.250 | −0.371185E + 04 | 0.397732E + 04 | 0.269869E + 03 | −0.878476E + 04 | 0.825721E + 03 | 0.8250E − 01 |
| 0.260 | −0.424835E + 04 | 0.454494E + 04 | 0.310798E + 03 | −0.878476E + 04 | 0.825721E + 03 | 0.8250E − 01 |
| 0.270 | −0.483369E + 04 | 0.516314E + 04 | 0.354345E + 03 | −0.112379E + 05 | 0.886757E + 03 | 0.7804E − 01 |
| 0.300 | −0.690287E + 04 | 0.734018E + 04 | 0.499591E + 03 | −0.157230E + 05 | 0.980329E + 03 | 0.7419E − 01 |

*Note:* E(··) is computer shorthand for powers, e.g., 1.2E + 04 = $1.2 \times 10^4$ and 1.2E − 04 = $1.2 \times 10^{-4}$.

Table 6.2. *Energy Budget for an Oscillating Vertical Film Flow* $(Re = 3.38, We = 0.00324, St = 2.0, d_x = 30)$

| α | RKINE | SURTE | REYNS | DISSI | SHEST | DISR |
|---|---|---|---|---|---|---|
| 0.001 | $-0.109750E-03$ | $0.343706E-08$ | $-0.109389E-04$ | $-0.272042E+02$ | $0.272041E+02$ | $0.2673E+03$ |
| 0.010 | $-0.120451E+00$ | $0.374859E-03$ | $-0.109353E-01$ | $-0.283781E+03$ | $0.283672E+03$ | $0.2695E+03$ |
| 0.035 | $-0.119116E+02$ | $0.427198E+00$ | $-0.688861E+00$ | $-0.165553E+04$ | $0.164388E+04$ | $0.2781E+03$ |
| 0.045 | $-0.430775E+02$ | $0.246012E+01$ | $-0.252438E+01$ | $-0.322188E+04$ | $0.317887E+04$ | $0.2722E+03$ |
| 0.050 | $-0.818877E+02$ | $0.565474E+01$ | $-0.506161E+01$ | $-0.471417E+04$ | $0.463169E+04$ | $0.2659E+03$ |
| 0.055 | $-0.160028E+03$ | $0.130809E+02$ | $-0.106468E+02$ | $-0.727012E+04$ | $0.710766E+04$ | $0.2573E+03$ |
| 0.100 | $-0.284214E+04$ | $0.615935E+03$ | $-0.401875E+03$ | $-0.307489E+05$ | $0.276929E+05$ | $0.1326E+03$ |
| 0.200 | $-0.183754E+06$ | $0.738101E+05$ | $-0.578412E+05$ | $-0.458944E+06$ | $0.259214E+06$ | $0.1885E+02$ |
| 0.300 | $-0.765016E+06$ | $0.503710E+06$ | $-0.521788E+06$ | $-0.282077E+07$ | $0.207233E+07$ | $0.5596E+01$ |

of oscillation. How are the long surface waves suppressed by dissipation? We conjecture that it is necessary, although it may not be sufficient, that more dissipation take place in an oscillating film near the free surface than at the wall in comparison with the case of the stationary plate. To verify this conjecture, we obtain DISR = DISF/DISW, i.e., the ratio of the energy dissipation per unit thickness at the free surface to that at the wall for both cases of stationary and oscillating plates. The results are given in the last column of Tables 6.1 and 6.2. It is seen that this ratio for an oscillating film is at least 1 order of magnitude larger than that for the case of a stationary plate. For the latter case this ratio is of the order of 1. Therefore the dissipation takes place mainly at the free surface to suppress the surface waves in contrast to the unstable steady film flow in which the wall and free-surface shear layers dissipate approximately the same amount of energy. Energy budgets for other flow parameters have also been calculated. The same mechanism is found to be responsible for the stabilization (Lin and Chen 1998).

## 6.5. Conclusion

The inherently unstable viscous liquid film flow down a vertical plate can be stabilized by oscillating the plate at appropriate amplitudes and frequencies, although the stabilization is obtainable over a relatively small range of Re. The range of Re and the forcing amplitude and frequency required depend on other flow parameters. The mechanism of stabilization is the enhancement of the mechanical energy dissipation through viscosity by the plate oscillation that promotes the concentration of the energy dissipation near the free surface. The numerical results also show that the surface tension does not contribute to the stabilization of the inherently unstable film flow. The work done by the bulk fluid on the surface leads to an increase in only the surface energy, since no surface viscosity is considered.

## Acknowledgment

This work was supported by U.S. National Science Foundation grant CTS-9616135.

## References

Benjamin, T. B. 1957. J. Fluid Mech. **2**, 554.
Bauer, R. J. and von Kerczek, C. H. 1991. J. Appl. Mech. **58**, 278.
Coward, A. V. and Renardy, Y. Y. 1995. Bull. Am. Phys. Soc. **40**, 12, 1950.
DeBruin, G. J. 1974. J. Eng. Math. **8**, 259.
Floryan, J. M., Davis, S. H., and Kelly, R. E. 1987. Phys. Fluids **30**, 983.
Kelly, R. E., Goussis, D. A., Lin, S. P., and Hsu, F. K. 1989. Phys. Fluids A **1**, 819.
Kelly, R. E. and Hu, H. C. 1993. J. Fluid Mech. **249**, 373.
Lin, S. P., Chen, J. N., and Woods, D. R. 1996. Phys. Fluids **8**, 3247.
Lin, S. P. and Chen, J. N. 1997. Phys. Fluids **9**, 3926.
Lin, S. P. and Chen, J. N. 1998. Phys. Fluids **9**, 1787.
Or, A. C. 1997. J. Fluid Mech. **335**, 213.
Or, A. C. and Kelly, R. E. 1996. Bull. Am. Phys. Soc. **40**, 12, 1949.
Squire, H. B. 1933. Proc. R. Soc. London S. A **142**, 621.
Woods, D. R. and Lin, S. P. 1995. J. Fluid Mech. **294**, 391.
Woods, D. R. and Lin, S. P. 1996. J. Appl. Mech. **63**, 1051.
Yih, C. S. 1968. J. Fluid Mech. **31**, 737.
Yih, C. S. 1963. Phys. Fluids **6**, 321.
Yih, C. S. 1955. Q. Appl. Math. **12**, 434.

# 7

# Three-Dimensional Waves in Thin Liquid Films

ALEXANDER A. NEPOMNYASHCHY

## 7.1. Introduction

The spontaneous generation of waves on the surface of a liquid viscous film was first studied by P. L. Kapitsa and S. P. Kapitsa in the late 1940s (Kapitsa, 1948; Kapitsa and Kapitsa, 1949). The theoretical explanation of this phenomenon on the basis of the linear stability theory was done by Benjamin (1957) and Yih (1963). The waves in the falling film were a subject of numerous experimental and theoretical investigations (for a survey, see the review paper by Chang, 1994). A remarkable feature of wave evolution is the breakdown of two-dimensional waves and the appearance of three-dimensional structures (Chang et al., 1994; Liu et al., 1995).

This chapter is devoted to the consideration of the latter phenomenon. In Section 7.2 we summarize the main results of the linear stability theory for a flow on an inclined plane. In subsequent sections, we develop the amplitude equations that describe weakly nonlinear waves. In the case of a small inclination angle we obtain the perturbed Korteweg–de Vries (KdV) equation and investigate the stability of waves with an inclined front of propagation (Section 7.3). In Section 7.4 we discuss the special case of a vertical plane and a moderate surface tension in which the waves are governed by the perturbed Zakharov–Kuznetsov (ZK) equation. The physically realistic limit of a strong surface tension governed by the anisotropic Kuramoto–Sivashinsky equation is considered in Section 7.5. Section 7.6 contains some concluding remarks.

## 7.2. Linear Stability of the Parallel Flow on an Inclined Plane

Let us consider a viscous incompressible liquid flowing under the action of the gravity force (gravity acceleration is $g$) along the inclined rigid surface. The density of the liquid is $\rho$, the coefficient of kinematic viscosity is $\nu$, and the surface-tension coefficient is $\sigma$, the angle between the surface and the horizontal direction is $\beta$, and the mean thickness of the layer is $d$. We use $d$, $\nu/d$, $d^2/\nu$, and $\rho(\nu/d)^2$ as units of length, velocity, time, and pressure, respectively. The motion of the fluid is governed by the following system of equations:

$$\frac{\partial \mathbf{v}}{\partial t} + (\mathbf{v} \cdot \nabla)\mathbf{v} = -\nabla p + \Delta \mathbf{v} + G\mathbf{e}, \tag{1}$$

$$\nabla \cdot \mathbf{v} = 0; \tag{2}$$

where $\mathbf{e} = (\sin\beta, 0, -\cos\beta)$ and $G = ga^3/\nu^2$ is Galileo number. The boundary conditions are

$$y = 0 : \quad \mathbf{v} = 0; \tag{3}$$

$$y = 1 + h : \quad (p - \gamma G^{1/3} K)n_i - \sigma'_{ik}n_k = p_a n_i, \tag{4}$$

$$\frac{\partial h}{\partial t} + u\frac{\partial h}{\partial x} + w\frac{\partial h}{\partial z} = v, \tag{5}$$

where $\sigma'_{ik} = \partial v_i/\partial x_k + \partial v_k/\partial x_i$, $\gamma = (\sigma/\rho)(v^4 g)^{-1/3} K$ is the mean curvature of the free surface, $u$, $v$, and $w$ are the $x$, $y$, and $z$ components of $\mathbf{v}$, respectively, and $h$ is the deviation of the layer thickness from its mean value.

The boundary problem of Eqs. (1)–(5) always has a solution describing a stationary parallel flow with a flat surface:

$$h_0 = 0, \quad v_0 = 0, \quad w_0 = 0, \quad u_0 = G\,\sin\beta\,U_0(z), \quad p_0 = p_a + G\,\cos\beta\,P_0(z), \tag{6}$$

where

$$U_0(z) = z - \frac{z^2}{2}, \qquad P_0(z) = 1 - z.$$

Let us impose a disturbance,

$$\mathbf{v} = \mathbf{v}_0 + \tilde{\mathbf{v}}, \qquad p = p_0 + \tilde{p}, \qquad h = h_0 + \tilde{h}, \tag{7}$$

substitute Eqs. (7) into Eqs. (1)–(5), and linearize. Assuming that

$$(\tilde{h}, \tilde{\mathbf{v}}, \tilde{p}) = [1, \mathbf{V}(z), P(z)]\exp(ik_x x + ik_y y + \lambda t), \tag{8}$$

we obtain an eigenvalue problem that determines the spectrum of growth rates $\lambda(k_x, k_y, G, \gamma)$. We are interested in only the deformational mode generating a long-wave instability that is characterized by the following property: $\lambda(0, 0, G, \gamma) = 0$. For long waves, the solution can be obtained by means of expansions in powers of the small parameters $k_x$, $k_y$:

$$\mathbf{V} = \sum_{n,m=0}^{\infty} k_x^n k_y^m \mathbf{V}_{nm}, \qquad P = \sum_{n,m=0}^{\infty} k_x^n k_y^m P_{nm}, \qquad \lambda = \sum_{n,m=0}^{\infty} k_x^n k_y^m \lambda_{nm}. \tag{9}$$

Omitting the calculations, we present the expressions for $\lambda_{nm}$ up to the fourth order (Nepomnyashchy, 1974a; Nepomnyashchy, 1976):

$$\lambda_{00} = 0; \quad \lambda_{10} = -iG\,\sin\beta; \quad \lambda_{20} = \frac{2}{15}(G\,\sin\beta)^2 - \frac{1}{3}G\,\cos\beta;$$

$$\lambda_{30} = i\left[G\,\sin\beta + \frac{4}{63}(G\,\sin\beta)^3 - \frac{10}{63}G\,\cos\beta G\,\sin\beta\right];$$

$$\lambda_{40} = -\frac{1}{3}\gamma G^{1/3} - \frac{75872}{2027025}(G\,\sin\beta)^4 + \frac{17363}{155925}(G\,\sin\beta)^2 G\,\cos\beta$$

$$\qquad - \frac{157}{224}(G\,\sin\beta)^2 - \frac{2}{45}(G\,\cos\beta)^2 + \frac{3}{5}G\,\cos\beta;$$

$$\lambda_{02} = -\frac{1}{3}G\cos\beta; \quad \lambda_{12} = i\left[G\sin\beta - \frac{10}{63}G\cos\beta G\sin\beta\right];$$

$$\lambda_{22} = -\frac{2}{3}\gamma G^{1/3} + \frac{17363}{155925}(G\sin\beta)^2 G\cos\beta - \frac{157}{224}(G\sin\beta)^2$$

$$- \frac{4}{45}(G\cos\beta)^2 + \frac{6}{5}G\cos\beta;$$

$$\lambda_{04} = -\frac{1}{3}\gamma G^{1/3} - \frac{2}{45}(G\cos\beta)^2 + \frac{3}{5}G\cos\beta.$$

One can see that the flow of Eqs. (6) is unstable ($\lambda > 0$) with respect to long-wavelength disturbances ($k = |\mathbf{k}| \to 0$), if $G > G_0$, where

$$G_0 = \frac{5\cos\beta}{2\sin^2\beta}. \tag{10}$$

If $\beta \neq \pi/2$, the neutral surface $\lambda(G, k_x, k_y) = 0$ has the following form at small $k$:

$$G = G_0\left[1 + \frac{k_y^2}{k_x^2} + H_0(\gamma, \beta)k_x^2\right] + O\left(k_x^4, k_y^4, k_y^4/k_x^2\right),$$

where

$$H_0(\gamma, \beta) = \gamma\left(\frac{2}{5}\right)^{2/3}\frac{\sin^{4/3}\beta}{\cos^{5/3}\beta} + \frac{7743}{2240} - \cot^2\beta\frac{1}{36036}. \tag{11}$$

The function $H_0(\gamma, \beta)$ decreases monotonically when $\beta$ decreases. If $\beta < \beta_0$, where $\beta_0$ is defined by the equation $H_0(\gamma, \beta_0) = 0$, the point $k = 0$ becomes a point of the maximum of the neutral curve. The angle $\beta_0$ is very small ($\beta_0 \approx 9'45''$ in the limit $\gamma \to 0$; $\beta_0 \approx 9'7''$ for water, $\gamma = 2850$).

If $H_0(\gamma, \beta)$ is positive, there is a long-wavelength instability. Let us define

$$k_m = \left[\frac{G - G_0}{G_0 H_0(\gamma, \beta)}\right]^{1/2} \tag{12}$$

and estimate the terms $\Lambda_{n0} = \lambda_{n0}k_x^n$ in expansion (9) for $k_x \sim k_m$. It is obvious that $|\Lambda_{20}| \sim |\Lambda_{40}|$. If $H_0 \gg 1(\gamma \gg 1$, and inclination angles are moderate), then $|\Lambda_{20}|$ and $|\Lambda_{40}|$ are much larger than $|\Lambda_{30}|$ in the whole region where the long-wavelength expansions are applicable (i.e., for $k_m G \sin\beta \ll 1$), except a small region near the threshold:

$$G - G_0 \sim \frac{1}{G_0 H_0 \sin^2\beta} \sim G_0^{2/3}\gamma^{-1}. \tag{13}$$

Outside of the latter region, the characteristic time of the evolution of the wave amplitude is much less than the time scale connected with the dispersion of waves; hence the dispersion is negligible. For small inclination angles ($H_0 \sim 1$), in the whole region where long-wavelength expansions are applicable, we obtain the opposite inequality:

$$|\Lambda_{30}| \gg |\Lambda_{20}|, \quad |\Lambda_{40}|,$$

and the dispersion of waves is essential. A special analysis is necessary in the case of a vertical plane ($G_0 = 0$).

We consider the nonlinear development of instability in each case separately.

## 7.3. Small Inclination Angle

We are going to derive amplitude equations by using the method of multiscale expansions. First, let us consider the generic case, $H_0 = O(1)$, $\beta \neq \pi/2$. Note that for small $G - G_0$, the instability region has a size $O[(G - G_0)^{1/2}]$ in the $k_x$ direction and a size $O(G - G_0)$ in the $k_y$ direction. Let us introduce new spatial variables $x_1 = \epsilon x$ and $y_2 = \epsilon^2 y$ and an infinite number of time variables $t_n = \epsilon^n t$, $n = 1, 2, \ldots$. The deviation of the solution $F = (\mathbf{v}, p, h)$ from the basic flow $F_0 = (\mathbf{v}_0, p_0, h_0)$ is expanded into an asymptotic series in powers of $\epsilon$. Putting $G = G_c + G_2\epsilon^2$, we collect the terms of the same order in $\epsilon$.

In the lowest order, one can find that

$$F - F_0 = \epsilon^m a_1(x_1, y_2, t_1, t_2, \ldots) f(z; 0, 0, G_0) + o(\epsilon^m), \tag{14}$$

where $f(z; k_x, k_y, G)$ is the eigenfunction of the linear stability theory corresponding to the wave vector $(k_x, k_y)$, $m$ is a certain number, and $a_1$ is the amplitude function (the scaled deformation of the interface). Considering solvability conditions in each power of $\epsilon$, one gets an infinite system of equations for amplitude functions $a_n$, $n = 1, 2, \ldots$. This system can be written formally as a unique equation in the form

$$\frac{\partial a}{\partial t} = \sum_{n=1}^{\infty} \epsilon^n L_n(a), \tag{15}$$

where

$$\frac{\partial}{\partial t} = \epsilon \frac{\partial}{\partial t_1} + \epsilon^2 \frac{\partial}{\partial t_2} + \cdots, \qquad a = a_1 + \epsilon a_2 + \cdots, \tag{16}$$

$L_n(a)$ is a certain function of $a$ and its spatial derivatives. Because of the conservation law, each term in amplitude equation (15) has the form

$$L_n(a) = \frac{\partial}{\partial x_1} Q_n(a) + \frac{\partial}{\partial y_2} P_n(a). \tag{17}$$

In the second order, one obtains the term

$$L_1(a) = \sigma_{10} \frac{\partial a}{\partial x_1}, \qquad \sigma_{10} = \left( \frac{\partial \lambda}{\partial k_x} \right)_c = -i(\lambda_{10})_c = -G_0 \sin \beta, \tag{18}$$

which describes the propagation of waves with their group velocity that can be eliminated by using a moving reference frame:

$$\xi = x_1 - \sigma_{10} t. \tag{19}$$

In higher orders, we find that one can get bounded smooth solutions as $t \to \infty$ if $m = 2$.

Finally, we obtain the following sequence of terms:

$$L_2(a) = 0,\tag{20}$$

$$L_3(a) = \sigma'_{10}\frac{\partial a}{\partial \xi} + \sigma_{30}\frac{\partial^3 a}{\partial \xi^3} + \gamma_1\frac{\partial}{\partial \xi}(a^2),\tag{21}$$

$$L_4(a) = \sigma'_{20}\frac{\partial^2 a}{\partial \xi^2} + \sigma_{40}\frac{\partial^4 a}{\partial \xi^4} + \sigma_{02}\frac{\partial^2 a}{\partial y_2^2} + \gamma_2\frac{\partial^2}{\partial \xi^2}(a^2),\tag{22}$$

$$L_5(a) = \sigma''_{10}\frac{\partial a}{\partial \xi} + \sigma'_{30}\frac{\partial^3 a}{\partial \xi^3} + \sigma_{50}\frac{\partial^5 a}{\partial \xi^5} + \sigma_{12}\frac{\partial^3 a}{\partial \xi \partial y_2^2} + \gamma'_1\frac{\partial}{\partial \xi}(a^2) + \gamma_3\frac{\partial}{\partial \xi}\left(a\frac{\partial^2 a}{\partial \xi^2}\right)$$
$$+ \gamma_4\frac{\partial}{\partial \xi}\left[\left(\frac{\partial a}{\partial \xi}\right)^2\right] + \gamma_5\frac{\partial}{\partial \xi}(a^3),\tag{23}$$

$$L_6(a) = \sigma''_{20}\frac{\partial^2 a}{\partial \xi^2} + \sigma'_{40}\frac{\partial^4 a}{\partial \xi^4} + \sigma_{60}\frac{\partial^6 a}{\partial \xi^6} + \sigma_{22}\frac{\partial^4 a}{\partial \xi^2 \partial y_2^2} + \gamma'_2\frac{\partial^2}{\partial \xi^2}(a^2)$$
$$+ \gamma_6\frac{\partial^2}{\partial y_2^2}(a^2) + \gamma_7\frac{\partial}{\partial \xi}\left(a\frac{\partial^3 a}{\partial \xi^3}\right) + \gamma_8\frac{\partial}{\partial \xi}\left(\frac{\partial a}{\partial \xi}\frac{\partial^2 a}{\partial \xi^2}\right) + \gamma_9\frac{\partial^2}{\partial \xi^2}(a^3),\tag{24}$$

etc., where the coefficients in linear terms are determined by derivatives of $\lambda(k_x, k_y, G)$ in the critical point and nonlinear interaction coefficients $\gamma_j, \gamma'_j, (j = 1, 2, \ldots)$ are real constants. By using the expressions for $\lambda_{mn}$, we find that

$$\sigma'_{10} = -G_2 \sin\beta; \quad \sigma_{30} = -G_0 \sin\beta; \quad \gamma_1 = -G_0 \sin\beta;$$

$$\sigma'_{20} = -\frac{1}{3}G_2 \cot\beta; \quad \sigma_{40} = -\frac{1}{3}G_0 H_0 \cot\beta; \quad \gamma_2 = -\frac{1}{2}G_0 \cos\beta,$$

etc. Because the mean value of $a$ does not change with time, the class of solutions with zero mean value should be chosen.

By means of a scale transformation and an additional change of the reference frame we can rewrite amplitude equation (15) in the following form:

$$U_T + 3(U^2)_X + U_{XXX} + \bar{\epsilon}[U_{XX} + U_{XXXX} + D(U^2)_{XX} - U_{YY}]$$
$$+ \bar{\epsilon}^2\left[\alpha_1\frac{\partial^5 U}{\partial X^5} + \alpha_2 U_{XYY} + \alpha_3(U^2)_{XXX} + \alpha_4(U_X^2)_X + \alpha_5(U^3)_X\right]$$
$$+ \bar{\epsilon}^3\left[\beta_1\frac{\partial^6 U}{\partial X^6} + \beta_2 U_{XXYY} + \beta_3(U^2)_{YY} + \beta_4(U^2)_{XXXX} + \beta_5(U_X^2)_{XX} + \beta_6(U^3)_{XX}\right]$$
$$= O(\epsilon^4),\tag{25}$$

where

$$\bar{\epsilon} = \epsilon\frac{(\sigma'_{20}\sigma_{40})^{1/2}}{|\sigma_{30}|} + o(\epsilon); \quad D = 3\frac{\sigma_{30}\gamma_2}{\gamma_1\sigma_{40}} + o(1),\tag{26}$$

etc. Thus, we obtain the perturbed KdV equation.

The $1 + 1$-dimensional version of problem (25) (without any dependence on $Y$) was considered formerly by Nepomnyashchy (1976) and Bar and Nepomnyashchy (1995). The traveling-wave solutions

$$U = u(\Xi - \Xi_0), \qquad \Xi = X - cT, \qquad \Xi_0 = \text{const.}, \tag{27}$$

which satisfy conditions

$$u(\Xi + 2\pi/q) = u(\Xi), \qquad \int_0^{2\pi/q} u(\Xi)d\Xi = 0, \tag{28}$$

were constructed by means of the asymptotic expansion

$$u = u_0 + \bar{\epsilon}u_1 + \bar{\epsilon}^2 u_2 + \cdots, \qquad c = c_0 + \bar{\epsilon}c_1 + \bar{\epsilon}^2 c_2 + \cdots. \tag{29}$$

In the leading order in $\bar{\epsilon}$, one obtains:

$$u_0 = u_0(\Xi - \Xi_0; q, k) = \frac{2q^2 K^2}{\pi^2}\left[\text{dn}^2\left(\frac{(\Xi - \Xi_0)qK}{\pi}\right) - \frac{E}{K}\right], \tag{30}$$

$$c_0 = \frac{4q^2 K^2}{\pi^2}\left(2 - k^2 - \frac{3E}{K}\right), \tag{31}$$

where dn is Jacobi's delta amplitude function with the modulus $k$ and $E = E(k)$ and $K = K(k)$ are complete elliptic integrals. Contrary to the case of the nonperturbed KdV equation, the parameter $k$ is not arbitrary but is connected with the spatial period $2\pi/q$ by the following formula, which can be obtained from the solvability condition for the equation in the next order:

$$\frac{\pi^2}{4q^2 K^2} = -\left(2 - k^2 - \frac{3E}{K}\right) - \frac{(D - 3)}{7}$$

$$\times \frac{-14(1 - k^2 + k^4)E^2/K^2 + 2(10 - 15k^2 + 13k^4 - 4k^6)E/K - 2(3 - 6k^2 + 5k^4 - 2k^6)}{2(1 - k^2 + k^4)E/K - (2 - 3k^2 + k^4)}. \tag{32}$$

The stability analysis of the solutions showed that the waves are stable as $D < 5/4$ in a certain interval of wave numbers, e.g., inside a certain Busse stability balloon (for details, see Bar and Nepomnyashchy, 1995).

The general investigation of the $2 + 1$-dimensional problem (25) is beyond the scope of this chapter. Here we restrict ourselves to the investigation of the stability of spatially periodic waves with a plane front with respect to three-dimensional disturbances (see also Bar, 1996).

First, let us note that the stability criterion for one-dimensional traveling waves (27) is unchanged compared with the case of the one-dimensional equation. Indeed, the normal disturbances have the form $\tilde{u} = \Phi(\Xi)e^{ik_y Y + \Omega T}$ and are governed by the following equation:

$$\Omega\Phi - c\Phi' + 6(u\Phi)' + \Phi''' + \bar{\epsilon}[\Phi'' + \Phi'''' + 2D(u\Phi)'' + k_y^2\Phi] = 0, \tag{33}$$

where the prime corresponds to differentiation with respect to $\Xi$. Obviously, for any eigenvalue

$\Omega(k) = \Omega(0) - \bar{\epsilon} k_y^2$, so that the two-dimensional disturbances are less dangerous than the one-dimensional ones.

We consider now a more wide class of wavy solutions (oblique waves):

$$U = u(\eta), \qquad \eta = X + \alpha Y - cT, \qquad |\alpha| < 1. \tag{34}$$

Let us perform the following transformation of variables:

$$U(X, Y, T) = \beta^2 V(\xi, y, t), \qquad t = \beta^3 T, \qquad \xi = \beta \eta, \qquad y = \beta Y, \tag{35}$$

where $\beta^2 = 1 - \alpha^2$. We obtain

$$V_t - \hat{c} V_\xi + 3(V^2)_\xi + V_{\xi\xi\xi} + \hat{\epsilon} \left[ V_{\xi\xi} + V_{\xi\xi\xi\xi} + D(V^2)_{\xi\xi} - \frac{1}{\beta^2}(V_{yy} + 2\alpha V_{y\xi}) \right] = O(\hat{\epsilon}^2), \tag{36}$$

where $\beta^2 \hat{c} = c$ and $\hat{\epsilon} = \beta \bar{\epsilon}$.

The stationary solution satisfying the conditions

$$u = u(\xi) = u\left(\xi + \frac{2\pi}{q}\right), \qquad \langle u \rangle = \int_0^{\frac{2\pi}{q}} u(\xi) d\xi = 0 \tag{37}$$

is constructed by means of asymptotic expansions:

$$u = u_0 + \hat{\epsilon} u_1 + \cdots, \qquad \hat{c} = \hat{c}_0 + \hat{\epsilon} \hat{c}_1 + \cdots \tag{38}$$

(actually $\hat{c}_1 = 0$).

Although the solutions describing the oblique waves are rather similar to those describing the one-dimensional waves, the stability properties of these two kinds of waves turn out to be completely different. We will show that the oblique waves are unstable for any $\alpha \neq 0$ and any $q$.

To this end, it is sufficient to consider a special kind of disturbance:

$$V = u(\xi) + \Phi(\xi) e^{\lambda t + i k_y y}, \qquad \Phi(\xi + 2\pi/q) = \Phi(\xi); \tag{39}$$

$$\lambda = \lambda_0 + \hat{\epsilon} \lambda_1 + \cdots; \qquad k_y = \hat{\epsilon} K, \qquad K = O(1). \tag{40}$$

The stability analysis is fully similar to that described in Appendix A of the paper by Bar and Nepomnyashchy (1995). The only new circumstance is the additional term $\hat{\epsilon}[-(1/\beta^2)2\alpha V_{y\xi}]$ in Eq. (36), which gives the contribution into the solvability condition in second order in $\hat{\epsilon}$, which has the form

$$I_3 \lambda_1^3 + I_2 \lambda_1^2 - i I_1 \lambda_1 = 0, \tag{41}$$

where $I_j$ are real coefficients and $I_1 \propto K\alpha$, $I_2/I_3 > 0$ if $D < 5/4$. Thus the three roots are

$$\lambda_1 = 0, \qquad \frac{-I_2}{2I_3}\left(1 \pm \sqrt{1 + i\frac{4I_3 I_1}{I_2^2}}\right). \tag{42}$$

For any real nonzero $I_1$, one of the roots always has a positive real part, which corresponds to an instability.

A more detailed investigation of problem (25) is presented elsewhere (Bar and Nepomnyashchy, 1998).

### 7.4. Vertical Plane, Moderate Surface Tension

The expansions leading to the amplitude equation should be changed in the case of a vertical plane ($\beta = \pi/2$), because the coefficient $\lambda_{02} = 0$ and $G_0 = 0$. Let us take $G \ll 1$, $\gamma \sim 1$ (the latter condition corresponds to a moderate surface tension). By using the expressions for $\lambda_{mn}$, we find that the width of the instability interval

$$k_m = \left[ -\frac{\lambda_{20}(G)}{\lambda_{40}(G)} \right]^{1/2} = (2/5\gamma)^{1/2} G^{5/6}.$$

Let us note that because of the equality $\lambda_{02}(G) = 0$, the characteristic spatial scales in the $x$ and the $y$ directions are equal. Estimating different terms in the amplitude equation, we find that the orders of two groups of terms are relatively close:

$$\sigma_{30}(G)\frac{\partial^3 a}{\partial x^3}, \qquad \sigma_{12}(G)\frac{\partial^3 a}{\partial x \partial y^2}, \qquad \gamma_1(G)\frac{\partial}{\partial x}(a^2) = O(G^{31/6}),$$

$$\sigma_{20}(G)\frac{\partial^2 h}{\partial x^2}, \qquad \sigma_{40}(G)\frac{\partial^4 h}{\partial x^4}, \qquad \sigma_{22}(G)\frac{\partial^4 h}{\partial x^2 \partial y^2}, \qquad \sigma_{04}(G)\frac{\partial^4 h}{\partial y^4} = O(G^{16/3}),$$

while all other terms are much smaller. After scaling, a perturbed Zakharov–Kuznetsov (ZK) equation is obtained:

$$U_T + U U_X + \Delta U_X + \epsilon(U_{XX} + \Delta^2 U) = o(\epsilon), \tag{43}$$

where $\epsilon \sim G^{1/6}$ and $\Delta = \partial^2/\partial X^2 + \partial^2/\partial Y^2$.

Unlike the case considered in Section 7.3, the one-dimensional solitary and cnoidal waves $U = U(X - cT)$ governed by the nonperturbed ZK equation are unstable with respect to transverse disturbances (Zakharov and Kuznetsov, 1974; Spektor, 1988; Allen and Rowlands, 1995). In the case of a primary one-dimensional soliton, the instability leads to formation of a chain of two-dimensional solitons (Zakharov and Kuznetsov, 1974; Frycz and Infeld, 1989).

It is relatively easy to analyze the nonlinear evolution for the long-wavelength disturbances in the case of the nonperturbed ZK equation,

$$U_T + U U_X + \Delta U_X = 0. \tag{44}$$

Let us consider a class of solutions that can be presented in the form

$$u = u[X - \phi(Y, T), \bar{\epsilon}Y, \bar{\epsilon}T], \qquad \bar{\epsilon} \ll 1 \tag{45}$$

($\bar{\epsilon}$ is a small parameter characterizing the spatial scale of disturbances). Assume that $\phi_Y$ and

$\phi_T$ are slow functions of the transverse coordinate $y$ and time:

$$\phi_Y = k(\bar{\epsilon}Y, \bar{\epsilon}T), \qquad \phi_T = \omega(\bar{\epsilon}Y, \bar{\epsilon}T). \tag{46}$$

Define

$$Z = x - \phi(Y, T), \qquad \eta = \bar{\epsilon}Y, \qquad \tau = \bar{\epsilon}T. \tag{47}$$

Obviously,

$$k_\tau - \omega_\eta = 0. \tag{48}$$

Let us represent the solution in the form

$$u = u_0 + \bar{\epsilon}u_1 + \cdots, \qquad k = k_0 + \bar{\epsilon}k_1 + \cdots, \qquad \omega = \omega_0 + \bar{\epsilon}\omega_1 + \cdots \tag{49}$$

and collect the terms in each order in $\bar{\epsilon}$.

In the zeroth order, we obtain the following equation:

$$-\omega_0 u_{0Z} + u_0 u_{0Z} + \left(1 + k_0^2\right) u_{0ZZZ} = 0. \tag{50}$$

Recall that $\omega_0$ and $k_0$ are functions of slow variables $\eta$ and $\tau$. We are interested in modulations of the solitary wave solution

$$u_0 = 3\omega_0 \operatorname{sech}^2 \left( \frac{\omega_0^{1/2}}{2\sqrt{1 + k_0^2}} Z \right). \tag{51}$$

However, to avoid mathematical complications connected with formation of the shelf and radiation (Kodama and Ablowitz, 1981), we prefer to consider a cnoidal wave with a very large period $L \gg 1$. Thus we assume that our solution $u_0$ is exponentially close to Eq. (51) in the region $-L/2 \leq Z \leq L/2$ but satisfies conditions

$$u_0(Z + L) = u_0(Z), \qquad \int_{-L/2}^{L/2} u_0(Z)\, dZ = 0. \tag{52}$$

Let us note that, according to Eq. (48),

$$k_{0\tau} - \omega_{0\eta} = 0. \tag{53}$$

In the first order in $\bar{\epsilon}$, the solvability condition gives the equation

$$\frac{1}{2}\langle u_0^2 \rangle_\tau + \left(\langle u_{0Z}^2 \rangle k_0\right)_\eta = 0, \tag{54}$$

where

$$\langle f \rangle \equiv \int_{-L/2}^{L/2} f(Z)\, dZ.$$

Calculating the integrals in the limit $L \to \infty$, we obtain the following equation:

$$\left( \omega_0^{3/2} \sqrt{1 + k_0^2} \right)_\tau + \frac{2}{5}\left( \omega_0^{5/2} \frac{k_0}{\sqrt{1 + k_0^2}} \right)_\eta = 0. \tag{55}$$

Here we consider some particular solutions of the system of Eqs. (53) and (55).

First, there is a family of solutions with constant values of $\omega_0$ and $k_0$ that describe solitons with an oblique front. Linearizing Eqs. (53) and (55) near such a solution, we find that the long-wave instability disappears as $k_0^2 > k_*^2 = 3/5$. In other words, the oblique soliton is unstable if the inclination angle $\alpha$ is smaller than $\alpha_* = \tan^{-1}\sqrt{3/5} \approx 0.659$, in coincidence with the result by Allen and Rowlands (1995). Let us emphasize, however, that for $\alpha > \alpha_*$ there is a short-wave instability of solitons, so that the long-wave analysis is incomplete.

Now let us consider the simple wave solutions characterized by the univalued connection:

$$\omega_0 = \omega_0(k_0). \tag{56}$$

Substituting Eq. (56) into Eq. (55) and taking into account Eq. (53), we find that

$$\frac{d\omega_0}{dk_0} = -\frac{2}{3} \cdot \frac{k_0 \pm \sqrt{k_0^2 - 3/5}}{1 + k_0^2}. \tag{57}$$

Equation (57) has real solutions as $k_0^2 > 3/5$ (i.e., in the stable case). For any solution (56) of Eq. (57), Eq. (53) has the form

$$\frac{\partial k_0}{\partial \tau} - \frac{d\omega_0}{dk_0}\frac{\partial k_0}{\partial \eta} = 0.$$

For any initial conditions of the type $k_0 = k_0(\eta)$ we obtain a simple wave solution

$$k_0 = k_0[\eta + \omega_0(k_0)\tau].$$

Obviously, the appearance of a shock wave is inavoidable if $d\omega_0/dX < 0$ somewhere as $t = 0$. The description of a finite-amplitude shock wave cannot be performed in frames of assumptions (46). However, one can expect that some waves with piecewise smooth front will appear, which will contain fragments of oblique solitons with different inclination angles.

The theory of waves governed by the perturbed ZK is still far from completion. Numerical simulations (Toh et al., 1989; Frenkel and Indireshkumar, 1996; Ogawa and Liu, 1997) reveal the existence of oblique V-shaped chains of two-dimensional solitons on the background of weak chaotic waves.

## 7.5. Strong Surface Tension

Another kind of expansion is used in the problem of film-flow instability of a liquid with a large surface tension (Nepomnyashchy, 1974b).

In both cases considered in previous sections the appearance of long waves ($k_m \ll 1$) is expected only near the threshold of instability: $G - G_0 \ll 1$. The instability interval $k < k_m$, where

$$k_m = \left[\frac{2\sin^2\beta G^{2/3}(G - G_0)}{5\gamma}\right]^{1/2}, \tag{58}$$

is narrow for any $G = O(1)$ if the parameter $\gamma \gg 1$. The small parameter $\epsilon \equiv k_m$ may be used for constructing an amplitude equation. It is assumed that $G - G_0 = O(1)$, but the region of wave vectors of unstable disturbances is narrow because the coefficients $\lambda_{40}$, $\lambda_{22}$, and $\lambda_{04}$ connected with the surface tension are large [$O(\epsilon^{-2})$ by definition]. In this case we find $m = 1$,

$$L_2(a) = \sigma_{20}\frac{\partial^2 a}{\partial \xi^2} + \sigma_{02}\frac{\partial^2 a}{\partial y_1^2} + \sigma_{40}\frac{\partial^4 a}{\partial \xi^4} + \sigma_{22}\frac{\partial^4 a}{\partial \xi^2 \partial y_1^2} + \sigma_{04}\frac{\partial^4 a}{\partial y_1^4} + \gamma_1\frac{\partial}{\partial \xi}(a^2), \tag{59}$$

etc. (Nepomnyashchy, 1974a; Nepomnyashchy, 1974c).

After a rescaling, we obtained a perturbed anisotropic Kuramoto–Sivashinsky equation:

$$U_T + (U^2)_X + U_{XX} - nU_{YY} + \Delta^2 U + \epsilon[\alpha_1 U_{XXX} + \alpha_2 U_{XYY} + \alpha_3(U^2)_{XX}$$
$$+ \alpha_4(U^2)_{YY} + \beta_1\Delta^2 U_X + \beta_2\nabla \cdot (U\nabla\Delta U)] = o(\epsilon), \tag{60}$$

where

$$\nabla = \mathbf{e}_X\frac{\partial}{\partial X} + \mathbf{e}_Y\frac{\partial}{\partial Y}; \quad \Delta = \nabla^2; \quad n = G_0/(G - G_0).$$

The term $\Delta^2 U$ and the terms with coefficients $\beta_1$ and $\beta_2$ are connected with the surface tension.

The stability of one-dimensional spatially periodic waves with the plane front

$$U = U_0(X); \quad U_0(X + 2\pi/q) = U_0(X) \tag{61}$$

governed by the nondispersive anisotropic Kuramoto–Sivashinsky equation was investigated by Nepomnyashchy (1974a) and Chang et al. (1994). The growth rate $\lambda$ of the disturbance

$$\tilde{U}(X, Y, T) = F(X)\exp(i\tilde{q}_x X + i\tilde{q}_y Y)\exp\lambda T, \tag{62}$$

where $F(\xi)$ is a $2\pi/q$ periodic function, is obtained from the following boundary value problem:

$$\mu F + \left(\frac{d}{dX} + i\tilde{q}_x\right)^4 F + \left(1 - 2\tilde{q}_y^2\right)\left(\frac{d}{dX} + i\tilde{q}_x\right)^2 F + 2\left(\frac{d}{dX} + i\tilde{q}_x\right)(U_0 F) = 0, \tag{63}$$

where $\mu = \lambda + n\tilde{q}_y^2 + \tilde{q}_y^4$.

For long-wave disturbances the expansions

$$F = \sum_{n=0}^{\infty} F_n\delta^n, \qquad \mu = \sum_{n=0}^{\infty} \mu_n\delta^n, \tag{64}$$

where

$$\delta = \left(\tilde{q}_x^2 + \tilde{q}_y^4\right)^{1/2}, \qquad \tilde{q}_x = \delta\cos\phi, \qquad \tilde{q}_y^2 = \delta\sin\phi,$$

may be used.

If only one-dimensional disturbances are taken into account ($\phi = 0$), it is known that solutions (61) are stable within the interval of wave numbers $q_2 < q < q_1$, where $q_1 \approx 0.837$ and $q_2 \approx 0.77$ (Nepomnyashchy, 1974b; Cohen et al., 1976; Frisch et al., 1986). Both stability boundaries are connected with long-wave disturbances. If $q > q_1$, then $\mu_1^2 > 0$; in this case, there are two real roots of opposite sign, one of which is always positive. If $q < q_1$, then $\mu_1^2 < 0$ (two imaginary roots), while the coefficient $\mu_2$ is negative if $q > q_2$ and positive otherwise.

For two-dimensional disturbances ($\phi \neq 0$) one obtains the following approximate expression for $\mu$:

$$\mu \approx (1 - \zeta \sin\phi\, \delta)\big(-\alpha \sin\phi\delta + \Lambda_2(\phi)\delta^2 \pm \{[\alpha \sin\phi\delta - \Lambda_2(\phi)\delta^2]^2$$

$$+ \big(\Lambda_1^2 \cos^2\phi\delta^2 + 2\alpha \sin\phi \cos^2\phi\delta^3\big)(1 + \zeta \sin\phi\delta)\}^{1/2}\big), \tag{65}$$

$$\lambda = \mu - n\delta \sin\phi - \delta^2 \sin^2\phi, \tag{66}$$

where $\alpha$, $\Lambda_1$, $\Lambda_2$, and $\zeta$ are numbers that depend only on $q$.

One can find that the oscillatory instability boundary $q = q_2$ is not changed (and does not depend on $n$).

In the region $q \approx q_1$ the boundary between monotonically growing and decaying disturbances in the plane $(\tilde{q}_x, \tilde{q}_y)$ in the vicinity of the point $(0, 0)$ in the case $n = 0$ ($\beta = \pi/2$) is described by the formula

$$\tilde{q}_x^2 \approx \frac{\tilde{q}_y^6}{\tilde{q}_y^2 + \Lambda_1^2/2\alpha}. \tag{67}$$

In the region $q < q_1$ ($\Lambda_1^2 < 0$) where the periodic wave is stable with respect to one-dimensional disturbances, there are monotonically growing two-dimensional modes. The region of growing disturbances is separated from the axis $\tilde{q}_y = 0$ by a gap. Inside the gap $\tilde{q}_y^2 < \tilde{q}_y^2 = -\Lambda_1^2/2\alpha$, all the disturbances decay.

The numerical calculations performed for finite values of $\tilde{q}_x$ and $\tilde{q}_y$ show that the two-dimensional disturbances with the maximal growth rate are characterized by the quasi-wave number $\tilde{q}_x = q/2$. When $q$ decreases, the motion becomes more stable. As $q < q_4 \approx 0.74$, the region of the stationary two-dimensional instability disappears.

When $n$ grows, the instability region shrinks. For each value of $q$ in the interval $q_4 < q < q_1$ there is a certain value $n = n_*(q)$ such that the wave is stable as $n > n_*(q)$. Because $q_4 < q_2$, we find that, for $n$ small enough, all the one-dimensional periodic wavy motions are unstable. This conclusion is particularly valid for the wavy motions on the vertical plane ($n = 0$). Because the quantity $n = 1/(G/G_0 - 1)$ decreases when $G$ grows, the stability interval for one-dimensional periodic waves shrinks and disappears completely as $G \approx 9.3G_0$.

The two-dimensional nonlinear waves generated by the described instability were studied numerically by Nepomnyashchy (1974c) and Petviashvili and Tsvelodub (1978). Because of the symmetry of the system with respect to the transformation $x \rightarrow -x$, an appearance of heteroclinic loops connecting unstable one-dimensional solutions is typical.

The influence of the disturbance $O(\epsilon)$ in Eq. (60) on the two-dimensional waves is not fully clear. In the case of one-dimensional waves, the linear dispersion term suppresses the spatial chaos (Chang, 1994). In the two-dimensional case, one can expect that the dispersion terms violate the reflection symmetry, the heteroclinic loops will be destroyed, and some limit cycles will appear (Demekhin and Shkadov, 1984; Trifonov 1990; see also Chang et al., 1994).

## 7.6. Conclusions

The long waves in a thin inclined film can be described by rather different mathematical models, depending on the relations among characteristic time scales of the main physical effects (dispersion, nonlinearity, instability, and dissipation). In the case in which the influence of the surface tension is moderate, one obtains weakly nonconservative perturbed KdV or ZK equations. If the influence of the surface tension is strong, the waves are governed by a strongly dissipative anisotropic Kuramoto–Sivashinsky equation. The manifestation of the resonance phenomena leading to a breakdown of the waves with planar front and appearance of three-dimensional structures is rather different in different cases. The analysis given in this chapter can be used to select an adequate mathematical model for a description of real three-dimensional waves in thin films.

# References

Allen, M. A. and Rowlands, G. 1995. Stability of obliquely propagating plane solitons of the Zakharov–Kuznetsov equation. J. Plasma Phys. **53**, 63–73.

Bar, D. E. 1996. The weakly nonlinear theory of long waves generated by instabilities. PhD thesis, Technion, Haifa.

Bar, D. E. and Nepomnyashchy, A. A. 1995. Stability of periodic waves governed by the modified Kawahara equation. Physica D **86**, 586–602.

Bar, D. E. and Nepomnyashchy, A. A. 1998. Stability of periodic waves generated by long-wavelength instabilities in isotropic and anisotropic systems, submitted to *Physica D*.

Benjamin, T. B. 1957. Wave formation in laminar flow down an inclined plane. J. Fluid Mech. **2**, 554–574.

Chang, H.-C. 1994. Wave evolution on a falling film. Ann. Rev. Fluid Mech. **26**, 103–136.

Chang, H.-C., Cheng, M., Demekhin, E. A., and Kopelevich, D. I. 1994. Secondary and tertiary excitation of three-dimensional patterns on a falling film. J. Fluid Mech. **270**, 251–275.

Cohen, B. I., Krommes, J. A., Tang, W. M., and Rosenbluth, M. N. 1976. Non-linear saturation of the dissipative trapped-ion mode by mode coupling. Nucl. Fusion **16**, 971–992.

Demekhin, E. A. and Shkadov, V. Ya. 1984. Three-dimensional waves in a liquid flowing down a wall. Fluid Dyn. **19**, 689–695.

Frenkel, A. L. and Indireshkumar, I. 1996. Derivations and simulations of evolution equations of wavy film flows. In *Mathematical Modeling and Simulation in Hydrodynamic Stability*, D. N. Riahi, ed., World Scientific, Singapore, pp. 35–81.

Frisch, U., She, Z. S., and Thual, O. 1986. Viscoelastic behaviour of cellular solutions to the Kuramoto–Sivashinsky model. J. Fluid Mech. **168**, 221–240.

Frycz, P. and Infeld, E. 1989. Spontaneous transition from flat to cylindrical solitons. Phys. Rev. Lett. **63**, 384–385.

Kapitsa, P. L. 1948. Wave flow on thin viscous fluid layers. Zh. Eksp. Teor. Fiz. **18**, 3–28.

Kapitsa, P. L. and Kapitsa, S. P. 1949. Wave flow on thin layers of liquid. Zh. Eksp. Teor. Fiz. **19**, 105–120.

Kodama, Y. and Ablowitz, M. J. 1981. Perturbations of solitons and solitary waves. Stud. Appl. Math. **64**, 225–245.

Liu, J., Schneider, J., and Gollub, J. P. 1995. Three-dimensional instabilities of film flows. Phys. Fluids **7**, 53–67.

Nepomnyashchy, A. A. 1974a. Stability of wavy regimes in a liquid film with respect to three-dimensional disturbances. Tr. Permsk. Gos. Univ. **316**, 91–104.

Nepomnyashchy, A. A. 1974b. Stability of wavy regimes in a film flowing down an inclined plane. Fluid Dyn. **9**, 354–359.

Nepomnyashchy, A. A. 1974c. Three-dimensional spatially periodic motions in a liquid film flowing down a vertical plane. In *Hydrodynamics*, pt. 7, Perm, 43–52.

Nepomnyashchy, A. A. 1976. Wavy motions in a layer of viscous liquid flowing down an inclined plane. Tr. Permsk. Gos. Univ. **362**, 114–124.

Ogawa, T. and Liu C.-F. 1997. Two-dimensional patterns of pulses appearing in a thin viscous film flow. Physica D **108**, 277–290.

Petviashvili, V. I. and Tsvelodub, O. Yu. 1978. Horseshoe-shaped solitons on a flowing viscous film of fluid. Sov. Phys. Dokl. **23**, 117–118.

Spektor, M. D. 1988. Stability of cnoidal waves in media with positive and negative dispersion. Sov. Phys. JETP **61**, 104–112.

Toh, S., Iwasaki, H., and Kawahara, T. 1989. Two-dimensionally localized pulses of a nonlinear equation with dissipation and dispersion. Phys. Rev. A **40**, 5472–5475.

Trifonov, Yu. Ya. 1990. Bifurcations of two-dimensional wavy regimes to three-dimensional ones for a vertically flowing liquid film. Izv. Akad. Nauk SSSR, Mekh. Zhidk. Gaza **5**, 109–114.

Yih, C. S. 1963. Stability of liquid flow down an inclined plane. Phys. Fluids **6**, 321–334.

Zakharov, V. E. and Kuznetsov, E. A. 1974. Three-dimensional solitons. Sov. Phys. JETP **39**, 285–286.

# 8

# Modulation Wave Dynamics of Kinematic Interfacial Waves

H.-C. CHANG, E. A. DEMEKHIN, R. M. ROBERTS, AND Y. YE

## 8.1. Introduction

C.-S. Yih was responsible for many of the pioneering linear stability theories for thin-film wave dynamics (Yih, 1963, 1967). The introduction of these classical hydrodynamic theories $\sim$40 years ago has greatly improved our understanding of interfacial phenomena in many industrially important multiphase flows. They have also inspired later efforts to extend his linear theories, valid for small-amplitude primary waves at inception, and to capture subsequent nonlinear wave dynamics of finite-amplitude waves. One particularly important direction is the onset of nonlinear modulation wave dynamics (Lin, 1974). Recent experiments (Liu and Gollub, 1993) for falling-film waves have shown that the modulation wave dynamics induces the primary waves to coarsen their texture and form unique localized wave structures – solitary waves. The wavelength and the speed of these solitary waves are very different from the primary waves and, for a vertically falling film, they rapidly become the dominant wave structure beyond 20 cm from the inlet. We have also found empirical evidence that this modulation instability causes slug formation in stratified gas–liquid flow (McCready and Chang, 1996) and precedes atomization from the crests of the solitary waves (Roberts, 1998). It is hence very important to analyze the modulation transition from relatively short and slow primary waves to well-separated, faster, and much larger wave structures. It is the dominant weakly nonlinear instability of most primary interfacial waves.

Modulation instability of thin-film waves is related to the Eckhaus instability of periodic rolls in Rayleigh–Benard instability and the Benjamin–Feir instability of deep-water waves. It involves a slow modulation of the wave amplitude over a distance much longer than the primary wavelength. However, the long-wave nature of many interfacial instabilities implies that the primary growth rate does not resemble a quadratic nose near onset as in short-wave instabilities. The usual Ginzburg–Landau multiscale expansion about the nose, which yields the Eckhaus and the Lange–Newell bounds, would be invalid. Considerable effort has then been expended to derive a different expansion formalism about the neutral wave number instead of the maximum-growing one at the nose (Cheng and Chang, 1990, 1995). However, there is an important and unique physical modulation mechanism that has been uncovered only recently.

Unlike deep-water free-surface problems, the local flow rate $q$ of a thin film scales nonlinearly with respect to the film height $h$ because of viscous effects. The parabolic velocity profile of a freely falling film, for example, yields $q \sim h^3$. A two-layer Couette flow, on the other hand, yields $q \sim h^2$ for each layer. Kinematics (mass conservation) for films, however, dictate that $[(\partial h)/(\partial t)] = [-(\partial q)/(\partial x)]$. On linearizing about a flat film of thickness $h_0$ and flow rate $q_0$, the above nonlinear scaling $q \sim h^n$ immediately yields the phase speed of long thin-film waves of small amplitude, $c_0 = nq_0/h_0$. This is consistent with the classical result that the phase speed of long falling-film waves is three times the average fluid velocity of a flat film and twice that of a two-layer Couette flow (Yih, 1963, 1967).

A more important implication of the nonlinear correlation between $q$ and $h$ concerns finite-amplitude periodic waves – to leading order in the long-wave expansion, larger waves will carry more liquid (have higher flow rate) than smaller ones. This can introduce a positive feedback mechanism through kinematics to destabilize a uniform wave field that is stationary in a moving frame. If there is a small and gradual variation in the wave amplitude, the corresponding variation in the flow rate will increase the average thickness below a patch of large waves since the flow rate in front of the patch is lower. This increase in thickness can, in turn, further enlarge the local wave amplitude and speed and hence the flow gradient. Not surprisingly, modulation theory for secondary instability of uniform wave fields on these films requires the inclusion of this important kinematic effect (Frisch et al., 1986; Renardy and Renardy, 1993; Cheng and Chang, 1995; Chang et al., 1997). Since the average thickness corresponds to a Fourier mode with zero wave number, this effect is often known as zero- (Goldstone) mode coupling/excitation.

In some earlier theories, however, the zero mode is included as only a slave mode – the wave field is allowed to change its average thickness but spatial variation in the thickness is not permitted (Cheng and Chang, 1995). In another theory, spatial variation of the zero mode is allowed but only when this variation is stabilizing (Renardy and Renardy, 1993). The most severe modulation arises when spatial thickness variation exists and is destabilizing. Frisch et al. (1986) show that this variation drives a persistent viscoelastic modulation of the primary waves of the Kuromoto–Sivashinsky (KS) equation. Chang et al. (1997) show that the modulation causes the formation of wave sinks in the generalized Kuramoto–Sivashinsky (gKS) equation, which includes wave dispersion. These wave sinks dilate the wave field in a nonuniform manner, and they can drive the evolution from a uniform primary wave field to isolated solitary waves.

However, all these earlier analyses show that, within the band of primary unstable wave modes with a normalized wave number $\alpha$ between 0 and 1, only a small band near the unit-neutral wave number is modulationally unstable. This band excludes the dominant primary wave with dimensional maximum-growing wave number $\alpha_m = 1/\sqrt{2}$ scaled with respect to the neutral one. As a result, modulation wave dynamics is not expected for naturally excited waves and would be observable only with artificial external forcing at high frequency. This contradicts the recent experimental observation of kinematic wave modulation by Gallagher et al. (1996) for a two-layer Couette flow. Modulation instability is observed for a subset of forced or naturally excited primary waves with $\alpha = \alpha_m = 1/\sqrt{2}$. This implies that higher-order effects omitted by the low-order KS and gKS evolution equations must be retained to predict whether modulation exists for naturally excited waves with the dominant wave number $\alpha_m$.

We present such a high-order modulation theory here for low-inertia thin-film waves near inception by analyzing a modified Kuramoto–Sivashinsky (mKS) equation. We obtain general criteria, analogous to the classical Eckhaus and Lange–Newell bounds, for modulation instabilities of long interfacial waves and favorably compare them with experimental data for falling films and two-layer Couette flow.

## 8.2. Evolution Equation and Periodic Waves

Formal long-wave and small-amplitude expansions of the local flow rate yield the form

$$q \sim q_0 + c_0\hat{h} + c_1\hat{h}_x + c_2\hat{h}_{xx} + c_3\hat{h}_{xxx} + c_4\hat{h}_{xxxx} + d_0(\hat{h}^2) + d_1(\hat{h}^2)_x, \qquad (1)$$

where $\hat{h} = (h - h_0)$ is the deviation height from the flat-film thickness $h_0$. The flat-film correlation $(q/q_0) = (h/h_0)^n$ yields the phase speed $c_0 = nq_0/h_0$ and the dominant nonlinear term $d_0(\hat{h})^2$. The other linear and nonlinear terms capture higher-order terms because of the waves. The KS equation retains these two and the nondispersive terms $c_1\hat{h}_x$ and $c_3\hat{h}_{xxx}$. In the frame moving with speed $c_0$, the equation is valid when these two growth and dissipation terms are of the same order and hence $(c_1/c_3) \sim \hat{\alpha}_0^2$, where $\hat{\alpha}_0$ is a typical dimensional wave number representing the long-wave expansion parameter (see, for example, Benney, 1966). One must hence be sufficiently close to criticality for the growth coefficient $c_1$ to be much smaller than the dissipation coefficient $c_3$. In most thin films, both instability and dispersion are introduced by inertia. As a result, the dispersion coefficients $c_2$ and $c_4$ are of the same order as $c_1$ and hence the linear dispersion terms are higher order than the linear growth and dissipation terms. In fact, the fifth-order dispersion term is negligible in the present resolution. The dominant nonlinear term $d_0\hat{h}^2$ arises from flat-film kinematics,

$$d_0 = \frac{1}{2}\left(\frac{d^2q}{dh^2}\right)_0 = \frac{n(n-1)}{2}\frac{q_0}{h_0^2}.$$

When it is balanced to the other two KS terms to ensure parity among nonlinearity growth and dissipation for finite-amplitude wave dynamics, one obtains the relative scaling between the amplitude and the long-wave expansion, $\hat{h} \sim 0(\hat{\alpha}_0^3)$. The second nonlinear term is a kinematic term arising from wave motion and its coefficient $d_1$ is typically unit order.

With these typical orders for a thin film, the formal flow rate expansion of approximation (1) in the kinematic condition yields

$$H_\tau + H_{\xi\xi} + H_{\xi\xi\xi\xi} + (H^2)_\xi + \delta H_{\xi\xi\xi} + \epsilon(H^2)_{\xi\xi} = 0 \qquad (2)$$

for low-inertia thin films. This mKS equation, valid up to $O(\hat{\alpha}_0^7)$, represents a higher-order evolution equation than the KS and gKS equations for nondispersive and dispersive thin-film waves. We have also used KS scalings in Eq. (2): $t = c_3\tau/c_1^2$, $x - c_0t = (c_3/c_1)^{1/2}\xi$ and $(h - h_0) = (c_1^3/c_3)^{1/2}(H/d_0)$ such that the normalized wave number $\alpha$ is unstable between zero and unity and the maximum-growing wave number is $\alpha_m = 1/\sqrt{2}$. The corrections arise from the dominant dispersive coefficient, $\delta = c_2/(c_1c_3)^{1/2}$, and the secondary nonlinear coefficient, $\epsilon = (d_1/d_0)/(c_1c_3)^{1/2}$. The KS equation results when both $\epsilon$ and $\delta$ vanish and the gKS when only $\epsilon$ vanishes.

Periodic solutions $U$ to Eq. (2) that are stationary in a frame moving with a constant wave speed $C$ are known as traveling waves or stationary waves. They are defined by

$$\frac{d^4U}{d\xi^4} + \delta\frac{d^3U}{d\xi^3} + \frac{d^2U}{d\xi^2} - C\frac{dU}{d\xi} + (U^2)_\xi + \epsilon(U^2)_{\xi\xi} = \frac{dQ}{d\xi}, \qquad (3a)$$

or

$$U_{\xi\xi\xi} + \delta U_{\xi\xi} + U_\xi - CU + U^2 + \epsilon(U^2)_\xi = Q, \qquad (3b)$$

$$U(\xi) = U(\xi + 2\pi/\alpha) \qquad (3c)$$

where $\xi$ is the moving coordinate, $2\pi/\alpha$ is the wavelength, and $Q$ is the flow rate in the moving frame. It is convenient to rescale the wavelength to $2\pi$ by defining $\zeta = \alpha\xi$ such that Eqs. (3)

become

$$\alpha^4 \frac{d^4 U}{d\zeta^4} + \delta\alpha^3 \frac{d^3 U}{d\zeta^3} + \alpha^2 \frac{d^2 U}{d\zeta^2} - C\alpha \frac{dU}{d\zeta} + \alpha(U^2)_\zeta + \epsilon\alpha^2 (U^2)_{\zeta\zeta} = 0, \tag{4a}$$

$$\alpha^3 U_{\zeta\zeta\zeta} + \delta\alpha^2 U_{\zeta\zeta} + \alpha U_\zeta - CU + U^2 + \epsilon\alpha(U^2)_\zeta = Q. \tag{4b}$$

With the flow rate $Q$, the wave speed $C$ can be determined only by the imposition of an additional constraint. This can be either the zero-flow-rate condition or the zero-mean-thickness condition (Chang et al., 1993). The two resulting stationary wave families are related, and we hence focus on only the zero-mean-thickness family $U(\zeta; \alpha, \delta, \epsilon)$ described by Eq. (4b) and

$$\langle U \rangle = \frac{1}{2\pi} \int_0^{2\pi} U \, d\zeta = 0. \tag{5}$$

The linear terms in Eq. (4b) show that $U = e^{i\zeta}$ and $C = -\delta$ are a solution at $\alpha = 1$. This indicates that the $U$ family bifurcates from the trivial solution at this neutral wave number. A simple bifurcation analysis then shows that it is a pitchfork bifurcation such that the solution exists only for $\alpha < 1$. The two branches of the pitchfork correspond to a shift of $\pi$ and hence correspond to the same member of the periodic wave family. The analysis yields very simple expressions for the near-neutral $(1 - \alpha \ll 1)$ members of the $U$ family for small $\epsilon$ and $\delta$; $(1 - \alpha) \ll \epsilon, \delta \ll (1 - \alpha)^{1/2}$:

$$\begin{aligned} U \sim {} & \sqrt{(24 + 48\epsilon^2 + 36\delta\epsilon + 6\delta^2)(1 - \alpha) - 124(1 - \alpha)^2} \sin\alpha x \\ & + (-2 + 2\alpha)\sin 2\alpha x + (-4\epsilon - \delta)(1 - \alpha)\cos 2\alpha x, \\ C \sim {} & -\delta + (6\epsilon + 3\delta)(1 - \alpha) - (13\epsilon + \delta/2)(1 - \alpha)^2, \\ \langle Q \rangle \sim {} & (12 + 24\epsilon^2 + 18\epsilon\delta + 3\delta^2)(1 - \alpha) - 60(1 - \alpha)^2. \end{aligned} \tag{6}$$

We note that this near-neutral expansion cannot capture many secondary instabilities. Oscillatory overtone instability due to 1–2 resonance occurs at $\alpha = 0.52$ (Chang et al., 1993) and subharmonic instability also occurs at a low wave number beyond the validity of approximations (6). However, the dominant primary wave with $\alpha_m = 1/\sqrt{2}$ is well described by approximations (6). Its dominant instability is then the modulation instability since it is stable to both the overtone and the subharmonic instability.

## 8.3. Symmetries and Modulation Instability

The stability of $U(\xi)$ must be determined from Eq. (2) in the same moving frame

$$H_\tau - CH_\xi + H_{\xi\xi} + H_{\xi\xi\xi\xi} + (H^2)_\xi + \delta H_{\xi\xi\xi} + \epsilon(H^2)_{\xi\xi} = 0 \tag{7}$$

and linearizing about $U$, $H = U + \Psi$, we obtain

$$\Psi_\tau = L\Psi, \tag{8}$$

where

$$L \cdot = C \frac{d}{d\xi} \cdot - \frac{d^2}{d\xi^2} \cdot - \frac{d^4}{d\xi^4} \cdot - \delta \frac{d^3}{d\xi^3} \cdot - (2U\cdot)_\xi - \epsilon(2U\cdot)_{\xi\xi}.$$

The stability of $U$ to disturbances of the same wavelength is then determined by the spectrum of $L$ with periodic boundary conditions. The eigenvalues must all be stable or neutrally stable, otherwise the periodic wave would never have been formed.

There is, however, one neutral eigenvalue for $L$ from the symmetries of thin-film kinematic waves. The first symmetry is a translational symmetry – $U$ is defined up to an arbitrary translation. If $U(\xi)$ is a stationary wave, so is $U(\xi + a)$. For a small translation, $U(\xi + a) \sim U(\xi) + aU_\xi(\xi)$. Substituting this into the linearized version of Eq. (7), one concludes that $U_\xi$ is a null eigenfunction corresponding to a zero eigenvalue:

$$LU_\xi = 0. \tag{9}$$

Hence, the spectrum of $L$ always contains one zero eigenvalue.

For the gKS equation with $\epsilon = 0$, Eq. (7) possesses another Galilean symmetry. If $U(\xi)$ is a stationary wave, so is $U(\xi - 2\chi\tau) + \chi$, where $\chi$ is an arbitrary constant corresponding to a shift in the baseline – a nonzero-mean-thickness correction. This symmetry is also captured by the gKS version of Eqs. (3) at $\epsilon = 0$ by the transformations $U \to U + \chi, C \to C + 2\chi$, and $Q \to Q - \chi^2 - C\chi$. This Galilean invariance originates from the $q \sim h^n$ scaling of a flat thin viscous film and involves the flat-film terms in Eq. (3b) without derivatives. However, only the second-order expansion in the wave amplitude $\hat{h}$ is retained here and the flow rate is in the moving frame (and hence the correction $C\chi$). Note that the perturbation term with the parameter $\epsilon$ breaks this invariance since this nonlinear term does not correspond to a flat film.

Since $U(\xi - 2\chi\tau) + \chi$ is also a stationary wave solution, but one with nonzero mean thickness, one can again expand it for small $\chi$ as we have done for the translation $a$. Hence, expanding about $\chi = 0$, one gets $(\partial/\partial\chi)[U(\xi - 2\chi\tau) + \chi]_{\chi=0} = -2\tau U_\xi + 1$. This must be a solution to the linearized version of Eq. (7) but because of the appearance of $\tau$, it is now a time-dependent solution of Eq. (8), $(\partial/\partial\tau)(-2\tau U_\xi + 1) = L(-2\tau U_\xi + 1)$ at $\epsilon = 0$. This then shows that the spectrum has a generalized null eigenvalue such that

$$L\left(-\frac{1}{2}\right) = U_\xi. \tag{10}$$

When $\epsilon \neq 0$, this generalized eigenvalue deviates from zero but for small $\epsilon$, it remains a near-neutral eigenvalue.

To capture the shift due to $\epsilon$, we need the adjoint operator at $\epsilon = 0$ with respect to the $L_2$ inner product $(\cdot, \cdot)$ over the wavelength $2\pi/\alpha$,

$$L^+ = -C\frac{d}{d\xi} - \frac{d^2}{d\xi^2} - \frac{d^4}{d\xi^4} + \delta\frac{d^3}{d\xi^3} + 2U\frac{d}{d\xi}, \tag{11}$$

which also has a simple zero and a generalized zero:

$$L^+ 1 = 0, \qquad L^+ \Phi = 1. \tag{12}$$

That $\Psi$ exists is easily verifiable by the solvability condition $(1, U_\xi) = 0$. The generalized null eigenfunction $\Phi$ must be constructed numerically. However, in our multiscale expansion of the modulation theory, the solvability condition involves only 1 and hence $\Phi$ is inconsequential. The inhomogeneous terms to any equation involving $L$ with $\epsilon = 0$ must have a zero mean over the periodic domain.

The two zero eigenvalues of the gKS stationary wave in Eqs. (9) and (11), because of translational and Galilean symmetries, begin to interact and move away from the origin when we break both symmetries by introducing long-wave modulation and nonzero $\epsilon$. We choose the long length scale for the modulation to be $y = \epsilon^{-1/2}\zeta$ and show that it gives rise to an instability at the long time scale $\epsilon^{-1/2}\tau$. Letting $\theta = \epsilon^{-1/2}\tau$ and carrying out the multiscale expansion,

$$\frac{\partial}{\partial \zeta} \rightarrow \frac{\partial}{\partial \zeta} + \epsilon^{1/2}\frac{\partial}{\partial y}, \qquad H \rightarrow W_0 + \epsilon^{1/2}W_1 + \epsilon W_2,$$

of Eq. (7) in the normalized $\zeta$ coordinate, we obtain

$$\mathcal{L}_0 W_0 = 0,$$
$$\mathcal{L}_0 W_1 = \frac{\partial W_0}{\partial \theta} - \mathcal{L}_1\frac{\partial W_0}{\partial y},$$
$$\mathcal{L}_0 W_2 = \frac{\partial W_1}{\partial \theta} - \mathcal{L}_2\frac{\partial^2 W_0}{\partial y^2} - \mathcal{L}_1\frac{\partial W_1}{\partial y} - \alpha^2(W_0^2)_{\zeta\zeta}, \tag{13}$$

where

$$\mathcal{L}_0 = \alpha C\frac{d}{d\zeta} - \alpha^2\frac{d^2}{d\zeta^2} - \alpha^4\frac{d^4}{d\zeta^4} - \alpha^3\frac{d^3}{d\zeta^3} - \alpha(2U\cdot)_\zeta$$

is the scaled version of the gKS linear operator $L$ of Eq. (8) with $\epsilon = 0$ and the other two operators arise from multiscale expansion:

$$\mathcal{L}_1 = C\alpha - 2\alpha^2\frac{d}{d\zeta} - 4\alpha^4\frac{d^3}{d\zeta^3} - 3\delta\alpha^3\frac{d^2}{d\zeta^2} - 2\alpha U,$$

$$\mathcal{L}_2 = -\alpha^2 - 6\alpha^4\frac{d^2}{d\zeta^2} - 3\delta\alpha^3\frac{d}{d\zeta}.$$

From Eq. (9), it is clear that $W_0$ is related to the simple null eigenfunction

$$W_0 = S(y, \theta)\alpha U_\zeta. \tag{14}$$

The generalized zero in Eq. (11), however, manifests itself in a more subtle way. Taking the derivative with respect to $\alpha$ of the gKS version ($\epsilon = 0$) of Eq. (4a), we obtain the relationship due to Galilean invariance:

$$\mathcal{L}_0\frac{\partial U}{\partial \alpha} = -\alpha\frac{dC}{d\alpha}U_\zeta - \frac{1}{\alpha^2}\mathcal{L}_1(\alpha U_\zeta), \tag{15}$$

and the thickness shift $\chi$ is now transformed into a change in wave number $\alpha$.

Since $\langle W_0 \rangle = 0$, the second equation for $W_1$ is solvable, and from Eq. (15), it is clear that $W_1$ must be of the form $aU_\alpha - b/2$. This yields

$$\begin{aligned}
\mathcal{L}_0 W_1 &= a\mathcal{L}_0 U_\alpha + b\alpha U_\zeta \\
&= -\frac{a}{\alpha^2}\mathcal{L}_1(\alpha U_\zeta) - a\alpha\frac{dC}{d\alpha}U_\zeta + b\alpha U_\zeta \\
&= \frac{\partial W_0}{\partial \theta} - \mathcal{L}_1\frac{\partial W_0}{\partial y} \\
&= \frac{\partial S}{\partial \theta}\alpha U_\zeta - \alpha\frac{\partial S}{\partial y}\mathcal{L}_1 U_\zeta,
\end{aligned} \tag{16}$$

and hence $a = [\alpha^2(\partial S/\partial y)]$ and $b = a(dC/d\alpha) + (\partial S/\partial \theta)$

$$W_1 = \alpha^2 \frac{\partial S}{\partial y} U_\alpha - \frac{1}{2}\left(\alpha^2 \frac{\partial S}{\partial y}\frac{dC}{d\alpha} + \frac{\partial S}{\partial \theta}\right). \tag{17}$$

Substituting $W_0$ and $W_1$ from Eqs. (14) and (17) into the inhomogeneous part of the equation for $W_2$ in Eqs. (13) and integrating over $\zeta$ to ensure solvability, we get

$$\frac{1}{2}S_{\theta\theta} + BS_{\theta y} + CS_{yy} = 0. \tag{18}$$

For the case of zero mean thickness, $\langle U \rangle = \langle U \rangle_\alpha = 0$, the coefficients are

$$B = \frac{C\alpha}{2} - \frac{\alpha^2}{2}\frac{dC}{d\alpha},$$

$$c = -\frac{\alpha^3 C}{2}\frac{dC}{d\alpha} - \alpha^3 \frac{d\langle U^2 \rangle}{d\alpha}. \tag{19}$$

The long-wave modulation $S(y, \theta)$ grows if $D = B^2 - 2C$ is negative. From Eq. (4b), we see that

$$\langle Q \rangle = \langle U^2 - CU \rangle = \langle U^2 \rangle,$$

and if we define $\Omega = C\alpha$ as the nonlinear wave frequency of the stationary wave, the determinant $D$ can be expressed simply as

$$D = \frac{\alpha^2}{4}\left[\left(\frac{d\Omega}{d\alpha}\right)^2 + 8\alpha \frac{d\langle Q \rangle}{d\alpha}\right] < 0 \tag{20}$$

in terms of the gradient in the flow rate $\langle Q \rangle$ and wave frequency $\Omega$ with respect to $\alpha$ along the periodic wave family $U$.

This simple result is consistent with that derived by Chang et al. (1997) for the gKS stationary waves. It has a simple geometric interpretation as conservation of wave nodes and mass:

$$\frac{\partial \alpha}{\partial \theta} = -\frac{\partial \Omega}{\partial \alpha}\frac{\partial \alpha}{\partial y} - 2\alpha \frac{\partial \chi}{\partial y},$$

$$\frac{\partial \chi}{\partial \theta} = -\frac{\partial \langle Q \rangle}{\partial \alpha}\frac{\partial \alpha}{\partial y}, \tag{21}$$

with the coupling $2\alpha(\partial \chi/\partial y)$ term corresponding to the generalized zero in Eqs. (12) and (17). Extension to include the $\epsilon$ term of the mKS equation yields the same result, provided $\epsilon$ is small. If one linearizes Eqs. (21) about a particular stationary wave of wave number $\alpha$, exponential growth results when the linearized equation becomes elliptic when its Jacobian

$$\underset{\approx}{J} = \begin{pmatrix} \frac{d\Omega}{d\alpha} & 2\alpha \\ \frac{d\langle Q \rangle}{d\alpha} & 0 \end{pmatrix}$$

yields nonimaginary eigenvalues for negative $(4/\alpha^2)D = (\text{trace})^2 - 4\det$, viz., Eqs. (20).

Condition (20) is best viewed from Eqs. (21) with the variation in the local thickness $\chi$, driven by the gradient in the flow rate $\langle Q \rangle$ carried by waves of different wave number. For nondispersive waves with $(d\Omega/d\alpha) = 0$, this flow rate gradient of kinematic waves is solely responsible for the modulation instability. With the introduction of dispersion such that $\delta \neq 0$ in approximations (6), the nonlinear dilation effect of nonlinear dispersion can relieve the compression effect of the modulation and hence provide a stabilizing mechanism.

It is clear from approximations (6) that $d\langle Q \rangle/d\alpha$ is negative for all near-neutral primary waves with $\alpha$ close to 1. Since $\delta$ and hence $d\Omega/d\alpha$ are typically small, this implies that all near-neutral waves are modulationally unstable. One expects $d\langle Q \rangle/d\alpha$ to remain negative for very small $\alpha$ since the large solitary waves should carry more flow rate than the smaller periodic waves. However, these solitary waves are beyond the description of the near-neutral expansion in approximations (6). Nevertheless there is an intermediate $\alpha$ region with positive $d\langle Q \rangle/d\alpha$ that is still within the region of validity of approximations (6). This sign change in $d\langle Q \rangle/d\alpha$ is due to curvature effects that escapes the flat-film scaling arguments. Simple substitution of approximations (6) into Eq. (20) then yields the lower bound of a modulationally unstable near-neutral wave, for small $\epsilon$ and $\delta$:

$$\alpha_c = 0.9 - 0.15\epsilon^2 - 0.09\epsilon\delta - 0.01\delta^2. \tag{22}$$

All waves between $\alpha_c$ and 1 are modulationally unstable. Alternatively, the dominant primary wave with wave number $\alpha_m$ becomes unstable at the critical condition defined by

$$\epsilon_c = \pm 1.1 - 0.3\delta. \tag{23}$$

For thin films, dispersion due to inertia typically slows down the shorter waves and $\delta$ is positive. However, $\epsilon$ can have both signs, and the $\pm$ signs of Eq. (23) represent the critical condition for corresponding signs of $\epsilon$.

## 8.4. Comparison with Modulation Experiments

The best-studied example of thin-film modulation wave dynamics is for a falling film on a plane inclined at $\theta$. Liu and Gollub (1993) have carefully recorded the primary wave frequencies $\omega_c = c_0\alpha_c$ that can trigger modulation instability. We use Nakaya's (1975) lubrication equation for inclined films to derive the following expressions for $\delta$ and $\epsilon$:

$$\delta = \left(2 - \frac{40}{63}\bar{R}\cot\theta + \frac{32}{63}\bar{R}^2\right)\left[\left(-\frac{2}{3}\cot\theta + \frac{8}{15}\bar{R}\right)\left(\frac{2}{3}\frac{\bar{W}}{\sin\theta} - \frac{6}{5}\cot\theta\right.\right.$$
$$\left.\left. + \frac{157}{56}\bar{R} + \frac{8}{45}\bar{R}\cot^2\theta - \frac{138904}{155925}\bar{R}^2\cot\theta + \frac{1213952}{2027025}\bar{R}^3\right)\right]^{-1/2},$$

$$\epsilon = \frac{-\cot\theta + \frac{24}{15}\bar{R}}{2}\left[\left(-\frac{2}{3}\cot\theta + \frac{8}{15}\bar{R}\right)\left(\frac{2}{3}\frac{\bar{W}}{\sin\theta} - \frac{6}{5}\cot\theta + \frac{157}{56}\bar{R}\right.\right.$$
$$\left.\left. + \frac{8}{45}\bar{R}\cot^2\theta - \frac{138904}{155925}\cot\theta + \frac{1213952}{2027025}\bar{R}^3\right)^{-1}\right]^{1/2}, \tag{24}$$

where $\bar{R} = [(gh_0^3\sin\theta)/(2v^2)]$, $\bar{W} = [\sigma/(\rho gh_0^2)]$, $\theta = 6.4°$, $v = 0.00000616$ m$^2$/s, $\sigma = 0.067$ N/m, and $\rho = 1150$ kg/m$^3$ for Liu and Gollub's (1993) experiment. We obtain $\alpha_c$ from the above expressions and $\omega_c = \alpha_c(2 - c)$ to get the critical dimensionless frequency. However, Nakaya's

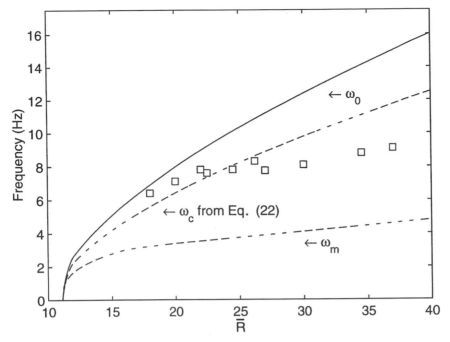

Figure 8.1. Critical frequency $\omega_c$ measured for modulation below the neutral frequency $\omega_0 = c_0\alpha_0$ for a falling film by Liu and Gollub (1993) by varying the film thickness $h_0$. Primary waves with frequency higher than $\omega_c$ are modulationally unstable. The prediction from Eq. (22) and the maximum-growing frequency $\omega_m$ are also shown.

(1975) equation does not predict an accurate dimensional $\hat{\alpha}_0$, and we hence convert these values to dimensional form by using a dimensional-neutral wave number from the full Orr–Sommerfeld analysis of Liu and Gollub (1993). Thus adjusted, the predicted critical frequency for modulation instability is seen to be in good agreement with their data for critical modulation frequency obtained by varying $h_0$, as shown in Fig. 8.1.

Gallagher et al. (1996) have recently explored modulation wave dynamics in two-layer Couette flow in a device whose moving outer cylinder is in contact with a very viscous liquid. The inner layer, on the other hand, consists of a less viscous fluid that is neutrally buoyant with respect to the outer fluid. To determine the parameters $\epsilon$ and $\delta$ for this two-layer Couette flow, we neglect curvature and assume an infinitely viscous but deformable upper fluid translating at velocity $U_p$. A quadratic polynomial is then used to describe the Couette–Poiseuille velocity profile of the less viscous lower fluid. The usual integral formulation yields an approximate depth-averaged equation of motion for the lower fluid (Roberts, 1998):

$$\frac{\partial q}{\partial t} + \frac{\partial q}{\partial x}\left(2.5\frac{q}{h} - 0.25\right) - \frac{\partial h}{\partial x}\left(1.326\frac{q^2}{h^2} - 0.0773\frac{q}{h} - 0.125\right)$$
$$= 0.82h\left(Wh_{xxx} - \frac{\partial\Pi}{\partial x}\right) + \frac{4.94}{R}\left(\frac{1}{h} - \frac{2q}{h^2}\right). \tag{25}$$

This equation must be solved in conjunction with the kinematic equation $(\partial h/\partial t) = [-(\partial q/\partial x)]$. The Couette geometry allows for different flat-film solutions with different film height, flow rate, and viscous dissipation, provided a pressure gradient is imposed such that the linear Couette velocity profile is changed to a quadratic Poiseuille–Couette profile. For long waves, a family of such profiles exists in the lower film and each is locally sustained by the lubrication pressure

$$\frac{\partial \Pi}{\partial x} = \frac{3}{mR(L-h)} \left[ \frac{q - (L - 1/2)}{(L-h)^2} + \frac{1}{(L-h)} \right], \tag{26}$$

where $x$ is scaled by the depth of the inner fluid $l$, $t$ by $l/U_p$, $L$ is the total height of both layers scaled by $l$, $R = U_p l/\nu$ is the Reynolds number of the lower fluid, $W = \sigma/(\rho U_p^2 l)$ is the Weber number of the lower fluid, and $m$ is a small number representing the ratio in the viscosity of the lower fluid to the upper one. Gallagher et al. (1996) have used two neutrally buoyant fluids with $\rho = 1110 \, \text{kg/m}^3$ and $m = 0.0159$. The physical properties of the lower fluid are $\nu = 0.0005$ m$^2$/s and $\sigma = 0.01$ N/m, and the apparatus gap height is 1 cm, such that $L = (1 \, \text{cm}/l)$. The only remaining parameters are the plate speed $U_p$, which is varied between 0 and 50 cm/s, and the inner fluid depth $l$, which ranges from 0.3 to 0.7 cm. Both parameters are varied experimentally.

We have also simulated these equations for several conditions, as shown in Fig. 8.2. The boundary between steady and modulation wave dynamics in the $U_p - l$ parameter space obtained from our simulation is in rough agreement with the data of Gallagher et al. (1996). To obtain an explicit modulation condition, we have also performed a long-wave and weakly

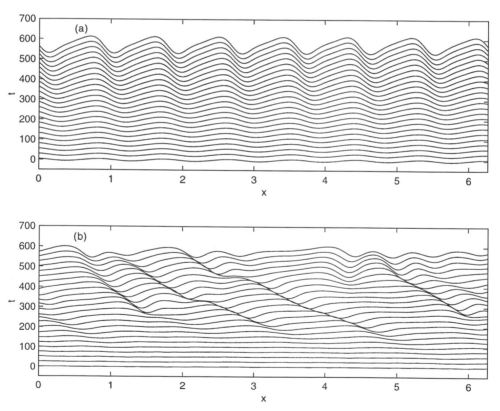

Figure 8.2. Simulation of the averaged equations with random initial conditions showing (a) steady traveling periodic waves at $R = 50$, $W = 0.225$, $m = 0.0159$, $L = 2.5$; (b) modulation wave dynamics at $R = 125$, $W = 0.036$, $m = 0.0159$, $L = 2.5$.

nonlinear expansion of Eqs. (9) and (11) near criticality ($l = l_c = 0.32$ cm) in the limit of vanishing $m$ to yield the linear and the nonlinear coefficients:

$$c_0 = \frac{1 + 2mH^3}{1 + 4mH^3},$$

$$c_1 = \frac{mH^3 R}{3(1 + 4mH^3)} (c_0^2 - c_0 + 1/6),$$

$$c_2 = \left[ \frac{mH^3 R}{3(1 + 4mH^3)} \right]^2 (2c_0 - 1)(c_0^2 - c_0 + 1/6),$$

$$c_3 = \frac{mH^3 R}{3(1 + 4mH^3)} W, \tag{27}$$

$$d_0 = \left( 24 - \frac{3}{H} \right) \Big/ R,$$

$$d_1 = -0.6 + 0.433/H,$$

where $H = L - 1$ and $\hat{\alpha}_m \sim [(c_0^2 - c_0 + 1/6)/2W]^{1/2}$. The onset of primary waves occurs at the critical dimensionless gap height of $L_c = 3.1$ or, equivalently, a critical inner film thickness of $l_c = 0.32$ cm for the 1-cm gap in the device of Gallagher et al. (1996).

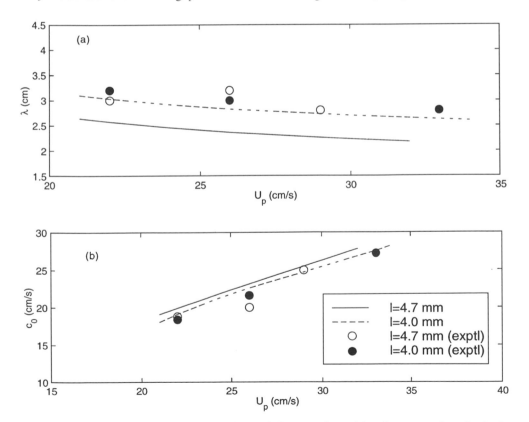

Figure 8.3. Simulated and measured wavelengths $\lambda$ and phase speeds $c_0$ of the primary wave from the depth-averaged equations with random initial conditions.

We have verified from our simulations in Fig. 8.2 with random initial conditions that the stable stationary waves indeed have dimensional wave numbers $\hat{\alpha}_m$ corresponding to the maximum-growing one from linear theory. We also compare $2\pi/\hat{\alpha}_m$ and $c_0(\hat{\alpha}_m)$ from our linear theory with the measured wavelengths and speeds of Gallagher et al. (1996) in Fig. 8.3. The excellent agreement seen indicates that all their primary waves are dominant ones with wave number $\hat{\alpha}_m$.

The coefficients in Eqs. (27) yield $\epsilon$ and $\delta$ in terms of $l$ and $U_p$,

$$\epsilon = -0.025R\left[\left(c_0^2 - c_0 + 1/6\right)/W\right]^{1/2},$$

$$\delta = \frac{mH^3R}{3(1+4mH^3)}(2c_0 - 1)\left[\left(c_0^2 - c_0 + 1/6\right)/W\right]^{1/2}, \tag{28}$$

and substitution into Eq. (23) yields the critical condition when the dominant primary waves become modulationally unstable,

$$U_p \sim 15.0(l - l_c)^{-1/4}, \tag{29}$$

beyond criticality ($l > l_c$). This bound is seen to be in satisfactory agreement with the critical data of with Gallagher et al. (1996) (see Fig. 8.4).

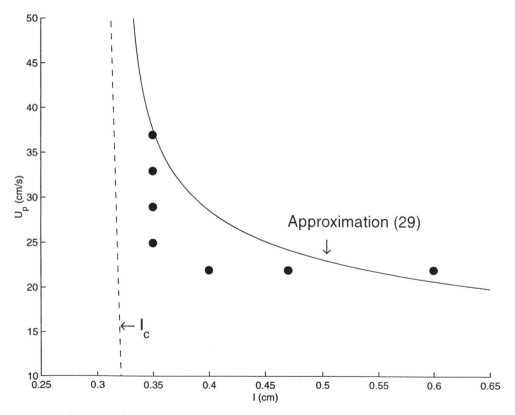

Figure 8.4. The onset of primary waves occur at $l_c = 0.32$ cm. The boundary for modulation instability measured by Gallagher et al. (1996) as compared with the prediction of approximation (29).

## Acknowledgments

The work is partially supported by a NASA grant and an U.S. National Science Foundation grant. Y. Ye received a fellowship from the Center for Applied Mathematics at Notre Dame.

## References

Benney, D. J. 1966. Long waves on liquid films. *J. Math. Phys.* **45**, 150.

Chang, H.-C., Demekhin, E. A., and Kopelevich, D. I. 1993. Laminarizing effects of dispersion in an active-dissipative nonlinear medium. *Phys. D* **63**, 299.

Chang, H.-C., Demekhin, E. A., Kopelevich, D. I., and Ye, Y. 1997. Nonlinear wavenumber selection in gradient flow systems. *Phys. Rev. E* **55**, 2818.

Cheng, M. and Chang, H.-C. 1990. A generalized sideband stability theory via center manifold projection. *Phys. Fluids A* **2**, 1364.

Cheng, M. and Chang, H.-C. 1995. Competition between subharmonic and sideband secondary instabilities on a falling film. *Phys. Fluids A* **7**, 34.

Frisch, U., She, Z. S., and Thual, O. 1986. Viscoelastic behavior of cellular solutions to the Kuramoto–Sivashinsky model. *J. Fluid Mech.* **168**, 221.

Gallagher, C. T., Leighton, D. T., and McCready, M. J. 1996. Experimental investigation of a two-layer shearing instability in a cylindrical Couette cell. *Phys. Fluids A* **8**, 2385.

Lin, S. P. 1974. Finite-amplitude sideband stability of a viscous fluid. *J. Fluid Mech.* **63**, 417.

Liu, J. and Gollub, J. P. 1993, Onset of spatially chaotic waves on flowing films. *Phys. Rev. Lett.* **70**, 2289.

McCready, M. J. and Chang, H.-C. 1996. Formation of large disturbances on sheared and falling films. *Chem. Eng. Commun.* **141**, 347.

Nakaya, C. 1975. Long waves on a thin fluid layer flowing down an inclined plane. *Phys. Fluids* **18**, 1407.

Renardy, M. and Renardy, Y. 1993. Derivation of amplitude equations and analysis of sideband instabilities. *Phys. Fluids A* **5**, 2738.

Roberts, R. M. 1998. Ph.D. Dissertation, University of Notre Dame.

Yih, C.-S. 1963. Stability of liquid flow down an inclined plane. *Phys. Fluids* **6**, 321.

Yih, C.-S. 1967. Instability due to viscosity stratification. *J. Fluid Mech.* **27**, 337.

# 9

# Multifilm Flow Down an Inclined Plane

## Simulations Based on the Lubrication Approximation and Normal-Mode Decomposition of Linear Waves

C. POZRIKIDIS

## 9.1. Introduction

In a seminal paper, Yih (1963) studied the growth of small-amplitude waves on the free surface of a liquid film flowing down an inclined plane, under the auspices of the normal-mode linear stability theory. In the 35 years that have elapsed, his analysis was extended and generalized in several ways. Subsequent authors studied the effects of film heating, surface-active agents, and non-Newtonian fluid properties, and discussed the evolution of finite-amplitude and three-dimensional waves in the context of weakly nonlinear stability theories and by numerical simulation at zero and nonzero Reynolds numbers. Reviews can be found in the articles by Weinstein and Kurz (1991), Chen (1993), Chang (1994), Oron et al. (1997), and Pozrikidis (1998b).

In one important extension, the single-film flow was generalized to multifilm flow of immiscible fluids with different physical properties and interfacial tensions. This generalization is by no means an academic exercise, for it is relevant and has been studied with reference to multifilm coating in photographic technology (e.g., Kistler and Schweizer, 1997). Figures 9.1(a) and 9.1(b) display drawings of patented multilayer slide-coating and curtain-coating apparatus showing the formation and merging of several liquid films on an inclined plane and their subsequent coating onto a moving support (U.S. patents 2,761,791 and 3,508,947). In practice, as many as 13 superimposed layers of gelatin emulsions may flow down the inclined surface and then be coated onto the support to produce a color film. To prevent uneven thicknesses leading to a defective product, the coating engineer must secure the flatness of the interfaces and thus the stability of the multifilm flow.

In this chapter, we make two contributions to the theory of multifilm flow. First, we formulate a system of equations governing the evolution of film thicknesses in the absence of fluid inertia, working under the auspices of the lubrication approximation; we derive a system of nonlinear partial differential equations governing the evolution of interfacial waves; and we implement a finite-difference method of solution. For the sake of generality, we allow the inclined wall to have periodic undulations, but the wavelength of the interfacial perturbations is restricted to be equal to the period of the wall. Numerical solutions confirm that the three-layer configuration is unstable under certain conditions in agreement with linear theory and, furthermore, illustrate the steepening of finite-amplitude interfacial waves and suggest the development of interfacial folding.

Second, we discuss a method of extracting the properties of the linear normal modes from the results of a numerical simulation with arbitrary small-amplitude interfacial waves. When the fluid inertia is negligible, the $N$-layer configuration has $N$ normal modes. A complete description of each normal mode requires the specification of (a) the phase and the amplitude of each interfacial wave relative to those of the interface that is closest to the wall, (b) the phase

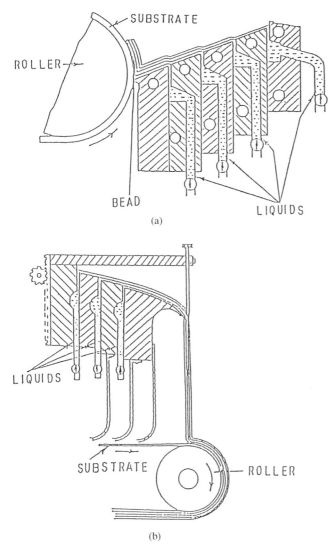

Figure 9.1. Technical drawings of (a) a slide-coating, (b) a curtain-coating apparatus used in the photographic industry, showing the formation and subsequent flow of several superimposed films down an inclined plane (U.S. patents 2,761,791 and 3,508,947).

velocity, and (c) the growth rate of the perturbation. These properties have been derived in closed form only for the simple case of single-film flow; selected properties have been graphed for particular cases of two-film and three-film flow. We present a general method for extracting the complete properties of the normal modes from the results of a numerical simulation in the limit of small deformations based on the Prony fitting of a signal with a sum of complex exponentials. The method requires solving systems of linear equations with small size and thus bypasses the formulation and solution of a generalized eigenvalue problem. The numerical procedure is tested with success using the results of simulations based on the lubrication approximation.

The above-mentioned two developments establish a point of departure for further computational studies and theoretical analyses of multifilm flow.

## 9.2. Lubrication Model for Multifilm Flow Down an Inclined Wall

Consider the gravity-driven flow of $N$ superposed liquid films down an inclined wall with small-amplitude periodic undulations of wavelength $L$, as shown in Fig. 9.2. The position of the wall is described by the equation $y = y_0(x)$, and the position of the $i$th interface is described by the equation $y = y_i(x, t)$, $i = 1, \ldots, N$. All interfaces are assumed to have periodic shapes that conform to the period of the wall. The $i$th film is confined between the $i - 1$ interface and the $i$th interface, with the understanding that the zeroth interface represents the wall.

In this section, we develop an approximate theory that allows us to simulate the evolution of the interfaces from specified initial shapes, in the limit as (a) the slopes $\partial y_i / \partial x$ are uniformly small, for $i = 0, \ldots, N$, (b) the fluid inertia is negligible, and (c) the ratio $H/L$ is sufficiently small; $H$ is the total layer thickness in unidirectional flow. Under these conditions, we may work under the auspices of the lubrication theory and assume that, to leading-order approximation, the flow within each film is locally unidirectional and parallel to the wall (e.g., Pozrikidis, 1997). The governing equations of Stokes flow simplify to

$$\frac{\partial p^{(i)}}{\partial x} = \mu_i \frac{\partial^2 u^{(i)}}{\partial y^2} + \rho_i g \sin \theta_0, \tag{2.1a}$$

$$\frac{\partial p^{(i)}}{\partial y} = -\rho_i \cos \theta_0, \tag{2.1b}$$

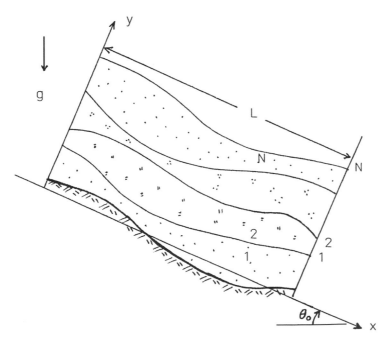

Figure 9.2. Schematic illustration of multifilm flow down an inclined wall with periodic undulations.

where the subscript or superscript $i$, taking values $1, \ldots, N$, designates the layer number, as shown in Fig. 9.2, and $\rho_i$ and $\mu_i$ are the density and the viscosity, respectively, of the $i$th fluid. Continuity of velocity across the interfaces requires that

$$u^{(i)}(y = y_i) = u^{(i+1)}(y = y_i) \tag{2.2a}$$

for $i = 1, \ldots, N - 1$. The no-slip boundary condition at the wall requires that

$$u^{(1)}(y = y_0) = 0. \tag{2.2b}$$

Continuity of shear stress at the interfaces requires that

$$\left[ \frac{\partial u^{(i)}}{\partial y} \right]_{y=y_i} = \lambda_i \left[ \frac{\partial u^{(i+1)}}{\partial y} \right]_{y=y_i} \tag{2.2c}$$

for $i = 1, \ldots, N$, where $\lambda_i = \mu_{i+1}/\mu_i$ are the viscosity ratios, with the understanding that $\mu_{N+1} = 0$ and thus $\lambda_N = 0$. Finally, balancing the normal stresses on either side of the $i$th interface and the surface tension yields

$$[p^{(i)}]_{y=y_i} = [p^{(i+1)}]_{y=y_i} + \gamma_i \kappa_i \tag{2.2d}$$

for $i = 1, \ldots, N$, where $\gamma_i$ is the tension of the $i$th interface and $\kappa_i = -y_i''/(1 + y_i'^2)^{3/2}$ is the curvature of the $i$th interface; a prime designates a derivative with respect to $x$.

The statement of the problem is now complete, and we proceed to formulate the solution. For simplicity, the time $t$ is omitted from the arguments of the dependent variables. Integrating Eq. (2.1b) with respect to $y$ and using condition (2.2d), we find that the pressure distribution within the $i$th layer is given by

$$p^{(i)}(x, y) = p^{(i+1)}[x, y = y_i(x)] + \gamma_i \kappa_i(x) + \rho_i g \cos \theta_0 \, [y_i(x) - y] \tag{2.3}$$

for $i = 1, \ldots N$, where $p^{(N+1)} = P_0$ is the constant ambient pressure. Differentiating Eq. (2.3) with respect to $x$, we find

$$\frac{\partial p^{(i)}}{\partial x} = \frac{\partial p^{(i+1)}}{\partial x} + \frac{\partial p^{(i+1)}}{\partial y} \frac{\partial y_i}{\partial x} + \gamma_i \frac{\partial \kappa_i}{\partial x} + \rho_i g \cos \theta_0 \frac{\partial y_i}{\partial x}. \tag{2.4}$$

Using Eq. (2.3) to evaluate $\partial p^{(i+1)}/\partial y$, we find

$$\frac{\partial p^{(i)}}{\partial x} = \frac{\partial p^{(i+1)}}{\partial x} + \gamma_i \frac{\partial \kappa_i}{\partial x} + (\rho_i - \rho_{i+1}) g \cos \theta_0 \frac{\partial y_i}{\partial x} \tag{2.5}$$

with the understanding that $\rho_{N+1} = 0$. Using relation (2.5), we find

$$-\frac{\partial p^{(i)}}{\partial x} = -\sum_{j=i}^{N} \gamma_j \frac{\partial \kappa_j}{\partial x} - g \cos \theta_0 \sum_{j=i}^{N} (\rho_j - \rho_{j+1}) \frac{\partial y_j}{\partial x}. \tag{2.6}$$

Finally, we introduce the approximation $\kappa_i = -y_i''$ and obtain

$$-\frac{\partial p^{(i)}}{\partial x} = \sum_{j=i}^{N} \gamma_j \frac{\partial^3 y_j}{\partial x^3} - g \cos \theta_0 \sum_{j=i}^{N} (\rho_j - \rho_{j+1}) \frac{\partial y_j}{\partial x}. \tag{2.7}$$

The velocity profile within the $i$th layer arises by integration of Eq. (2.1a), yielding

$$u^{(i)}(x, y) = A_i(x) + B_i(x)y - G_i(x)y^2, \tag{2.8}$$

where we have defined

$$G_i(x) \equiv \frac{1}{2\mu_i}\left[-\frac{\partial p^{(i)}}{\partial x} + \rho_i g \sin\theta_0\right]. \tag{2.9}$$

The right-hand side of Eq. (2.9) may be evaluated from a knowledge of the instantaneous interfacial profiles by use of Eq. (2.7). When the interfaces are flat, the functions $G_i(x)$ are constant. The flow rate along the $x$ axis within the $i$th layer is given by

$$Q_i(x) \equiv \int_{y_i}^{y_{i+1}} u^{(i)}dx = A_i(x)(y_i - y_{i-1}) + \frac{1}{2}B_i(x)\left(y_i^2 - y_{i-1}^2\right) - \frac{1}{3}G_i(x)\left(y_i^3 - y_{i-1}^3\right). \tag{2.10}$$

To evaluate the functions $A_i(x)$ and $B_i(x)$, we use the interfacial and wall conditions expressed by Eqs. (2.2a)–(2.2c), obtaining

$$A_i(x) + B_i(x)y_i - G_i(x)y_i^2 = A_{i+1}(x) + B_{i+1}(x)y_i - G_{i+1}(x)y_i^2 \tag{2.11a}$$

for $i = 1, \ldots, N - 1$,

$$A_1(x) + B_1(x)y_0 - G_1(x)y_0^2 = 0, \tag{2.11b}$$

$$B_i(x) - 2G_i(x)y_i = \lambda_i[B_{i+1}(x) - 2G_{i+1}(x)y_i] \tag{2.11c}$$

for $i = 1, \ldots, N - 1$, and

$$B_N(x) = 2G_N(x)y_N. \tag{2.11d}$$

A straightforward rearrangement can be made to replace the recursive relation (2.11c) by the explicit relation

$$B_i(x) = 2G_i(x)y_i + 2\sum_{k=i+1}^{N}\frac{\mu_k}{\mu_i}G_k(x)(y_k - y_{k-1}) \tag{2.11e}$$

for $i = 1, \ldots, N - 1$. Equations (2.11d) and (2.11e) provide us with expressions for the evaluation of the functions $B_i(x)$. Once these functions are available, $A_1(x)$ follows from Eq. (2.11b) and the rest of the functions $A_i(x)$ follow from Eq. (2.11a).

Now, a mass balance for each film requires that

$$\frac{\partial Q_i}{\partial x} = -\frac{\partial y_i}{\partial t} + \frac{\partial y_{i-1}}{\partial t} \tag{2.12}$$

for $i = 1, \ldots, N$. Combining these expressions, we find

$$\frac{\partial y_i}{\partial t} = -\sum_{j=1}^{i}\frac{\partial Q_j}{\partial x}. \tag{2.13}$$

Substituting Eq. (2.7) into Eq. (2.9), the result into Eq. (2.10), and the outcome into Eq. (2.13), we derive a system of fourth-order nonlinear partial differential equations governing the evolution of the interfaces and of the free surface.

### 9.2.1. *One Film*

In the case of one film, $N = 1$, and a flat support located at $y_0 = 0$, we readily compute $B_1(x) = 2G_1(x)y_1$, $A_1(x) = 0$ and derive the nonlinear partial differential equation

$$\frac{\partial y_1}{\partial t} = -\frac{\rho_1 g \sin \theta_0}{3\mu_1} \frac{\partial}{\partial x}\left[ y_1^3 \left( 1 - \cot \theta_0 \frac{\partial y_1}{\partial x} + \Gamma_1 H_R^2 \frac{\partial^3 y_1}{\partial x^3} \right) \right]. \tag{2.14}$$

The dimensionless group

$$\Gamma_1 \equiv \frac{\gamma_1}{\rho_1 g H_R^2 \sin \theta_0} \tag{2.15}$$

is the inverse Bond number defined with respect to a specified constant reference film thickness $H_R$.

### 9.2.2. *Two Films*

In the case of two films, $N = 2$, and a flat support located at $y_0 = 0$, we compute

$$B_2(x) = 2G_2(x)y_2, \tag{2.16a}$$

$$B_1(x) = 2G_1(x)y_1 + 2\lambda_1 G_2(x)(y_2 - y_1), \tag{2.16b}$$

$$A_1(x) = 0, \tag{2.16c}$$

$$A_2(x) = G_1(x)y_1^2 + G_2(x)y_1[2\lambda_1(y_2 - y_1) - 2y_2 + y_1], \tag{2.16d}$$

and then find

$$Q_1 = y_1^2 \left[ \frac{2}{3} G_1 y_1 + \lambda_1 G_2 (y_2 - y_1) \right], \tag{2.17a}$$

$$Q_2 = (y_2 - y_1) \left\{ G_1 y_1^2 + 2G_2(y_2 - y_1) \left[ \lambda_1 y_1 + \frac{1}{3}(y_2 - y_1) \right] \right\}, \tag{2.17b}$$

where

$$G_1 \equiv \frac{\rho_2 g \sin \theta_0}{2\mu_2} \lambda \left\{ \Gamma_2 \left( \frac{\gamma_1}{\gamma_2} H_R^2 \frac{\partial^3 y_1}{\partial x^3} + \frac{\partial^3 y_2}{\partial x^3} \right) - \cot \theta_0 \left[ \frac{\partial y_2}{\partial x} + \left( \frac{1}{\beta} - 1 \right) \frac{\partial y_1}{\partial x} \right] + \frac{1}{\beta} \right\}, \tag{2.18a}$$

$$G_2 \equiv \frac{\rho_2 g \sin \theta_0}{2\mu_2} \left( \Gamma_2 H_R^2 \frac{\partial^3 y_2}{\partial x^3} - \cot \theta_0 \frac{\partial y_2}{\partial x} + 1 \right). \tag{2.18b}$$

We have introduced the dimensionless numbers

$$\beta \equiv \frac{\rho_2}{\rho_1}, \tag{2.19a}$$

$$\lambda \equiv \frac{\mu_2}{\mu_1}, \tag{2.19b}$$

$$\Gamma_2 \equiv \frac{\gamma_2}{\rho_2 g H_R^2 \sin \theta_0} \tag{2.19c}$$

The evolution of the interfaces is described by the equations

$$\frac{\partial y_1}{\partial t} = -\frac{\partial Q_1}{\partial x},$$  (2.20a)

$$\frac{\partial y_2}{\partial t} = -\frac{\partial Q_1}{\partial x} - \frac{\partial Q_2}{\partial x}.$$  (2.20b)

### 9.2.3. *Numerical Solutions*

A standard explicit finite-difference method was implemented for solving the system of equations (2.13) on a one-dimensional uniform grid by using the first-order Euler forward differentiation method to approximate the time derivatives. The first spatial derivative of the flow rates on the right-hand side was approximated with either central second-order or backward first-order differences; the second choice amounts to upwind differentiation.

In Fig. 9.3(a), we present typical stages in the evolution of the interfaces of a three-layer system flowing down a vertical wall, for unperturbed film thicknesses $H_1/H = 0.6$, $H_2/H = 0.1$, $H_3/H = 0.3$, viscosity ratios $\mu_2/\mu_1 = 0.2$, $\mu_3/\mu_1 = 1.0$, all density ratios equal to unity, and in the absence of interfacial and free surface tensions. At the initial instant, the interfaces have wavy shapes with amplitudes $A_1/H = 0.02$, $A_2/H = 0.0164$, $A_3/H = 0.0118$, phase lags with respect to the origin $\varphi_1 = 0$, $\varphi_2 = 0.23\pi$, $\varphi_3 = -0.18\pi$, and wavelength $L/H = 4$. These conditions roughly correspond to an unstable normal mode according to linear theory, as is discussed in Section 9.3. We observe that the amplitudes of the interfaces grow in time and the sinusoidal waves steepen. At long times, numerical instabilities develop, indicating the tendency of the waves to overturn and the interfaces to fold.

The simulations presented in Fig. 9.3(a) were carried out by the central-difference method. The corresponding evolution simulated by the upwind differentiation method is shown in Fig. 9.3(b); the important effect of numerical diffusivity is apparent. Similar behaviors are observed for a broad range of systems and flow conditions.

## 9.3. Linear Waves

Consider now flow down an inclined plane, and assume that the distortions of the interfaces are small compared with the unperturbed layer thicknesses as well as with the wavelength of a disturbance $L$. The ratio $H/L$ is allowed to be arbitrary. These assumptions allow us to apply the method of domain perturbation and effectively derive modified interfacial conditions at the position of the unperturbed interfaces.

### 9.3.1. *Normal-Mode Analysis*

Carrying out a normal-mode analysis, we find that exponentially growing or decaying interfacial and free surface waves are described by the equations

$$y = y_j(x, t) = Y_j + A_j^{NM,l}(t) \cos\left[k\left(x - c_P^{NM,l}t\right) - \Delta\phi_j^{NM,l}\right]$$  (3.1)

for $j = 1, \ldots, N$, where $Y_j$ are the unperturbed interface positions, $k = 2\pi/L$ is the wave number, the superscript *NM* stands for normal mode, $c_P^{NM,l}$ is the phase velocity, and $\Delta\phi_j^{NM,l}$ is the

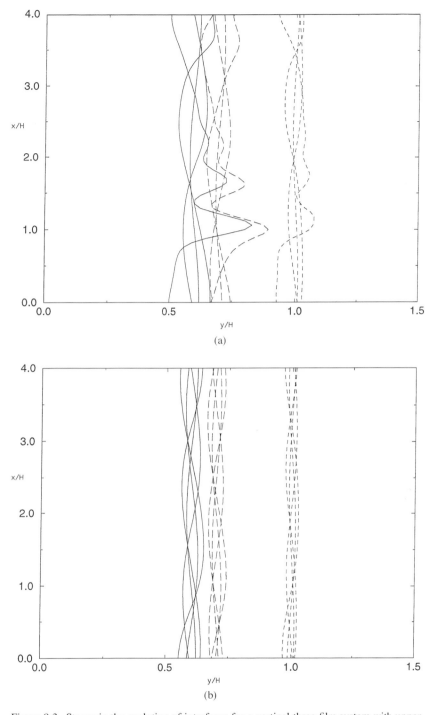

Figure 9.3. Stages in the evolution of interfaces for a vertical three-film system with unperturbed film thicknesses $H_1/H = 0.6$, $H_2/H = 0.1$, $H_3/H = 0.1$, viscosity ratios $\mu_2/\mu_1 = 0.2$, $\mu_3/\mu_1 = 1.0$, all density ratios equal to unity, in the absence of interfacial and free surface tensions. At the initial instant, the interfaces have sinusoidal shapes with amplitudes $A_1/H = 0.02$, $A_2/H = 0.0164$, $A_3/H = 0.0118$, phase lags with respect to the origin $\varphi_1 = 0$, $\varphi_2 = 0.23\pi$, $\varphi_3 = -0.18\pi$, and wavelength $L/H = 4$. (a) Results obtained with the centered differentiation method at times $\hat{t} = t\rho_1 g H/\mu_1 = 0, 5, 10, 20$; (b) results obtained with the upwind differentiation method at times $\hat{t} = 0, 5, 10, 15, 20$.

phase lag of the wave on the $j$th interface with the respect to the wave on the first interface, all corresponding to the $l$th normal mode; by definition, $\Delta\phi_1^{NM,l} = 0$. The amplitudes $A_j^{NM,l}(t)$ are exponential functions of time given by

$$A_j^{NM,l}(t) = A_j^{NM,l}(t = 0)\exp\left(\sigma_l^{NM,l}t\right),\tag{3.2}$$

where $A_j^{NM,l}(t = 0)$ is the initial amplitude and $\sigma_l^{NM,l}$ is the growth rate of the $l$th normal mode.

For each unperturbed $N$-layer configuration, there are $N$ normal modes with negative, zero, or positive values of the growth rate $\sigma_l^{NM,l}$ and corresponding stable, neutrally stable, and unstable behavior. A complete description of the $l$th normal mode requires the specification of the following $2N$ quantities: (a) the ratio of the amplitudes of the interfacial waves $r_j^{NM,l} \equiv A_j^{NM,l}(t)/A_1^{NM,l}(t)$, for $j = 2, \ldots, N$, (b) the phase lags $\Delta\phi^{NM,l} = 2, \ldots, N$, (c) the phase velocity $c_P^{NM,l}$, and (d) the growth rate $\sigma_l^{NM,l}$. For single-film flow, only the phase velocity and the growth rate are required.

Yih (1963) studied the stability of the single-film flow and obtained the following closed-form expressions for the phase velocity and the growth rate in the limit of Stokes flow:

$$c_P^{NM,l} = \frac{\rho_1 g \sin\theta_0 H^2}{2\mu_1}\left(1 + \frac{2}{1 + \cosh 2\hat{k} + 2\hat{k}^2}\right),\tag{3.3a}$$

$$\sigma_l^{NM,l} = -\frac{\rho_1 g \sin\theta_0 H}{\mu_1}\left(\frac{\sinh 2\hat{k}}{2\hat{k}} - 1\right)\frac{\hat{k}^2\Gamma_1 + \cot\theta_0}{1 + \cosh 2\hat{k} + 2\hat{k}^2},\tag{3.3b}$$

where $\hat{k} = kH$, $H$ is the unperturbed film thickness, and the inverse Bond number $\Gamma_1$ is defined in Eq. (2.15) with $H_R = H$.

A number of authors studied the stability of the two-film flow and presented graphs of selected properties of unstable normal modes, including Kao (1965a, 1965b, 1968), Kobayashi and Scriven (1981, AIChE Annual Meeting; 1982 AIChE Winter Meeting), Lin (1983), Kobayashi et al. (1986, Spring AIChE Meeting), Loewenherz and Lawrence (1989), Weinstein (1990), Kobayashi (1986, AIChE Spring Meeting), Kobayashi (1992), and Chen (1993). Their results showed that the flow can be unstable when the less viscous fluid is adjacent to the wall and the surface tension is sufficiently small.

Akhtaruzzaman et al. (1978), Wang et al. (1978), and Lin (1983), formulated the linear stability problem for an arbitrary number of layers and presented results for three-film flow, but did not identify unstable behaviors. The three-film flow was discussed in detail by Weinstein and Kurz (1991), who pointed out that instability is predicted even in the context of the lubrication approximation when the viscosity of the middle film is lower than that of the films at the top and next to the wall. In contrast, the simplified formulation for two-film flow erroneously predicts that the motion is stable under any conditions. More recently, Kobayashi (1995) compared theoretical predictions with laboratory observations, and Kliakhandler and Sivashinsky (1997) studied the weakly nonlinear instability.

### 9.3.2. *Extraction of Normal Modes by Exponential Fitting*

In computational studies of nonlinear instability, it is both expedient and physically meaningful to use an initial condition that corresponds to the most unstable normal mode. But since the complete properties of this mode are available only in the case of the single-film flow, we develop a method of extracting them from the results of a numerical simulation with arbitrary initial conditions, in the limit of small interface deformations.

We begin by considering the motion when the interfaces and free surface are perturbed with arbitrary sinusoidal small-amplitude waves whose relative amplitudes and phase lags do not necessarily correspond to a normal mode. Without loss of generality, we assume that, at the origin of time, the interfaces and the free surface are described by the equation

$$y = y_j(x, t = 0) = Y_j + A_j \cos(kx - \varphi_j) \tag{3.4}$$

for $j = 1, \ldots, N$, where $Y_j$ are the unperturbed layer positions and $\varphi_j$ are the phase lags with respect to the designated origin of the $x$ axis. One way to describe the evolution of these waves is to express them in terms of a linear combination of the $N$ normal modes, writing

$$y_j(x, t = 0) = Y_j + \sum_{l=1}^{N} A_j^{NM,l}(t = 0) \cos(kx - \varphi_j^{NM,l}). \tag{3.5}$$

The arguments of the trigonometric functions in Eq. (3.5) evolve according to Eq. (3.1), and the functions $A_j^{NM,l}(t)$ evolve according to Eq. (3.2). Combining these equations, we derive evolution equations for the arbitrary linear waves in terms of the properties of the normal modes,

$$y_j(x, t) = Y_j + \sum_{l=1}^{N} A_j^{NM,l}(t = 0) \cos\left[k\left(x - c_p^{NM,l}t\right) - \varphi_j^{NM,l}\right] \exp(\sigma_I^{NM,l}t). \tag{3.6}$$

To compute the $2N^2$ constants $A_j^{NM,l}(t = 0), \varphi_j^{NM,l}, l = 1, \ldots, N, j = 1, \ldots, N$, we set the right-hand side of Eq. (3.4) equal to the right-hand side of Eq. (3.5) and require that the sums of coefficients of like trigonometric functions balance to zero, thereby obtaining a system of $2N$ algebraic–trigonometric equations. If the normal-mode wave amplitude ratios $r_j^{NM,l}$ and phase lags $\Delta\phi_j^{NM,l}$ were available for $l = 1, \ldots, N, j = 2, \ldots, N$, we could have invoked their definitions to obtain an additional set of $2N(N-1)$ equations, and this would raise the total number of equations to the number of the unknowns.

Faced with the unavailability of the complete properties of the normal modes, we develop a method of extracting them from the numerical solution subject to an arbitrary initial condition, in the limit of small perturbations. As a first step, we express Eq. (3.5) in the form

$$y_j(x, t) = Y_j + F_{j,c}(t) \cos kx + F_{j,s}(t) \sin kx, \tag{3.7}$$

where

$$F_{j,c}(t) = \sum_{l=1}^{N} A_j^{NM,l}(t = 0) \cos\left(\sigma_R^{NM,l}t + \varphi_j^{NM,l}\right) \exp(\sigma_I^{NM,l}t), \tag{3.8a}$$

$$F_{j,s}(t) = \sum_{l=1}^{N} A_j^{NM,l}(t = 0) \sin\left(\sigma_R^{NM,l}t + \varphi_j^{NM,l}\right) \exp(\sigma_I^{NM,l}t) \tag{3.8b}$$

are the first-order cosine and sine coefficients of the complete Fourier series of $y_j(x, t)$ with respect to $x$; we have introduced the real part of the complex growth rate $\sigma_R^{NM,l} = kc_P^{NM,l}$. The key idea is that we may express the left-hand sides of Eqs. (3.8a) and (3.8b) as sums of complex exponentials and we may recover the $2N^2$ real unknowns $A_j^{NM,l}(t = 0), \varphi_j^{NM,l}$, by performing an $N$-mode complex exponential fitting by using a method developed by Prony (e.g., Hildebrand,

1974, pp. 457–463; Kay and Marple, 1981; Marple, 1987, pp. 303–349). To implement the method, we write

$$F_{j,c}(t) = \frac{1}{2} \sum_{l=1}^{N} [c_{j,c}^{(l)} e^{-i\sigma^{NM,l}t} + c_{j,c}^{(l)*} e^{i\sigma^{NM,l*}t}], \tag{3.9a}$$

$$F_{j,s}(t) = \frac{1}{2} \sum_{l=1}^{N} [c_{j,s}^{(l)} e^{-i\sigma^{NM,l}t} + c_{j,s}^{(l)*} e^{i\sigma^{NM,l*}t}], \tag{3.9b}$$

where $\sigma^{NM,l} = \sigma_R^{NM,l} + i\sigma_I^{NM,l}$, are the complex growth rates, $i$ is the imaginary unit, an asterisk designates the complex conjugate, and $c_{p,q}^{(r)}$ are complex coefficients. Setting the right-hand side of Eq. (3.9a) or Eq. (3.9b) equal to the right-hand side of Eq. (3.8a) or Eq. (3.8b), we find

$$c_{j,c}^{(l)} = A_j^{NM,l}(t=0)e^{-i\varphi_j^{(l)}}, \tag{3.10a}$$

$$c_{j,s}^{(l)} = A_j^{NM,l}(t=0)e^{-i[\varphi_j^{(l)} - \pi/2]} \tag{3.10b}$$

Assume now that we have available an $(M + 4)$-long time series for $F_{j,c}(t)$ with constant sampling time $\Delta t$, and denote, for brevity, $\zeta_q \equiv F_{j,c}(q\Delta t)$, where $q = 0, 1, 2, \ldots$. Prony's method proceeds in four stages: At the first stage, we solve the generally overdetermined $M \times 2N$ linear system of equations,

$$\begin{bmatrix} \zeta_1 & \zeta_2 & \cdots & \zeta_{2N} \\ \zeta_2 & \zeta_3 & \cdots & \zeta_{2N+1} \\ \cdots & \cdots & \cdots & \cdots \\ \zeta_M & \zeta_{M+1} & \cdots & \zeta_{2N+M-1} \end{bmatrix} \begin{bmatrix} \alpha_{2N} \\ \cdots \\ \alpha_2 \\ \alpha_1 \end{bmatrix} = - \begin{bmatrix} \zeta_{2N+1} \\ \zeta_{2N+2} \\ . \\ . \\ \zeta_{2N+M} \end{bmatrix}, \tag{3.11}$$

for the $2N$ real unknowns $\alpha_1, \alpha_2, \ldots, \alpha_{2N}$ where $M \geq 2N$. At the second stage, we compute the roots of the $2N$-degree characteristic polynomial,

$$P_{2N}(z) = z^{2N} + \alpha_1 z^{2N-1} + \alpha_2 z^{2N-2} + \cdots + \alpha_{2N-1}z + \alpha_{2N}, \tag{3.12}$$

and call them $z_1, z_2, z_3, \ldots, z_{2N}$. Since the coefficients of the characteristic polynomial are real, complex roots appear in pairs of complex conjugates: $(z_1, z_2 = z_1^*)$, $(z_3, z_4 = z_3^*)$, ..., $(z_{2N-1}, z_{2N} = z_{2N-1}^*)$. At the third stage, we extract the complex growth rates from the relations

$$\begin{aligned} z_1 &= \exp(-i\sigma^{NM,l}\Delta t), & z_2 &= \exp(i\sigma^{NM,l*}\Delta t), \\ z_3 &= \exp(-i\sigma^{NM,2}\Delta t), & z_4 &= \exp(i\sigma^{NM,2*}\Delta t), \\ &\cdots \\ z_{2N-1} &= \exp(-i\sigma^{NM,N}\Delta t), & z_{2N} &= \exp(i\sigma^{NM,N*}\Delta t). \end{aligned} \tag{3.13}$$

Finally, we recover the coefficients $c_{j,c}^{(l)}$ by solving a generally overdetermined linear system of equations that arises by writing (3.9a) for $t = q\Delta t$, $q = 0, 1, 2, \ldots$, subject to the computed values of $\sigma^{NM,l}$, and then compute $A_j^{NM,l}(t=0)$, $\varphi_j^{(l)}$, by using Eq. (3.10a).

There is an ambiguity in the definition of $\sigma^{NM,l}$, stemming from our freedom to interchange the complex conjugate roots $z_1$ and $z_2$ on the left-hand sides of Eqs. (3.13), which amounts to replacing $\sigma^{NM,l}$ with its complex conjugate on the right-hand side, and similarly for the rest of the complex growth rates. To resolve this ambiguity, we also perform the Prony fitting of the sine coefficient $F_{j,s}(t)$: We introduce a time series for $F_{j,s}(t)$, set $\zeta_q \equiv F_{j,s}(q \Delta t)$, and work in a similar fashion with Eq. (3.9b) in place of Eq. (3.9a) to recover a $2N$-tuple of values $A_j^{NM,l}(t = 0)$, $\varphi_j^{(l)}$ for each interface. The proper values of $\sigma^{NM,l}$ are the ones that give the same – in practice, nearly the same – values of $A_j^{NM,l}(t = 0)$, $\varphi_j^{(l)}$ computed from the fitting of $F_{j,c}(t)$ or $F_{j,s}(t)$. Combining the results for the interfaces and the free surface, we deduce the wave amplitude ratio and phase lag of the normal modes.

### 9.3.3. *Implementation of Prony's Method*

In the present implementation of the Prony method, the Fourier coefficients $F_{j,c}(t)$ and $F_{j,s}(t)$ are computed in terms of Fourier integrals by use of the Simpson 1/3 rule with overlapping parabolas (e.g., Pozrikidis, 1998a). The overdetermined systems of equations for the polynomial coefficients and for the coefficients of the Prony expansion are solved by the normal equation least-squares method followed by Cholesky decomposition.

Weinstein and Kurz (1991, Fig. 4) presented graphs of the complete properties of unstable normal modes for three superimposed films, computed on the basis of the lubrication approximation. For $H_1/H = 0.6$, $H_2/H = 0.1$, $H_3/H = 0.3$, viscosity ratios $\mu_2/\mu_1 = 0.2$, $\mu_3/\mu_1 = 1.0$, all density ratios equal to unity, and in the absence of interfacial and free surface tensions, their graphs show the occurrence of an unstable normal mode with relative interface amplitudes $A_2/A_1 = 0.82$, $A_3/A_1 = 0.58$ and phase lags $\Delta\phi_2^{NM} = 0.23\pi$, $\Delta\phi_3^{NM} = -0.18\pi$. Using these values, along with $A_1/H = 0.01$ and $L/H = 4$, we generated initial interface profiles, and computed their evolution down a vertical plane wall by using the numerical method described in Section 9.2.

In Figs. 9.4(a)–9.4(c), we show the normalized Fourier coefficients $\hat{F}_{j,c} \equiv F_{j,c}/H$ and $\hat{F}_{j,s} \equiv F_{j,s}/H$ for the two interfaces and the free surface. The circles and squares correspond to the Prony fitting of a single normal mode for the cosine and the sine coefficients, respectively, yielding a complex growth rate $\sigma\mu_1/(\rho_1 gH) = 0.806 + i0.078$, where $i$ is the imaginary unit. The symbols mark the time series used to generate the overdetermined systems of linear equations. The ratio of the phase velocity $k\sigma_I$ to the velocity of the fluid at the unperturbed position of the first interface is equal to 1.22, which is in excellent agreement with the value read off the graph in Fig. 4(c) of Weinstein and Kurz (1991).

In Fig. 9.4(d), we plot on a linear-log scale the amplitudes of the interfaces $\text{Amp}(t)$ – defined as half the difference between the maximum and the minimum displacement of an interface from the unperturbed position – reduced by their initial values, and obtain a nearly straight line with decaying undulations. The slope of the straight line provides us with a growth rate that is consistent with that deduced from the single-mode exponential fitting, $\sigma_I\mu_1/(\rho_1 gH) = 0.078$. The undulations are attributed to the weak presence of the two stable normal modes.

Next we consider an initial condition that corresponds to a combination of stable and unstable normal modes. The relative amplitudes of the interfacial waves at the initial instant are $A_2/A_1 = 1.00$ and $A_3/A_1 = 1.00$; all interfaces are perturbed in phase. In Figs. 9.5(a)–(c), we show the normalized Fourier coefficients $\hat{F}_{j,c} \equiv F_{j,c}/H$ and $\hat{F}_{j,s} \equiv F_{j,s}/H$ for an initial perturbation with $A_1/H = 0.01$ and $L/H = 4$. The symbols mark the trace of a three-mode exponential fitting, revealing the existence of two stable and one unstable normal mode in agreement with linear theory. The results suggest that the relative interface amplitudes corresponding to the unstable normal mode are approximately equal to $A_2/A_1 = 0.88$ and $A_3/A_1 = 0.43$, and the corresponding phase lags are $\Delta\phi_2^{NM} = 0.08\pi$ and $\Delta\phi_3^{NM} = -0.21\pi$. Considering that the

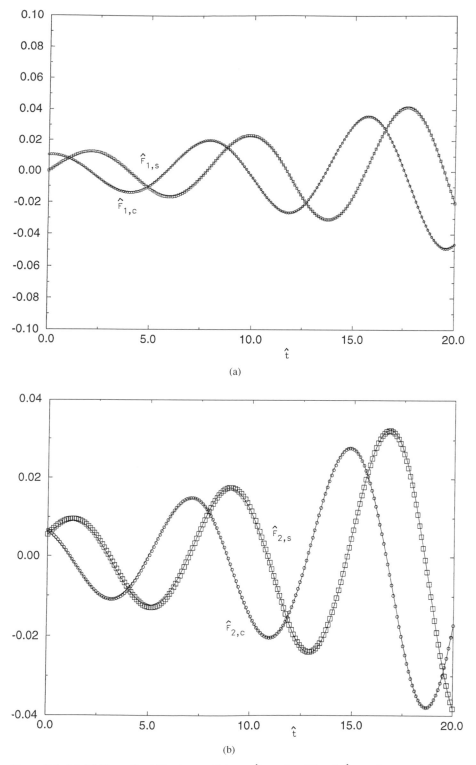

(a)

(b)

Figure 9.4. (a)–(c) Normalized Fourier coefficients $\hat{F}_{j,c} \equiv F_{j,c}/H$ and $\hat{F}_{j,s} \equiv F_{j,s}/H$ for the two interfaces and for the free surface, subject to a normal-mode disturbance on a three-film system, as described in the text, displayed against the dimensionless time $\hat{t} = t\rho_1 g H/\mu_1$. (d) Evolution of the amplitude of the interfaces; the solid curve is for the first interface, the long-dashed curve is for the second interface, and the short-dashed curve is for the free surface.

124

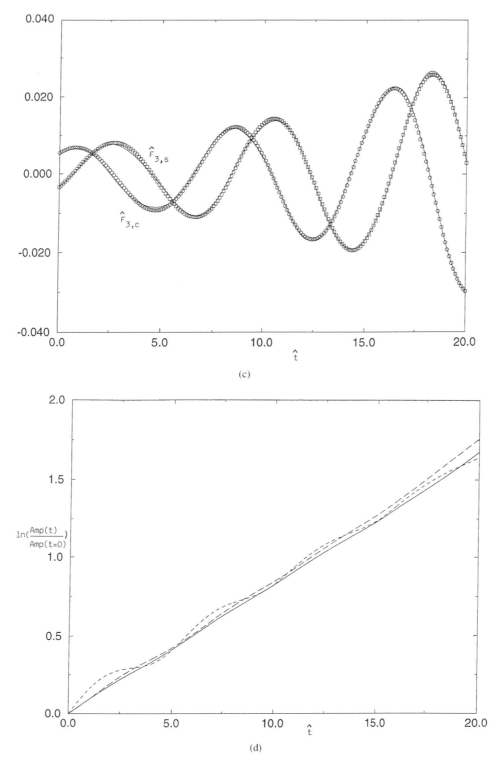

Figure 9.4. *(continued)*

125

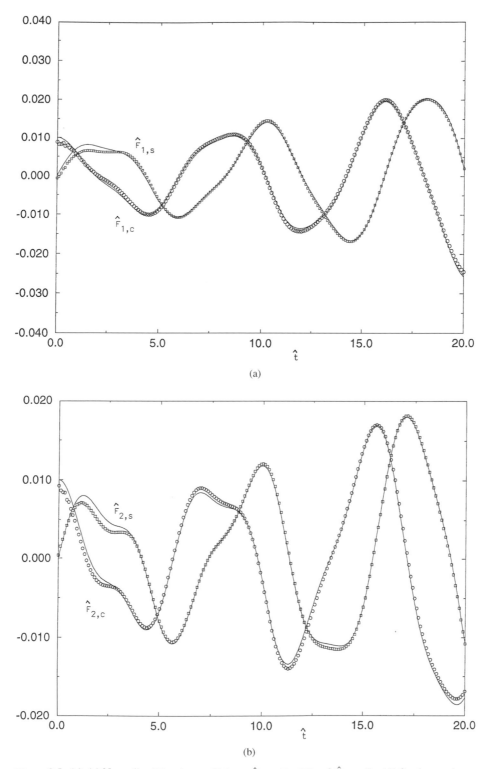

(a)

(b)

Figure 9.5. (a)–(c) Normalized Fourier coefficients $\hat{F}_{j,c} \equiv F_{j,c}/H$ and $\hat{F}_{j,s} \equiv F_{j,s}/H$ for the two interfaces and for the free surface, subject to a non-normal-mode disturbance on a three-film system, as described in the text, displayed against the dimensionless time $\hat{t} = t\rho_1 g H/\mu_1$. (d) Evolution of the amplitude of the interfaces; the solid curve is for the first interface, the long-dashed curve is for the second interface, and the short-dashed curve is for the free surface.

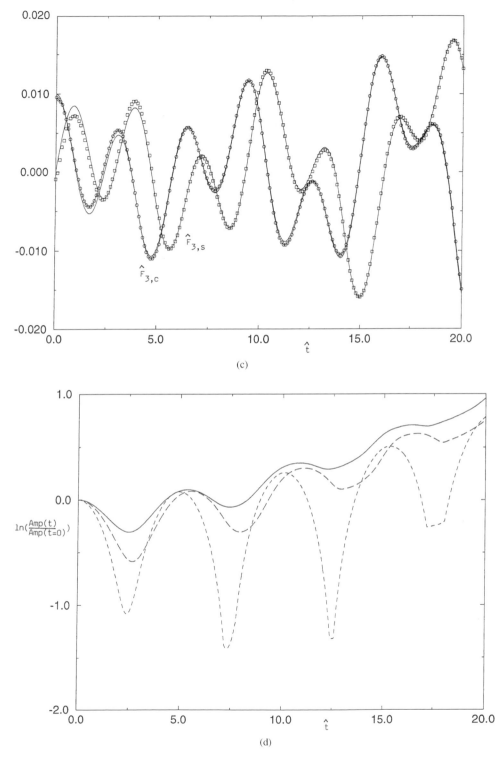

Figure 9.5. *(continued)*

127

initial amplitude of the interfaces is 10% of the unperturbed middle film thickness, the numerical results are in satisfactory agreement with the predictions of linear theory stated in a previous paragraph. In Fig. 9.5(d), we show the evolution of the amplitude of the interfaces and of the free surface on a linear-log scale, and observe strongly oscillatory and discontinuous behaviors, a result of the competition of the normal modes. It would certainly have been difficult to extract the properties of the normal modes by analyzing the amplitudes of the interfacial profiles.

## Acknowledgments

I am indebted to Bhaskar D. Rao of the Department of ECE, University of California, San Diego, for bringing to my attention Prony's method. This research was supported by the U.S. National Science Foundation and the SUN Microsystems Corporation. Acknowledgment is made to the donors of the Petroleum Research Fund, administered by the American Chemical Society, for partial support.

## References

Akhtaruzzaman, A. F. M., Wang, C. K., and Lin, S. P. 1978. Wave motion in multilayered liquid films. *J. Appl. Mech.* **45**, 25–31.

Chang, H.-C. 1994. Wave evolution on a falling film. *Ann. Rev. Fluid Mech.* **26**, 103–136.

Chen, K. 1993. Wave formation in the gravity-driven low-Reynolds-number flow of two liquid films down an inclined plane. *Phys. Fluids A* **5**, 3038–3048.

Hildebrand, F. B., 1974. *Introduction to Numerical Analysis*. Dover, New York.

Kao, T. W. 1965a. Stability of two-layer viscous stratified flow down an inclined plane. *Phys. Fluids* **8**, 812–820.

Kao, T. W. 1965b. Role of the interface in the stability of stratified flow down an inclined plane. *Phys. Fluids* **8**, 2190–2194.

Kao, T. W. 1968. Role of viscosity stratification in the stability of two-layer flow down an incline. *J. Fluid Mech.* **33**, 561–572.

Kay, S. M. and Marple, L. Jr. 1981. Spectrum analysis – a modern perspective. *Proc. IEEE* **69**, 1380–1419.

Kistler, S. F. and Schweizer, P. M., eds. 1997. *Liquid Film Coating*. Chapman & Hall, London.

Kliakhandler, I. L. and Sivashinsky, G. I. 1997. Viscous damping and instabilities in stratified liquid film flowing down a slightly inclined plane. *Phys. Fluids* **9**, 23–30.

Kobayashi, C. 1992. Stability analysis of film flow on an inclined plane. I. One layer, two layer flow. *Ind. Coating Res.* **2**, 65–88.

Kobayashi, C. 1995. Stability analysis of film flow on an inclined plane. II. Multi-layer flow. *Ind. Coating Res.* **3**, 103–125.

Lin, S. P. 1983. Effects of surface solidification on the stability of multi-layered liquid films. *J. Fluids Eng.* **105**, 119–121.

Loewenherz, D. S. and Lawrence, C. J. 1989. The effect of viscosity stratification on the stability of a free surface flow at low Reynolds number. *Phys. Fluids A* **1**, 1686–1693.

Marple, S. L. 1987. *Digital Spectral Analysis with Applications*. Prentice-Hall, Englewood Cliffs, NJ.

Oron, A., Davis, S. H., and Bankoff, S. G. 1997. Long-scale evolution of thin liquid films. *Rev. Mod. Phys.* **69**, 931–980.

Pozrikidis, C. 1997. *Introduction to Theoretical and Computational Fluid Dynamics*. Oxford U. Press, Oxford.

Pozrikidis, C. 1998a. *Numerical Computation in Science and Engineering*. Oxford U. Press, Oxford.

Pozrikidis, C. 1998b. Gravity-driven creeping flow of two adjacent layers through a channel and down a plane wall. *J. Fluid Mech.* **371**, 345–376.

Wang, C. K., Seaborg, J. J., and Lin, S. P. 1978. Instability of multi-layered liquid films. *Phys. Fluids* **21**, 1669–1673.

Weinstein, S. J. 1990. Wave propagation in the flow of shear-thinning fluids down an incline. *AIChE J.* **36**, 1873–1889.

Weinstein, S. J. and Kurz, M. R. 1991. Long-wavelength instabilities in three-layer flow down an incline. *Phys. Fluids* **3**, 2680–2687.

Yih, C.-S., 1963. Stability of liquid flow down an inclined plane. *Phys. Fluids* **6**, 321–334.

# 10

# Spatial Evolution of Interfacial Waves in Gas–Liquid Flows

M. J. McCREADY

## 10.1. Introduction

The most prominent problem in the field of gas–liquid flows has always been, and continues to be, prediction of the flow regime. Early work consisted of observing the flow regime and noting the changes as boundaries on a two-dimensional map. Mandhane et al. (1974) give one of the most popular for horizontal flows. Other work has considered mechanisms necessary for transitions from a chosen base state (e.g., stratified flow) and has used linear stability analyses to predict when the stratified state was unstable. Examples of this type of work are by Kordyban and Ranov (1970) and Taitel and Dukler (1976), who used an inviscid formulation, and Lin and Hanratty (1986), Brauner and Maron (1991), Barnea (1987, 1991), and Crowley et al. (1992), who used equations averaged normal to the interface, which is similar to a long-wave approximation. Both of these types of theories are severe approximations of reality and a full numerical solution to the stability problem, including a suitable turbulence model, should be used as a basis for the predictions. However, even when this is done, there is an experimental difficulty that should be looked out for, that is, waves and larger disturbances in flow channels and pipes are convectively unstable, meaning that they evolve and grow with distance. If the growth is very rapid, then perhaps the final state will be observed in a very short distance. However, close to the neutral stability curve, growth will be slow and a distance of hundreds or thousands of characteristic length scales could be needed for development. Thus great care is needed, but has not always been given, to predicting the transition boundaries for either wave onset or flow regimes.

A further complication is that the state of the flow at any location could also depend on the specific conditions that occur at the channel inlet. That different inlet configurations give different regime behavior has been noted by many investigators (e.g., Fan et al., 1993). A subtler but still interesting effect is that very small amplitude noise present in the inlet can influence the macroscopic properties of the disturbances. Kuru et al. (1997) attribute inlet noise to defects in the interfacial wave field at different conditions.

In this chapter, wave evolution with distance in a horizontal gas–liquid flow is examined on different length scales with the intention of demonstrating the issues that arise for the general problem of disturbance propagation in multifluid systems. For flow situations close to neutral stability, in which only a band of short waves is linearly unstable, two different kinds of wave dynamics are identified. For sufficiently deep layers, individual waves evolve only slowly and the wave field can be described as a family of steady waves that has a definable relation between amplitude and wave period. For waves in this region it is argued that a distribution of steady amplitude–period waves exists because of the details of the noise that existed at the channel inlet when the waves began to grow. In contrast, for thinner layers the waves can evolve very rapidly and no relation between amplitude and wave period exists. The evolution of large-wavelength

129

waves, which have been directly implicated in flow regime transition (Fan et al., 1993) or roll wave formation (Bruno and McCready, 1988), is also discussed. A necessary condition for waves growing in the long-wave region appears to be that long waves must be linearly unstable. Finally, the value of understanding the convective nature of disturbances is demonstrated by a description of how disturbances in cocurrent, gas–liquid packed bed reactors can be used to study and influence the reaction behavior.

## 10.2. Flow Geometry

The flow geometry of interest to this chapter is shown in Fig. 10.1. The channel is 7 m long, 2.54 cm high, and 30.5 cm wide and is oriented horizontally. Air is the gas and the liquid is water and glycerin–water solutions with viscosities up to 20 cP. All wave measurements shown in this chapter were obtained by parallel wire conductance methods. Jurman (1990) gives details in a dissertation.

## 10.3. Effect of Noise on Waves

Although difficult to create in an experiment, ideal wave growth for a two-layer flow occurs starting at an inlet where both the gas and liquid are injected without finite disturbances. The infinitesimal noise, which preferably is flat over the range of possible waves, is amplified with distance to form waves. Because growth is exponential, only wavelengths in a narrow range near the peak of the growth curve are observed. Figure 10.2 shows a wave tracing for conditions

Figure 10.1. Two-layer, horizontal gas–liquid flow.

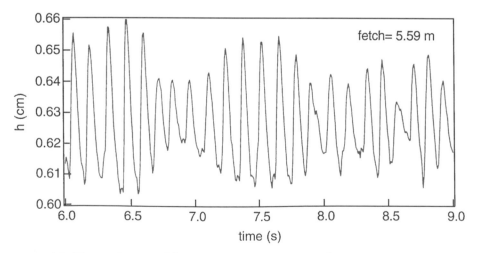

Figure 10.2. Wave tracing at conditions close to neutral stability at a distance where wave growth has saturated.

close to neutral stability where this idealization is essentially achieved. Figure 10.3 shows the corresponding wave spectrum and the linear growth curve. It is seen that there is a clearly identifiable dominant frequency at 8 Hz. This agrees with the linear stability prediction that the wave peak should be close to 9 Hz. The other frequency content is a weak overtone and very weak low-frequency peaks associated with the slight beating of the tracing. The measurements were taken 5.6 m from the channel inlet (fetch = 5.6 m), which, based on these and other observations, lets us conclude that the average wave amplitude has grown to its maximum value and the linear growth has been saturated. However, this does not mean that there are not

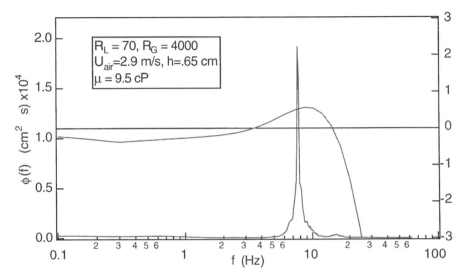

Figure 10.3. Interfacial wave spectrum for conditions of Fig. 10.2.

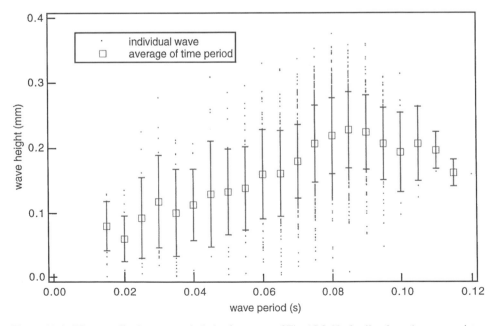

Figure 10.4. Wave amplitude–wave period plot for waves of Fig. 10.2. No family of steady waves exists.

still changes in the waves. The beating is indicative of dynamics and thus one would expect continuing evolution. A measure of this is given in Fig. 10.4 that plots individual wave amplitude versus the corresponding wave period. It is seen that there is no preferred value for amplitude (compare with Fig. 10.8 below) for a given period. Thus there is no family of steady waves. An even more dramatic example of an oscillating wave field is shown in Figs. 10.5–10.7. The wave field comprises wave packets; consequently Fig. 10.7 shows no preferred relation between amplitude and wave period.

In contrast, if the liquid depth is increased, the oscillating wave field gives way to a more stable wave field that exists as a family of steady waves. A possible explanation for this is given by

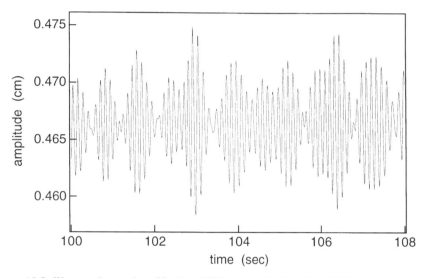

Figure 10.5. Wave tracing at $R_L = 30$, $R_G = 3930$, $\mu_L = 16$ cP, and $h = 0.47$ cm. Very severe beating is present. Local dynamics prevent any chance of steady waves.

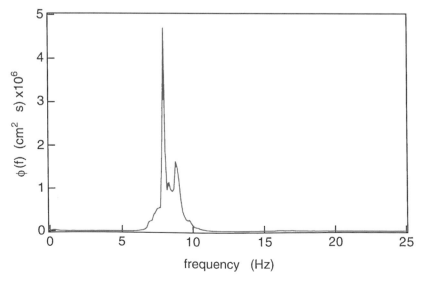

Figure 10.6. Wave spectrum of Fig. 10.5. A (rare) split peak is seen.

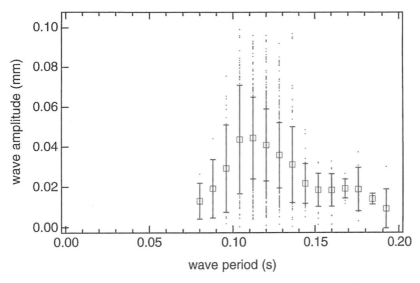

Figure 10.7. Amplitude–period plot for conditions of Fig. 10.4. No relation between amplitude and time period exists.

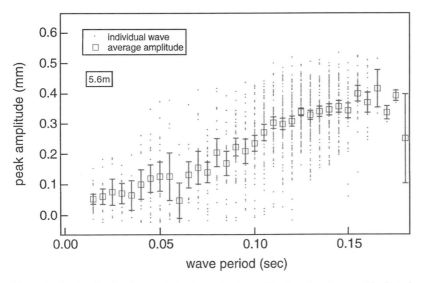

Figure 10.8. Amplitude–time period relation for $R_L = 78$, $R_G = 3640$, $\mu_L = 10$ cP, and $h = 0.7$ cm.

Sangalli et al. (1997), who explain that the depth change alters the relative speed of the fundamental compared with the first overtone; a deeper layer gives a more stable fundamental–overtone state. Figure 10.8 shows the amplitude–time period for a deeper liquid. It is seen that there is a very definite relation between the wave amplitude and the wave period. Presumably the noise present at the inception of the wave growth determines, to a large extent, the details of the wave field. There is some slow wave evolution, and eventually the influence of noise will be lost. However, for conditions of steady waves, it is not possible to predict this *a priori*. Sangalli et al. (1992) give a calculation procedure for quantifying the rate of evolution. They represent

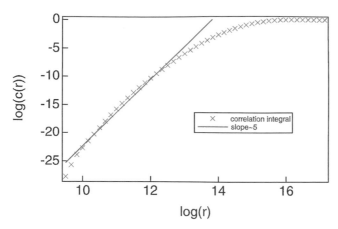

Figure 10.9. The correlation dimension of the data of Fig. 10.5. The embedding dimension is 8 (many different ones were tried). This calculation does not converge.

wave tracings at two locations as vectors and calculate how much the inner product changes between the displaced signals (vectors) to determine how fast the wave field is evolving. They found that the rate of evolution increased as the liquid depth decreased, consistent with the arguments presented here.

It is worthwhile to address one last point of similarity for both the steady wave and oscillatory wave conditions, which is shared with all convectively unstable flows. Even complete knowledge of the wave and velocity field at a single location at the present and all previous times is not enough to predict the future wave field. Thus a calculation of the dimension of the interface time series will not converge to a small value. Such calculations have been proposed for characterizing turbulence or predicting flow regimes but these are not correct. Either the lasting effect of noise, or dynamics, that in earlier times was caused by noise makes the future state (at a fixed location) independent from the current state for distances greater than approximately one wavelength. Figure 10.9 shows the correlation dimension (Grassberger and Procaccia, 1983) plot for the data of Fig. 10.5. The embedding dimension shown was 8 but this had been varied over the range of 4 to 16. Even the local dynamics that prevents a relation between amplitude and time period does not lead to a low dimension attractor. However, it is possible that a low dimensional attractor could exist in a reference frame moving with the wave field.

## 10.4. Evolution of Large-Wavelength Waves

Sufficiently far above criticality, the waves do not remain of small amplitude with wavelengths in the range of the peak growth rate. Waves with wavelengths that are long compared with the channel height will grow with distance. Figure 10.10 shows wave spectra as functions of distance for an air–water flow and the corresponding linear growth curve. Since the gas flow is turbulent, a turbulence model (Uphold, 1997) is used and so there is some uncertainty about its applicability to a gas–liquid flow. Further, once some waves reach finite amplitude, the stability calculation should be altered because the base state will be changing. The data show that a low-frequency peak is growing, and there is no indication if it will ever saturate before the waves begin to break. The linear stability prediction for the growth rate shows that waves up to more

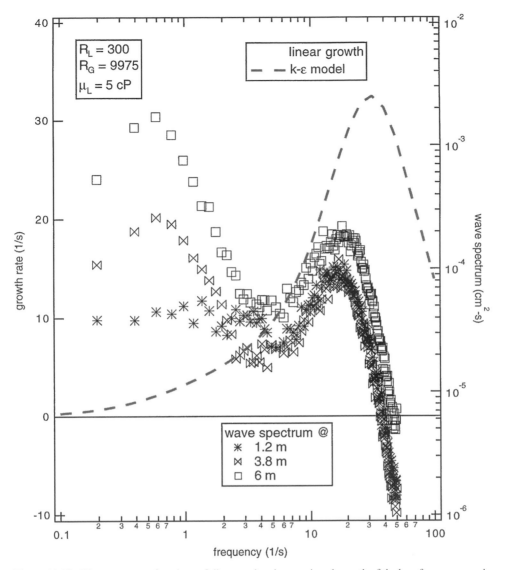

Figure 10.10. Wave spectra as functions of distance showing continued growth of the low-frequency peak. The linear stability prediction uses a $k$–$\varepsilon$ turbulence model that is described in a dissertation by Uphold.

than 100 Hz are unstable, but other than the short-wave peak, there is no suggestion of how the long-wave peak is selected. We might presume nonlinear effects, and McCready and Chang (1996) have given one mechanism. They suggest that a weak nonlinear interaction between the short-wave peak and a long wave that travels at the same speed could be responsible for selecting the low-frequency peak. However, this is by no means confirmed. Another possibility is that flow and pressure oscillations in the gas flow caused by the finite-amplitude short waves, which is really an alteration in the base state, act to select the low-frequency peak.

It is almost certain that any low-frequency peak will not grow large unless the low frequency region is linearly unstable. Figures 10.11 and 10.12 show wave spectra for air–water at fixed fetch and $R_L = 2180$ and increasing $R_G$. It should be emphasized that the since each of the data sets of Figs. 10.11 and 10.12 is shown at only one location, we are viewing each of

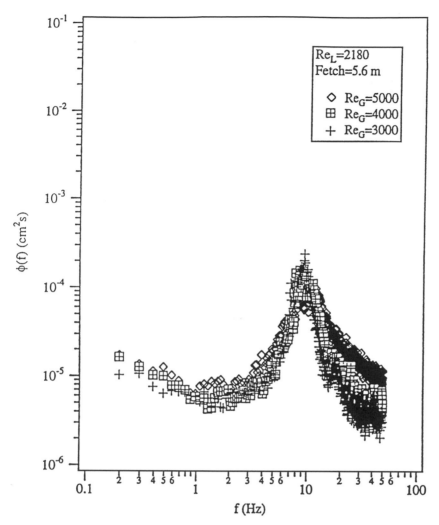

Figure 10.11. Wave spectra at conditions below $Re_G$ where long waves become unstable.

the conditions at different relative stages of evolution. However, it is possible to make some general observations. It is seen that for $R_G < 6000$, there is no definable low-frequency peak. At $R_G = 6000$, a little low-frequency peak occurs; for $R_G > 6000$ the low-frequency region increases significantly with increasing $R_G$. The prediction for when long waves should be unstable is shown in Fig. 10.13. For a liquid Reynolds number of 2180, long waves are stable until $R_G > 5000$. This is entirely consistent with the data of Figs. 10.11 and 10.12. It is also interesting to note that the popular one-dimensional model (Barnea, 1987, 1991; Brauner and Maron, 1991) does not give reliable stability results, at least in this region.

## 10.5. Removing the Effect of Fetch on Wave Measurements

Since there is seldom unlimited flow distance in experimental situations and the data above show a clear effect of flow distance on waves of very different length scales, a solution is needed. The obvious one is to do experiments in a circular, rotating geometry. Jahne et al.

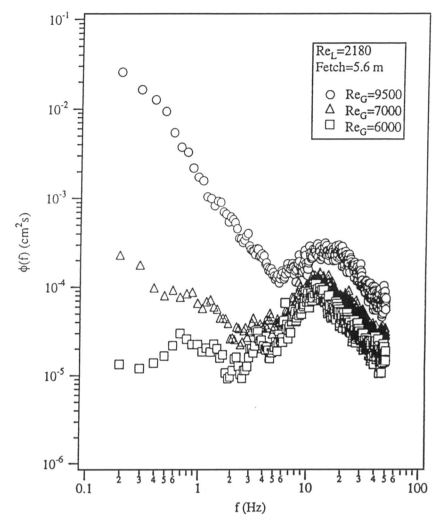

Figure 10.12. Wave spectra at conditions where long waves become unstable. Significant growth in the low-frequency region is seen.

(1979) report the use of a circular wind-wave tunnel for interfacial transport measurements. However, there is a pronounced effect of centrifugal acceleration on both the wave field and the liquid flow field if the interface is perpendicular to gravity. Thus this orientation is of limited usefulness for the measurement of wave properties.

Sangalli et al. (1995) and Gallagher et al. (1996) report on a novel experiment that removes the effect of flow distance and does not create large curvature effects. By use of matched density (immiscible) liquids in a cylindrical Couette flow, in which the outside cylinder is rotated, an interface oriented parallel to gravity is constructed. With a nominal diameter of 20 cm, curvature effects are minimal and the matched liquid density and rotation of the outside cylinder prevents any significant centrifugal effects. Sangalli et al. (1995) show the first confirming evidence of a supercritical interfacial instability and also verify the weakly nonlinear theory of Blennerhassett (1980) and Renardy and Renardy (1993) for the Landau constant for a two-layer flow. The experiments of Sangalli et al. were done under conditions for which only a

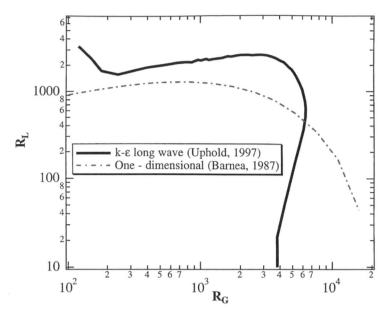

Figure 10.13. Stability boundary air–water flow in $R_L$–$R_G$ space.

range of short waves was unstable so the effect of finite length was not expected to be too important. However, Gallagher et al. (1996) conducted their experiments for a depth ratio that was always unstable to long waves. From all previous studies, which were done in channels and in light of the above-mentioned uncertainty in the linear stability calculations, a sensible conclusion was that large wavelength and amplitude disturbances would always form at large distances if long wavelength waves were unstable. This is the premise that underlies the use of linear stability theory for prediction of flow regime transitions (Lin and Hanratty, 1986; Barnea, 1987, 1991). Surprisingly, Gallagher et al. (1996) found a range of rotation rates within which either no visible waves formed or for which only short-wavelength, small-amplitude waves were found – even when long waves were unstable. Thus the following premise should be reviewed – if long waves are unstable in channel flows, the result must be large-amplitude, long-wavelength waves.

## 10.6. Closing Comments

A question that arises in response to the first part of this chapter is why is such strong emphasis is placed on inlet noise as compared with the effect of noise away from the inlet. Certainly, there is a continuous level of noise from the turbulent gas flow. However, noise will have the biggest effect at the channel inlet where the wave amplitude is very small. As the waves grow larger, they are less susceptible to small fluctuations. The existence of a family of steady waves in Fig. 10.8 shows that turbulent fluctuations may not be disrupting finite-amplitude waves.

Most multifluid flows are unstable to spatially growing disturbances, particularly if the flow rates are at values of reasonable process conditions. Thus it is extremely important to determine if the flow is unstable to disturbances that could cause regime transitions or strong disruptions of the flow. There are many examples of incorrect knowledge of the flow regime that have led to process failure. While for a given flow rates slugs may not have yet formed in a 15-m laboratory pipe, they could form in a process pipeline that is many miles long.

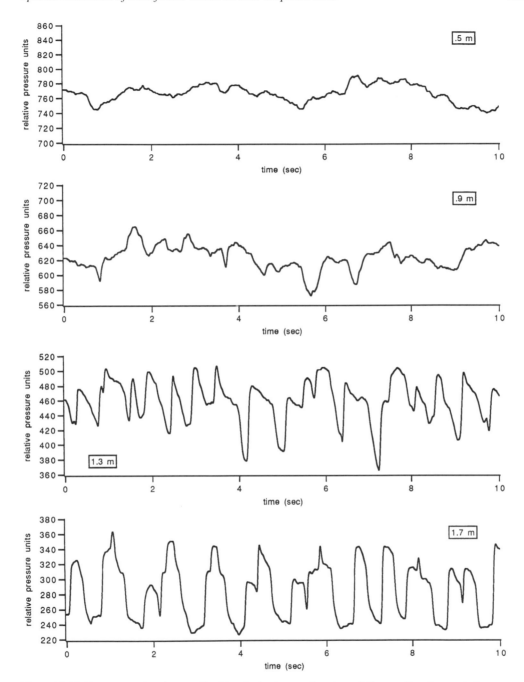

Figure 10.14. Pressure tracings in a gas–liquid packed column as functions of distance. Development of strong disturbances (pulses) with distance is seen.

That large disturbances grow with distance is not always a difficulty and can be taken advantage of in certain situations. Figure 10.14 shows local measurements of pressure in a gas–liquid, cocurrent downflow, packed bed column (trickle bed). (More details about the experiment are available in the works of Helwick et al., 1992 and Krieg et al., 1995). In these measurements the strength and coherence of the disturbances are increasing with distance. For these conditions, the top of the column is trickling and the bottom is "pulsing." Pulses in packed columns are strong coherent disturbances that are analogous to slugs in pipe flow. Wu et al. (1995, 1998) exploited this effect to conduct reaction experiments for a sequential reaction under trickling and pulsing, but at identical flow rates. This was possible because the column was packed primarily with inert particles but with active catalyst particles confined to regions either near the top (trickling) or bottom (pulsing). They found substantial changes in the outcome of the reaction that can be attributed to the change in the local fluctuations in the mass transfer rate for the different regimes.

## Acknowledgments

This work has been supported by the U.S. Department of Energy under grant DE-FG02-96ER14637 and NASA, Microgravity Science Division, under grant NCC-466.

## References

Barnea, D. 1987. A unified model for predicting flow-pattern transitions for the whole range of pipe inclinations. *Int. J. Multiphase Flow* **13**, 1–12.

Barnea, D. 1991. On the effect of viscosity on stability of stratified gas–liquid flow – application to flow pattern transition at various pipe inclinations. *Chem. Eng. Sci.* **46**, 2123–2131.

Blennerhassett, P. J. 1980. On the generation of waves by wind. *Trans. R. Soc. London* **298**, 451–494.

Brauner, N. and Maron, D. M. 1991. Analysis of stratified/non stratified transitional boundaries in horizontal gas–liquid flows. *Chem. Eng. Sci.* **46**, 1849–1859.

Bruno, K. and McCready, M. J. 1988. Origin of roll waves in horizontal gas–liquid flows. *AIChE J.* **34**, 1431–1440.

Crowley, C. J., Wallis, G. B., and Barry, J. J. 1992. Validation of a one-dimensional wave model for the stratified to slug flow regime transition, with consequences for wave growth and slug frequency. *Int. J. Multiphase Flow* **18**, 249–271.

Fan, Z., Lusseyran, F., and Hanratty, T. J. 1993. Initiation of slugs in horizontal gas–liquid flows. *AIChE J.* **39**, 1741–1753.

Gallagher, C. T., McCready, M. J., and Leighton, D. T. 1996. Experimental investigation of a two-layer shearing instability in a cylindrical Couette cell. *Phys. Fluids* **8**, 2385–2392.

Grassberger, P. and Procaccia, I. 1983. Characterizing strange attractors. *Phys. Rev. Lett.* **50**, 346–349.

Helwick, J. A., Dillon, P. O., and McCready, M. J. 1992. Time varying behavior of cocurrent gas–liquid flows in packed beds. *Chem. Eng. Sci.* **47**, 3249–3256.

Jahne, B., Munnich, K. O., and Siegenthaler. 1979. Measurements of gas exchange and momentum transfer in a circular wind-water tunnel. *Tellus* **31**, 321–329.

Jurman, L. A. 1990. Interfacial waves on sheared, thin liquid films. Ph.D. dissertation, Department of Chemical Engineering, University of Notre Dame.

Kordyban, E. S. and Ranov, T. 1970. Mechanism of slug formation in horizontal two-phase flow. *J. Basic Eng.* **92**, 857–864.

Krieg, D. A., Helwick, J. A., Dillon, P. O., and McCready, M. J. 1995. Origin of disturbances in cocurrent gas–liquid packed bed flows. *AIChE J.* **41**, 1653–1666.

Kuru, W. C., Sangalli, M., and McCready, M. J. 1996. Transition to three-dimensional waves in concurrent gas–liquid flows, in Advances in Multi-Fluid Flows, 52–69, SIAM, Philadelphia, PA.

Lin, P. Y. and Hanratty, T. J. 1986. Prediction of the initiation of slugs with linear stability theory. *Int. J. Multiphase Flow* **12**, 79–98.

Mandhane, J. M., Gregory, G. A., and Aziz, K. 1974. A flow pattern map for gas–liquid flow in horizontal pipes. *Int. J. Multiphase Flow* **1**, 537–553.

McCready, M. J. and Chang, H.-C. 1996. Formation of large disturbances on sheared and falling liquid films. *Chem. Eng. Commun.* **141-142**, 347–358.

Renardy, Y. and Renardy, M. 1993. Derivation of amplitude equations and analysis of sideband instabilities in two-layer flows. *Phys. Fluids A* **5**, 2738–2762.

Sangalli, M., McCready, M. J., and Chang, H.-C. 1997. Stabilization mechanisms of short waves in gas–liquid flow, *Phys. Fluids*, **9**, 919–939.

Sangalli, M., Prokopiou, Th., McCready, M. J., and Chang, H.-C. 1992. Observed transitions in two-phase stratified gas–liquid flow. *Chem. Eng. Sci.* **47**, 3289–3296.

Sangalli, M., Gallagher, C. T., Leighton, D. T., Chang, H.-C., and McCready, M. J. 1995. Finite amplitude wave evolution at the interface between two viscous fluids. *Phys. Rev. Lett.* **75**, 77–80.

Taitel, Y. and Dukler, A. E. 1976. A model for predicting flow regime transitions in horizontal and near horizontal gas–liquid flow. *AIChE J.* **22**, 47–55.

Uphold, D. D. 1997. Wave behavior in two-layer channel flows. Ph.D. dissertation, Department of Chemical Engineering, University of Notre Dame.

Wu, R., McCready, M. J., and Varma, A. 1995. Influence of mass transfer coefficient fluctuation frequency on performance of three-phase packed bed reactors. Chem. Eng. Sci. **50**, 3333–3344.

Wu, R., McCready, M. J., and Varma, A. 1998. Enhancing performance of three-phase packed-bed reactors: experiments and model. *AIChE J.* (in review).

# 11

# The Shear Breakup of an Immiscible Fluid Interface

GRÉTAR TRYGGVASON AND SALIH OZEN UNVERDI

## 11.1. Introduction

In this chapter the two-dimensional Kelvin–Helmholtz instability of a sheared fluid interface separating immiscible fluids of different densities and viscosities is studied by numerical simulations of the Navier–Stokes equations. The evolution is determined by the density ratio of the fluids, the Reynolds number in each fluid, and the Weber number. Unlike the Kelvin–Helmholtz instability for miscible fluids, in which the sheared interface evolves into well-defined concentrate vortices if the Reynolds number is high enough, the presence of surface tension leads to the generation of fingers of interpenetrating fluids. In the limit of a small density ratio the evolution is symmetric, but for large density stratification the large-amplitude stage consists of narrow fingers of the denser fluid penetrating into the less dense one. The dependency on the density difference is explained in terms of the advection of interfacial vorticity by the density-weighted mean velocity. Even though the simulations are confined to only the two-dimensional aspect of the problem, it is found that the fingers can break up into isolated drops. While the initial growth rate is well predicted by inviscid theory, once the Reynolds numbers are sufficiently high, the large-amplitude behavior is strongly affected by viscosity and the mode that eventually leads to fingers is longer than the most unstable one.

The Kelvin–Helmholtz instability of an initially flat interface separating fluids moving in the opposite direction is a classical problem in fluid mechanics. In its simplest form, the fluids are assumed to be inviscid and the flow on either side of the interface to be irrotational. Linear stability analysis dates back to the past century, and computations of the nonlinear formation of a concentrated vortex are among the earliest examples of computational fluid dynamics studies (Rosenhead, 1931). As computers became widely available, studies progressed along two paths: Several investigators examined the evolution of an infinitely thin vortex sheet, separating two potential flow regions. This turned out to be a difficult problem because of the ill posedness of the vortex sheet (and the rapid growth of small perturbations) as well as the formation of a singularity at the point where the vortex should appear. The difficulty was eventually resolved by Krasny (1986), who showed that a regularization of the vortex sheet resulted in both a well-posed problem and eliminated the singularity formation. Other investigators examined the nonlinear evolution by solving the full Navier–Stokes equations (see, for example, Patnaik et al., 1976; Corcos and Sherman, 1984) and showed that a perturbed shear layer develops into a row of vortices. Tryggvason et al. (1991) examined the limit of high Reynolds numbers and a small initial thickness and compared full Navier–Stokes simulations with the regularized inviscid model of Krasny (1986). They showed that while it was not possible to equate the regularization directly to a physical effect, the limit of small initial thickness and high Reynolds numbers appeared to approach the same limit as the inviscid model when the regularization was reduced.

The Kelvin–Helmholtz instability of two identical fluids is now a fairly well-studied problem (for early three-dimensional simulations see, e.g., Metcalfe et al., 1987). The corresponding problem for a finite density stratification is much less studied, particularly when the fluids are immiscible and it is necessary to include finite surface tension. The importance of the breakup of an interface separating immiscible fluids of different material properties is considerable. Liquid fuels are usually burned by first atomizing a fuel jet to increase the surface area and hence the evaporation rate. In engineering modeling of spray combustion, the initial atomization is by far the least-understood aspect. Indeed, there are not completely unfounded reasons to believe that if the initial atomization was understood, current models would lead to a reasonably accurate prediction of the rest of the spray (CRAY Research, 1995). The importance of the initial breakup is demonstrated by the large number of atomizers that have been proposed. The books by Lefebvre (1989) and Bayvel and Orzechowski (1993), for example, discuss many practical aspect of atomization and the design of devices to atomize liquids.

Fluid jets break up in different ways, depending on the governing nondimensional parameters. When capillary effects are large, a jet undergoes a Rayleigh instability due to waves that are longer than its diameter and breaks up into a stream of relatively large drops. When capillary effects are small the jet is unstable to shorter waves that are generally enhanced by aerodynamic effects, resulting in smaller drops. At very large Weber numbers the jet breaks up into ligaments (or fibers) that then break up into drops much smaller than the jet diameter. Under atmospheric conditions, in which the external flow has small effects, the jet breakup is believed to be strongly affected by the turbulence level in the jet. However, for high-pressure combustors, the density ratio is often much smaller and the injected jet laminar. Experimentally it is found that the effect of the ambient fluid is small if the density ratio of the fuel to gas is large. For experimental investigations of the different breakup modes see, for example, Farago and Chigier, 1992, and Wu et al., 1994.

For jets, several authors have examined the Rayleigh breakup of an axisymmetric jet, both by inviscid boundary integral methods as well as by methods that include viscous effects. See, for example, the article by Spangler et al. (1995), who use a boundary integral method to simulate the axisymmetric breakup of a periodic jet. These authors also simulated, in addition to the Rayleigh breakup, the breakup of a jet in the so-called first and second wind-induced breakup regimes, in which aerodynamic effects are important. Their results show that short-wave disturbances grow into a ring of fluid encircling the main jet. The breakup of this ring would lead to the drops that are observed to be ejected from a jet in the second wind-induced breakup regime. The same authors have also recently simulated the axisymmetric, low Weber number breakup of a jet emerging from a nozzle (Hilbing and Heister, 1995), and a fully viscous simulation of a similar problem has been presented by Richards et al. (1994). While no numerical simulations have been published of the breakup of a jet when capillary effects are small, the breakup of a flat sheared interface has been examined by a number of authors. The effect of surface tension for fluids of the same density was examined by Hou et al. (1997), who assumed that viscous effects could be ignored completely. They found that large surface tension suppressed rollup and that the linear instability instead evolved into long fingers of interpenetrating fluids. When surface tension was lowered, the fingers folded over and at low enough surface tension the interface rolled up, as in the classical nonlinear Kelvin–Helmholtz instability. Viscosity as well as surface tension has been included by Zaleski et al. (1996), who also found the development of long fingers. They also examined the effect of three dimensionality and found that the two-dimensional folds that develop initially can evolve into a fully three-dimensional finger of one fluid pointing into the other fluid. In the present study, we examine the two-dimensional aspect of interface breakup in more detail than Zaleski et al. did.

## 11.2. Problem Formulation and Numerical Method

We examine a horizontal interface separating two different fluids. The fluid above the interface is moving to the left with velocity $U_1$ and the fluid on the bottom is moving to the right with velocity $U_2$. The density of the top fluid is $\rho_1$ and the bottom fluid has density $\rho_2$. The computational domain is a rectangle with periodic horizontal boundaries and rigid, moving walls at the top and the bottom. The interface is nearly flat, initially, and its evolution is determined by the velocity difference across the interface, the surface tension, and the density and the viscosity of either fluid. When the viscosities are low enough, we expect the growth rate to be well predicted by linear stability analysis for inviscid flows (see Drazin and Reid, 1981, for example). For perturbations of the form

$$A = A_0 e^{st+ikx},$$

it is found that

$$s = -ik\frac{\rho_1 U_1 + \rho_2 U_2}{\rho_1 + \rho_2} \pm \sqrt{\frac{k^2 \rho_1 \rho_2 (U_1 - U_2)^2}{(\rho_1 + \rho_2)^2} - \frac{Tk^3}{\rho_1 + \rho_2}}. \tag{1}$$

Here, $T$ is the surface-tension and $k$ is the wave number. The first part of this expression is the phase velocity, $c = -\mathrm{Im}(s)/k$, and the second part is the growth rate $\sigma = \mathrm{Re}(s)$. In nondimensional form we have

$$\tilde{\sigma} = \frac{\sigma T}{\rho_2 \Delta U^3} = \frac{1}{\mathrm{We}} \sqrt{\frac{r}{1+r}\left(\frac{1}{1+r} - \frac{1}{\mathrm{We}}\right)}, \tag{2}$$

$$\tilde{c} = \frac{c}{\Delta U} = \frac{1}{\Delta U}\frac{U_1 + rU_2}{(1+r)}, \tag{3}$$

where we have constructed a time and a velocity scale by

$$\frac{T}{\rho_2 \Delta U^3}, \quad \Delta U.$$

The Weber number is defined as

$$\mathrm{We} = \frac{\rho_2 \Delta U^2}{Tk},$$

and we have introduced $\Delta U = U_2 - U_1$ and $r = \rho_1/\rho_2$. The expression for the phase velocity simply shows that the initial perturbation is advected by the density-weighted average velocity, but the growth rate is positive if the expression under the square root is positive:

$$\mathrm{We} > 1 + r. \tag{4}$$

The growth rate is maximum for

$$\mathrm{We} = \frac{3}{2}(1+r). \tag{5}$$

Although the computations presented in this chapter are done by solving the full Navier–Stokes equations, it is useful to remind the reader that the evolution of the vortex sheet strength, $\gamma =$

$(\mathbf{u}_2 - \mathbf{u}_1) \cdot \mathbf{s}$, of an interface separating two regions of irrotational flow of inviscid fluid is given by

$$\frac{D\gamma}{Dt} + \gamma \frac{\partial \mathbf{U} \cdot \mathbf{s}}{\partial s} = \frac{\rho_2 - \rho_1}{\rho_2 + \rho_1} \left( \frac{D\mathbf{U}}{Dt} \cdot \mathbf{s} + \gamma \frac{\partial \gamma}{\partial s} \right) - \frac{2T}{\rho_2 + \rho_1} \frac{\partial \kappa}{\partial s}. \tag{6}$$

The first term on the left-hand side of Eq. (6) is the rate of change of the vortex sheet strength following a point that moves with the average of the velocities on either side of the sheet, $\mathbf{U}$; the second term on the left represents change in vortex sheet strength due to stretching of the interface. If we integrate the left-hand side over a small segment of the interface, bounded by Lagrangian points, we find that it represents the change of circulation of this segment. The first term on the right-hand side represents the baroclinic generation of vorticity due to local acceleration, and the second term in the parenthesis represents transport of vorticity by the density-weighted average velocity. If the convective derivative on the left-hand side is written in terms of an density-weighted average velocity, this terms disappears. The last term represents generation of vorticity by surface tension. Here, $\kappa$ is the curvature. If we add gravity effects [which have been left out in Eq. (6)], we obtain an additional term on the right-hand side that shows that baroclinic vorticity production is proportional to the slope of the interface. We note that it is the second term on the right that is responsible for any asymmetry in the evolution of an initially symmetric perturbation for finite density differences. For the Rayleigh–Taylor instability, in which a heavy fluid falls into a lighter fluid, this asymmetry manifests itself in the penetration of big bubbles of lighter fluid upward while the heavy fluid falls down in narrow spikes (see, for example, Tryggvason, 1988). For small density differences the evolution is, on the other hand, symmetric. While the effect of density differences is perhaps less well known for the Kelvin–Helmholtz instability, a finite density difference will result in a similar asymmetry. The presence of surface tension also leads to changes in the vortex sheet strength. This effect is present even for zero stratification, and if we integrate Eq. (6) over a small Lagrangian segment of the interface (putting $\rho_2 = \rho_1$), we find that the rate of change of the circulation, $\Gamma$, is

$$\frac{D\Gamma}{Dt} = -\frac{2T}{\rho_2 + \rho_1} (\kappa_2 - \kappa_1), \tag{7}$$

where the subscripts 1 and 2 refer to the end points of the small segment.

The numerical method used for the computations presented here is based on writing one set of equations for the entire computational domain, independently of how many different fluids are involved. This is possible by allowing for different material properties in the formulation and adding singular terms at the phase boundaries to ensure that the correct boundary conditions are satisfied. The resulting one-field Navier–Stokes equations are

$$\frac{\partial \rho \bar{u}}{\partial t} + \nabla \cdot \rho \bar{u} \bar{u} = -\nabla p + \nabla \cdot \mu (\nabla \bar{u} + \nabla \bar{u}^T) + \int_F \bar{F}_\sigma \delta(\bar{x} - \bar{x}_f) \, da. \tag{8}$$

Here, $\bar{u}$ is the velocity vector, $p$ is the pressure, and $\rho$ and $\mu$ are the discontinuous density and viscosity fields, respectively. The surface forces $\bar{F}_\sigma$ act on only the interface between the different fluids and appears in the current formulation multiplied by a two- or a three-dimensional delta function $\delta$. The integral is over the entire front. It is important to note that this equation contains no approximations beyond those in the usual Navier–Stokes equations. In particular, it contains implicitly the proper stress conditions for the fluid interface. Since the density and the viscosity are different for the different fluids, it is necessary to track the evolution of these fields by solving the equations of state that simply specify that each fluid particle retains its original density and viscosity:

$$\frac{D\rho}{Dt} = 0, \qquad \frac{D\mu}{Dt} = 0. \tag{9}$$

The momentum equations are also supplemented by an equation of mass conservation, which for incompressible flows is simply

$$\nabla \cdot \bar{u} = 0. \tag{10}$$

Combining this equation with the momentum equation leads to an elliptic equation for the pressure.

The momentum equation is discretized on a regular staggered grid by second-order centered finite differences for the spatial derivatives and a second-order time integration scheme. The continuity equation, when combined with the momentum equation, results in a pressure equation that is not separable as for homogeneous flow and is solved by a multigrid package (MUDPACK from NCAR). To advect the material properties and to evaluate the surface-tension term in the momentum equation, we track the interface between the different phases explicitly by using a moving grid of lower dimension than what we use for the conservation equations. This grid is usually referred to as a front. The one-field formulation used here is common to other techniques for multifluid flows such as the volume-of-fluid and level-set methods. In these methods, however, the phase boundary is not tracked explicitly, but reconstructed from a marker function. Explicitly tracking the interface avoids the difficulty of advecting such a marker function and allows accurate evaluation of surface forces.

Computing surface tension accurately is one of the most difficult part of methods intended for simulations of multifluid flows. Our current algorithm, which appears to be very satisfactory, is based on computing directly the force on each surface element by

$$\bar{F}_\sigma = \int_{\Delta S} T \frac{\partial \bar{t}}{\partial s} \, \mathrm{d}s = T(\bar{t}_2 - \bar{t}_1) \tag{11}$$

in a semi-implicit way. Here $\bar{t}$ is a tangent to the boundary of the surface element and $T$ is the surface-tension coefficient. A comparable expression is available for the three-dimensional case. By computing the surface forces in this way, we explicitly enforce that the integral over any portion of the surface gives the right value, and for closed surfaces, in particular, we enforce that the integral of the surface tension over the whole surface is zero. This is particularly important for long time simulations, in which a failure to enforce this constraint can lead to unphysical motion.

The method has been implemented for two- and three-dimensional flows on regular grids and for axisymmetric geometries in which stretched grids are used to allow local grid refinement. It has been applied to a number of multifluid problems and tested and validated in a number of ways, not only to check the implementation, but also its accuracy. Those tests include comparisons with analytical solutions for simple problems, other numerical computations, and experiments. The actual resolution requirement varies with the parameters of the problem. High Reynolds numbers, for example, generally require finer resolution than lower ones, as in other numerical calculations. We have also found that for problems for which the surface tension varies, such as for contaminated bubbles and drops moving by thermal migration we generally require finer resolution than for flows for which the surface tension is constant. However, in all cases we have found that the methods converge rapidly under grid refinement, and in those cases in which we have other solutions we have found excellent agreement, even for modest resolutions. The method has also been extended in various ways to allow for simulations of heat transfer, surfactant effects, thermal migration, and phase changes. For a more detailed description of the method, see Unverdi and Tryggvason (1992) and for validations by comparison with analytical results see Nobari et al. (1996) and Esmaeeli and Tryggvason (1996), for example. For comparison with experiments see Qian et al. (1997), for example.

## 11.3. Results

The breakup of an initially flat interface is studied here. For high-pressure sprays the density difference between the liquid fuel and the ambient gas is usually much less than at normal pressure. Since small density differences also make the numerical computations easier, we examine relatively modest density ratios here. If the density difference is very small, then the effect of stratification will be small and results for zero stratification will apply. Here, we compare the evolution of an immiscible interface with zero stratification with the case of a density ratio of 10.

### 11.3.1. *Linear Instability*

Since we assume that the velocity field is initially discontinuous and since the Reynolds numbers are high, we expect the initial growth to be well predicted by inviscid analysis, Eq. (1). To examine if that is the case, several simulations were done with the eigenmode from the stability analysis as the initial condition. The nondimensional vortex sheet strength, as computed from linear stability analysis, is given by

$$\tilde{\gamma} = -\frac{1}{\text{We}} + \frac{r-1}{r+1}\frac{\tilde{\zeta}}{\text{We}}\cos(\tilde{x}) \pm 2\tilde{\zeta}\tilde{\sigma}\,\sin(\tilde{x}), \tag{12}$$

where $\tilde{\zeta} = k\zeta$ is the nondimensional initial amplitude, taken to be 0.05 times the wavelength, $\tilde{\sigma}$ is the linear growth rate, given by Eq. (2), and $\tilde{x} = kx$ is the nondimensional horizontal space coordinate. In Fig. 11.1 we show the amplitude versus time for Weber numbers equal to 3, no density stratification, and three different grid resolutions. The amplitude has been divided by $e^{\sigma t}$. If the growth rate remained constant, this curve would be parallel to the time axis. As the amplitude of the wave increases, nonlinear effects reduce the growth, so the line slopes

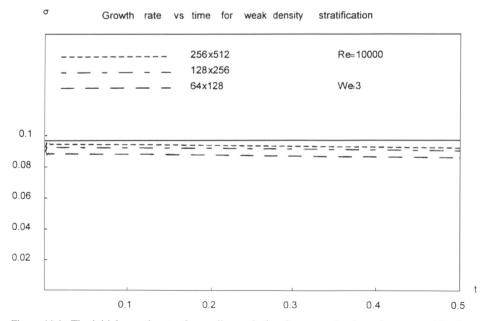

Figure 11.1. The initial growth rate of a small perturbation for a zero density ratio computed by three different grid resolutions. The solid line is the inviscid theoretical predictions. The amplitude has been divided by $e^{\sigma t}$.

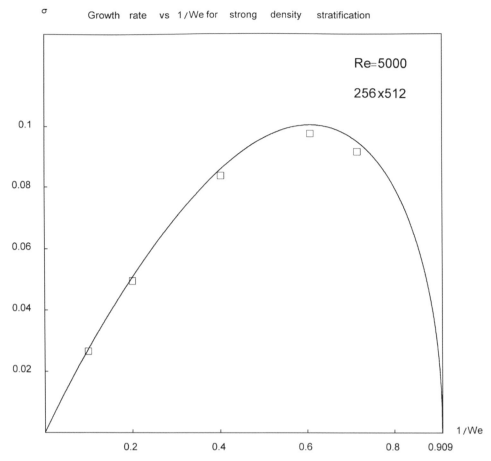

Figure 11.2. The linear growth rate for a density ratio of 10 for different nondimensional wavelengths. The solid curve is the prediction of linear inviscid theory.

slightly downward. The solid line is the theoretical prediction, and it is clear that the numerical computations are converging to that value. The Reynolds number, based on the velocity jump and the wavelength, is 10,000. We have repeated these calculations for several different Reynolds numbers and find almost no variations for values ranging from 600 to 10,000, suggesting that viscous effects are indeed small at the earliest time. In Fig. 11.2, the computed nondimensional growth rate (open squares) is plotted versus the inverse Weber number for Re = 5000 and a density ratio of 10, along with the growth rate predicted by linear stability theory (continuous curve). The growth rate is well predicted, although the finite resolution used leads to a slight underprediction. Similar results are obtained for the zero stratification case.

### 11.3.2. *Nonlinear Evolution for Zero Density Differences*

The evolution of an interface between two fluids with equal densities and viscosities is shown in Fig. 11.3 for two Weber numbers and Re = 10,000. In the column on the left, We = 3, which corresponds to a wavelength equal to the most unstable one. In the column on the right, We = 6.0, and the wavelength is twice the most unstable one. We show three periods, although only one was actually computed. The initial conditions are the same as those in Subsection 11.3.1. In both cases, the initial disturbance grows rapidly and the wave becomes steeper on the

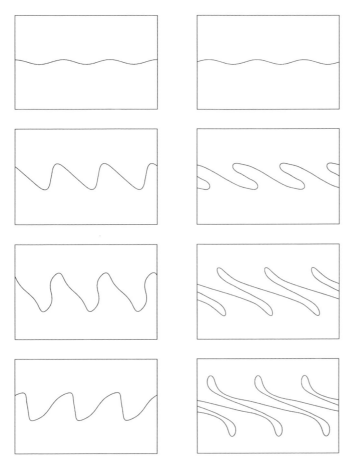

Figure 11.3. The evolution of the material interface for We = 3 (left column) and We = 6 (right column) and zero density stratification. The interface is shown at four times.

downwind side. The most unstable wave reaches a finite amplitude and is then pulled back into a symmetric shape that then decays, but in the We = 6 case, the wave folds over, resulting in a tongue or a finger of one fluid penetrating into the other fluid. As the finger grows longer, its tips bend outward. Hou et al. (1997) observed similar fingers, but in their inviscid simulations the length of the finger increased continuously. Here, viscous effects eventually lead to an increased thickness of the shear layer and therefore a reduction in the local shear strength. Once the shear is low enough, the interface is pulled back by surface tension. A somewhat similar evolution is seen for very viscous Kelvin–Helmholtz instabilities between immiscible fluids, in which the shear layer is initially thin enough so that the perturbations grow, but viscous diffusion is rapid enough to eliminate the vorticity concentration in the eye of the vortex. Here, it is surface tension that prevents rollup and the effect of viscosity is felt only late in the evolution as is evident by both comparing with the results of Hou et al. (1997) and by the fact that we see rollup for these Reynolds numbers when the surface tension is set to zero.

The effect of viscosity is not, however, simply to increase the thickness of the shear layer by viscous diffusion. In Fig. 11.4 the vorticity (for two periods) at four times for the We = 3 case is shown. Initially, the vorticity at the interface is advected toward the middle of the steep part of the interface in the same way as the vorticity forming a vortex sheet with zero surface

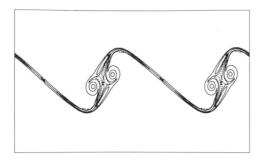

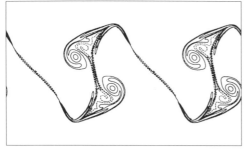

Figure 11.4. The vorticity in the fluid at four times for the We = 3 case in Fig. 11.3.

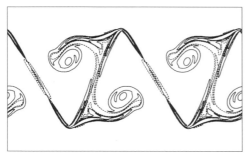

tension. Here, however, the vorticity peak divides in two as the interface becomes steeper, and in the first frame in Fig. 11.4, the vorticity has separated from the crest of the wave. The appearance of two peaks in the vorticity was already predicted by Hou et al. (1997), who used inviscid simulations. As vorticity is shed from the interface, the local pressure minimum above the crest is reduced, and as the driving mechanism for the growth of the instability is removed, surface tension pulls the interface back. For higher Weber numbers, we see a similar shedding of vorticity at early times, but the vorticity is weaker and is left behind when the fingers grow. As surface tension eventually pulls the fingers back, vorticity is shed from their tips.

We have done several additional computations for various different Weber numbers and find that for low Weber numbers and early times, the length of the finger increases with increasing Weber number, but the slope at which it penetrates the other fluid decreases, so the highest Weber number fingers are nearly parallel to the interface. Since a large number of modes are unstable for high Weber numbers, the interface is generally wavy because of short capillary waves that are exited during the rollup. As the Weber number is increased further (and the relative effect of surface tension decreased) the initial wave rolls up once before starting to be drawn out into a finger parallel to the interface. This transition takes place between We $= 10$ and We $= 15$ for the initial conditions used here. When the interface folds over once, the interface appears to pinch off near the fold. Here, the interface is not allowed to change its topology, so the interface stays connected. At even higher Weber numbers, the interface rolls up, as we would expect for miscible fluids. We have not examined the transition from the folding over to rollup, but refer the reader to Hou et al. (1997) for inviscid simulations in the higher Weber number range and an examination of the pinch off. Since separation is likely to be less important at very high Weber numbers, we expect their results to give a very good picture of the evolution of real systems in that parameter range. Our results for small density differences appear to be in reasonably good agreement with the inviscid computations of Hou et al. (1997), but we have not done a detailed comparison yet.

### 11.3.3. *Nonlinear Evolution for a Density Ratio of 10*

Figure 11.5 shows the evolution of the interface for a Reynolds numbers of 5000 (based on the heavier bottom fluid) and two different Weber numbers. In the left column the Weber number is equal to 1.7, which is close to the most unstable wavelength (We $= 1.65$). In the right column, We $= 5$. The initial conditions are the same as for the low density ratio case and in the figure, the lighter fluid is on the top. As we saw for the smaller density ratio, the initial perturbation grows and the waves become steeper. Here, however, the evolution is strongly asymmetric and while the trough is smooth, the crest, in which the heavy fluid reaches into the lighter one, is much sharper. The wave moves to the right with the heavy bottom fluid, and the left side becomes steeper. While it is very similar to a breaking water wave at this point, surface tension eventually stabilizes the wave.

In Fig. 11.6, we continue the higher Weber number simulation in Fig. 11.5 to a larger amplitude. After surface tension rounds the crest, a thin finger or a ligament is pulled off the crest by the lighter fluid. The tip of this finger eventually develops a circular drop and pinches off. While somewhat reminiscent of the formation of drops from a round ligament, the two dimensionality of this problem makes the usual explanation, that the azimuthal component of the surface tension causes the collapse of the ligament, not applicable. However, similar necking has been seen in other two-dimensional simulations of fluid interfaces, such as in Hele–Shaw cells (Tryggvason and Aref, 1983) and for vortex sheets separating inviscid, immiscible fluids (Hou et al., 1997). We offer the following conjecture to explain its formation: Consider a stationary liquid filament ending in a rounded end. Because of surface tension, the pressure in the rounded end is higher than in the rest of the filament and this causes an axial flow from the end to the straight part of the filament. Since there is no axial flow in the straight part, conservation of mass causes the filament to bulge out as it becomes shorter. Where the bulge starts, the curvature is opposite to what it is at the end of the filament and the pressure there is therefore lower than in the straight part of the filament. Thus there will not only be a flow from the bulge at the end into this low-pressure region, but from the straight part of the filament as well. This draining of fluid from the straight part makes it thinner, eventually leading to necking as the bulge moves toward the straight part.

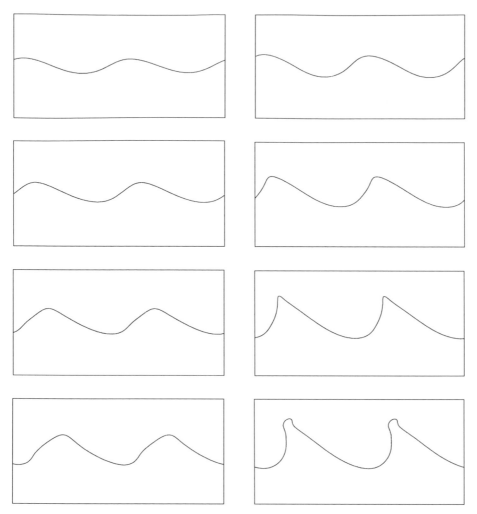

Figure 11.5. The evolution of the material interface for We = 1.7 (left column) and We = 5 (right column) and a density ratio of 10. The interface is shown at four times.

The vorticity for the We = 5 case is shown at three times in Fig. 11.7. As the face of the wave becomes steeper, vorticity is shed from the crest as for the low density ratio case. The differences with the weak density stratification case are even more pronounced in the vorticity distribution than in the interface shape. Here, the shed vorticity is nearly all in the light fluid and little vorticity is deposited into the heavy fluid. The diffusion and the shedding of vorticity reduce the shear, and eventually surface tension pulls the crest back. Since there is now relative motion between the drop and the top fluid, vorticity is deposited into the light fluid from the drop surface.

Even though the density ratio is only 10 here, the differences with the constant density case are striking and much more reminiscent of what we would expect to see for an air/water interface, for example. The reason is the advection of vorticity along the interface by the density-weighted velocity, and, in the limit of a completely free surface, the vorticity moves simply with the heavy fluid. For inviscid flows, the interface vorticity (the vortex sheet strength) never leaves the interface, but for viscous flows we generally expect separation, particularly for a nonzero surface tension.

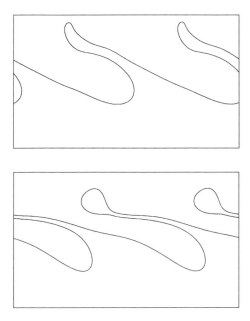

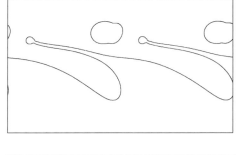

Figure 11.6. The continuation of the We = 5 case in Fig. 11.5 to larger amplitudes.

## 11.4. Conclusion

To start looking at the shear breakup of jets, we have done several two-dimensional simulations of immiscible periodic shear layers. The Reynolds numbers have been selected to be sufficiently high so that the initial instability is well predicted by inviscid theory, but viscous effects become important at larger amplitude. The linear stability analysis predicts that surface tension stabilizes short waves and yields a wavelength with the largest linear growth rate (most unstable wave). Generally we find that the inviscidly most unstable mode saturates quickly and perturbations of longer wavelength are the ones that grow to larger amplitude. Exactly which wavelength is the most dangerous one depends on the Reynolds number. We have, so

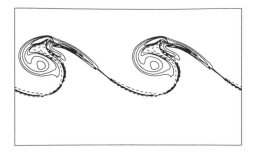

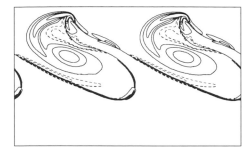

Figure 11.7. The vorticity in the fluid at three times for the We = 5 case in Fig. 11.6.

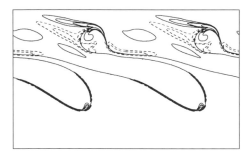

far, conducted two sets of simulations, one for zero density differences and the other at density ratios of 10. For zero stratification, surface tension prevents Kelvin–Helmholtz rollup, as seen for miscible fluids, and fingers of one fluid penetrate the other fluid. The slope of these fingers depends on the nondimensional wavelength (Weber number). While viscous effects limit the growth of the fingers at low Weber numbers, high Weber number fingers can become very long. At even higher Weber numbers the interface start to exhibit a behavior more similar to the classical nonlinear Kelvin–Helmholtz instability and rollup. The transition is complex, however, and we find intermediate states in which the interface folds over once before being stretched into a long finger, for example. At larger density ratios, the evolution is no longer symmetric and we find waves of the heavy liquid that grow into the lighter one. As for zero stratification, waves with wavelengths close to the most unstable one are stabilized at large amplitude by viscous effects but longer wavelengths lead to a wave breaking in which a finger of the heavy fluid is pulled into the lighter fluid. Even in the two-dimensional simulations presented here, these fingers eventually break down into drops. The slope of the fingers and the size of the resulting drops depend on the nondimensional wavelength (Weber number).

At larger amplitudes, we expect three-dimensional effects to take over and control the actual breakup process. We have not examined the three-dimensional aspects here, but preliminary computational studies by Zaleski et al. (1996) show that three-dimensional perturbations can alter the evolution in a significant way and lead to a breakup of the ridges or folds that run perpendicularly to the flow direction into round fingers that run parallel to the flow direction. Such fingers have been observed experimentally in the fiber breakup mode of jets.

## 11.5. Acknowledgments and Disclaimer

Effort sponsored by the U.S. Air Force Office of Scientific Research, Air Force Material Command, USAF, under grant number F49620-96-1-0356. The U.S. Government is authorized to reproduce and distribute reprints for Governmental purpose notwithstanding any copyright notation thereon. The views and conclusions contained herein are those of the authors and should not be interpreted as necessarily representing the official policies or endorsements, either expressed or implied, of the Air Force Office of Scientific Research or the U.S. Government.

## References

L. Bayvel and Z. Orzechowski, *Liquid Atomization,* Taylor & Francis, Washington D.C. (1993).

G. M. Corcos and F. S. Sherman, "The mixing layer: deterministic models of turbulent flow. Part 1. Introduction and the two-dimensional Flow," *J. Fluid Mech.* **139**, 29–65 (1984).

CRAY Research. KIVA Users Meeting, Detroit (1995).

P. G. Drazin and W. H. Reid, *Hydrodynamic Stability*, Cambridge U. Press, Cambridge (1981).

A. Esmaeeli and G. Tryggvason, "An inverse energy cascade in two-dimensional, low Reynolds number bubbly flows," *J. Fluid Mech.* **314**, 315–330 (1996).

Z. Farago and N. Chigier, "Morphological classification of disintegration of round liquid jets in a coaxial air stream," *Atomization Sprays* **2**, 137–153 (1992).

J. H. Hilbing and S. D. Heister, "Developments in nonlinear modeling of atomization processes," in *ILASS-AMERICAS 95. 8th Annual Conference on Liquid Atomization and Spray Systems*, Troy, MI, May 21–24 (1995).

T. Y. Hou, J. S. Lowengrub, and M. J. Shelley, "The long-time motion of vortex sheets with surface tension," *Phys. Fluids* **9**, 1933 (1997).

R. Krasny, "Desingularization of periodic vortex sheet roll-up," *J. Comput. Phys.* **65**, 292–313 (1986).

A. Lefebvre, *Atomization and Sprays,* Taylor & Francis, Washington, D.C. (1989).

R. W. Metcalfe, S. A. Orszag, M. E. Brachet, S. Menon, and J. J. Riley, "Secondary instability of a temporally growing mixing layer," *J. Fluid Mech.* **184**, 207–243 (1987).

M. R. Nobari, Y.-J. Jan, and G. Tryggvason, "Head-on collision of drops – a numerical investigation," *Phys. Fluids* **8**, 29–42 (1996).

P. C. Patnaik, F. S. Shearman, and G. M. Corcos, "A numerical simulation of Kelvin–Helmholtz waves of finite amplitude," *J. Fluid Mech.* **73**, 215–240 (1976).

J. Qian, G. Tryggvason, and C. K. Law, "An experimental and computational study of bouncing and deforming droplet collision," Submitted for publication.

J. R. Richards, A. M. Lenhoff, and A. N. Beris, "Dynamic breakup of liquid–liquid jets," *Phys. Fluids* **6**, 2640–2655 (1994).

L. Rosenhead, "The formation of vortices from a surface of discontinuity," *Proc. R. Soc. London* A **234**, 170 (1931).

C. A. Spangler, J. H. Hilbing, and S. D. Heister, "Nonlinear modeling of jet atomization in the wind-induced regime," *Phys. Fluids* A **7**, 964–971 (1995).

G. Tryggvason and H. Aref, "Numerical experiments on Hele–Shaw flow with a sharp interface," *J. Fluid Mech.* **136**, 1–30 (1983).

G. Tryggvason, "Numerical simulation of the Rayleigh–Taylor instability," *J. Comput. Phys.* **75**, 253–282 (1988).

G. Tryggvason, W. J. A. Dahm, and K. Sbeih, "Fine structure of vortex sheet rollup by viscous and inviscid simulations," *ASME J. Fluid Eng.* **113**, 31–36 (1991).

S. O. Unverdi and G. Tryggvason, "A front tracking method for viscous incompressible flows," *J. Comput. Phys.* **100**, 25–37 (1992).

P.-K. Wu, R. F. Miranda, and G. M. Faeth, "Effects of initial flow conditions on primary breakup of nonturbulent and turbulent liquid jets," *AIAA* paper 94-0561, AIAA, Washington, D.C. (1994).

S. Zaleski, J. Li, R. Scardovelli, and G. Zanetti, "Direct simulation of multiphase flows with density variations," *Proceedings of IUTAM symposium*, Marseille, July 8–10 (1996). F. Anselmet and L. Fulachier (eds.), Kluwer, Dordecht, The Netherlands.

# 12

# Two-Fluid-Layer Flow Stability

S. ÖZGEN, G. S. R. SARMA, G. DEGREZ, AND M. CARBONARO

## 12.1. Introduction

Because of its many industrial applications, the stability problem of a two-fluid system has attracted considerable interest. Among these applications one may count lubrication, coating and polymer processing procedures, photographic development, and deicing/anti-icing fluid applications on aircraft wings in aeronautics. The latter example provides the essential motivation of this study. The main objective is to determine the effects of a layer of deicing/anti-icing fluid on the flow and performance characteristics of an airplane wing.

Among the previous works that are most relevant to this problem, the pioneering works of Miles (1962) and Yih (1967) should be mentioned. Miles has studied the instability of a liquid film sheared by a turbulent boundary layer. The effect of the air flow is taken into account by calculation of the stresses that the airflow exerts on a dry wavy surface and the imposition of these as conditions at the liquid/gas interface. Yih has studied the stability of two-layer Couette and plane Poiseuille flows for very long waves. He has reported that, apart from the Tollmien–Schlichting (T-S) mode (or hard mode) found in the classical parallel flow stability theory, a second mode (interfacial or soft mode) exists that is brought up by the viscosity difference between the two fluids.

Experimental works of Cohen and Hanratty (1965) and Craik (1965) are important in this context. Cohen and Hanratty have studied the instabilities of liquid films several millimeters thick by using water and other liquids that are 4 to 10 times more viscous than water. Craik has investigated much thinner water films (less than 1 mm).

Hooper and Boyd (1983) have investigated an unbounded flow of two fluids with different viscosities for very short waves. The configuration has been found to be unstable at equal densities and zero surface tension. The instability found is relevant to any flow geometry in which a shear flow acts in the neighbourhood of a viscosity jump.

Renardy (1985) has considered a channel flow of two stratified viscous fluids and has revealed the existence of unstable regions missed out by the long- or the short-wave asymptotic analyses by performing a full linear analysis.

Hooper (1985) has investigated a semi-infinite flow geometry. Here the lower fluid is bounded by the wall and the interface, whereas the upper fluid is bounded only by the interface. It has been shown that if the lower fluid is also the more viscous one, the configuration is unstable to long waves, a phenomena termed the thin-layer effect.

Hooper (1989) has also studied the stability problem of Couette and Poiseuille flows of two viscous fluids in a channel. It has been shown that if a thin layer of less viscous fluid is placed next to the two walls, the effect is to lubricate and hence to stabilize the flow.

Yih (1990) has investigated a two-fluid boundary layer with a motivation similar to ours. He has considered a configuration in which a boundary layer shears a dense, viscous fluid (typical of deicing fluids). He has represented the boundary-layer velocity profile with a Blasius profile and has further approximated it by a piecewise linear profile.

The work of Miesen and Boersma (MB) (1995) is also very important, as they consider a two-fluid boundary layer in which the shearing layer is taken to be turbulent instead of being laminar. This configuration is practically more interesting, as turbulent boundary layers are encountered more often in two-fluid systems. Because a larger portion of the flow over a wing is turbulent, this work provides a good example on that case. They have reported two modes of instability, namely the interfacial mode and the internal mode. The interfacial mode has propagation speeds larger than the mean interfacial speed, whereas the internal mode propagation speeds are smaller than this value.

In this chapter, mainly the results of a parametric study for a two-fluid boundary-layer stability problem are presented although there are a few remarks and results for the Couette and plane Poiseuille flow configurations, especially regarding the non-Newtonian aspects. In Section 12.2, the description and the formulation of the problem are outlined. The method of solution is briefly explained in Section 12.3. Results are presented and discussed in Section 12.4, and, finally, the last section is devoted to conclusions.

## 12.2. Problem Definition and Formulation

Flow geometry is shown in Fig. 12.1. Viscosities are denoted by $\mu^*$, densities by $\rho^*$, and depths of the fluids by $d^*$. Subscripts 1 and 2 denote the upper (shearing) and the lower (sheared) layers respectively. Power-law behavior of the lower fluid is depicted by the power-law index $n$ and the dimensional consistency factor $k^*$. Variables with an asterisk denote dimensional quantities. Velocity profile for the upper layer is the Blasius profile and is denoted by $U_1$, while the profile for the lower layer is $U_2 = a_2 y$. Velocity gradient of the lower layer is denoted by $a_2 = U_0/l$, where $U_0$ is the mean interfacial velocity given as $U_0 = (a_1/m)^{1/n} * l$ with

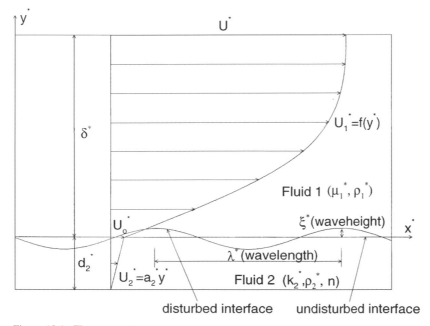

Figure 12.1. Flow geometry.

$m = (k_2^*/\mu_1^*)(U^*/d_1^*)^{n-1}$ and $l = d_2^*/d_1^*$. Parameter $a_1$ is the velocity gradient at the interface for the upper fluid. All distances are normalized by $d_1^*$ (Blasius length scale) and velocities by $U^*$ (free-stream velocity relative to the interface).

A simple non-Newtonian model is the power-law model that assumes dependence of viscosity on shear rate as well as on temperature. This model has been shown to represent the stress–strain relationship to a good approximation for some anti-icing fluids for which $n < 1$ exclusively (shear-thinning behavior). Accordingly, the viscosity is defined as

$$\mu^* = k^*(I_2^*)^{(n-1)/2}, \tag{1}$$

where $I_2^*$ is an invariant of the rate of strain tensor and is defined as $I_2^* = 2e_{lm}^* e_{lm}^*$, in which $e_{lm}^*$ denote the elements of the rate of strain tensor. If the lower fluid is Newtonian ($n = 1$), $k^*$ becomes the viscosity and $m$ simply becomes the ratio of viscosities.

Linear parallel stability theory with temporal formulation is used. The mean flow is a steady, two-dimensional, and incompressible boundary-layer flow. Because of Squire's theorem it is sufficient to consider two-dimensional disturbances:

$$\Phi(x, y, t) = \phi(y) \exp[i\alpha(x - ct)], \tag{2}$$

in which $\alpha$ is the dimensionless wave number (real), $c$ is the complex wave speed ($c = c_r + i c_i$), the real part being the propagation speed and the imaginary part being the amplification factor. The flow is unstable when $c_i > 0$, stable when $c_i < 0$, and neutrally stable when $c_i = 0$. The parameter $\sigma_i = \alpha * c_i$ is the amplification rate, and the dimensionless wavelength is $\lambda = 2\pi/\alpha$. The stability problem is governed by the Orr–Sommerfeld equation. For the upper layer (a prime denotes a differentiation with respect to $y$) it is

$$(U_1 - c)(\phi'' - \alpha^2\phi) - U_1''\phi = (1/i\alpha R)(\phi'''' - 2\alpha^2\phi'' + \alpha^4\phi). \tag{3}$$

Although the equation is similar for the lower layer, the density and the viscosity stratifications and the power-law behavior need to be taken into account:

$$(U_2 - c)(\chi'' - \alpha^2\chi) = \left(m a_2^{n-1}/i\alpha Rr\right)[n\chi'''' + 2(n-2)\alpha^2\chi'' + n\alpha^4\chi]. \tag{4}$$

In Eq. (4) $r = \rho_2^*/\rho_1^*$ is the density ratio and $R = \rho_1^* U^* d_1^*/\mu_1^*$ is the Reynolds number. For the amplitude function, $\phi$ and $\chi$ are used for the two layers to avoid confusion. The boundary and the free-stream conditions are given as follows:

$$\chi = \chi' = 0 \quad \text{at} \quad y = -l; \quad \phi \to 0, \quad \phi' \to 0 \quad \text{when} \quad y \to \infty. \tag{5}$$

Also, four interface conditions are necessary to satisfy the continuity of velocity and stress components at the interface. Derivations of these conditions can be found in Yih (1967):

$$\phi(0) = \chi(0), \tag{6}$$

$$\phi'(0) - \chi'(0) = \frac{\phi(0)}{\tilde{c}}(a_2 - a_1), \tag{7}$$

$$\alpha^2\phi(0) + \phi''(0) = m n a_2^{n-1}[\chi''(0) + \alpha^2\chi(0)], \tag{8}$$

$$-i\alpha R(\tilde{c}\phi' + a_1\phi) - (\phi''' - 3\alpha^2\phi') + i\alpha Rr(\tilde{c}\chi' + a_2\chi)$$
$$+ m a_2^{n-1}[n\chi''' - (4-n)\alpha^2\chi'] = i\alpha R[(r-1)F^{-2} + \alpha^2 S]\phi/\tilde{c}. \tag{9}$$

The parameter $\tilde{c} = c - U_0$ and is the wave speed relative to the interface. Equation (9) is the normal stress condition in which the effect of gravity is given by the Froude number

$F = U^*/(g^*d_1^*)^{1/2}$ and the surface tension by surface-tension parameter $S = T^*/(\rho_1^* U^{*2}d_1^*)$, where $g^*$ is the gravitational acceleration and $T^*$ is the surface-tension coefficient.

## 12.3. Solution Method

The solution method is the backward integration method outlined by Drazin and Reid (1981). It consists of choosing a sufficiently large distance from the interface $y = y_e$, so that $U(y_e) = 1$ and $U''(y_e) = 0$ to some degree of accuracy. Equation (3) is integrated starting at $y = y_e$ toward the wall by use of the asymptotic solutions $\phi_1$ and $\phi_2$ as the starting values of the integral, which are given as

$$
\begin{aligned}
\begin{bmatrix} \phi_1 & \phi_1' & \phi_1'' & \phi_1''' \end{bmatrix} &= \begin{bmatrix} 1 & -\alpha & \alpha^2 & -\alpha^3 \end{bmatrix}, \\
\begin{bmatrix} \phi_2 & \phi_2' & \phi_2'' & \phi_2''' \end{bmatrix} &= \begin{bmatrix} 1 & -\beta & \beta^2 & -\beta^3 \end{bmatrix},
\end{aligned}
\tag{10}
$$

with $\beta^2 = \alpha^2 + i\alpha R(1 - c)$. In order to maintain the linear independence of the solutions during integration, a Gram–Schmidt orthonormalization procedure has to be used. When the interface, that is, $y = 0$, is reached the equation to be integrated is switched from Eq. (3) to Eq. (4). Equation (4) is also a fourth-order ordinary differential equation and hence has a general solution of the form

$$
\chi = B_1\chi_1 + B_2\chi_2 + B_3\chi_3 + B_4\chi_4.
\tag{11}
$$

As we cannot say a priori anything about the solution in the finite domain $-l \le y \le 0$, four linearly independent solutions $\chi_i$ are chosen as those corresponding to an orthonormal basis of initial vectors with $[\chi, \chi', \chi'', \chi''']_{y=0} = \bar{\varepsilon}_i$, with the unit vectors $i = 1, 2, 3, 4$:

$$
\begin{aligned}
\bar{\varepsilon}_1 &= \begin{bmatrix} 1 & 0 & 0 & 0 \end{bmatrix}, \quad \bar{\varepsilon}_2 = \begin{bmatrix} 0 & 1 & 0 & 0 \end{bmatrix}, \quad \bar{\varepsilon}_3 = \begin{bmatrix} 0 & 0 & 1 & 0 \end{bmatrix}, \\
\bar{\varepsilon}_4 &= \begin{bmatrix} 0 & 0 & 0 & 1 \end{bmatrix}.
\end{aligned}
\tag{12}
$$

Equation (4) is integrated with Eqs. (12) as the starting value of the integral. When the wall is reached, boundary conditions given in Eq. (5) are to be satisfied. Eigenvalues of the problem are $\alpha$, $c_r$, $c_i$, and $R$, whereas the eigenfunction is $\phi$ (or $\chi$ in the lower layer). Two of these parameters are fixed ($c_i$, $R$) and the remaining two ($\alpha$, $c_r$) are searched. For the details of the solution procedure one may refer to Özgen et al. (1998).

## 12.4. Results and Discussion

### 12.4.1. *Boundary Layer with Laminar Profile*

Before the effects of various parameters on the stability characteristics are discussed. It is feasible to characterize the two instability modes. Figure 12.2(a) shows the neutral stability curves in the $R - \alpha$ plane for various viscosity ratios with $r = 1$, $l = 0.5$, $S = 0$. The thick curves are for the T-S mode, whereas the thinner curves are for the interfacial mode, a representation that holds throughout the paper. Looking at the curves for $m = 10$ (solid curves), we see that the region enclosed by the T-S mode neutral stability curve is unstable, whereas the remaining regions are stable. However, the two neutral stability curves of the interfacial mode enclose a stable domain whereas the surrounding regions are unstable. Hence there are regions in the $R - \alpha$ plane that are stable for the T-S mode but unstable for the interfacial mode. Another point that merits mentioning is that long waves (low wave numbers) are unstable for all Reynolds numbers, which is the thin-layer effect reported by Hooper (1985). In order to understand better

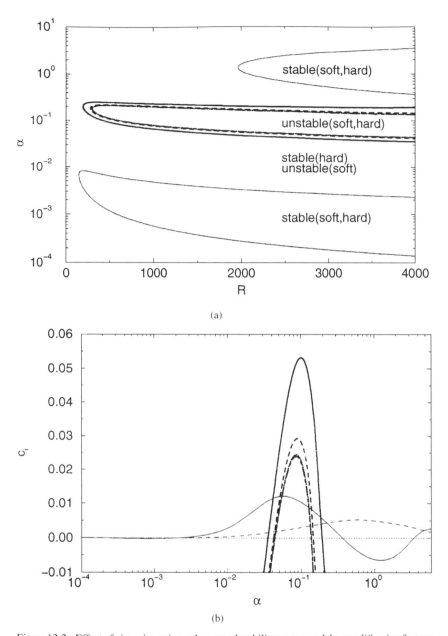

Figure 12.2.  Effect of viscosity ratio on the neutral stability curves and the amplification factors of the two modes, ($r = 1, l = 0.5$, $S = 0$): (—), $m = 10$; (– –), $m = 100$; ($\cdots$), $m = 1000$; (–·–), single layer.

the difference between the two modes, one may refer to Fig. 12.2(b). This figure shows an $\alpha - c_i$ plot for the same configuration as in Fig. 12.2(a) at $R = 4000$. The highest amplification factors for the configurations treated occur close to this Reynolds number, so it has been chosen to contrast the various configurations better. Looking at the curves for $m = 10$, we see that the most striking feature that can be seen is the difference in the magnitudes of the amplification factors of the two modes.

Viscosity stratification is responsible for the presence of the interfacial mode of instability; therefore it is natural to emphasize the effect of this parameter on the stability characteristics. The effect of increasing the viscosity of the lower layer on the amplification factors of the two modes can be seen again in Fig. 12.2. At smaller values of $m$, not only the unstable domain spans a larger wave-number range but the corresponding amplification factors are larger as well. The magnitudes of the amplification factors of the T-S mode approach those of a single-layer flow and they become almost indistinguishable for the interfacial mode as $m$ increases. Therefore, in the limit $m \to \infty$, we have the eigenvalues belonging to the T-S mode that are identical to those for a single-layer flow and eigenvalues belonging to an interfacial mode that are asymptotically neutrally stable. This suggests that, in the limit $m \to \infty$, the lower layer remains idle in terms of contribution to the stability characteristics of the flow. Since $m$ for deicing/anti-icing fluid air combination is very large $\left(\approx 5 \times 10^5\right)$ it is the T-S mode that manifests itself first.

Sensitivity of the interfacial mode against the viscosity stratification has been tested with a less drastic variation of this parameter ($2 < m < 10$) (not shown). The effect on the neutral stability curves of the hard mode and the soft mode at very low wave numbers was seen to be negligible. However, a considerable growth in size of the stable region of the soft mode at moderate and high wave numbers has been observed with decreasing viscosity ratio. Still not shown, viscosity ratios less than 1 have also been treated, and it has been observed that the long waves are now stable, regardless of the Reynolds number. This is the lubrication effect previously reported by Hooper (1989).

The effect of increasing the thickness of the lower layer (moving the interface away from the wall) is shown in Fig. 12.3. The figure is complicated so we investigate it starting from small wave numbers to larger ones. At very small wave numbers we see a region of stability embedded in an unstable region for the interfacial mode. The variation of the thickness of the

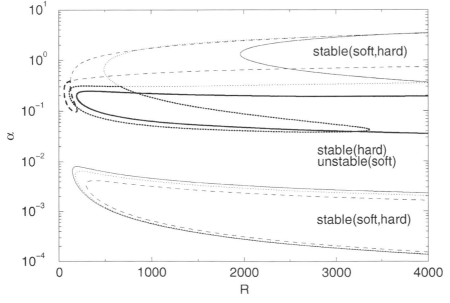

Figure 12.3. Effect of thickness of the lower layer on the neutral stability curves for the two instability modes ($m = 10, r = 1, S = 0$): (—), $l = 0.5$; ($\cdots$), $l = 1.0$; ($--$), $l = 2.0$.

lower layer has almost no effect on this region because these waves are almost flat ($\approx 10^4$ times the boundary-layer height) and the wall would have hardly any effect on them. Looking at moderate and high wave numbers from the curve for $l = 1.0$, we see that the lower branches of the hard and the soft modes coalesce at a point (at $R = \sim 3400$ in this case). This suggests that this intersection may occur at very high values of the Reynolds number for smaller values of $l$, which happens at a value of 40,000 for $l = 0.5$. Another point is that the region where both the hard and the soft modes are stable has expanded and starts at $R \approx 600$ for $l = 1.0$. On the other hand, the unstable region of the hard mode has shrunk and now spans a region $100 < R < 3400$.

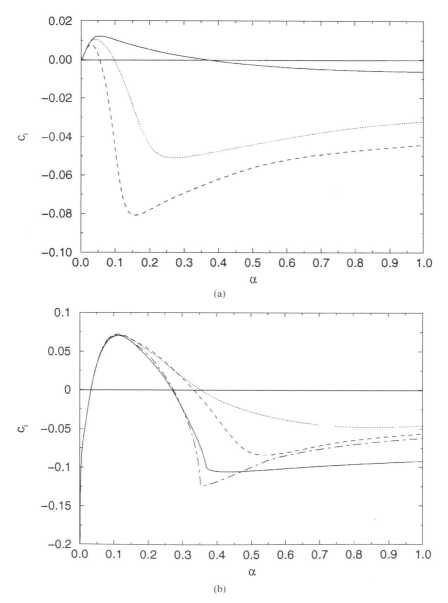

Figure 12.4. Effect of surface tension on the amplification factors of the two modes ($m = 10$, $l = 0.5$, $F^{-2} = 0$, $R = 4000$): (a) (—), $S = 0$; ($\cdots$), $S = 1$; (——), $S = 5$ ($r = 1$). (b) (—), $S = 0$; ($\cdots$), $S = 1$; (——), $S = 3$; (—·—), $S = 10$ ($r = 0.5$).

In the case for $l = 2.0$ these effects have become more pronounced; the hard mode has almost disappeared, giving space for a large stable region that is present at $R > 150$.

The effect of surface tension on the soft and the hard modes is shown in Fig. 12.4. In Fig. 12.4(a) we see the stabilizing effect of surface tension on the interfacial mode. This effect is more obvious for short waves and there is a cutoff wave number above which the configuration is unconditionally stable for $S > 0$ ($S = 0$ curve becomes unstable again at higher values of $\alpha$). The value cutoff wave number decreases as the surface-tension parameter increases, meaning that more waves are stable for a larger surface-tension parameter. The reason why the effect is more pronounced for short waves lies in the fact that the surface-tension force is directly proportional to the interface curvature. The curvature is higher for short waves; therefore the stabilizing effect is stronger for them. In Fig. 12.4(b) we observe a rather peculiar behavior. Here, the effect of surface tension on the hard mode of instability is shown for a density ratio of $r = 0.5$ for a surface-tension parameter $S$ between 0 and 10. In contrast to the behavior observed for the interfacial mode, low and moderate values of the surface-tension parameter have a destabilizing effect, especially on high wave numbers. It is only at higher values of this parameter that we observe a stabilizing behavior. This behavior was also observed for other ranges of the parameters and hence is not due to adverse density stratification.

The effect of gravity is given by the Froude number $F$ and is stabilizing for a favorable density stratification (heavier fluid at the bottom) and destabilizing for an adverse one. We have investigated the effect of density stratifications on the amplification factors of the soft mode at zero gravity and nonzero gravity. Interestingly, the behavior is not monotonic. Until a value of $r = \sim5$ the flow is destabilized with increasing density stratification, and only beyond this value do we see a stabilizing effect of this parameter. Although not shown, for nonzero gravity (Froude number) a cutoff wave number is seen above which the configuration is stable for all Reynolds numbers.

The effect of the non-Newtonian behavior of the lower fluid on stability characteristics is displayed in Fig. 12.5. Here, decreasing $n$ means increasing departure from Newtonian behavior. Looking at Fig. 12.5(a), we see that the boundary for the stable pocket at very low wave numbers for the soft mode remains the same for all the $n$ values considered. On the other hand, the pockets at high wave numbers have disappeared for $n \neq 1$. The boundaries of the unstable regions for the hard mode change slightly, suggesting at first glance that the non-Newtonian power-law nature of the lower fluid does not have a significant effect on stability characteristics. However, when we examine Fig. 12.5(b), which is for the same configuration as that for Fig. 12.4 at $R = 4000$, some obvious effects can be noted. To start with, the amplification factors for the hard mode become smaller with decreasing $n$. From the curve for $n = 0.2$ we see that the maximum amplification factor is approximately half of the value for $n = 1.0$. A similar stabilizing effect is evident for the soft mode as well. With decreasing power-law index the amplification factors decrease progressively, and for the value of $n = 0.2$ these have almost vanished. This behavior is very similar to the behavior observed when the viscosity of the lower fluid is increased (see discussion for Fig. 12.2). When we look at Eqs. (4), (8), and (9) we see that the viscosity ratio $m$ is always multiplied by a factor $a_2^{n-1}$. From the definition of $U_0$ and $a_2$, we note that $a_2^{n-1}$ becomes a very large number when the power-law index is decreased; thus, when multiplied by $m$, a very large effective viscosity ratio is attained. This means that, although the viscosity ratio $m$ remains the same, the boundary layer shears an effectively more viscous fluid at smaller values of $n$. This is why the effect of decreasing $n$ has the same effect as increasing the viscosity ratio.

So far, hypothetical configurations have been treated in order to demonstrate the effect of each parameter on the stability characteristics. In this paragraph, a real case, which was reported

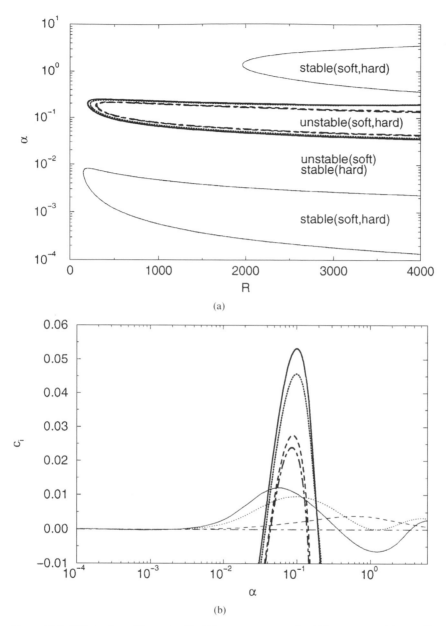

Figure 12.5. Effect of non-Newtonian liquid layer on the neutral stability curves and the amplification factors of the two instability modes ($m = 10$, $r = 1$, $l = 0.5$, $S = 0$): (——), $n = 1.0$; (···), $n = 0.8$; (– –), $n = 0.5$; (–·–), $n = 0.2$.

by Hendrickson and Hill (HH) (1987) and was also treated by Yih (1990) is investigated. It is a deicing fluid/air system, and the parameters of the problem are $m = 598802$, $r = 972$, $S = 0.01013$, and $F = 496$. For this configuration, the free-stream velocity $U^*$ is given to be 27.28 m/s, the thickness of the fluid is $d_2^* = 1.1$ mm, and the Blasius length scale is $d_1^* = 3.08 \times 10^{-4}$ m, which makes $l = 3.5668$. The Reynolds number based on the Blasius length scale is 496. Both the hard (T-S) mode and the soft (interfacial) mode were captured by our

Table 12.1. *Comparison of the Results for a Special Case Treated by Yih*

| Author | $\alpha_{\text{low}}$ | $\alpha_{\text{up}}$ | $\alpha_{\sigma_{i,\max}}$ | $\sigma_{i,\max}$ | $(c_r - U_0)_{\sigma_{i,\max}}$ | Mode |
|---|---|---|---|---|---|---|
| Yih | 0.0040 | 0.3666 | 0.2008 | $4.58 \times 10^{-6}$ | $2.08 \times 10^{-5}$ | Interfacial |
| Boelens and Hoeijmakers | 0.0054 | 0.3767 | 0.2057 | $5.17 \times 10^{-6}$ | $2.64 \times 10^{-5}$ | Interfacial |
| Present study | 0.0053 | 0.3754 | 0.2022 | $5.13 \times 10^{-6}$ | $2.55 \times 10^{-5}$ | Interfacial |
| HH | n.a. | n.a. | 0.1402 | n.a. | n.a. | T-S |
| Present study | 0.0913 | 0.2008 | 0.1504 | $2.06 \times 10^{-3}$ | 0.3449 | T-S |

computations. The results of the computations together with the analytical and numerical results of other scientists have been given in Table 12.1. Subscripts low and up are the lower and the higher values of $\alpha$ where the $\alpha - c_i$ curve intersects the $c_i = 0$ line. As can be seen, there are 3 orders of magnitude difference between the amplification rates ($\sigma_i$) of the hard and the soft modes. Our numerical results agree remarkably well with the numerical results of Boelens and Hoeijmakers (1997) (who use a spectral technique) and the analytical results of Yih (who uses an asymptotic technique for very small wave numbers). However, there remains one more point to be discussed on Yih's paper. He estimates that the waves observed in the experiments have a wave number $\alpha$ equal to 0.5 based on the liquid thickness $d_2^*$, which corresponds to $\alpha = 0.14$ based on the Blasius length scale $d_1^*$. This agrees very well with the wave-number value we obtain for the hard mode for this configuration. Therefore there is evidence that the waves observed in the experiments of HH belong to the T-S mode but not the interfacial mode, as the mode with the highest amplification rate would most likely be observed. Yet when Yih compares his results with those of HH, he mentions that the most unstable wave for the T-S mode for a Blasius boundary layer has a wave number $\alpha = 0.26$ based on the displacement thickness $\delta^*$, which he gives as 0.534 mm. However, when he normalizes the dimensional wave number $\alpha^*$, he divides (instead of multiplying) it by the reference length and thereby finds that the most unstable mode has $\alpha = 0.126$ based on the liquid thickness $d_2^*$. Comparing this with the experimental value of $\alpha = 0.5$, he concludes that the mode observed in the experiment is not the T-S mode but is the interfacial mode. However, with the correct algebra, a value of $\alpha = 0.536$ (based on liquid thickness) is found, which is very close to the observed experimental value.

### 12.4.2. *Boundary Layer with Turbulent Profile*

In this subsection, the laminar boundary-layer profile is replaced by a turbulent one. For this purpose, the following approximations are used for the velocity profile:

$$U(y) = (\tau^*/\mu_1^*)y d_2^*/U^* \quad \text{for } 0 \leq y \leq \left[s\mu_1^*/(\rho_1^*\tau^*)^{1/2}\right]/d_2^*,$$
$$U(y) = ay^c + b \quad \text{for } y \geq \left[s\mu_1^*/(\rho_1^*\tau^*)^{1/2}\right]/d_2^*, \tag{13}$$

where $s$ defines the thickness of the sublayer and its value is chosen such that $y_s \ll 1/\alpha$, as given by Miles (1962). The parameters $a$, $b$, and $c$ are constants that are functions of the specific flow configuration. For this subsection only, all lengths are normalized by $d_2^*$, which is a more convenient length scale. The method of solution will be applied to an air/water system given by MB (1995). The parameters are $m = 56.4$, $r = 824$, $S = 0.2636$, and $F = 1061$. Free-stream velocity is 40 m/s, and the thickness of the liquid layer $d_2^*$ is given as $0.145 \times 10^{-3}$ m. Friction coefficient $C_f$ is given as 0.02, which yields a shear stress at the interface equal to 19.4 Pa from

Table 12.2. *Comparison of the Results for a Turbulent Boundary-Layer Configuration*

| Author | $\alpha_{\text{low}}$ | $\alpha_{\text{up}}$ | $\alpha_{\sigma_{i,\max}}$ | $\sigma_{i,\max}$ | $(c_r - U_0)_{\sigma_{i,\max}}$ | Mode |
|---|---|---|---|---|---|---|
| MB | 0.0 | 5.2 | 1.85 | 0.0146 | n.a. | Interfacial |
| Present study | 0.0 | 5.12 | 1.91 | 0.0112 | $6.48 \times 10^{-3}$ | Interfacial |
| MB | 0.5 | 1.30 | 0.95 | $1.53 \times 10^{-3}$ | n.a. | Internal |
| Present study | 0.67 | 1.37 | 1.04 | $1.14 \times 10^{-3}$ | $-0.041$ | Internal |

$\tau^* = (1/2)C_f \rho_1^* U^{*2}$. A summary of the analysis is given in Table 12.2 together with the results of MB. As can be seen, both the interfacial and the internal mode were captured by our code. There is an order of magnitude difference between the amplification rates of the two modes. The results show very good agreement for all the parameters that could be extracted from Fig. 7 of MB.

We have also tested our code for the cases treated in the experiments of Cohen and Hanratty (1965) and Craik (1965). The agreement of our results with the experimental ones turned out to be good. Still better agreement was achieved with the numerical results of MB (1995) for the same cases. For more details of this analysis, Özgen et al. (1998) may be referred to.

### 12.4.3. *Two-Layer Couette Flow*

With a few alterations, the code is capable of handling various flow geometries such as two-layer Couette and Poiseuille flows. Equations (3), (8), and (9) now have terms involving the power-law index of the upper layer. Only some results regarding the non-Newtonian aspects are treated here for the two-fluid Couette flow. Figure 12.6(a) shows the results for a case with a power-law fluid in the upper layer and Newtonian fluid in the lower. The Reynolds numbers are based on the properties of the upper fluid and are defined as $R = \rho_1 U^{*2-n_1} d^{*n_1} / k_1^*$ for power-law fluids, where $d^*$ is the channel height. The effect of changing the power-law index of the lower fluid is shown in this figure. In addition to a few shear-thinning fluids ($n < 1$), a shear-thickening (dilatant) ($n > 1$) fluid has also been added. When we look at the curve for $n_1 = 0.8$, $n_2 = 1.0$, we see that there is an unstable domain embedded in a stable domain, which itself is surrounded by an unstable domain. When we decrease the power-law index of the upper fluid the stable domain becomes smaller. In all of these cases very small wavelengths are unstable (thin-layer effect) except for the $n_2 = 1.2$ case. It means that the effect of a high power-law index is to decrease the effective viscosity ratio (in this case it seems it becomes less than 1). Another noteworthy feature is that the unstable subdomain ceases to exist at $n$ values less than 0.8.

Figure 12.6(b) shows a configuration in which both layers are non-Newtonian. Here the power-law index of the upper fluid is fixed and the lower fluid is varied. Here the effect of decreasing the power-law index is to enlarge the stable domain and at a value of $n$ smaller than 1.0 we see that very long waves become unconditionally stable. The unstable subdomain exists for some cases ($0.6 < n < 1.0$) but does not for others. Apparently, the non-Newtonian nature of either one of the fluids has a great influence on the stability characteristics.

For the Poiseuille flow case, several test cases reported by Yiantsios and Higgins (1988) have been treated. The agreement of the results with those of Yiantsios and Higgins was seen to be perfect, which again shows the versatility and the feasibility of our method and code.

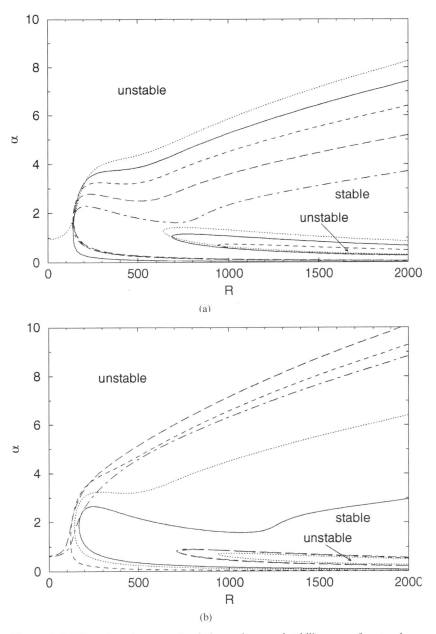

Figure 12.6. Effect of varying power-law index on the neutral stability curves for a two-layer Couette flow ($m = 2, r = 1, l = 1, S = 0$): (a)($\cdots$), $n_1 = 1.2$; (—), $n_1 = 1.0$; (— —), $n_1 = 0.8$; (— —), $n_1 = 0.6$; (—·—), $n_1 = 0.4$ ($n_2 = 1.0$). (b)(—), $n_2 = 1.2$; ($\cdots$), $n_2 = 1.0$; (— —), $n_2 = 0.8$; (— —), $n_2 = 0.6$; (—·—), $n_2 = 0.4$ ($n_1 = 0.8$).

## 12.5. Conclusions

The presence and importance of the interfacial (soft:Yih) mode of a two-fluid flow configuration discovered by Yih has been ascertained for non-Newtonian fluids as well. Agreement of the computed results with the experimental, numerical, and theoretical work reported in the literature is encouraging, and the numerical algorithm and the code developed have shown a capability for handling a wide range of flow geometries and parameter ranges. The results for the non-Newtonian fluids need to be elaborated further because there is evidence that the instabilities formed in such systems are different in nature from those formed with Newtonian fluids. An experimental database for deicing/anti-icing fluids is currently under development at the Von Karman Institute and will be used for further validation purposes in future studies.

## References

Boelens, O. J. and Hoeijmakers, H. W. M. 1997. Wave formation on de-/anti-icing fluids. In *EUROMECH, 3rd European Fluid Mechanics Conference*, Göttingen, Germany.

Cohen, L. S. and Hanratty, T. J. 1965. Generation of waves in the concurrent flow of air and a liquid. *AIChEJ.* **11**, 138–144.

Craik, A. D. D. 1965. Wind generated waves in thin liquid films. *J. Fluid Mech.* **26**, 369–392.

Drazin, P. G. and Reid, W. H. 1981. *Hydrodynamic Stability*. Cambridge U. Press, Cambridge.

Hendrickson, G. S. and Hill, E. G. 1987. Effects of de/anti-icing fluids on airfoil characteristics. Boeing Rep.

Hooper, A. P. 1985. Long-wave instability at the interface between two viscous fluids: thin layer effects. *Phys. Fluids* **28**, 1613–1618.

Hooper, A. P. 1989. The stability of two superposed viscous fluids in a channel. *Phys. Fluids A* **1**, 1133–1142.

Hooper, A. P. and Boyd, W. G. C. 1983. Shear-flow instability at the interface between two viscous fluids. *J. Fluid Mech.* **128**, 507–528.

Miesen, R. and Boersma, B. J. 1995. Hydrodynamic stability of a sheared liquid film. *J. Fluid Mech.* **301**, 175–202.

Miles, J. W. 1962. On the generation of surface waves by shear flows. *J. Fluid Mech.* **13**, 443–448.

Özgen, S., Sarma, G. S. R., and Degrez, G. 1998. Two-fluid boundary layer stability. *Phys. Fluids* **10**, 2746–2757.

Renardy, Y. 1985. Instability at the interface between two shearing fluids in a channel. *Phys. Fluids* **28**, 3441–3443.

Yiantsios, S. G. and Higgins, B. G. 1988. Linear stability of plane Poiseuille flow of two superposed fluids. *Phys. Fluids* **31**, 3225–3238.

Yih, C. S. 1967. Instability due to viscosity stratification. *J. Fluid Mech.* **27**, 337–352.

Yih, C. S. 1990. Wave formation on a liquid layer for de-icing airplane wings. *J. Fluid Mech.* **212**, 41–53.

# PART THREE

# Waves and Dispersion

# 13

# On Modeling Unsteady Fully Nonlinear Dispersive Interfacial Waves

THEODORE YAOTSU WU

## 13.1. Introduction

This study on nonlinear, dispersive, unsteady interfacial waves in layered fluids is intended first to develop a theory for modeling such fully nonlinear fully dispersive (FNFD) wave phenomena. Such waves are ubiquitous in the open ocean and in the atmosphere locally marked with certain topographical characteristics. Previous investigations have found that these waves may have hazardous effects on ocean engineering, coastal infrastructures and environmental quality, longshore current and sediment transport, with possible degrees and extents still unclear (see, e.g., Osborne et al., 1978; the 5-year, 3-State Congress *Chesapeake Bay Program*, 1983, that investigated the mid-1960s drastic drop of the previously rich Chesapeake marine products). To improve our knowledge on the physical processes involving oceanic internal waves, it is therefore essential to secure theoretical models capable of accurate prediction of the wave performance to their exact roles of nonlinear and dispersive effects, such as wave breaking, run up on seabed, and turbulent mixing in coastal waters.

The approach adopted in pursuing this study is inspired by the highly scholarly spirit of Chia-Shun Yih, to whose honor this special Symposium is dedicated. Dr. Yih won international esteem for his contributions to fluid mechanics, which are important, enlightening, and extensive, spanning over a variety of physical phenomena and a broad scope of applications. At the Third U.S. Congress of Applied Mechanics, Yih (1958; 1991, Selected Papers, A3) published the powerful transformation that introduced his associated-flow velocity (equal to the real fluid velocity multiplied by the square root of variable density of the fluid). In terms of this new velocity, he showed that the equation of steady two-dimensional flow of stratified fluids can be reduced to exactly linear form for the associated flow under suitable upstream conditions. This method has thus benefited many researchers in finding exact solutions to their original nonlinear challenging problems.

In the present study, we attempt to develop a system of basic equations, in $(1 + 2)$ dimensions (1 time plus 2 horizontal space), for modeling unsteady, FNFD interfacial waves in two-layered fluids under the action of gravity and surface tension, with the viscous effects neglected. For simplicity, we consider here the interfacial waves in two-layered fluids as bounded top and bottom by two horizontal flat plates of infinite extent. The theoretical results subject to this confinement are generally useful to provide accurate simulation of analogous motion of the *internal wave mode* of the associated two-layered fluid systems, having the top surface also free to move, as pointed out theoretically and experimentally by Zhu et al. (1986, 1987), except for the special case in which the top fluid layer is exceedingly thinner than the lower one. In the field, giant internal waves found in the Andman Sea (Osborne, 1978) have been observed to run in a single group of five to ten solitarylike waves, with a typical length of

10 km, stretching straight spanwise to 200 km and reaching to a magnitude of 200 m along the pycnocline $\sim 100$ m below the ocean surface. Even with such great amplitude, the ocean surface would remain hardly displaced, other than carrying a signature of the giant internal waves submerged beneath in the form of belts of white-capped rip waves of $\sim 1$ m in height. Thus, the rigid-top surface would serve in general as a good approximation in reality. It is this slight idealization that actually renders the exact solution possible to our nonlinear problem at hand. Another factor that appears to facilitate the exact solution is the adoption of the horizontal velocity at the interface together with the interface elevation as the basic unknown variables exclusively. In comparison, there are alternatives in taking the velocities at the solid boundary (as in Rayleigh, 1876, for the single-layer case), or the layer-mean velocity (as in Boussinesq, 1871; Korteweg and de Vries, 1895; Green and Naghdi, 1976; Wu, 1981), or making joint uses of boundary integrals (as in Choi, 1995; Choi and Camassa, 1996; Grue et al., 1997). However, these alternative methods do not seem to excel in simplicity and accuracy over the present one, as will be shown in Section 13.2 with the mathematical formulation.

From his life and work, Chia-Shun Yih left us with a vivid image that exhibited him so well as possessing an intense interest in what he chose to study, a crisp clear physical concept in making an incisive penetration to the core of the problem being pursued, and emerging with a solution marked with beauty in simplicity and elegance and the power in drawing conclusions of essence. In life, he showed a bountiful love for nature. In work, he had vigor to spare in always trying his very best. To friends, he might have done just as well as a painter or poet, with just as warm a sincerity as the devoted friend he was. It is with such a feeling that this work is pursued with an intent of emulating the high quality he so well demonstrated.

## 13.2. A Theory for Unsteady Nonlinear Interfacial Waves in Two-Layered Fluids

In developing a theory for modeling fully nonlinear and dispersive wave phenomena, it is important to ascertain that the nonlinear and dispersive effects do play their full exact roles in the theory. These effects can be closely estimated by two key parameters, namely

$$\alpha = a/h, \qquad \epsilon = h/\lambda, \tag{1}$$

representing, respectively, the nonlinear and the dispersive effects for characterizing waves of amplitude $a$ and typical length $\lambda$ in water of rest depth $h$. They are both assumed small for weakly nonlinear and weakly dispersive waves, with $\alpha = O(\epsilon^2) \ll 1$ for the Boussinesq family and both are of the order of unity for waves of finite amplitude.

To develop the theory, we proceed to adopt Euler's equations for describing three-dimensional inviscid wave motion in two layers of immiscible fluids of infinite horizontal extent under the action of gravity and surface tension. The two fluid layers have rest thickness $h_j$ and constant density $\rho_j$, with $j = 1$ for the lower and $j = 2$ for the upper layer, and with $\rho_2 < \rho_1$ for the static stability. The two fluids are bounded below by a rigid horizontal bottom at $z = -h_1$ and on top by another horizontal flat plate of infinite extent at $h = h_2$ so the system has only the interface free to move (see Fig. 13.1). In motion, the fluids move with velocity $(\mathbf{u}_j, w_j) = (u_j, v_j, w_j)$ in each layer, with the interface displaced from its rest position at $z = 0$ to elevation $z = \zeta(\mathbf{r}, t)$ as a function of the horizontal position

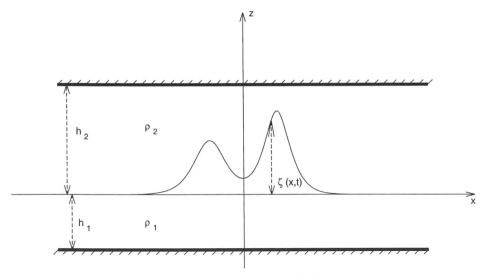

Figure 13.1. A sketch of interfacial wave motion in two-layered fluids.

vector $\mathbf{r} = (x, y, 0)$ and time $t$. Assuming that the fluid is incompressible and inviscid, we have the Euler equations of continuity, horizontal and vertical momentum for the two fluids ($j = 1, 2$) as

$$\nabla \cdot \mathbf{u}_j + \frac{\partial w_j}{\partial z} = 0, \tag{2}$$

$$\frac{d\mathbf{u}_j}{dt} = \frac{\partial \mathbf{u}_j}{\partial t} + \mathbf{u}_j \cdot \nabla \mathbf{u}_j + w_j \frac{\partial \mathbf{u}_j}{\partial z} = -\frac{1}{\rho_j} \nabla p_j, \tag{3}$$

$$\frac{dw_j}{dt} = \frac{\partial w_j}{\partial t} + \mathbf{u}_j \cdot \nabla w_j + w_j \frac{\partial w_j}{\partial z} = -\frac{1}{\rho_j} \frac{\partial p_j}{\partial z} - g, \tag{4}$$

where $\nabla = (\partial_x, \partial_y, 0)$ ($\partial_x = \partial/\partial x$, etc.) is the horizontal gradient operator, $p_j$ is the pressure, and $g$ is the gravitational acceleration. The boundary conditions are

$$\tilde{w}_j = \tilde{D}_j \zeta \quad [\tilde{D}_j = \partial_t + \tilde{\mathbf{u}}_j \cdot \nabla, \quad \text{on } z = \zeta(\mathbf{r}, t)], \tag{5}$$

$$p_1 = p_2 + \rho_1 \gamma \nabla \cdot \mathbf{n} \quad [z = \zeta(\mathbf{r}, t)], \tag{6}$$

$$w_2 = 0 \quad (z = h_2), \tag{7}$$

$$w_1 = 0 \quad (z = -h_1), \tag{8}$$

where $\tilde{\mathbf{u}}_j(\mathbf{r}, t) = \mathbf{u}_j[\mathbf{r}, \zeta(\mathbf{r}, t), t]$ is the value of $\mathbf{u}_j$ at the interface; likewise for $\tilde{w}_j(\mathbf{r}, t)$, $\rho_1 \gamma$ is the uniform surface tension of the interface and $\mathbf{n}$ is the upward unit vector normal to the interface. The subscripts $t$ and $z$ denote partial differentiation.

We adopt the variables $\tilde{\mathbf{u}}_1$, $\tilde{\mathbf{u}}_2$, $\tilde{w}_1$, $\tilde{w}_2$, and $\zeta$ as the unknowns in formulating the theory, for the reason primarily due to the relative simplicity in establishing an exact theory. To proceed, we first project the momentum equations under conditions (2), (5), and (6) onto the free interface to obtain the projected equation in terms of the unknowns. For an arbitrary flow variable $f_j(\mathbf{r}, z, t)$, it approaches, as $z \to \zeta(\mathbf{r}, t)$ from within the $j$-side fluid domain, the interfacial value $\tilde{f}_j(\mathbf{r}, t)$:

$$f[\mathbf{r}, \zeta(\mathbf{r}, t), t] = \tilde{f}(\mathbf{r}, t), \tag{9}$$

such as $\tilde{\mathbf{u}}_j(\mathbf{r}, t)$ and $\tilde{w}_j(\mathbf{r}, t)$. Their time derivatives are related, by the chain rule, as

$$\partial_t \tilde{f}(\mathbf{r}, t) = [\partial_t f(\mathbf{r}, z, t) + \zeta_t \partial_z f]|_{z=\zeta}, \tag{10}$$

and similarly for their derivatives with respect to $x$ and $y$. Using these relations, we find

$$\left. \frac{df_j}{dt} \right|_{z=\zeta} = \tilde{D}_j \tilde{f} \quad (\tilde{D}_j = \partial_t + \tilde{\mathbf{u}}_j \cdot \nabla). \tag{11}$$

Making use of these formulas, we can derive straightforwardly from Eqs. (2)–(6) the equation

$$\tilde{D}_j \tilde{\mathbf{u}}_j + \left[ g(t) + \tilde{D}_j^2 \zeta \right] \nabla \zeta = -(1/\rho_j) \nabla \tilde{p}_j, \qquad (j = 1, 2). \tag{12}$$

Here, we have extended the case of constant gravity (e.g., Choi 1995) to include the more general case of Faraday's waves produced in a horizontal water tank under vertical oscillation, a case that is equivalent to having a time-dependent gravity acceleration with reference to the tank frame. Now applying dynamic condition (6) to Eq. (12) yields

$$\tilde{D}_1 \tilde{\mathbf{u}}_1 - \mu \tilde{D}_2 \tilde{\mathbf{u}}_2 + \left[ g_e + \tilde{D}_1^2 \zeta - \mu \tilde{D}_2^2 \zeta \right] \nabla \zeta = -\gamma \nabla \nabla \cdot \mathbf{n}, \tag{13}$$

$$\mu = \rho_2/\rho_1, \qquad g_e(t) = (1 - \mu)g(t), \qquad \mathbf{n} = (-\zeta_x, -\zeta_y, 1)/(1 + |\nabla \zeta|)^{1/2}. \tag{14}$$

This resulting equation, although superficially involving only $(\tilde{\mathbf{u}}_1, \tilde{\mathbf{u}}_2, \zeta)$, all pertaining to the interface, actually has incorporated the vertical momentum equation as well as the kinematic and dynamic conditions at the free interface to yield this equation of an overall equilibrium. Furthermore, it is exact.

We note that Eq. (13) and the two kinematic equations in (5) are three equations for five unknowns, $(\tilde{\mathbf{u}}_1, \tilde{\mathbf{u}}_2, \tilde{w}_1, \tilde{w}_2, \zeta)$, and that these equations are all exact. We can accomplish closure of the system by further seeking from the general solution to the velocity field equations for the two fluids two more exact equations relating the five unknown variables.

Since the three basic equations assembled so far are exact, we may ignore the nonlinearity parameter $\alpha$ by regarding it as arbitrary and consider first the special case of inviscid long waves in shallow water by assuming only the dispersion parameter $\epsilon = h_1/\lambda$ to be small, with $h_1$ and $h_2$ assumed equal in order of magnitude. [It turns out that this assumption can also be relaxed eventually; see the discussion following Eq. (30)].

From here onward, we assume the flow to be continuous, continuously differentiable, and irrotational in the flow domain of the two fluids except for the interface, across which any discontinuities of the tangential velocity will render the interface a vortex sheet. Therefore the velocity fields have potentials, $\phi_j(\mathbf{r}, z, t)$ $(j = 1, 2)$, such that

$$\mathbf{u}_j = \nabla \phi_j, \qquad w_j = \partial \phi_j/\partial z \quad (-h_1 \le z \le \zeta \text{ for } j = 1; \quad \zeta \le z \le h_2 \text{ for } j = 2). \tag{15}$$

To pursue analysis for small $\epsilon$, we shall have the vertical lengths scaled by $h_1$ and horizontal lengths by a typical wavelength $\lambda$; the three-dimensional Laplace equation satisfied by the velocity potential $\phi_j$ then involves the parameter $\epsilon = h_1/\lambda$ as

$$(\phi_j)_{zz} + \epsilon^2 \nabla^2 \phi_j = 0 \quad (j = 1, 2). \tag{16}$$

Further, with $\phi_j$ scaled by $c\lambda$, where $c = \sqrt{g_o h_1}$ is a linear wave speed ($g_o$ is a reference value of $g_e$), $\phi_j$ satisfying Eqs. (16), (7), and (8) may assume a series expansion of the form

$$\phi_j(\mathbf{r}, z, t; \epsilon) = \sum_{n=0}^{\infty} \epsilon^{2n} \phi_{jn}(\mathbf{r}, z, t) = \sum_{n=0}^{\infty} \frac{(-1)^n}{(2n)!} [\epsilon H_j(z)]^{2n} \nabla^{2n} \phi_{jo}(\mathbf{r}, t; \epsilon), \tag{17}$$

$$H_1(z) = z + h_1 = z + 1, \qquad H_2(z) = z - h_2. \tag{18}$$

Here, the horizontal velocity $\mathbf{u}_j = \nabla \phi_j$ (scaled by $c$) and the elevation $\zeta$ (scaled by $h_1$) are assumed to be of the order of $\alpha$, which is arbitrary and hence not explicitly designated. The function $\phi_{jo}(\mathbf{r}, z, t; \epsilon)$, which is the only unknown involved in $\phi_j$, may depend on the parameter $\epsilon$ as a result of taking an appropriate regrouping of the complementary solutions of the higher-order equations for $\phi_{jn}$ such that $\nabla \phi_{j0}(\mathbf{r}, z, t; \epsilon) = O(\alpha)$ as $\epsilon \to 0$. This regrouping is admissible provided the medium is uniform ($h_j = \text{const.}$, $j = 1, 2$) and horizontally unbounded in the absence of any boundary effects of specific orders in magnitude.

From this expansion of $\phi_j$, we deduce the horizontal and the vertical velocity components $\mathbf{u}_j$ and $w_j$, both scaled by $c = \sqrt{g_o h_1}$, from $\mathbf{u}_j = \nabla \phi_j$, $w_j = \epsilon^{-1} \partial \phi_j/\partial z$, and using Eq. (17) we can readily obtain the on-surface horizontal and vertical velocity components $\tilde{\mathbf{u}}_j$ and $\tilde{w}_j$. The result can be written in operator form as

$$\tilde{\mathbf{u}}_j(\mathbf{r}, t) = A_j[\mathbf{u}_{jo}(\mathbf{r}, t)] = \sum_{n=0}^{\infty} \epsilon^{2n} A_{jn} \mathbf{u}_{jo}(\mathbf{r}, t), \tag{19}$$

$$\tilde{w}_j(\mathbf{r}, t) = B_j[\mathbf{u}_{jo}(\mathbf{r}, t)] = \sum_{n=0}^{\infty} \epsilon^{2n+1} B_{jn} \mathbf{u}_{jo}(\mathbf{r}, t), \tag{20}$$

where

$$A_{jn} = \frac{(-1)^n}{(2n)!} \eta_j^{2n} \nabla^{2n}, \qquad B_{jn} = \frac{(-1)^{n+1}}{(2n+1)!} \eta_j^{2n+1} \nabla^{2n} \nabla \quad (n = 0, 1, 2, \ldots). \tag{21}$$

Having attained the above expressions for $\tilde{\mathbf{u}}_j$ and $\tilde{w}_j$ in terms of $\mathbf{u}_{jo}$, we have actually accomplished a theory for modeling FNFD gravity-capillary internal waves in two-layered fluid systems of this family, which we classify as:

(A) the $(\mathbf{u}_{0j}, \zeta)$ system – the boundary velocity basis.

For the $(\mathbf{u}_{0j}, \zeta)$ system, the basic equations consist of kinematic conditions (5), the interface-projected momentum equation (13) and series relations (19) and (20):

$$\zeta_t = (1/\epsilon) \tilde{w}_j - \tilde{\mathbf{u}}_j \cdot \nabla \zeta, \qquad (j = 1, 2), \tag{22}$$

$$\tilde{D}_1 \tilde{\mathbf{u}}_1 - \mu \tilde{D}_2 \tilde{\mathbf{u}}_2 + \left[ g_e + \epsilon^2 (\tilde{D}_1^2 \zeta - \mu \tilde{D}_2^2 \zeta) \right] \nabla \zeta = -\gamma \nabla \nabla \cdot \mathbf{n}, \tag{23}$$

$$\tilde{\mathbf{u}}_j(\mathbf{r}, t) = A_j[\mathbf{u}_{jo}(\mathbf{r}, t)], \qquad \tilde{w}_j = B_j[\mathbf{u}_{jo}(\mathbf{r}, t)], \qquad (j = 1, 2) \tag{24}$$

where the operators $A_j$ and $B_j$ are specified in Eqs. (19)–(21) and the parameter $\epsilon$ is retained in Eq. (22) and installed in Eq. (23) by the same scaling rule together with the time scaled by $\lambda/c$. After substituting Eqs. (24) for $\tilde{\mathbf{u}}_j$ and $\tilde{w}_j$ into Eqs. (22) and (23), we obtain Eqs. (22) and (23) for the three unknowns $(\mathbf{u}_{1o}, \mathbf{u}_{2o}, \zeta)$ as the complete system of basic equations for modeling FNFD gravity-capillary internal waves in the two-layered fluids. However, it is seen that this model is still quite complicated for applications owing to the repeated appearance of the series and the series products in the equations. Such complications can be curtailed if $(\tilde{\mathbf{u}}_j, \zeta)$ are adopted as the basic variables.

To achieve this new objective, we first obtain the inverse of Eq. (19) as

$$\mathbf{u}_{jo}(\mathbf{r}, t) = J_j[\tilde{\mathbf{u}}_j(\mathbf{r}, t)] = \sum_{n=0}^{\infty} \epsilon^{2n} J_{jn} \tilde{\mathbf{u}}_j(\mathbf{r}, t), \tag{25}$$

$$J_{jo} = 1, \qquad J_{jn} = -\sum_{m=0}^{(n-1)} A_{j(n-m)} J_{jm} \quad (n = 1, 2, \ldots), \tag{26}$$

which determines all the $J_{jn}$, with the leading few given as

$$J_{j1} = -A_{j1}, \qquad J_{j2} = A_{j1}^2 - A_{j2}, \qquad J_{j3} = A_{j1}A_{j2} + A_{j2}A_{j1} - A_{j3} - A_{j1}^3, \quad \text{etc.} \tag{26a}$$

Here we note that in general the operators $A_{jn}$ and $J_{jm}$ are noncommutative, i.e., $A_{jn}J_{jm} \neq J_{jm}A_{jn}$ $(m + n > 2)$; it is nevertheless true that the inversion is identical on $\mathbf{u}_{jo}$ or on $\tilde{\mathbf{u}}_j$, i.e., $A_jJ_j = J_jA_j = 1$.

We complete the conversion by substituting Eq. (25) for $\mathbf{u}_{jo}$ into Eq. (20), which gives

$$\tilde{w}_j(\mathbf{r}, t) = B_j[\mathbf{u}_{jo}] = B_jJ_j[\tilde{\mathbf{u}}_j] = K_j[\tilde{\mathbf{u}}_j] = \sum_{n=0}^{\infty} \epsilon^{2n+1} K_{jn} \tilde{\mathbf{u}}_j(\mathbf{r}, t), \tag{27}$$

$$K_{jn} = \sum_{m=0}^{n} B_{jm} J_{j(n-m)} \quad (n = 0, 1, 2, \ldots), \tag{28}$$

which determines all the $K_{jn}$, with the leading few given as

$$K_{jo} = B_{jo}, \qquad K_{j1} = -B_{jo}A_{j1} + B_{j1}, \qquad K_{j2} = B_{jo}(A_{j1}^2 - A_{j2}) - B_{j1}A_{j1} + B_{j2}, \quad \text{etc.} \tag{28a}$$

Having obtained this expression for $\tilde{w}_j$ in terms of $\tilde{\mathbf{u}}_j$, we now have established another theoretical model that we designate as:

(B) the $(\tilde{\mathbf{u}}_j, \zeta)$ system – the interface velocity basis.

For the $(\tilde{\mathbf{u}}_j, \zeta)$ system, the basic equations consist of Eqs. (5) and (13) with the conversion of Eqs. (27) and (28):

$$\zeta_t + \tilde{\mathbf{u}}_j \cdot \nabla \zeta = \sum_{n=0}^{\infty} \epsilon^{2n} K_{jn} \tilde{\mathbf{u}}_j, \qquad (j = 1, 2) \tag{29}$$

$$\tilde{D}_1 \tilde{\mathbf{u}}_1 - \mu \tilde{D}_2 \tilde{\mathbf{u}}_2 + \left[ g_e + \epsilon^2 (\tilde{D}_1^2 \zeta - \mu \tilde{D}_2^2 \zeta) \right] \nabla \zeta = -\gamma \nabla \nabla \cdot \mathbf{n}, \tag{30}$$

where the operators $K_{jn}$ are given in Eq. (28), and here again the dispersion parameter $\epsilon = h_1/\lambda$ is retained for order indication.

We note that system (B), compared with model system (A), gains in simplicity and elegance since in (B) there is only one term involving a single series expansion.

We note that if $\mathbf{u}_o$ and $\zeta$ are analytic everywhere in the flow domain, the original series of Eqs. (19) and (20) are convergent within their radius of convergence, which is infinite. The inverted series of Eqs. (25)–(28) then define the inverse functions that are analytic within the flow domain since these inverted series are noted to possess a finite radius of convergence. It is therefore important to realize that it is no longer necessary to require $\epsilon$ to be small, and we may indeed set $\epsilon = 1$ by rescaling all the lengths, horizontal as well as vertical, by $h_1$.

In principle, these two models, both being exact, are therefore entirely equivalent in modeling this class of interfacial waves without limitation to the order of magnitude of nonlinearity and dispersion.

### 13.3. Conclusion and Discussion

Summarizing, we have now obtained two sets of models for describing FNFD, time-varying gravity-capillary waves on two-layered liquids of uniform depth in terms of the two sets of basic variables. Concerning the prospects of their further development and needs of efficient computational methods in implementing applications, we can make the following expository comments and discussion.

1. In the absence of the upper layer fluid, $\rho_2 = 0$ and $\mu = 0$, the above basic equations for the two model systems reduce to the case of a single layer of fluid, as recently reported by Wu (1997, 1998).

2. In addition, an alternative model can be derived based on the variables $(\bar{\mathbf{u}}_1, \bar{\mathbf{u}}_2, \zeta)$, where $\bar{\mathbf{u}}_j$ is the depth mean of $\mathbf{u}_j$ over each layer depth. The basic equations can be obtained by following the work of Wu (1998) for the single-layer case. This set of basic variables is perhaps the most commonly used in literature. However, it remains to be seen if this model is as simple as system (B) in practical applications.

3. For modeling fully nonlinear but weakly dispersive interfacial waves on the layered shallow fluids within the parametric regime $[\alpha = O(1), \epsilon \ll 1]$, in the spirit of Green and Naghdi (1976) and Ertekin et al. (1986), the $(\bar{\mathbf{u}}_1, \bar{\mathbf{u}}_2, \zeta)$ system, Eqs. (29) and (30), actually provides most directly the desired model equations simply by truncating the series in Eq. (29) to $n = N(\geq 1)$ terms, with a resulting accuracy up to $O(\epsilon^{2N})$. For $N = 0$, the basic equations reduce to the shallow water equations of Airy's class, which implies, as in the single-layer case, the nonexistence of permanent waves. For the corresponding model equation based on the $(\bar{\mathbf{u}}_j, \zeta)$ system, reference may be made to Choi (1995), Choi and Camassa (1996), and Wu (1998).

4. In somewhat more restricted cases, the present FNFD models can serve as the basis for deducing the family of Boussinesq model, the Kortweg–de Vries-type model for unidirectional waves, the bidirectional wave model of Wu (1994) for studying wave–wave interactions (Yih and Wu, 1995; Wu, 1995), the higher-order nonlinear and dispersive effects (Wu and Zhang, 1996a) and the related processes of transport of mass and energy (Wu and Zhang, 1996b).

5. These mathematical models can be further extended to evaluate the response of nonlinear dispersive systems physically open to external excitation at resonance of the system, such as the remarkable phenomenon of periodic generation of upstream-radiating solitons by disturbances moving steadily at transcritical velocity, as first discovered by Wu and Wu (1982) and Huang et al. (1982), as well as the interesting effects due to variations in channel shapes (Teng and Wu, 1992).

### Acknowledgment

This study has been sponsored by the U.S. National Science Foundation through the Hazard Mitigation Program under grant CMS-9503620.

# References

Boussinesq, J. 1871. *C. R. Acad. Sci.* Paris **72**, 755–759.

Choi, W. 1995. *J. Fluid Mech.* **295**, 381–394.

Choi, W. and Camassa, R. 1996. *J. Fluid Mech.* **313**, 83–103.

Ertekin, R. C., Webster, W. C., and Wehausen, J. V. 1986. *J. Fluid Mech.* **169**, 275–292.

Green, A. E. and Naghdi, P. M. 1976. *J. Fluid Mech.* **78**, 237–246.

Grue, J., Friis, H. A., Palm, E., and Rusas, P. O. 1997. *J. Fluid Mech.* **351**, 223–252.

Huang, D. B., Sibul. O. J, Webster, W. C., Webausen, J. V., Wu. D. M., and Wu. T. Y. 1982. Ships moving in the transcritical range, in *Proceedings of the Conference on the Behavior of Ships in Restricted Waters*, Vol 2, 26/1–26/10. Varna: Bulgarian Ship Hydrodynamics Center.

Korteweg, D. J. and de Vries, G. 1895. *Philos. Mag.* **39**, 422–443.

Osborne, A. R., Burch, T. L., and Scarlet, R. I. 1978. *J. Pet. Technol.* **30**, 1497–1504.

Rayleigh, Lord, 1876. *Phil. Mag.* **1**(5), 257–279.

Teng, M. H. and Wu, T. Y. 1992. *J. Fluid Mech.* **242**, 211–233.

Wu, D. M. and Wu, T. Y. 1982. Three-dimensional nonlinear long waves due to moving surface pressure, in *Proceedings of the Fourteenth Symposium on Naval Hydrodynamics*, pp. 102–125. Washington, DC: National Academy Press.

Wu, T. Y. 1981. *J. Eng. Mech. Div. ASCE* **107**, 501–522.

Wu, T. Y. 1994. *Methods Appl. Anal.* **1**(1), 108–117.

Wu, T. Y. 1995. *Acta Mech. Sin.* **11**, 289–306.

Wu, T. Y. 1997. In *Proceedings of the Twelfth International Workshop on Water Waves and Floating Bodies*, pp. 321–324 (at Centennial Celebration of Georg Weinblum, March 16–20, 1997, Marseilles, France).

Wu, T. Y. 1998. Nonlinear waves and solitons in water. *Phys. D* **123**, 48–63.

Wu, T. Y. and Zhang, J. E. 1996a. In *Mathematics is for Solving Problems: A Volume in Honor of Julian Cole on his 70th Birthday*, pp. 233–249. Philadelphia: Society for Industrial and Applied Mathematics.

Wu, T. Y. and Zhang, J. E. 1996b. Mass and energy transfer between unidirectional interacting solitons (A tribute to Prof. C. C. Yu in honor of his 80th Anniversary). *Chin. J. Mech.* **12**(1), 79–84.

Yih, C.-S. 1958. In *Proceedings of the Third U.S. National Congress on Applied Mechanics*, pp. 857–861.

Yih, C.-S. 1991. *Selected Papers by Chia-Shun Yih*, Vol I. A3, pp. 22–26. Singapore: World Scientific.

Yih, C.-S. and Wu, T. Y. 1995. General solution for interaction of solitary waves including head-on collisions, *Acta. Mech. Sin.*, **11**, 193–199.

Zhu, J., Wu, T. Y., and Yates, G. T. 1986. In *Proceedings of the Sixteenth Symposium on Naval Hydrodynamics*, pp. 186–197. Washington, D.C.: National Academy Press.

Zhu, J., Wu, T. Y., and Yates, G. T. 1987. In *Proceedings of the Third International Symposium on Stratified Flows*, ASCE. pp. 74–83.

# 14

# Instabilities in the Coupled Equatorial Ocean–Atmosphere System

HENK A. DIJKSTRA AND PAUL C. F. VAN DER VAART

## 14.1. Introduction

The large-scale interaction between the ocean and the atmosphere is one of the important factors of natural climate variability. The El Niño/Southern Oscillation (ENSO) phenomenon in the tropical Pacific is one of the most prominent examples of climate variability on interannual time scales. ENSO has large effects on climate, even far outside the Pacific basin, and occurs on relatively short time scales. Therefore it is one of the best-studied climate phenomena, both observational and theoretical. ENSO can be described as an oscillatory mode in the coupled ocean–atmosphere system, arising through large-scale instabilities involving the sea surface temperature (SST), the low-level atmospheric winds, and movements of the ocean–atmosphere interface. After an introduction into the phenomenon and its basic physics, the remainder of this chapter focuses on the nature of the instability causing the oscillation.

### 14.1.1. *The El Niño/Southern Oscillation Phenomenon*

The equatorial tropical Pacific climate system is a complex dynamical system, involving strong coupling between the ocean and the atmosphere (Philander, 1990). The time-averaged SST is characterized by a cold tongue [Fig. 14.1(a)] of 24°C water in the eastern Pacific and a warm pool of 30°C near the western boundary of the basin. Once approximately every 4 years, the SST in the eastern Pacific increases by a few degrees over a period of ~1 year. These events are called El Niño (literally, the little boy), referring to the Christmas Child, since the maximum of the event is usually around December. The 1997/1998 El Niño had one of the largest amplitudes of this century. The SST anomaly pattern [with respect to the mean state in Fig. 1(a)] for December 1997 is plotted in Fig. 1(b). El Niño is seen as a basin-wide SST perturbation with a maximum amplitude near the South American coast of ~3°C.

One of the measures of the condition of the eastern Pacific Ocean is the NINO3 index, which gives the SST anomaly from the mean state averaged over the box [150W − 90W] × [5S − 5N]; this index is therefore positive during an El Niño. The drawn curve in Fig. 14.2 shows the course of this index from 1950 to 1997. El Niño episodes occur once every 3 to 7 years and last more than 1 year, with substantial variations in strength. The strongest El Niños were those of 1982/1983 and 1997/1998, which had approximately the same NINO3 amplitude.

Normally the equatorial Pacific surface winds, the trade winds, are directed westward and are driven by a pressure difference between a high-pressure region in the east (e.g., at Tahiti) and a low-pressure region in the west (e.g., at Darwin). During an El Niño, the pressure is lower than normal in Tahiti and higher than normal in Darwin. These variations in atmospheric pressure are known as the Southern Oscillation. The Southern Oscillation index (SOI), the pressure

179

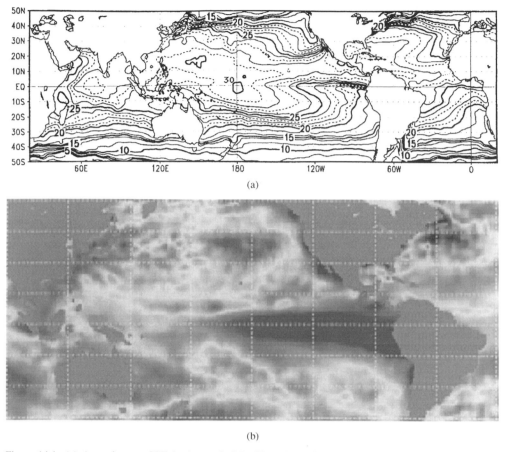

(a)

(b)

Figure 14.1. (a) Annual mean SST in the tropical Pacific. Along the equator, contour levels are in degrees centigrade. (b) SST anomaly pattern for December 1997. Light areas indicate zero anomalies, whereas darker shading indicates positive anomalies, with a maximum amplitude of $3°C$.

difference between Tahiti and Darwin, is plotted as the dotted curve in Fig. 14.2. When this index is negative (positive), the westward surface winds are weak (strong).

Although the strong negative correlation between SOI and NINO3 in Fig. 14.2 is obvious, it took until 1969 (Bjerknes, 1969) before it was realized that the changes occurring in the atmosphere and the ocean are closely related. Warm water in the east causes a weakening of the trade winds, which in turn drives changes in the oceanic circulation that influence the SST. The ENSO is therefore a coupled ocean–atmosphere phenomenon. The warm phase of the oscillation coincides with El Niño (positive NINO3) in the ocean and with weak trade winds (negative SOI) in the atmosphere. The cold phase (also called La Niña, the little girl) coincides with strong trade winds (positive SOI) and lower than normal SSTs (negative NINO3) in the eastern part of the equatorial Pacific.

### 14.1.2. *Basic Physics*

Over the years, measurements and modeling have led to a better understanding of ENSO, based on the knowledge of the behavior of the ocean and the atmosphere. Under normal conditions, the trade winds induce a slope in the sea level along the equator (Fig. 14.3). In the

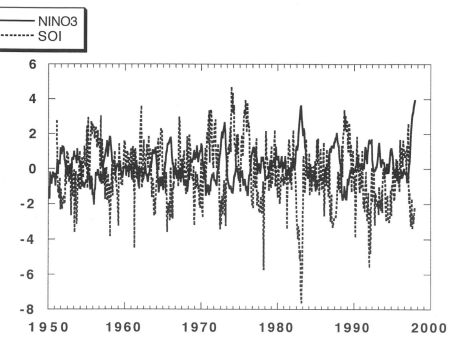

Figure 14.2. Time series of the SST anomaly averaged over a box in the eastern Pacific (NINO3) and the difference in sea level pressure [Southern Oscillation index (SOI)] between Tahiti (eastern Pacific) and Darwin (western Pacific).

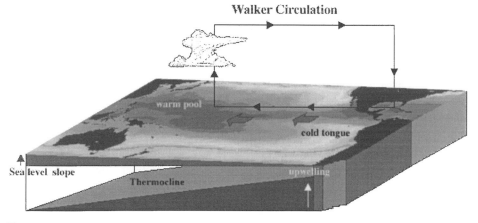

Figure 14.3. Sketch of normal ocean–atmosphere conditions. The slope in the boundary between warm surface water and the colder water below (the thermocline) is idealized as being constant. Westward surface winds induce a slope in sea level and thermocline and cause, together with equatorial upwelling, the characteristic cold tongue/warm pool SST structure.

west, the sea level is on average ~25 cm higher than in the east. Because of the higher hydrostatic pressure, the 20°C isotherm is pushed more downward in the west than in the east. The depth of this isotherm is a measure of the boundary between the warm surface water and the cold water below and is called the thermocline. The cold water is normally situated ~50 m from the surface in the east but is at a much larger depth (at ~200 m) in the west (Fig. 14.3).

Frictional processes in the surface layer of the ocean and the Coriolis effect at slightly off-equatorial latitudes induce a northward (southward) (Ekman) mass transport just north (south)

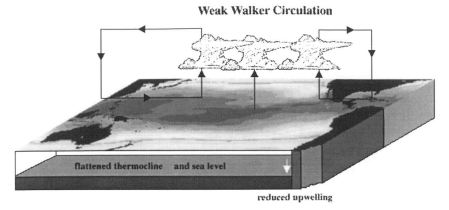

Figure 14.4. Ocean–atmosphere conditions during an El Niño event with a flattened sea level and thermocline, a weak Walker circulation, and a reduced upwelling causing the relatively warm SST in the Eastern Pacific.

of the equator. At the equator, this divergence is compensated for by an upward flow bringing relatively cold water to the surface. Together, the slope in the thermocline and the upwelling cause the cold tongue of water in the east and the warm pool in the west. Air warms up above warm water and then rises, which leads to convection and cloud formation while air descends over the cold water. Hence the east–west SST contrast induces surface winds that blow from east to west, thus reinforcing the trade winds. The circulation driven by this zonal SST gradient is called the Walker circulation.

A classical explanation of El Niño is that of a delayed oscillator (Suarez and Schopf, 1988). At the beginning of an El Niño, the surface winds weaken, which causes a change in the slope of the sea level and the thermocline. The thermocline tends to flatten and, together with a reduced upwelling, this leads to a warming of the eastern Pacific. This in turn causes a weakening of the Walker circulation, leading to a further weakening of the surface winds (Fig. 14.4). The ocean, however, does not immediately adapt to the changes in windstress. Long (Kelvin and Rossby) waves in the thermocline are excited to adapt to the changing forcing conditions at the surface. In the equatorial ocean, a Kelvin wave takes ∼2 months to travel from west to east and a Rossby wave ∼6 months to travel in the opposite direction. During an El Niño, the easterly thermocline lies deeper on the equator, but less deep in areas just to the north and the south of the equator. These off-equatorial shallower areas first propagate to the west, reflect at the west coast of the Pacific, and then travel to the east. Arriving at their easterly destination, they cause a shallowing of the thermocline which terminates El Niño.

In more modern terms, ENSO is viewed as an oscillatory mode in the coupled ocean–atmosphere system. During this oscillation, the surface winds strengthen and weaken because of variations in the SST, which induces changes in the ocean circulation. The latter affect the SST, and the phase differences needed for the oscillation are due to the adaptation processes of the ocean, involving a basin-wide reorganization of the thermocline.

## 14.2. The Zebiak–Cane Model

One of the first models that was able to simulate ENSO reasonably was that of Zebiak and Cane (1987). In its original version, an annual mean state and seasonal cycle of both ocean and atmosphere are obtained from observations, and the evolution of anomalies with respect to this reference state are computed. The model produces recurring warm events that are irregular in both amplitude and spacing, but favor a 3–4-year period.

The model describes large-scale motions in the tropical ocean and atmosphere in a domain of infinite extent in the latitudinal direction ($y$). The ocean is bounded by meridional walls representing the coasts of Asia ($x = 0$) and America ($x = L$). The ocean component of the model consists of a well-mixed layer of mean depth $H_1$ embedded in a shallow water layer of mean depth $H = H_1 + H_2$ that has a constant density $\rho_1$ (Fig. 14.5). Only long-wave motions above the thermocline are considered, and the deep ocean (having a constant density $\rho_2$) is assumed to be at rest. Deviations from the averaged thermocline depth $H$ are indicated by $h$, and the horizontal velocities $u$ and $v$ are the dynamical variables in this layer. The evolution of both velocity and thermocline anomalies is governed by the shallow water equations on an equatorial $\beta$-plane driven by a windstress $\tau = (\tau^x, \tau^y)$. The equations are (Philander, 1990)

$$\frac{\partial u}{\partial t} + \epsilon_o u - \beta y v + g'\frac{\partial h}{\partial x} = \frac{\tau^x}{\rho_1 H}, \tag{1a}$$

$$\beta y u + g'\frac{\partial h}{\partial y} = \frac{\tau^y}{\rho_1 H}, \tag{1b}$$

$$\frac{\partial h}{\partial t} + \epsilon_o h + H\left(\frac{\partial u}{\partial x} + \frac{\partial v}{\partial y}\right) = 0, \tag{1c}$$

where $g' = g(\rho_2 - \rho_1)/\rho_1$ is the reduced gravity and $\epsilon_o$ is a friction coefficient. The characteristic velocity in the ocean is given by $c_0 = (g'H)^{\frac{1}{2}}$, which is the phase speed of the first free

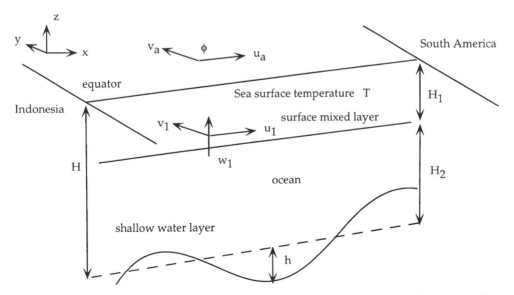

Figure 14.5. Schematic representation of the Zebiak–Cane model showing both surface layer and shallow water layer, the latter bounded below by the thermocline.

baroclinic Kelvin wave. The boundary conditions are

$$\int_{-\infty}^{\infty} u(0, y, t) \, dy = 0, \qquad u(L, y, t) = 0. \tag{2}$$

The first condition models a zero net mass flux through the western boundary ($x = 0$), while the second is a kinematic condition on the eastern boundary ($x = L$) corresponding to zero normal flow.

The intensification of wind-driven currents through rotation and frictional processes is modeled through a fixed-depth Ekman surface layer. The zonal and the meridional velocities in this layer are $u_s$ and $v_s$, respectively, and the governing equations are given by (Zebiak and Cane, 1987)

$$\epsilon_s u_s - \beta y v_s = \frac{H_2}{H_1} \frac{\tau^x}{\rho_1 H}, \tag{3a}$$

$$\epsilon_s v_s + \beta y u_s = \frac{H_2}{H_1} \frac{\tau^y}{\rho_1 H}, \tag{3b}$$

in which $\epsilon_s$ models the surface layer friction.

The horizontal velocities ($u_1, v_1$) influencing the heat transport in the mixed layer consist of two components, i.e.,

$$u_1 = u + u_s, \qquad v_1 = v + v_s. \tag{4}$$

The velocities without subscripts are due to the vertical mean currents above the thermocline and the other part is due to the vertical shear currents associated with the fixed-depth surface layer. The vertical velocity component, the upwelling, is determined from continuity:

$$w_1 = H_1 \left( \frac{\partial u_1}{\partial x} + \frac{\partial v_1}{\partial y} \right). \tag{5}$$

The evolution of the mixed-layer temperature $T$ is governed by the equation

$$\frac{\partial T}{\partial t} + \epsilon_w (T - T_0) + \frac{w_1}{H_1} \mathcal{H}(w_1)[T - T_{\text{sub}}(h)] + u_1 \partial_x T + v_1 \partial_y T = 0, \tag{6}$$

where $\mathcal{H}$ is a continuous approximation of the Heaviside function. The second term in Eq. (6) is the Newtonian cooling term, with inverse damping time $\epsilon_w$, representing all vertical processes as mixing, sensible and latent heat fluxes, and long-wave and short-wave radiation. $T_0$ is the temperature of radiative equilibrium that is realized in the absence of large-scale horizontal motion in the upper ocean and atmosphere. The next term models the heat flux due to upwelling through the total velocity $w_1$ and the approximate vertical temperature gradient $[T - T_{\text{sub}}(h)]/H_1$. The subsurface temperature ($T_{\text{sub}}$) depends on the thermocline deviations and models the effect in which heat is transported upward (if $w_1 > 0$) when the cold water is farther from the surface. It is parameterized by

$$T_{\text{sub}}(h) = T_{s0} + (T_0 - T_{s0}) \tanh \left( \frac{h + h_o}{H} \right), \tag{7}$$

where $h_o$ is some offset value. In this formulation, $T_{\text{sub}}$ cannot exceed $T_0$ and cannot decrease below $2T_{so} - T_0$. The last two terms in Eq. (6) represent horizontal advection.

The atmospheric zonal and meridional boundary-layer velocities ($u_a$, $v_a$) and geopotential $\Phi$ (proportional to sea-level pressure) are assumed to adapt instantaneously to the forcing. The atmospheric circulation is forced by SST anomalies with respect to $T_0$. The dynamics of the atmosphere component follows the work of Gill (1980), with the mechanical balances being among Coriolis force, pressure gradient, and boundary-layer friction, while the thermodynamical balances are between diabatic heating and divergent flow. The equations are

$$\epsilon_a u_a - \beta y v_a + \frac{\partial \Phi}{\partial x} = 0, \tag{8a}$$

$$\beta y u_a + \frac{\partial \Phi}{\partial y} = 0, \tag{8b}$$

$$\epsilon_a \Phi + c_a^2 \left( \frac{\partial u_a}{\partial x} + \frac{\partial v_a}{\partial y} \right) = -\alpha_T (T - T_0), \tag{8c}$$

where $\epsilon_a$ represents boundary-layer friction and $\alpha_T$ is a proportionality constant relating SST anomalies and heat forcing. The constant $c_a$ is the phase speed of the first baroclinic Kelvin wave in the atmosphere (Gill, 1980).

We close the model equations by prescribing how the windstress $\tau$ is determined from the atmosphere model. Following Dijkstra and Neelin (1995), we assume that the windstress is only zonal ($\tau^y = 0$) and part of the windstress, say $\tau_{\text{ext}}$, is unrelated to coupled processes within the basin. Even in the absence of zonal SST gradients, there still would be a weak zonal circulation, for example, driven by the zonally asymmetric land–sea contrast. The remainder is induced by the basin-wide zonal SST gradient and is assumed proportional the zonal surface velocity. Hence,

$$\tau^x = \tau_{\text{ext}} + \gamma u_a. \tag{9}$$

In the coupled model, the winds can be calculated once the SST deviations from $T_0$ are known. From the winds, the ocean response in both the shallow water layer and the surface layer is determined and subsequently the SST can be calculated from Eq. (6). The numerical details for solving the set of governing equations can be found in the work of Van der Vaart et al. (1998). Variables are expanded into spectral basis functions, with Chebyshev polynomials in the zonal direction and Hermite functions in the meridional direction. By use of collocation techniques, a set of nonlinear algebraic equations is obtained for the steady states of the model. The analysis of the stability of these steady states leads to a generalized eigenvalue problem. Both steady states and their linear stability are traced through parameter space by use of continuation techniques (Dijkstra et al., 1995).

## 14.3. Bifurcation Analysis of the Zebiak–Cane model

The Zebiak–Cane model turns out to be very well suited to determine the details of the evolution of ENSO events and has been widely used to study their physics. Many results of this model have been reviewed in the work of Neelin et al. (1994), in which many references also can be found. From these results, it appears that a large-scale instability of the coupled ocean–atmosphere plays a major role in the dynamics of ENSO. It is this instability that is the focus of the results below.

### 14.3.1. *The Warm Pool/Cold Tongue Mean State*

The control parameter in the system is the coupling parameter $\mu$, which is a dimensionless product of $\alpha_T$ and $\gamma$. Its physical interpretation is the amount of windstress per degree of SST anomaly. Other parameters are fixed at their best values as estimated from observations. At zero coupling ($\mu = 0$), the ocean circulation and consequently the SST is determined by the external zonal windstress $\tau_{\text{ext}}$. This windstress is assumed to have the form

$$\tau_{\text{ext}} = -F_0 e^{-\frac{(\alpha y)^2}{2}}, \tag{10}$$

where $\alpha$ controls the meridional extension of the external wind. The amplitude $F_0$ corresponds to a dimensional value of 0.01 Pa, which is $\sim$10% of the observed windstress. At each latitude, the external wind is zonally constant. In response to the external wind, the equatorial temperature, say $T_{\text{ext}}$, increases monotonically from $\sim$25.5°C in the east to $\sim$28.5°C in the west. The thermocline is approximately linear at the equator, its depth is increasing westward, and it has slight off-equatorial maxima near the western boundary.

At small $\mu$, the additional windstress due to coupling is approximately the atmospheric response to the cooling $T_{\text{ext}} - T_0$. This enhances the westward winds over most of the basin, leading to larger upwelling and a stronger thermocline slope and strengthening the cold tongue in the eastern part of the basin. The temperature $(T - T_0)_{EC}$ of the cold tongue and the vertical velocity just below the cold tongue $w_E$ [Fig. 14.6] demonstrate that there is a unique steady solution as a function of $\mu$, with more upwelling as coupling gets stronger.

At $\mu = 0.5$, the spatial structure of the mean state is as shown in Fig. 14.7. The zonal scale of the cold tongue [Fig. 14.7(a)] is set by a delicate balance of thermocline and surface layer feedbacks (Dijkstra and Neelin, 1995). The meridional extent of the cold tongue is determined both by the Ekman spreading length $(\epsilon_s/\beta)$ and by meridional advection. The thermocline field [Fig. 14.7(b)] shows the off-equatorial maxima and a deeper (shallower) equatorial thermocline in the west (east). This indicates that the reservoir of heat content lies off equatorial in the central and the western parts of the basin. The zonal wind response $u_a$ [Fig. 14.7(c)] shows the intensification of the westward winds, with a maximum west of the cold tongue. The vertical

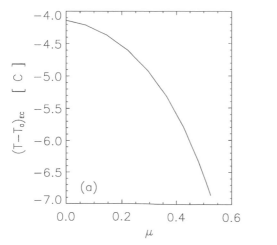
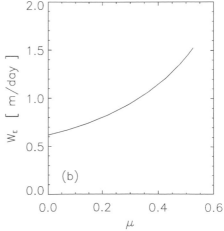

Figure 14.6. (a) Eastern Pacific ($x/L = 0.8$) equatorial SST deviation from $T_0 = 30$°C, (b) upwelling velocity as a function of the coupling strength $\mu$.

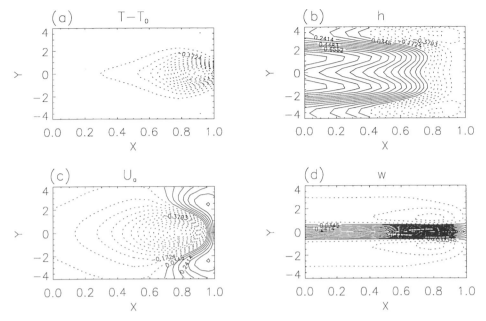

Figure 14.7. The mean state at standard parameter values and $\mu = 0.5$ in the $X = x/L$ and $Y = y/\lambda$ planes, where $\lambda = (c_0/\beta)^{\frac{1}{2}}$ is a characteristic meridional length scale. (a) $T - T_0$, maximum $6.6°$C; (b) thermocline depth, maximum 82.3 m; (c) zonal wind $u_a$, maximum 9.5 m/s; (d) vertical velocity $w_1$, maximum 1.44 m/day. In all panels, values are scaled with the maximum value of each field and the contour levels (with interval 0.069) are with respect to this maximum.

velocity structure [Fig. 14.7(d)] is clearly controlled by Ekman divergences. Upward velocities are restricted to an equatorial zone, and the maximum amplitude occurs in the eastern part of the basin.

### 14.3.2. *Linear Stability of the Coupled Climatology*

Along the branch of steady states in Fig. 14.6, we determine the linear stability simultaneously by writing the total solution vector $\phi$, consisting of ocean, atmosphere quantities, and the SST, as

$$\phi(x, y, t) = \bar{\phi}(x, y) + \tilde{\phi}(x, y)e^{\sigma t}. \tag{11}$$

Here, the vector $\bar{\phi}$ represents the mean state, $\tilde{\phi}$ are perturbations with respect to this mean state, and $\sigma$ is the complex growth rate of the perturbation. In Fig. 14.8, the path of six modes – which become leading eigenmodes at high coupling – is plotted as a function of the coupling strength $\mu$. In Fig. 14.8(a) a larger dot size indicates a larger value of $\mu$ and both period and growth rate of the modes are given in year$^{-1}$. In Fig. 14.8(b) only the growth rate is plotted against $\mu$. One oscillatory mode becomes unstable as $\mu$ is increased and a Hopf bifurcation occurs near $\mu = 0.5$.

At the critical value $\mu_c = 0.5$, for which the mean state was shown in Fig. 14.7, time–longitude diagrams of the equatorial thermocline, temperature, and zonal wind anomalies of this oscillatory mode are shown in Fig. 14.9. The SST pattern [Fig. 14.9(b)] displays a nearly standing oscillation for which the spatial scale is confined to the cold tongue of the mean state. There is a slight eastward propagation of the SST anomaly in the central equatorial Pacific. The

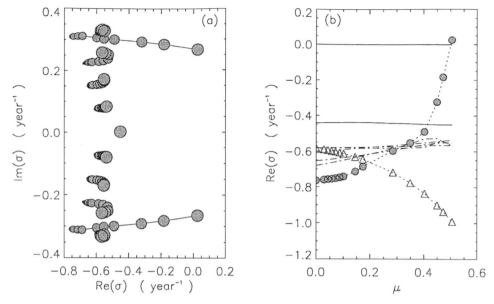

Figure 14.8. (a) Plot of the eigenvalues for the six leading eigenmodes in the [Re($\sigma$), Im ($\sigma$)] plane. Values of the coupling strength $\mu$ are represented by dot size [the smallest dot is the uncoupled case ($\mu = 0$) for each mode; the largest dot is the fully coupled case at the Hopf bifurcation ($\mu_c = 0.5$)]. The Hopf bifurcation that yields the ENSO mode occurs where the path of one eigenvalue first crosses Re($\sigma$) = 0. (b) The growth rate of the leading modes as a function of coupling strength.

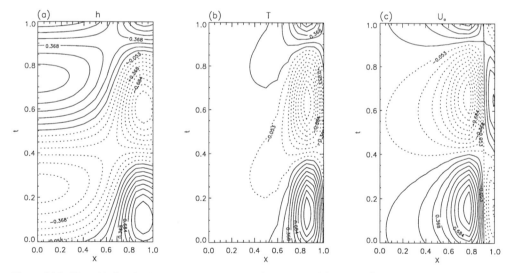

Figure 14.9. Time ($t$)–longitude ($X = x/L$) diagram at the equator of the anomalies of (a) SST (max = 1.4°C), (b) thermocline depth (max = 9.5 m), (c) zonal wind (max = 3.0 m/s). The period of the oscillation is 3.7 years. Note that the amplitude of the oscillation is not determined by the linear stability analysis. The maximum amplitudes are relative magnitudes of the different fields, i.e., the mode displays a thermocline deviation of ~10 m per degree SST anomaly.

thermocline anomaly [Fig. 14.9(a)] shows western anomalies in heat content leading those with the same sign at the eastern boundary. These anomalies are out of phase with the SST anomalies with a lag of ~5 months. The wind response [Fig. 14.9(c)] is much broader zonally and is in phase with the SST anomaly. Although the wind maximum in the central Pacific is more to the

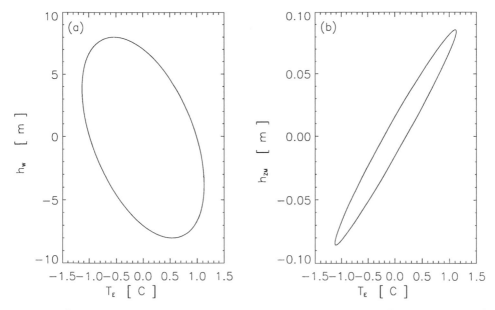

Figure 14.10. (a) Phase relation between the equatorial thermocline depth anomaly in the western part of the basin and the SST anomaly in the east, (b) the phase relation between the zonally averaged equatorial thermocline anomaly $h_{ZM}$ and SST anomaly in the east ($T_E$). Note that the amplitudes in both panels are arbitrary, but that their ratio is fixed. The direction of rotation is clockwise in both cases.

east than that in observations, the time–longitude diagrams do correspond reasonably to those observed (Neelin et al., 1994).

The phase relationships between wind/SST anomalies and thermocline anomalies can be seen more clearly in Fig. 14.10. Here, the thermocline depth in the western Pacific $h_W$ and the zonal mean equatorial thermocline displacement $h_{ZM}$ are related to the SST anomaly in the eastern Pacific (at $x/L = 0.92$) over one cycle of the oscillation. Figure 14.10(a) shows the characteristic ENSO phase relationship between SST and thermocline anomalies with a relatively shallow (deep) western thermocline in case of a warm (cold) event. As the closed curve is traversed clockwise over one cycle of the oscillation, it is seen that a cold event is followed by an extreme positive western thermocline anomaly. As the SST anomaly becomes zero, the western thermocline anomaly is still positive, and Fig. 14.10(b) shows that this also holds for the zonally averaged thermocline anomaly. Hence the equatorial heat content is slowly built up after the cold event by the increase of the trade winds. This sets the stage for the following warm event in which the equatorial heat content is discharged. After the warm event, the zonally averaged thermocline anomaly is negative as the SST anomaly goes through zero again and the equatorial heat content is low, which causes the next cold event.

The meridional structure of the ENSO mode is shown by plots of the different fields (Figs. 14.11–14.13) at several phases of the oscillation relative to the period, i.e., phase $t = 1/2$ indicates the fields after half a period. The starting point of the description is a positive SST anomaly in the eastern Pacific (early El Niño phase), as shown in Fig. 14.11 at $t = 0$. Eastward zonal wind anomalies to the west of the maximum in the SST anomaly (Fig. 14.11) are present, as can be seen in Fig. 14.13 at $t = 0$. The wind response amplifies the positive SST anomaly ($t = 1/16$ to $1/8$), and the spatial scale of the SST anomaly is controlled by the shape of the cold tongue (see Fig. 14.7). The equatorial thermocline response to the weaker surface winds up to $t = 1/8$ results in a negative anomaly (i.e., negative heat content) in the western Pacific

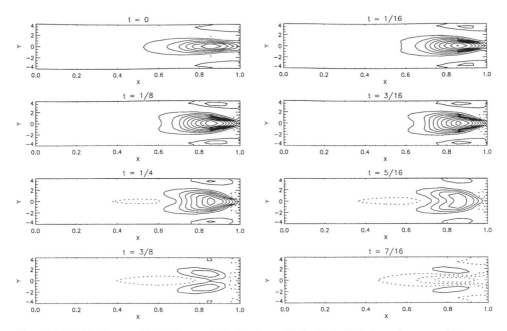

Figure 14.11. Plan forms of the SST anomaly during the oscillation in the $X$–$Y$ plane; times are with respect to the period of the oscillation. Solid (dotted) curves represent warm (cold) anomalies.

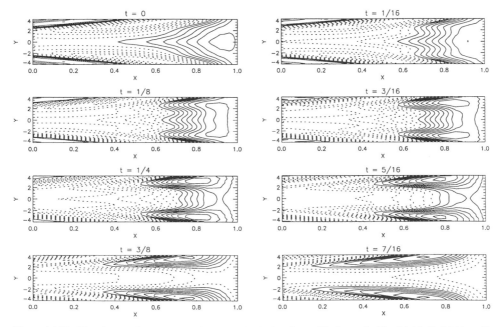

Figure 14.12. Plan forms of the thermocline anomaly during the oscillation as in Fig. 14.11. Solid (dotted) curves represent positive (negative) anomalies.

(Fig. 14.12, $t = 1/8$). This anomaly is at its minimum a few months later than the maximum of equatorial SST. As long as the positive thermocline/SST anomaly in the eastern part of the basin does not weaken, this negative anomaly cannot be discharged. However, because of ocean wave reflections at the eastern boundary, the mass fed along the equator to the eastern basin

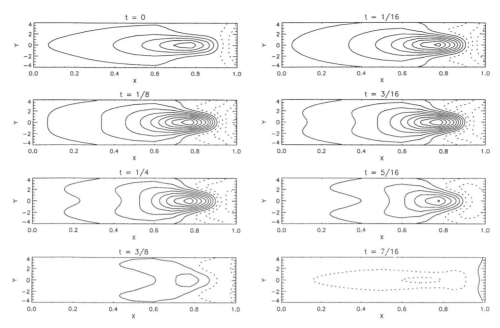

Figure 14.13. Plan forms of the zonal wind anomaly during the oscillation as in Fig. 14.11. Solid (dotted) curves represent anomalous eastward (westward) winds.

is transformed into a collective of long Rossby waves that propagates westward (Fig. 14.12, $t = 0–1/8$).

At the equator, the eastern positive thermocline anomaly and consequently the SST anomalies are weakened. This reduces the east–west SST gradient, causing the anomalous eastward winds to weaken (Fig. 14.13, $t = 1/8$ to $5/16$). Termination of the El Niño phase sets in as the western warm pool discharges its, previously builtup, negative heat content ($t = 3/8–7/16$ in Fig. 14.12). This is characteristic of the recharge oscillator (Jin, 1997a), showing a negative zonally mean thermocline anomaly at the equator during the transition from warm to cold SST anomalies. As the thermocline rises in the east, the SST anomaly becomes negative and, through coupled processes, its amplitude increases. The trade winds recover (Fig. 14.13, $t = 7/16$), and the positive off-equatorial thermocline anomalies propagate westward (Fig. 14.12, $t = 3/8–7/16$). Then the cycle starts over again, but with the signs of the perturbations reversed.

### 14.3.3. *Weakly nonlinear analysis*

The results of the linear stability analysis do not provide information on the finite amplitude of the fields for supercritical conditions. In this subsection, the equilibration of the perturbations to finite amplitude is studied in a weakly nonlinear context, i.e., for coupling values of $\mu$ just above the Hopf bifurcation. Let

$$\epsilon = \frac{\mu - \mu_c}{\mu_c} \ll 1 \tag{12}$$

be a measure of the distance beyond critical conditions, i.e., $\mu = \mu_c$ where $\text{Re}(\sigma) = 0$. For these values of the coupling strength $\mu$, equilibration of the unstable perturbations will occur on a time scale that is long compared with the time scale of growth. Therefore a new time variable is introduced:

$$\tau = \epsilon^2 t. \tag{13}$$

Coupling strength $\mu$, time, and the solution vector $\phi$ are expanded in terms of $\epsilon$ and the fundamental mode with time dependence $e^{i\omega_c t}$, where $\omega_c = \text{Im}(\sigma)$ at $\mu = \mu_c$,

$$\phi = \bar{\phi} + \epsilon A(\tau)\tilde{\phi}e^{i\omega_c t} + \epsilon^2\big(|A(\tau)|^2\tilde{\phi}_{02} + A^2(\tau)\tilde{\phi}_{22}e^{2i\omega_c t}\big) + \epsilon^3\tilde{\phi}_{13}e^{i\omega_c t} + c.c., \tag{14a}$$

$$\partial_t \rightarrow i\omega_c + \epsilon^2\partial_\tau, \tag{14b}$$

$$\mu = \mu_c(1 + \epsilon^2 m), \qquad m = \mathcal{O}(1). \tag{14c}$$

In these expansions, $c.c.$ denote complex conjugate, $m$ is the new control parameter, and $A(\tau)$ is the (complex) amplitude of the initially unstable mode with spatial structure $\tilde{\phi}$.

By substituting expansions (14) into the governing equations and collecting terms of like orders in $\epsilon$ and $e^{i\omega_c t}$, one can reduce the full equations to a scalar equation for the amplitude $A(\tau)$. This becomes a Landau equation:

$$\frac{\partial A}{\partial \tau} = m\frac{\partial \sigma}{\partial \mu}A - \Lambda A|A|^2, \tag{15}$$

where the coefficients are evaluated at $\mu = \mu_c$ and are calculated numerically within the spectral setup (Van der Vaart and Dijkstra, 1997). When Eq. (15) is solved for $A$, the solution for the SST field becomes

$$T(x, y, t) = \bar{T}(x, y) + \epsilon A(\tau)\tilde{T}(x, y)e^{i\omega_c t} + \mathcal{O}(\epsilon^2), \tag{16}$$

where the mean state is represented by $\bar{T}$ and the critical mode by $\tilde{T}$. If the coefficients $\partial\sigma/\partial\mu$ and $\Lambda$ satisfy the conditions for a supercritical Hopf bifurcation, $\text{Re}(\sigma_\mu) > 0$ and $\text{Re}(\Lambda) > 0$, finite-amplitude solutions to Eq. (15) exist of the form

$$A(\tau) = ge^{i\Omega\tau}, \qquad g = \sqrt{\frac{\text{Re}\big(\frac{\partial\sigma}{\partial\mu}\big)}{\text{Re}(\Lambda)}}, \qquad \Omega = \text{Im}\bigg(\frac{\partial\sigma}{\partial\mu}\bigg) - \frac{\text{Im}(\Lambda)}{\text{Re}(\Lambda)}. \tag{17}$$

Using these expressions, one can derive the amplitude and the total period of the stable limit cycle for coupling values beyond the Hopf bifurcation. For example, the period $P$ is given by

$$P = \frac{2\pi}{\omega_c + \epsilon^2\Omega}. \tag{18}$$

In Fig. 14.14(a) the period of the periodic orbit is plotted as a function of the coupling strength $\mu$. The period of ENSO is set by the critical period (at the Hopf bifurcation), rather than the frequency of the instability at supercritical coupling strength, which is also found in other studies (Battisti and Hirst, 1989; Neelin et al., 1994; Jin, 1997b). The amplitude of the SST anomaly in the cold tongue increases strongly with coupling being $\sim 3°\text{C}$ for $\epsilon = 0.1$.

### 14.3.4. *Irregular Behavior of ENSO*

Temporal variability in the tropical Pacific does not only occur on interannual time scales. One other major time-dependent phenomenon in the tropical Pacific is the seasonal cycle (Horel,

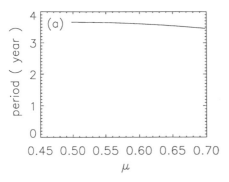 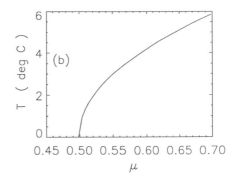

Figure 14.14. (a) The frequency, (b) the amplitude of the finite-amplitude limit cycle as obtained from the weakly nonlinear analysis within the fully coupled Zebiak–Cane model in a finite ocean basin. Shown is the maximum amplitude of the SST anomaly near the position of the cold tongue.

1982). At the equator, SSTs are warmest during winter and spring and are coldest during summer and fall when the northward winds are maximal. The annual component of the seasonal cycle in the eastern Pacific is a strange phenomenon, considering the semiannual component of the forcing. It is by now clear that coupled processes, in particular those in the surface layer, are involved to get an annual response to a semiannual forcing (Change et al., 1995). The interaction of seasonal cycle and ENSO variability can lead to very complicated temporal behavior (Jin et al., 1994; Tziperman et al., 1994; Chang et al., 1996). It was shown that the interactions of an externally forced seasonal cycle and the internal ENSO oscillation lead to subharmonic frequency locking and chaotic behavior.

Transient simulations in which variants of the Zebiak–Cane model are used in the strongly nonlinear regime and subjected to seasonal forcing with annual period and atmospheric (white) noise (Jin et al., 1996; Tziperman et al., 1995) lead to the understanding that the seasonal cycle controls the timing of warming events whereas the noise forcing is the most likely candidate for ENSO irregularity. An example of results of such a simulation in which the model above is used at supercritical conditions ($\mu = 0.7$) is shown in the time–longitude diagrams of Fig. 14.15. The deterministic period of the supercritical solution retains its identity over the influence of the seasonal and stochastic forcing [Fig. 14.16(a)]. The phase-space view of eastern Pacific SST anomaly versus western Pacific thermocline depth anomaly [Fig. 14.16(b)] shows the effect of the noise as a fuzzy signature of the otherwise stable periodic orbit. This broadening compares with results for which more complicated models were used (Blanke et al., 1997).

## 14.4. Summary

A bifurcation analysis of an intermediate complexity model has been performed to study the physics of the El Niño/Southern Oscillation phenomenon. Both the mean state and the instabilities are determined by coupled processes between the ocean and the atmosphere. An external wind induces a weak zonal SST gradient, with nonzero upwelling and a small thermocline slope. The SST gradient is amplified by the coupled processes to give the cold tongue/warm pool spatial structure of the mean state. The amplitude of the cold tongue depends on the coupling strength $\mu$.

This state becomes unstable to a single oscillatory mode as the coupling strength is increased. The mechanism of this mode is a large-scale instability, involving motion at the air–sea interface and in the thermocline and the strengthening/weakening of the trade winds due to anomalies in

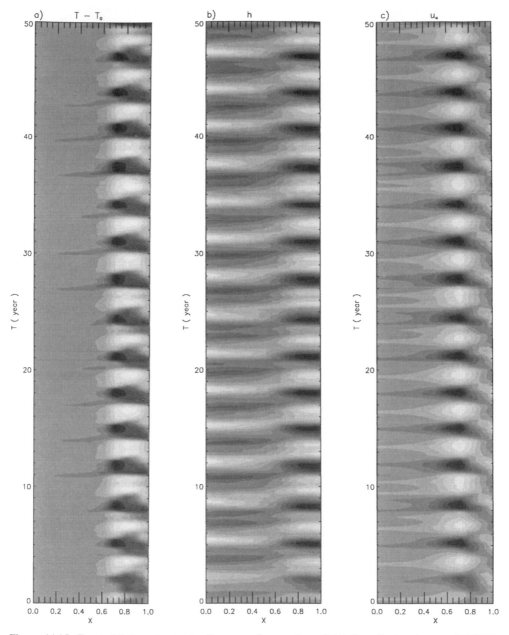

Figure 14.15. Equatorial time–longitude diagrams of anomalies of (a) $T - T_0$ (maximum = 5.732°C), (b) thermocline depth (maximum = 71 m), (c) zonal wind (maximum = 16.5 m s$^{-1}$). Simulation at super-critical conditions, with stochastic forcing and seasonal cycle included. Dark (light) colors represent negative (positive) anomalies.

SST. The ocean provides the memory of the oscillation because it not only responds to actual winds, but also to past winds because of the propagation of waves in the thermocline. The spatial pattern of the mean state is important to set the location where amplification of distur-bances can occur. As the cold tongue warms, the trade winds relax, the equatorial thermocline tilt reduces, and the deepening of the eastern Pacific thermocline further amplifies the SST

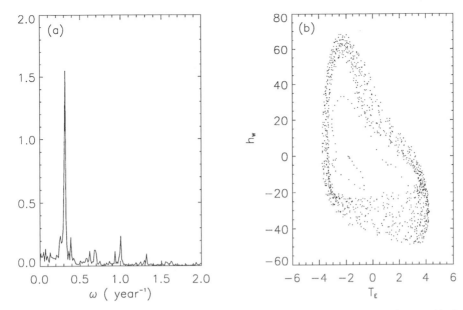

Figure 14.16. (a) Power spectrum of equatorial SST at $(x/L = 0.8)$ for the case of supercritical coupling strength (with respect to the annual mean state) with stochastic forcing and seasonal cycle included, (b) phase-space view of eastern Pacific SST anomaly $(T_E)$ versus western Pacific thermocline depth anomaly $(h_W)$.

warming within the cold tongue. Because of ocean wave reflections at the eastern boundary, mass is exchanged from the equatorial region toward off-equatorial latitudes. This causes the discharge of equatorial heat content, which weakens the SST anomaly and thereby the wind anomaly. When the SST and the wind anomalies diminish, this discharge process of equatorial ocean heat content continues, leading to an anomalously shallow thermocline over the entire equator. Upwelling is able to induce negative SST anomalies and therefore induces the cold phase of ENSO. This recharge oscillation mechanism, first envisioned by Wyrtki (1985) and Cane and Zebiak (1985), and later illustrated by Jin (1996, 1997a), is clearly consistent with the ENSO mode found here as the critical mode of the coupled model.

A weakly nonlinear analysis shows that the spatial pattern of the periodic orbit and its period are quite insensitive to the coupling strength. As soon as supercritical conditions are reached, hardly any change in period occurs, although the magnitude of the actual perturbation amplitude increases with $\mu$. When the system is forced with a seasonal signal and white noise, the spectral peak of the ENSO mode broadens. When the oscillatory mode is not unstable, i.e., $\mu$ is below the critical value for instability, the mode still may be excited by stochastic forcing from the atmosphere (Blanke et al., 1997). The degree of subcriticality or supercriticality, which is hard to establish from observations, is essential whether irregularity of the ENSO cycle can be attributed to deterministic processes, i.e., interaction with the seasonal cycle, or whether stochastic noise processes are essential.

Models like the Zebiak–Cane model have trouble simulating a correct seasonal cycle as obtained from semiannual forcing. On the path to a unified theory for the mean state, the seasonal cycle, and the interannual variability in the tropical Pacific, the next step in the analysis is to study the stability of such a seasonal cycle along similar lines as those in the work of Jin et al. (1996). The Zebiak–Cane model has to be modified since its off-equatorial response is not adequate to simulate the equatorial asymmetry of the seasonal cycle. Furthermore, the tropical climate system has to be studied within the global climate system to account for the influence of tropical–extratropical interactions. The behavior of ENSO is anything but regular when viewed on a longer time scale. Burning questions are whether the intensity of El Niño varies over decades. A knowledge of this natural variability is necessary in order to determine whether changes in ENSO behavior can be expected as a result of the increase of greenhouse gases in the atmosphere.

Much has been learned over the past decades. The 1982/1983 El Niño was a total surprise to all, and only with hindsight it could be seen that something unusual had happened (Philander, 1990). During the last El Niño, the situation was totally different, with measurements of the state of the tropical Pacific being distributed world wide by means of the Internet and various centers providing operational El Niño forecasts and advice. Models have shown very different prediction skills and have indicated that details in the dynamics can make a difference between successful prediction or a failure. The theoretical research in which intermediate complexity models such as the Zebiak–Cane model are used has been and is still very useful for understanding the physics of the results of more elaborate models and for suggesting improvements of these models.

## Acknowledgments

This work was supported by the Dutch National Research Programme on Global Air Pollution and Climate Change within project 951235. All computations were performed on the CRAY C916 at the Academic Computing Centre, Amsterdam, the Netherlands, within projects SC283 and SC498. Use of these computing facilities was sponsored by the National Computing Facilities Foundation with financial support from the Netherlands Organization for Scientific Research. Collaboration with David Neelin (UCLA, Los Angeles) and Fei Fei Jin (UH, Honolulu) over the years is much appreciated.

## References

Battisti, D. and Hirst, A. (1989) Interannual variability in a tropical atmosphere-ocean model: influence of the basic state, ocean geometry and nonlinearity. *J. Atmos. Sci.* **46**, 1687–1712.

Bjerknes, J. (1969) Atmospheric teleconnections from the equatorial Pacific. *Mon. Weather Rev.* **97**, 163–172.

Blanke, B., Neelin, J. D., and Gutzler, D. (1997) Estimating the effect of stochastic wind stress forcing on ENSO irregularity. *J. Climat.* **10**, 1473–1486.

Cane, M. and Zebiak, S. (1985) A theory for El Niño and the Southern Oscillation. *Science* **228**, 1084–1087.

Chang, P., Ji, L., Wang, B., and Li, T. (1995) Interactions between the seasonal cycle and El Niño–Southern Oscillation in an intermediate coupled ocean-atmosphere model. *J. Atmos. Sci.* **52**, 2353–2372.

Chang, P., Ji, L., Li, H., and Flügel, M. (1996) Chaotic systems versus stochastic processes in El Niño–Southern Oscillation in coupled ocean-atmosphere models. *Phys. D* **98**, 301–320.

Dijkstra, H. and Neelin, J. D. (1995) Coupled ocean-atmosphere interaction and the tropical climatology, part II: why the cold tongue is in the east. *J. Climat.* **8**, 1343–1359.

Dijkstra, H., Molemaker, M., van der Ploeg, A., and Botta, E. (1995) An efficient code to compute nonparallel flows and their linear stability. *Comp. Fluids* **24**, 415–434.

Gill, A. (1980) Some simple solutions for heat induced tropical circulation. *Q. J. R. Meteorol. Soc.* **106**, 447–462.

Horel, J. (1982) The annual cycle in the tropical Pacific atmosphere and ocean. *Mon. Weather Rev.* **110**, 1863–1878.

Jin, F.-F. (1996) Tropical ocean–atmosphere interaction, the Pacific cold tongue, and the El Niño/Southern Oscillation. *Science* **274**, 76–78.

Jin, F.-F. (1997a) An equatorial recharge paradigm for ENSO: part I: conceptual model. *J. Atmos. Sci.* **54**, 811–829.

Jin, F.-F. (1997b) An equatorial recharge paradigm for ENSO: part II: a stripped-down coupled model. *J. Atmos. Sci.* **54**, 830–847.

Jin, F.-F., Neelin, J. D., and Ghil, M. (1994) El Niño on the devil's staircase: annual subharmonic steps to chaos. *Science* **264**, 70–72.

Jin, F.-F., Neelin, J. D., and Ghil, M. (1996) El Niño/Southern Oscillation and the annual cycle: subharmonic frequency-locking and aperiodicity. *Phys. D* **98**, 442–465.

Neelin, J. D., Latif, M., and Jin, F. (1994) Dynamics of coupled ocean–atmosphere models: the tropical problem. *Ann. Rev. Fluid Mech.* **26**, 617–659.

Philander, S. (1990) *El-Niño and the Southern Oscillation*. Academic, New York.

Suarez, M. and Schopf, P. S. (1988) A delayed action oscillator for ENSO. *J. Atmos. Sci.* **45**, 3283–3287.

Tziperman, E., Stone, L., Cane, M., and Jarosh, H. (1994) El Niño chaos: overlapping of resonances between the seasonal cycle and the Pacific ocean–atmosphere oscillator. *Science* **264**, 72–74.

Tziperman, E., Cane, M., and Zebiak, S. (1995) Irregularity and locking to the seasonal cycle in an ENSO prediction model as explained by the quasi-periodicity route to chaos. *J. Atmos. Sci.* **52**, 293–306.

Van der Vaart, P. and Dijkstra, H. (1997) Sideband instabilities of mixed barotropic/baroclinic waves growing on a midlatitude zonal jet. *Phys. Fluids* **9**, 615–631.

Van der Vaart, P. C., Dijkstra, H., and Jin, F.-F. (1998) The Pacific cold tongue and the ENSO mode. *J. Atmos. Sci.* (submitted).

Wyrtki, K. (1985) Water displacements in the Pacific and the genesis of El Niño cycles. *J. Geophys. Res.* **91**, 7129–7132.

Zebiak, S. and Cane, M. (1987). A model El Niño-Southern Oscillation. *Mon. Weather Rev.* **115**, 2262–2278.

# 15

# Large-Amplitude Solitary Wave on a Pycnocline and Its Instability

DANIEL T. VALENTINE, BRIAN C. BARR, AND TIMOTHY W. KAO

## 15.1. Introduction

A two-layered fluid with an interface of continuous density stratification between the layers (i.e., a pycnocline) is common in natural bodies of water. Internal solitary waves on a pycnocline have been studied extensively, beginning with Keulegan (1953). More recent theoretical work concentrates on the Korteweg and de Vries (KdV) equation and its extensions; see, e.g., Helfrich et al. (1984) and Yih (1980). Kao et al. (1985) and Helfrich and Melville (1986) performed laboratory experiments on the generation and propagation of internal solitons and their interaction with changes in bottom topography of the slope-shelf type. These experiments demonstrated the importance of nonlinear and real-fluid effects, especially in the interaction of these waves with a slope-shelf topography. Kao et al. (1985) applied Benney's (1966) method to a pycnocline modeled by a hyperbolic tangent to help interpret their experimental observations. Helfrich and Melville (1986) extended the theory to compare predictions with their experiments. The comparisons with theoretical predictions indicated that the waves generated were solitary waves that behaved as first-order KdV waves when the amplitudes were sufficiently small.

When the amplitudes of internal solitary waves are relatively large, higher-order theory is required for modeling them. Miles (1980) presented a review of the development of higher-order theories for a variety of solitary wave problems. An alternative approach to seek a more comprehensive understanding of the various interaction processes involving solitary waves on a pycnocline is to apply numerical simulation techniques to solve the two-dimensional Navier–Stokes equations for a Boussinesq fluid. This approach is followed in the present investigation; it allows investigation of the effects of large amplitudes (strong nonlinearities) and diffusion. Numerical solutions were generated to study the features of propagating solitary waves over a range of amplitudes, the effect of wave–wave and wave–wall collisions, as well as the effect of a step-shelf change in bottom topography on the structure of the wave. The onset of breaking is examined to assess the instability mechanism that leads to breaking after a collision event or after the wave interaction with a step-shelf topography.

## 15.2. Mathematical Model

The initial configuration of the flow field examined is illustrated in Fig. 15.1(a). It is a system of two superposed layers of fluid with a free surface in a channel of depth $d$ and length $L$. The shallow upper layer of depth $h_1$ has density $\rho_1$ and the deeper lower layer of depth $h_2$ has density $\rho_2$ with $\rho_2 > \rho_1$. The shaded region of maximum depth $\Delta h$, approximate width $\Delta L$ ($x = \Delta L$ is the location where the depth of the pool is approximately 0.01 $\Delta h$), and density $\rho_1$ represents the initial sech$^2$ step pool. This step pool is the disturbance used to

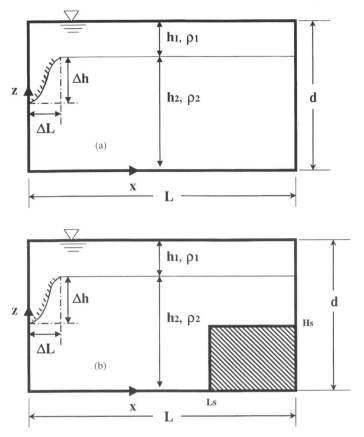

Figure 15.1. (a) Initial configuration of the model for the uniform depth case, (b) initial configuration of the model for the step-shelf topography.

generate a leading solitary wave. For all of the cases with $h_1 < h_2$, the leading solitary wave is a wave of depression. The modification of the flow-field geometry used to investigate the effect of a step-shelf topography is illustrated in Fig. 15.1(b). After establishing the initial density structure (as described below), we compute the motion that ensues from an initial state of rest.

With $d$, the depth of the channel, as the reference length and $\bar{C}_0$, the celerity of infinitesimal waves at the interface between two discrete layers, as the reference velocity, the scaled coordinates and components of velocity are

$$x = x^*/d, \qquad z = z^*/d, \qquad u = u^*/\bar{C}_0, \qquad w = w^*/\bar{C}_0,$$

where

$$\bar{C}_0 \equiv \left[ \frac{g(\rho_2 - \rho_1)}{\rho_2} \frac{h_1 h_2}{h_1 + h_2} \right]^{1/2}.$$

The model equations for the flow of an incompressible, viscous, diffusive, Boussinesq fluid are the continuity, Navier–Stokes, and diffusion equations. In two-dimensional, dimensionless form, they are given, respectively, by

$$\frac{\partial u}{\partial x} + \frac{\partial w}{\partial z} = 0, \tag{1}$$

$$\frac{\partial \zeta}{\partial t} + \frac{\partial u\zeta}{\partial x} + \frac{\partial w\zeta}{\partial z} = -\frac{1}{F_d^2}\frac{\partial \theta}{\partial x} + \frac{1}{\mathrm{Re}}\left(\frac{\partial^2 \zeta}{\partial x^2} + \frac{\partial^2 \zeta}{\partial z^2}\right), \tag{2}$$

$$\frac{\partial \theta}{\partial t} + \frac{\partial u\theta}{\partial x} + \frac{\partial w\theta}{\partial z} = \frac{1}{\mathrm{Re}\,\mathrm{Sc}}\left(\frac{\partial^2 \theta}{\partial x^2} + \frac{\partial^2 \theta}{\partial z^2}\right), \tag{3}$$

where

$$\zeta \equiv \frac{\partial u}{\partial z} - \frac{\partial w}{\partial x}, \qquad \theta \equiv \frac{\rho_2 - \rho}{\rho_2 - \rho_1},$$

$$u = \frac{\partial \psi}{\partial z}, \qquad w = -\frac{\partial \psi}{\partial x}, \tag{4}$$

$$\zeta = \frac{\partial^2 \psi}{\partial x^2} + \frac{\partial^2 \psi}{\partial z^2}, \tag{5}$$

where $\theta$, $\zeta$, and $\psi$ are the dimensionless density-difference ratio, the vorticity, and the stream function, respectively. The dimensionless time $t$ is defined by $t = t^*\bar{C}_0/d$. The dimensionless parameters of the system are the Reynolds number, $\mathrm{Re} = \bar{C}_0 d/\nu$, the densimetric Froude number, $F_d = \bar{C}_0/(g|\gamma_0|d)^{1/2}$, and the Schmidt number, $\mathrm{Sc} = \nu/D$, where $g$ is the gravitational acceleration, $\gamma_0 = (\rho_1 - \rho_2)/\rho_2$ is the reference density anomaly, $\nu$ is the kinematic viscosity, and $D$ is the diffusivity.

Equations (2), (3), and (5) are solved numerically subject to the following boundary and initial conditions. The tank bottom, left and right walls as well as the step face and shelf are no-slip, insulated walls. The free surface of the channel is a pure slip, rigid lid. These conditions lead to the following prescriptions for the parameters on the boundaries in Fig. 15.1(a):

$$0 \le x \le \frac{L}{d}, \qquad z = 0, \qquad \psi = 0, \qquad \zeta = \frac{\partial^2 \psi}{\partial z^2}, \qquad \frac{\partial \theta}{\partial z} = 0, \qquad u = w = 0;$$

$$0 \le x \le \frac{L}{d}, \qquad z = 1, \qquad \psi = 0, \qquad \zeta = 0, \qquad \frac{\partial \theta}{\partial z} = 0, \qquad u = \frac{\partial \psi}{\partial z}, \qquad w = 0;$$

$$x = 0, \qquad 0 \le z \le 1, \qquad \psi = 0, \qquad \zeta = \frac{\partial^2 \psi}{\partial x^2}, \qquad \frac{\partial \theta}{\partial x} = 0, \qquad u = w = 0;$$

$$x = \frac{L}{d}, \qquad 0 \le z \le 1, \qquad \psi = 0, \qquad \zeta = \frac{\partial^2 \psi}{\partial x^2}, \qquad \frac{\partial \theta}{\partial x} = 0, \qquad u = w = 0.$$

The initial condition is a state of rest. The initial depths of the upper and the lower layers were selected along with the size of the $\mathrm{sech}^2$ step pool illustrated in Fig. 15.1. The density-difference ratio in the upper layer and the $\mathrm{sech}^2$ step pool is set equal to 1.0 and in the lower layer is set equal to zero. The pycnocline between the two layers was established by the solution of a diffusion problem for $\theta$ over a time interval of 0.5 with the fluid in the basin held in a state of rest with an artificially large diffusion coefficient $(\mathrm{Re}\,\mathrm{Sc})^{-1}$ equal to 0.001. At $t = 0.5$ the characteristic parameters were set to $\mathrm{Re} = 10{,}000$ and $\mathrm{Sc} = 833$, which correspond to the small diffusivity of salt; these parameters are consistent with the experimental values of Kao et al. (1985). At the same time the $\mathrm{sech}^2$ step pool is released, causing motion to ensue.

## 15.3. Computational Method of Solution

The method applied to solve the equations described above is the ETUDE (explicit-in-time upstream-difference estimate) finite-difference method developed by Valentine (1995). It is a first-order-in-time and second-order-in-space method that eliminates second-derivative trunca-tion errors. It is an explicit scheme that possesses the transportive and conservative properties described by Roache (1972). Central differencing in space and forward differencing in time are applied except for the nonlinear terms in the equations of motion. An upwind method is used to model the nonlinear terms. The upwind scheme originally introduced by Torrence and Rockett (1969) has truncation errors that appear as false viscosity and false diffusion; see Kao et al. (1978) for a discussion of these issues. The ETUDE scheme was designed to eliminate these problems; see Valentine (1995) for details. The Poisson equation, Eq. (5), was solved by SOR (successive over-relaxation). The boundary conditions that must be updated in time are solved by application of second-order differences to predict unknowns along boundaries.

The finite-difference grid is defined as follows: $\Delta x = (L/d)/(M-1)$ and $\Delta z = 1/(N-1)$. The finite-difference grids used in the present investigation were $(M, N) = (301, 81)$ for a tank with $L/d = 15$ and $(601, 81)$ for a tank with $L/d = 30$. This is twice the number of grid points used in each of the coordinate directions by Saffarinia and Kao (1996) for the same flow-field parameters, i.e., the same Re, Sc, and $F_d$. The waves they generated were produced by a rectangular step pool disturbance. Their computational results were found to predict solitary wave behavior reasonably well based on comparisons with experiments and theory. Hence the resolution of the present computations is more than adequate. The computational time step

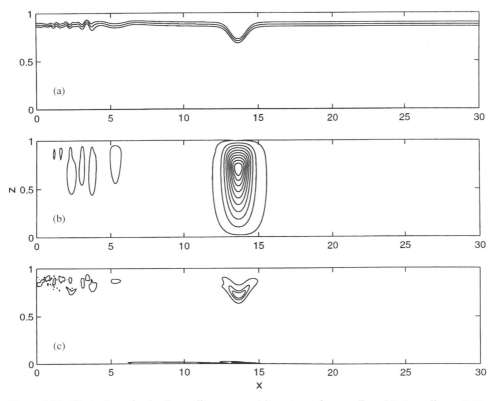

Figure 15.2. Illustration of a leading solitary wave: (a) contour of pycnocline, (b) streamline pattern, (c) vorticity contour.

used to compute the simulations described in this chapter is $\Delta t = 0.0010$. Double-precision arithmetic was also used.

## 15.4. Results and Discussion

The initial value problem described in the previous sections leads to the generation of a leading solitary wave of relatively large amplitude. Numerical simulations were computed to explore the behavior of the leading solitary wave for a range of amplitudes. Additional simulations were computed to examine the interaction of the leading solitary wave with another solitary wave, with a wall, or with a step shelf. In Subsection 15.4.1 we describe the properties of the leading solitary wave. In Subsections 15.4.2 and 15.4.3 we discuss the results of wave–wave and wave–wall collision simulations, respectively. In Subsection 15.4.4 we examine the creation of reflected and transmitted waves and the onset of breaking associated with a solitary wave encounter with a step-shelf topographical feature.

### 15.4.1. *Leading Solitary Wave*

The solitary waves computed in the present investigation are waves that satisfy the conditions of the Boussinesq approximation. Figure 15.2 shows the flow pattern induced by a leading

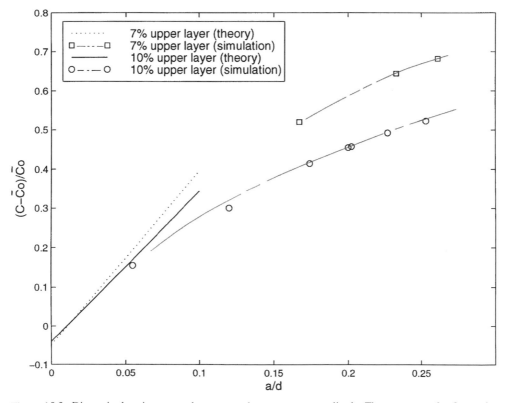

Figure 15.3. Dimensionless incremental wave speed versus wave amplitude. The curves are for first-order KdV waves as predicted for a hyperbolic-tangent pycnocline with a half-thickness measure of 0.018 $d$, as reported by Kao et al. (1985).

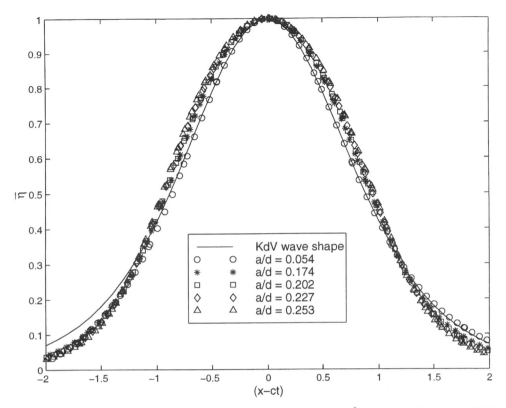

Figure 15.4. Normalized waveform for five cases compared with the sech$^2$ waveform of first-order KdV theory.

solitary wave sufficiently far from the initial disturbance used to create it, i.e., it is well beyond its sorting distance.* Figure 15.2(a) is a contour map of the pycnocline. This wave carries with it a closed streamline pattern, as illustrated in Fig. 15.2(b), and a patch of vorticity that is illustrated in Fig. 15.2(c). This patch of vorticity is quite important in sustaining the shape of the wave. It is the production and transport of vorticity that induces the velocity field induced by the wave. The production of vorticity by means of the $\partial\theta/\partial x$ term in Eq. (2), in turn, induces a velocity field by means of the stream function.

The leading solitary waves that were generated in the numerical experiments are waves that behave asymptotically as first-order KdV waves. The waves generated are similar to those generated and studied experimentally by Kao et al. (1985). To support this contention let us review the waves investigated by Kao et al.; they are internal waves that propagate on a finite-depth pycnocline. The maximum or reference value of the density anomaly for the experiments was $\gamma_0 = -0.01$; hence the Boussinesq approximation was invoked and found to be satisfied when theoretical predictions (based on first-order KdV theory) were compared with the experimental data. For waves with amplitudes equal to or less than the depth of the upper-layer fluid (i.e., for $a/h_1 < 1$), the ratio $u_{max}/c_0$ was shown to be equal to $a/h_1$, where $u_{max}$ is

---

* The sorting distance is the distance the leading wave must propagate to emerge from the initial disturbance. Hammack and Segur (1974) discuss the quantitative aspects of this concept.

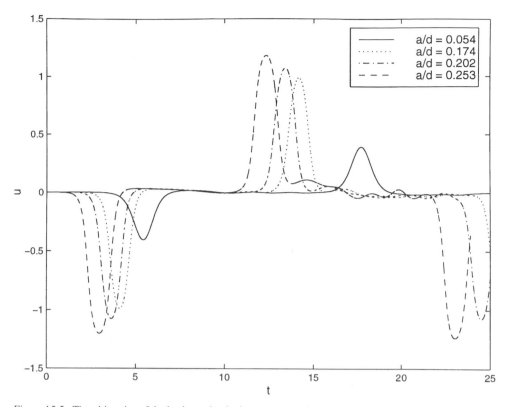

Figure 15.5. Time histories of the horizontal velocity as measured at $(x, z) = (22, 0.925)$ in a tank $L/d = 30$ for four wave–wave collision cases.

the maximum horizontal velocity in the upper layer and $c_0$ is the celerity of infinitesimal waves on a finite-width pycnocline. For $a/d$ less than approximately 0.05, the wavelength[*] $\lambda$ was shown to be proportional to $(a/d)^{-1/2}$, the incremental wave speed $\delta c/c_0 = (c - c_0)/c_0$, was shown to increase linearly with amplitude, i.e., $\delta c/c_0 = \frac{2}{3}ra/d$, where $r$ is a scaling parameter that arises from the theory and the shape of the wave was shown to be $\text{sech}^2$. These results are predicted by first-order KdV theory; hence the waves observed experimentally within the amplitude range $0 < a/d < 0.05$ are first-order KdV solitary waves. For $0.05 < a/d \leq 0.2$ the observed wavelengths were larger than first-order KdV, the difference increasing with increasing amplitude. For waves with $a/d > 0.1$, the wave speed, although it continually increases with wave amplitude, is significantly less than the predictions based on first-order theory. These findings are compared next with the computational results of the present work.

Figure 15.3 illustrates the effects of wave amplitude and upper-layer depth on wave celerity for the waves generated computationally. Also, speed increases with amplitude; however, for large-amplitude waves the increase in speed with amplitude in less than for small-amplitude waves. For a 10% upper-layer depth ($h_1/d = 0.1$), the scaling parameter $r$ reported by Kao et al. (1985) is approximately 5.2 for $a/d = 0.2$ (this parameter is a function of $a/d$, $h_1/d$ and the thickness of the pycnocline); this means that $(c - c_0)/rc_0 \approx 0.085$ for the wave computed in this investigation. Similarly, for the 7.5% upper layer the scaling parameter $r$ is approximately

---

[*] The wavelength of a solitary wave is defined as follows: $\lambda = (1/a) \int_0^\infty \eta(x - ct)\, d(x - ct)$, where $\eta$ is the shape of the $\theta = 0.5$ isopycnal; see Kao et al. (1985) for details.

7.0 for $a/d = 0.2$; this means that $(c - c_0)/rc_0 \approx 0.085$ as well. These predictions, when plotted on Fig. 12 in the work of Kao et al. (1985), fall precisely where expected within the experimental data reported. Figure 15.3 also shows that in a tank with a thinner upper layer a wave with the same amplitude, $a/d$, travels faster. This means that the waves of a given amplitude propagating on a thinner upper layer are more nonlinear.

Figure 15.4 is a plot of the $\theta = 0.5$ isopycnal normalized by the amplitude for several of the waves computed. The profiles are compared with the $\text{sech}^2$ function, which is the shape of first-order KdV waves. The smallest-amplitude wave ($a/d = 0.054$) compares closely with the $\text{sech}^2$ shape. The larger-amplitude waves ($a/d > 0.17$) have fuller profiles as compared with $\text{sech}^2$. Thus they have greater area under their profile for the same amplitude; hence, for the same amplitude as compared with a $\text{sech}^2$ wave they would be expected to have a larger wavelength. This is what the experiments reported by Kao et al. (1985) showed for waves with $a/d > 0.05$. The large-amplitude solitary waves generated in the present study deviate from first-order KdV theory with fuller profiles, slower speeds, and longer wavelengths.

### 15.4.2. Wave–Wave Collisions

We next examine the collision of a large-amplitude right-moving wave an identical left-moving wave. In a tank with $L/d = 30$, a wave is generated with the $\text{sech}^2$ step pool at $x = 0$ and a zero-shear insulated-wall symmetry condition at $x = 30$. The wave is allowed to propagate to a

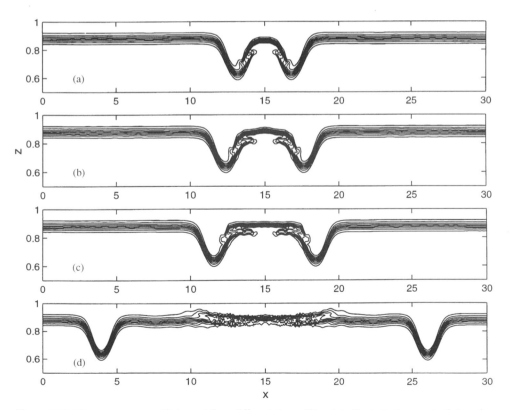

Figure 15.6. The wave–wave collisions at four different times. The plots illustrate the onset of shear instability and the generation of a mixed region at the back of the waves during the separation process. (a) $t = 9.0$, (b) $t = 9.5$, (c) $t = 10.0$, (d) $t = 15.0$.

point past the midpoint of the tank. At that time, the beginning half of the domain is truncated (up to $x = 15$ with this location identified as $x = 0$ in the new tank), removing most of the small linear waves following the leading solitary wave. This initial condition is then mirrored about the plane of symmetry (which is now at $x = 15$ in the new tank) to produce two fully developed isolated solitary waves propagating into quiescent fluid. This method allows the study of the effects of the wave–wave interaction without any disturbances interfering with the results. This procedure of capturing a wave is similar to that used by Segur and Hammack (1982). The modified tank is still with $L/d = 30$. The two waves thus generated are one coming from the direction of the new left wall at the new $x = 0$ and one from the new no-slip right wall at $x = 30$. At $(x, z) = (22, 0.925)$ of the new tank, the time history of the horizontal velocity was monitored for four simulations with different wave amplitudes. These time histories are compared in Fig. 15.5. Since the probe is located in the right half of the basin (a distance 8 units from the right wall), it first records the passage of the left-moving wave. It subsequently records the right-moving wave that emerges from the collision at $x = 15$. Finally, it records the left-moving wave reflected from the right wall. The oscillatory tail of the largest wave is quite pronounced. Any oscillatory tail for the smallest wave is insignificant. There is theoretical work that ascribes the loss of energy to other modes; see, e.g., Fenton and Rienecker (1982). The main point of this figure is to show that small-amplitude waves emerge relatively unscathed from collision. The large-amplitude waves shed small waves at the rear.

Figure 15.6 shows the shape of the pycnocline at four different times after collisions of two incident waves with amplitude $a/d = 0.253$, which is shown to shed small waves in Fig. 15.5. As the waves approach each other their interaction reduces the magnitude of the velocity field.

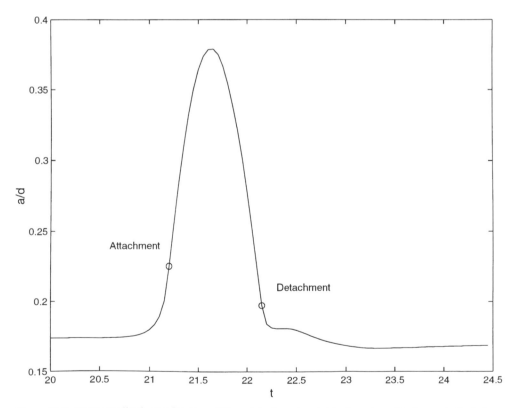

Figure 15.7. Wave amplitude as a function of time. Attachment and detachment amplitudes are indicated.

This results in an enhancement of the stability of the flow field. However, when the waves begin to separate, their interaction tends to enhance the upwelling of fluid between them as they begin to separate. This enhanced upward flow tends to distort the profile of the back of each wave, as illustrated in Fig. 15.6(a). The instability caused by enhanced shearing of the pycnocline at the back of the waves induces smaller-amplitude waves and a mixed region that are shed (or sorted) by the leading wave, as illustrated in Figs. 15.6(b), 15.6(c), and 15.6(d). Following Kao et al. (1985), we computed the local gradient Richardson number $J$ by applying the formula $JF_d^2 = (\partial\bar{\theta}/\partial z)/(\partial u/\partial z)^2$ in the region where the breaking is initiated. Values of $J < 0.25$ appeared in this region before the observed onset of instability; hence the Miles (1961, 1963) criterion for shear instability is indicative of the onset of breaking in this wave–wave collision. When the waves are sufficiently far apart and hence sorted from the collision process, the near balance between nonlinearity and dispersion is restored. The amplitude of the leading wave that emerges is only slightly smaller than the incident wave. In addition, the leading wave sorts itself from the mixed region well before the mixed region breaks up into unresolved fine-scale motions. Hence the waves that emerge from the collision process appear not to be affected by the small-scale mixing left behind.

### 15.4.3. *Wave–Wall Collisions*

The wave–wall collision can be characterized by the motion of the pycnocline and the location of its maximum amplitude $a$. Figure 15.7 is a plot of the wave amplitude for the onset wave with

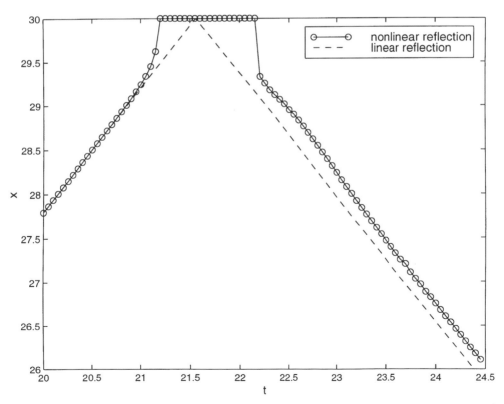

Figure 15.8. The trajectory of the horizontal location of the wave amplitude as a function of time, as compared with a linear wave reflection.

$a/d = 0.174$ of the $\theta = 0.5$ isopycnal (the center of the pycnocline) as it approaches and collides with the wall. The attachment and detachment points in the figure are when the wave trough attaches to and leaves from the wall, respectively. The incident wave increases in amplitude as it approaches the wall. The maximum amplitude of the wave during the collision event is more than twice the amplitude of the incident wave, evidence of the nonlinearity of the interaction as described by Su and Mirie (1980). The reflected wave amplitude is initially less than its ultimate value. Similar results for surface waves were recently reported by Cooker et al. (1997). Figure 15.8 is the trajectory of the incident and the reflected wave amplitudes shown in Fig. 15.7. The dotted line represents the trajectory of a linear wave traveling with the same speed as the onset wave. The vertical distance, at any point, between the linear wave trajectory and the solitary wave trajectory is called the phase shift. Note that it is not constant. A more easily measured value indicative of the nonlinear interaction is the wall residence time defined by Cooker et al.; they defined it to be the time interval between attachment and detachment. The wall residence times for the cases with $a/d$ equal to 0.054, 0.174, and 0.253 were calculated to be 0.095, 1.0, and 1.1, to within $\pm 0.05$, respectively. It is not possible to compare directly with the surface wave results of Cooker et al. For the largest-amplitude surface waves, they found that the wall residence time approached a constant. However, they did show that the detachment time first decreases and subsequently increases with wave amplitude while the attachment time decreases and appears to approach an asymptotic value. Thus, for surface waves larger than the ones they investigated, the residence time could eventually increase slightly. The results

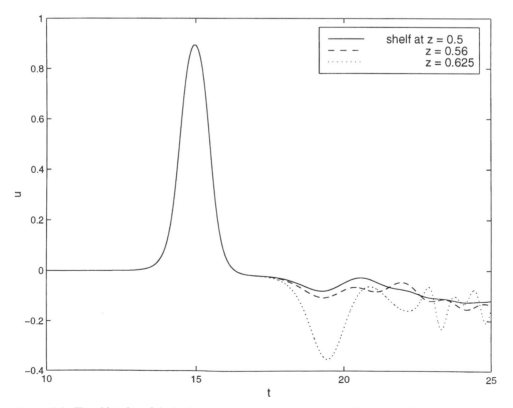

Figure 15.9. Time histories of the horizontal velocity of an onset wave with $a/d = 0.17$ that encounters a step shelf at $x = 22.5$ as measured with a probe at $(x, z) = (20, 0.925)$ in a tank $L/d = 30$.

reported here show a slight rise of wall residence time for large-amplitude solitary waves on a pycnocline. The work of Cooker at al. must be extended to handle the internal wave problem to compare directly with the present results.

### 15.4.4. *Interaction with Step-Shelf Topography*

Figure 15.9 shows the effect of shelf height on the size of the reflected wave. The higher the stepshelf, the larger the leading reflected wave. This is not unexpected, because the higher the shelf the larger the induced back flow around the corner. Figure 15.10 illustrates the changes in the isopycnals as the solitary wave encounters the highest step-shelf topography. The acceleration of fluid that is induced by the wave around the corner of the step is in the direction opposite the oncoming wave. This flow pulls the wave downward toward the face of the step, as shown somewhat in Figure 15.10(a). The subsequent rebound produces a transmitted and reflected wave that is clearly visible in Fig. 15.10. The region between the reflected and the transmitted waves is a mixed region created by the onset of shear instability in what appears to be a similar manner as the mixed region formed at the back of the reflected waves in the wave–wave collision. The reflected and the transmitted waves leave behind the mixed region. The production of the leading transmitted and reflected waves are observed to precede the breaking and mixing that occurs after the encounter event with the step shelf has taken place. Hence, like the wave–wave interaction processes, the waves that emerge do not appear to be

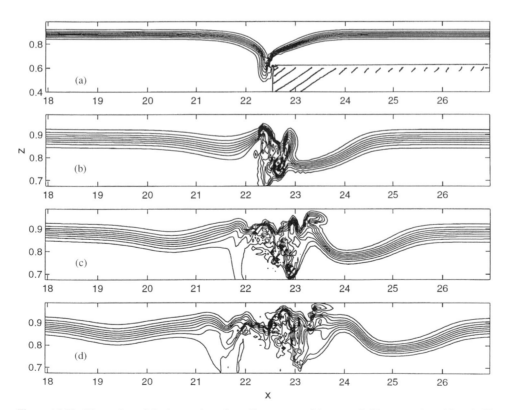

Figure 15.10. Illustration of the interaction of a solitary wave with a step-shelf topography with a shelf height of $z = 0.625$ and location $x = 22.5$. (a) $t = 17.0$, (b) $t = 17.5$, (c) $t = 18.0$, (d) $t = 18.5$.

influenced by the mixing. Future work on the energy contained in the leading transmitted and reflected waves should provide insight into coefficients of transmission and reflection, as well as energy lost to the generation of smaller waves and mixing.

## References

Benney, D. J. 1966. Long non-linear waves in fluid flows. *J. Math. Phys.* **45**, 52–63.

Cooker, M. J., Weidman, P. D., and Bale, D. S. 1997. Reflection of a high-amplitude solitary wave at a vertical wall. *J. Fluid Mech.* **342**, 141–158.

Fenton, J. D. and Rienecker, M. M. 1982. A Fourier method for solving nonlinear water-wave problems; application to solitary-wave interactions. *J. Fluid Mech.* **118**, 411–443.

Hammack, J. L. and Segur, H. 1974. The Korteweg–de Vries equation and water waves. Part 2. Comparison with experiments. *J. Fluid Mech.* **65**, 289–314.

Helfrich, K. R., Melville, W. K., and Miles, J. W. 1984. On interfacial solitary waves over slowly varying topography. *J. Fluid Mech.* **149**, 305–317.

Helfrich, K. R. and Melville, W. K. 1986. On long nonlinear internal waves over slope-shelf topography. *J. Fluid Mech.* **167**, 285–308.

Kao, T. W., Pan, F.-S., and Renourd, D. 1985. Internal solitons on the pycnocline: generation, propagation, and shoaling and breaking over a slope. *J. Fluid Mech.* **159**, 19–53.

Kao, T. W., Park, C., and Pao, H. P. 1978. Inflow, density currents and fronts. *Phys. Fluids* **21**, 1912–1922.

Keulegan, G. H. 1953. Characteristics of internal solitary waves. *J. Res. Natl. Bur. Stand.* **51**, 133–140.

Miles, J. W. 1980. Solitary waves. *Ann. Rev. Fluid Mech.* **12**, 11–43.

Miles, J. W. 1961. On the stability of heterogeneous shear flows. *J. Fluid Mech.* **10**, 496–508.

Miles, J. W. 1963. On the stability of heterogeneous shear flows. Part 2. *J. Fluid Mech.* **16**, 209–227.

Roache, P. J. (1972). *Computational Fluid Dynamics*. Hermosa Publishers, NM.

Saffarinia, K. and Kao, T. W. 1996. A numerical study of the breaking of an internal soliton and its interaction with a slope. *Dyn. Atmos. Oceans* **23**, 379–391.

Segur, H. and Hammack, J. L. 1982. Soliton models of long internal waves. *J. Fluid Mech.* **118**, 285–304.

Su, C. H. and Mirie, R. M. 1980. On head-on collisions between two solitary waves. *J. Fluid Mech.* **98**, 509–525.

Torrence, K. E. and Rockett, J. A. 1969. Numerical study of natural convection in an enclosure with localized heating from below: creeping flow to the onset of laminar instability. *J. Fluid Mech.* **36**, 33–54.

Valentine, D. T. 1995. Decay of confined, two-dimensional, spatially periodic arrays of vortices: a numerical investigation. *Int. J. Num. Methods Fluids* **21**, 155–180.

Yih, C.-S., 1980. *Stratified Flows*. Academic, NY.

# 16

# Stability and Pattern Selection in Parametrically Driven Surface Waves

PEILONG CHEN AND JORGE VIÑALS

## 16.1. Introduction

Parametrically driven surface waves (also known as Faraday waves) can be excited on the free surface of a fluid layer that is periodically vibrated in the direction normal to the surface at rest if the amplitude of the driving acceleration is large enough to overcome the dissipative effect of fluid viscosity [1–3]. Despite the simplicity of the configuration, this system displays a large number of features that are characteristics of strong nonlinearity and has served as a prototype of many nonlinear phenomena that are currently under active investigation. Among them we mention the discovery of stationary quasi-periodic patterns of surface standing waves (the analog of a quasi-crystal in solid state physics) [4–6], the transition to spatiotemporal chaos [7], the coexistence of chaotic and regular regions in an extended system, and a laboratory for detailed studies of turbulence [8].

We review recent developments in the study of Faraday waves for spatially extended systems in which the lateral dimensions of the system are much larger than the wavelength of the wave and sufficiently close to threshold of the primary instability so that the description can be made at the level of slow modulations of a base pattern of standing waves. Despite the fact that the linear stability analysis of the quiescent state was carried out over 40 years ago, it has been only recently that the stability boundaries for a viscous fluid have been determined both theoretically and experimentally. We first touch on the salient features and results of the linear regime above onset, and especially the interplay between two small parameters: the dimensionless distance away from threshold $\epsilon$ and the viscous damping parameter $\gamma$. Within a linear analysis, the stability results for a viscous fluid reduce to the inviscid case smoothly as $\gamma \to 0$. This is not the case for the weakly nonlinear case: For finite $\epsilon$ the limit $\gamma \to 0$ is singular.

We next discuss recent progress on the weakly nonlinear regime above onset, especially results on pattern selection. It is now known that different wave patterns can be excited depending on the fluid properties and the driving amplitude or frequency. At high viscous dissipation (a fluid of large viscosity or a low driving frequency), the observed wave pattern above threshold consists of parallel stripes [5, 9]. For lower dissipation, patterns of square symmetry (combinations of two perpendicular plane waves) are observed in the capillary regime (large frequencies) [4, 5, 7, 10–13]. At lower frequencies (the mixed gravity-capillary regime), higher symmetry patterns have been observed: hexagonal [6, 14], eightfold, and tenfold patterns [6]. The dominant mechanism leading to this rich phenomenology is triad resonant interactions through stable second-order waves. We address recent results concerning the calculation of cubic-order coefficients of the standing-wave amplitude equation in viscous fluids and the prediction of selected patterns of standing waves of various symmetries close to onset. Fairly comprehensive

211

sets of experiments have been recently conducted in the large aspect ratio systems, and their salient results are summarized.

## 16.2. Governing Equations and Linear Stability

We first briefly review the basic governing equations that correspond to an idealization of the classical configuration of the Faraday experiment: a semi-infinite fluid layer, unbounded in the $x$–$y$ direction, extending to $z = -\infty$, and with a planar free surface at $z = 0$ when at rest. The fluid is assumed incompressible and Newtonian. Under periodic vibration of the layer in the direction normal to the surface at rest, the equation governing fluid motion (in the comoving reference frame) is

$$\partial_t \mathbf{u} + (\mathbf{u} \cdot \nabla)\mathbf{u} = -\frac{1}{\rho}\nabla p + \nu\nabla^2\mathbf{u} + g_z(t)\hat{\mathbf{e}}_z, \tag{1}$$

where $\mathbf{u}$ is the velocity field, $p$ is the pressure, $\rho$ and $\nu$ are the density and the kinematic viscosity of the fluid, respectively, and $g_z(t) = -g - f\cos\omega_0 t$ is the effective gravity. The incompressibility condition is $\nabla \cdot \mathbf{u} = 0$. We note that these equations are invariant under time translation by a period of the driving force $t \to t + 2\pi/\omega_0$.

The base state is a quiescent fluid with a pressure $p = \rho g_z(t)z$. Pressure is redefined in what follows to absorb the body force, so that $p$ is the deviation from $\rho g_z(t)z$. Besides the null conditions at $z = -\infty$, there are four boundary conditions at the moving free surface [15]. Let $z = \zeta(x, y, t)$ be the instantaneous position of the surface. Then the normal velocity of an element of boundary equals the fluid velocity at that point:

$$\partial_t \zeta + [\mathbf{u}(z = \zeta) \cdot \nabla_H]\zeta = w(z = \zeta),$$

where $\nabla_H = \hat{\mathbf{e}}_x\partial_x + \hat{\mathbf{e}}_y\partial_y$ and we use $\mathbf{u} = (u, v, w)$ as the Cartesian components of the velocity. The air phase above the surface is assumed to be a passive phase of negligible density so that continuity of the tangential stress at the fluid surface reads $\hat{\mathbf{t}}_m \cdot \mathbf{T} \cdot \hat{\mathbf{n}}|_{z=\zeta} = 0$, $m = 1, 2$, where $\mathbf{T}$ is the stress tensor of components, $T_{ij} = [-p - \rho g_z(t)z]\delta_{ij} + \rho\nu(\partial_j u_i + \partial_i u_j)$, $\hat{\mathbf{t}}_1$ and $\hat{\mathbf{t}}_2$ are two unit tangent vectors (not mutually orthogonal), $\hat{\mathbf{t}}_1 = (1, 0, \partial_x\zeta)/[1 + (\partial_x\zeta)^2]^{1/2}$, $\hat{\mathbf{t}}_2 = (0, 1, \partial_y\zeta)/[1 + (\partial_y\zeta)^2]^{1/2}$, and $\hat{\mathbf{n}}$ is the outward-pointing normal, $\hat{\mathbf{n}} = (-\partial_x\zeta, -\partial_y\zeta, 1)/[1 + (\partial_x\zeta)^2 + (\partial_y\zeta)^2]^{1/2}$. The normal stress at the boundary is discontinuous because of capillarity, $\hat{\mathbf{n}} \cdot \mathbf{T} \cdot \hat{\mathbf{n}}|_{z=\zeta} = 2\sigma H$, where $\sigma$ is the interfacial tension and $2H$ is the mean curvature of the surface*:

$$2H = \{\partial_{xx}\zeta[1 + (\partial_y\zeta)^2] + \partial_{yy}\zeta[1 + (\partial_x\zeta)^2] - 2\partial_x\zeta\partial_y\zeta\partial_{xy}\zeta\}/[1 + (\partial_x\zeta)^2 + (\partial_y\zeta)^2]^{3/2}.$$

Most of the recent work has focused on the large aspect ratio limit in which the lateral dimensions of the fluid layer are much larger that the wavelength of the surface waves. Viscous dissipation along the container sidewalls, as well as other phenomena associated with the possible motion of the fluid meniscus, are thus neglected. Estimates of their relative contribution have been given in Refs. [16 and 17]. A separate issue concerns effects associated with a finite layer depth. For an inviscid fluid, it is well known that the linear dispersion relation for surface waves is modified. Additionally, viscous damping at the bottom of the fluid layer can introduce more

---

* In order to avoid excessive use of fences, we follow the convention in the remainder of the paper that the operator $\partial$ acts only on the function immediately following it.

subtle effects that lead to qualitatively new behavior. Kumar, for example, has suggested that in the case of a shallow layer in which viscous dissipation at the bottom of the fluid layer becomes dominant, the surface responds harmonically to the driving acceleration instead of subharmonically (as it does in the infinite depth limit). This point has been further elaborated on and demonstrated experimentally by Müller et al. [18]. Depth can also qualitatively affect pattern selection above onset, as shown recently in Ref. 19. In this case, depth affects damping of the primary wave through a modification of triad resonance with a weakly damped second-order wave.

Linearization of the governing equations and boundary conditions around the base state leads to the following equation for each Fourier component $\zeta_k(t)$ of the surface displacement [20]:

$$\partial_{tt}\zeta_k + 4\nu k^2 \partial_t \zeta_k + \left(\frac{\sigma k^3}{\rho} + gk + 4\nu^2 k^4 + kf\cos\omega_0 t\right)\zeta_k$$
$$- 4\nu k^2 \sqrt{\frac{\nu k^2}{\pi}} \int_0^t dt'[\partial_{t'}\zeta_k(t') + \nu k^2 \zeta_k(t')]\frac{\exp[-\nu k^2(t-t')]}{\sqrt{t-t'}} \tag{2}$$
$$= 4\nu k^2 \zeta_k(0)\mathrm{erfc}(\sqrt{\nu k^2 t}).$$

At long times, the transient term on the right-hand side of Eq. (2) can be neglected. The linear stability of the fluid layer was first addressed in the inviscid limit by Benjamin and Ursell [21]. Equation (2) with $\nu = 0$ reduces to the Mathieu equation for $\zeta_k(t)$, the solutions of which are well known. A set of instability regions (or tongues) appears in a $(k, f)$ diagram that intersects the $f = 0$ axis. Solutions within these regions are oscillatory with frequency $\omega_n = [(n+1)/2]\omega_0$, $n = 0, 1, 2, \ldots$, and an exponentially increasing amplitude. The fact that the regions of instability intersect the $f = 0$ line indicates that there is no threshold for instability in this limit. Benjamin and Ursell, however, argued that in practical cases (i.e., when the fluid is slightly dissipative) the higher-order modes (large $n$) would be suppressed and hence any appreciable surface disturbance should have low values of $n$.

The limit of small viscous dissipation was considered in Ref. 22 by the addition of a phenomenological damping term to the Mathieu equation [identical to the second term on the left-hand side of Eq. (2)]. Damping removes the degeneracy of the inviscid modes and introduces a finite threshold value of $f$ for instability that is a function of $n$. The lowest threshold corresponds to $n = 0$ and hence to subharmonic resonance. The threshold value in this approximation is $f_0 = \nu k_0 \omega_0/2$, where the wave number $k_0$ is determined by the resonance condition and the linear dispersion relation $\omega/2 = \omega_0 = gk_0 + \sigma k_0^3/\rho$. The range over which this approximation is valid can be obtained by examination of the scaling of the various terms in Eq. (2). For a wave of frequency $\omega_0$ and wave number $k_0$, the ratio of viscous dissipation to inertia is of the order of $\nu k_0^2/\omega_0$, which can be taken as a small parameter in the low dissipation regime. The term $\nu^2 k_0^4$ is then of higher order and can be neglected in this regime in front of $\omega_0$. The nonlocal term on the right-hand side of Eq. (2) describes the rectification of the wave that is due to vorticity diffusion in the vicinity of the moving interface. In the limit $\nu k_0^2/\omega_0 \ll 1$ the term $\nu k^2 \zeta_k(t')$ in the integrand is negligible in front of $\partial_{t'}\zeta_k(t)$, the latter being of the order of $\omega_0 \zeta_k$. The integral of the kernel $\{\exp[-\nu k^2(t-t')]/(\sqrt{t-t'})\}$ is then of the order of erf $(\sqrt{\nu k^2 t})$, where the time $t$ has to be taken as the characteristic time that is available for the vorticity being created at the moving interface to diffuse into the bulk before it reverses sign. This time is $1/\omega_0$. Therefore the leading contribution of the retarded term in Eq. (2) is of the relative order of $(\nu k_0^2/\omega_0)^{3/2}$ and hence negligible in front of the other viscous term [second term on the left-hand side of Eq. (2)]. In short, in a formal expansion in powers of both the interface displacement $\zeta$ and of $\nu k_0^2/\omega_0$, the damped Mathieu equation is the lowest order. In other words, the dominant dissipative force is viscous damping of the irrotational flow in the bulk. Damping arising from the rotational flow within the thin surface vortical layer is of the order of $\gamma^{3/2}$ [23].

Considerable progress has been achieved recently in extending the weakly viscous analyses described above. Corrections to the stability boundary predicted by the damped Mathieu equation were first obtained numerically by Kumar and Tuckerman [24] by a clever reformulation of the stability problem into an eigenvalue problem. Their results for the onset driving amplitude and the critical wavelength compared favorably with experiments in deep layers [25]. More recently, Müller et al. [18] obtained the stability boundaries analytically up to the order of $(\nu k_0^2/\omega_0)^{3/2}$ and presented evidence of harmonic response for shallow layers. Progress has also been achieved in the opposite limit of large viscous dissipation by Cerda and Tirapegui [26], who have derived an analytic approximation for the instability threshold.

An exact, albeit implicit, expression for the instability threshold for arbitrary viscous dissipation has been given in Ref. 27 by an extension of the eigenvalue analysis of Ref. 24. We mention here its limiting behavior at low viscosity. In this limit, one expects the critical wave number $k_{\mathrm{onset}}$ to be near $k_0$. It is then convenient to define dimensionless variables by using $1/\omega_0$ as the time scale and $1/k_0$ as the length scale. We also define a reduced wave number $\bar{k} = k/k_0$, a viscous damping coefficient $\gamma = 2\nu k_0^2/\omega_0$, the gravity wave $G = gk_0/\omega_0^2$, capillary wave $\Sigma = \sigma k_0^3/\rho\omega_0^2$ contributions to the dispersion relation, and the dimensionless amplitude of the driving acceleration $\Delta = fk_0/4\omega_0^2$. $G = 1$ corresponds to a pure gravity wave while $G = 0$ corresponds to a pure capillary wave. For $\gamma \ll 1$, $k_{\mathrm{onset}}$ and the dimensionless value of the critical driving amplitude $\Delta_{\mathrm{onset}}$ can be expanded as a power series of the damping coefficient $\gamma$:

$$\bar{k}_{\mathrm{onset}} = 1 + \frac{1}{3 - 2G}\gamma^{3/2} + \frac{-7 + 2G}{(3 - 2G)^2}\gamma^2 + \cdots,$$

$$\Delta_{\mathrm{onset}} = \gamma - \frac{1}{2}\gamma^{3/2} + \frac{11 - 2G}{8(3 - 2G)}\gamma^{5/2} + \cdots. \tag{3}$$

The first correction term is proportional to $\gamma^{3/2}$ and agrees with a low-viscosity expansion of the linearized equations given in Ref. 18.

## 16.3. Pattern Selection in Large Aspect Ratio Systems

We start by briefly reviewing recent experimental results on pattern selection and then proceed to outline the main results of the weakly nonlinear theory. Square patterns consisting of two sets of plane waves with wave vectors perpendicular to each other were generally observed in early experiments by Ezerskii et al. [11], Douady et al. [28], Tufillaro et al. [7], and Edwards and Fauve [5]. These experiments used low-viscosity fluids, in part because theories at the time were restricted to the low damping limit, in part because the lower value of the driving acceleration necessary to reach threshold allows for more precise measurements. These general indications of patterns of square symmetry in low-viscosity fluids, however, were to be taken with caution since, as Edwards and Fauve indicated [5], when viscosity $\nu \to 0$ the lateral boundaries can strongly influence the pattern because of the very large coherence length near onset. This is a direct consequence of the increasing curvature of the neutral stability curve with decreasing viscous damping [3]. Nevertheless, more recent observations in systems of aspect ratio of the order of 100 [6, 14] have confirmed the existence of stable square patterns for low damping.

Stripes (or lines) consisting of a single set of plane waves were first observed by Fauve et al. [29] in $CO_2$ very close to the liquid–vapor critical temperature. They were also observed later by Edwards and Fauve [5] in a glycerol/water mixture and by Daudet et al. [9] in glycerine. All three cases have the common feature of large viscous dissipation. In this limit, boundary effects are expected to be less relevant since the coherence length of the pattern is of the order of the wavelength. Again, these early observations have been later confirmed in experiments involving a much larger system size [14]. One may note here that stripe patterns in Faraday waves share some qualitative properties with roll states in Rayleigh–Bénard convection [5]. In particular, they both tend to be aligned normal to the sidewalls, especially at high forcing amplitudes. Also, if the driving amplitude is increased abruptly from below to above threshold, circular and spiral patterns have been observed in shallow circular experimental cells [30].

In parallel to the observation of stripes and square patterns, patterns of higher symmetry (consisting of three or more pairs of plane waves) were also observed. Christiansen et al. [4] first observed eightfold quasi-periodic patterns near threshold produced by four standing plane waves of equal amplitude and wave vectors at $45°$ angles. They studied a low-viscosity fluid (ethanol) in the capillary dominated region in a system with an aspect ratio of approximately 50. They also observed hexagonal patterns at higher driving amplitude. Interestingly, they reported that the eightfold and the hexagonal patterns were only observed in large enough systems (aspect ratio > 45). Later Torres et al. [31] observed tenfold patterns again in a low-viscosity fluid at a particular fluid depth, although the aspect ratio was relatively small (of the order of 10). Hexagons were also observed by Kumar and Bajaj [32]. Motivated by these reports, Kudrolli and Gollub [14] undertook a systematic survey of pattern selection at large aspect ratio. By using a cell with aspect ratio of the order of 100, they not only surveyed pattern selection at different driving frequencies but, more importantly, they scanned a wide range of fluid viscosities by using different silicon oils. Unfortunately the depth of the fluid layer used in the experiments was smaller than the surface wavelength. Their results concerning pattern selection can be summarized as follows: At intermediate and high frequencies (i.e., close to the capillary wave limit), squares appear at low viscosity and stripes at large viscosity. Results from previous experiments are qualitatively consistent with this picture. At low frequency (close to the gravity wave limit), hexagons are always the observed patterns. No quasi-periodic patterns with four or more plane waves were therefore observed. This lack of quasi-patterns was later confirmed by Binks et al. [19] in a detailed study of the effect of fluid depth on pattern selection.

The existence of quasi-periodic patterns in a system driven by a single frequency has been convincingly demonstrated in a fluid layer of depth that is large compared with the wavelength [6, 19]. The regions in which quasi-periodic patterns are observed are generally in agreement with earlier theoretical predictions by Zhang and Viñals [33, 34].

## 16.4. Weakly Nonlinear Theory and Amplitude Equations

The theoretical analysis of the weakly nonlinear regime was pioneered by Miles [35, 36] and Milner [17]. Their starting point is the simpler ideal or inviscid limit in which there is a Hamiltonian formulation of Faraday waves. The canonically conjugate variables are the surface displacement away from planarity and the velocity potential at the free surface [37]. Viscous effects were added as a perturbation. Milner explicitly calculated the interaction between waves of different orientation and predicted the existence of patterns of square symmetry above onset if the wave was allowed to detune away from resonance. He further analyzed secondary bifurcations from the base pattern of standing waves.

Surface shear stresses produce a viscous layer near the free surface of depth $\delta \sim (\nu/\omega_0)^{1/2}$. In the limit of low viscous dissipation that Milner considered, $\delta$ is much smaller than the wavelength. An argument of Landau and Lifshitz [38] states that even in a thin surface layer, velocity gradients will not be large. The implication is that energy dissipation within this region is small compared with that in the bulk where the flow is irrotational. Milner used this idea and developed a multiscale expansion based on the inviscid limit and treated viscous dissipation as a perturbation. The multiscale expansion [39] considers first a lowest-order solution that is a sum over marginally growing modes of frequency $\omega$ and wave vector $\{\mathbf{k}_j\}$:

$$\zeta_0 = \sum_j a_j(X, T)e^{i(\mathbf{k}_j \cdot \mathbf{x} - \omega t)} + c.c., \tag{4}$$

and a similar expression for the flow. Here $k = |\mathbf{k}_j|$ and $\mathbf{k}_{-j} = -\mathbf{k}_j$ and the variables $\mathbf{x}$ and $X$ are two dimensional. $X$ and $T$ are slowly varying and are proportional to an expansion parameter $\epsilon$, $X = \epsilon x$, $T = \epsilon t$. The surface and flow variables are expanded in power series, (e.g., $\zeta = \epsilon\zeta_0 + \epsilon^2\zeta_1 + \epsilon^3\zeta_2 + \cdots$), as well as the driving amplitude $f = \epsilon f_0 + \epsilon^2 f_1 + \cdots$. The governing equations and boundary conditions are then solved order by order in $\epsilon$. The equations at order $\epsilon^3$ yield an amplitude equation for $a_j$:

$$\begin{aligned}
\frac{\partial a_j}{\partial t} = & -\frac{ikf}{4\omega}a_{-j}^* - \frac{3\omega}{2k}(\hat{\mathbf{k}}_j \cdot \nabla)a_j + \frac{3i\omega}{4k^2}\nabla^2 a_j - \frac{3i\omega}{8k^2}(\hat{\mathbf{k}}_j \cdot \nabla)^2 a_j \\
& + iT_{jk}^{(1)}|a_k|^2 a_j + iT_{jk}^{(2)}a_k a_{-k}a_{-j}^*,
\end{aligned}$$

where $T_{jk}^{(i)}$ are known functions of the angle between $\mathbf{k}_j$ and $\mathbf{k}_k$ and the three terms that involve gradients of $a_j$ describe a small detuning of the wave away from subharmonic resonance. Milner argued that the absence of the quadratic terms (reflecting the symmetry $a_j \to -a_j$ induced by the symmetry under time translation) precluded the existence of hexagonal patterns above onset. As we will see below, hexagons can be selected for particular functional dependences of the cubic coefficients.

He then obtained dissipative terms order by order in $\epsilon$ by computing the rate of dissipation of the irrotational flow:

$$\dot{E} = -2\eta \int dV \left(\frac{\partial^2 \phi}{\partial x_i \partial x_j}\right)^2,$$

where $\phi$ is the velocity potential. The resulting expansion is of the form

$$\dot{E} = -D^{(0)}|a_j|^2 - D_{jk}^{(1)}|a_j|^2|a_k|^2 - D_{jk}^{(2)}a_j a_{-j}a_k^* a_{-k}^*.$$

When the contribution from damping is added, the final traveling-wave amplitude equation becomes [17]

$$\begin{aligned}
\frac{\partial a_j}{\partial t} = & -\gamma^{(0)}a_j - \frac{ikf}{4\omega}a_{-j}^* - \frac{3\omega}{2k}(\hat{\mathbf{k}}_j \cdot \nabla)a_j + \frac{3i\omega}{4k^2}\nabla^2 a_j - \frac{3i\omega}{8k^2}(\hat{\mathbf{k}}_j \cdot \nabla)^2 a_j \\
& - [\gamma_{jk}^{(1)} - iT_{jk}^{(1)}]|a_k|^2 a_j - [\gamma_{jk}^{(2)} - iT_{jk}^{(2)}]a_k a_{-k}a_{-j}^*. \tag{5}
\end{aligned}$$

The nonlinear coefficients $T_{jk}^{(i)}$ and $\gamma_{jk}^{(i)}$ are explicitly given in Ref. 17. It should be noted that there is a divergence in the nonlinear damping coefficient at $\theta_{jk} = \cos^{-1}(2^{1/3} - 1) \approx 74.9°$, the angle at which quadratic nonlinear waves of wave vector $\mathbf{k}_1 + \mathbf{k}_2$ have a frequency (according to the inviscid linear dispersion relation) that is in resonance with the driving frequency. This divergence is due to the fact that the equation of motion of this resonant mode was not considered, and hence its damping was not accounted for.

There is no simple argument based on Eq. (5) that allows a determination of the selected pattern above onset since the amplitude equation is not of gradient form. However, close to threshold ($f \approx \gamma$), for each pair of unstable traveling waves, only one standing wave (a combination of the two traveling waves) is unstable, whereas the other has a negative growth rate of $-2\gamma$. Hence the standing wave with growth rate $-2\gamma$ can be eliminated as a fast variable and a standing-wave amplitude equation derived. This is a recurring feature of parametric instabilities in weakly damped systems [3]. The resulting standing-wave amplitude equation was derived by Milner, who showed that it is of gradient form with an associated Lyapunov functional:

$$\mathcal{F} = \left(-\gamma\epsilon + \frac{\sigma^2}{2\gamma}\right)|A_j|^2 + \frac{1}{2}\left(\Gamma_{jk} + \frac{\sigma}{\gamma}T_{jk}\right)|A_j|^2|A_k|^2.$$

Here $\sigma$ is the detuning, $T$ is a combination of $T^{(1)}$ and $T^{(2)}$, and $\Gamma$ is a combination of $\gamma^{(1)}$ and $\gamma^{(2)}$. Milner restricted his attention to regular patterns comprising $N$ standing waves of equal amplitude $|A_j| = A$ and wave vectors forming a regular $N$-sided polygon. Minimization of the Lyapunov functional yields square patterns as the selected pattern if one allows for a continuous adjustment of the detuning away from resonance to an optimal value given by $\sigma^* = -\frac{1}{2}T|A|^2$ for the selected pattern ($T \equiv \sum_k T_{jk}$). In this case, $\mathcal{F}(A, \sigma^*) = -\epsilon\gamma N A^2 + (N/2)\Gamma A^4$ and the coefficient of nonlinear damping $\Gamma$ enters in an essential way in the determination of the selected pattern (note that $\Gamma$ depends on the angle between pairs of waves and hence on $N$). Given that viscous dissipation enters the theory originally as a perturbation, this result has been questioned by a number of authors [36, 40].

It was later recognized, however, that although the viscous boundary layer near the free surface is very thin and the bulk flow remains potential, the effect of the viscous layer still must be explicitly incorporated into the calculation of the potential flow. The first step in this direction is based on the so-called quasi-potential approximation [23], in which the boundary conditions are systematically expanded in powers of $\gamma$ and the effects of viscous dissipation taken into account by the introduction of effective boundary conditions on the otherwise irrotational flow. By using this approximation, Zhang and Viñals [33, 34] first expand the governing equations of motion and boundary conditions at the free surface in the small thickness of the viscous layer $1/\sqrt{\gamma}$. The resulting equations were recast into a nonlocal form that involves free-surface variables alone, thus eliminating the need to solve for the bulk flow:

$$\partial_t\zeta(\mathbf{x}, t) = \gamma\nabla^2\zeta + \hat{\mathcal{D}}\Phi - \nabla \cdot (\zeta\nabla\Phi)\frac{1}{2}\nabla^2(\zeta^2\hat{\mathcal{D}}\Phi) - \hat{\mathcal{D}}(\zeta\hat{\mathcal{D}}\Phi)$$
$$+ \hat{\mathcal{D}}\left[\zeta\hat{\mathcal{D}}(\zeta\hat{\mathcal{D}}\Phi) + \frac{1}{2}\zeta^2\nabla^2\Phi\right],$$

$$\partial_t\Phi(\mathbf{x}, t) = \gamma\nabla^2\Phi - (G - \Sigma\nabla^2)\zeta - 4f\zeta\sin(2t)\frac{1}{2}(\hat{\mathcal{D}}\Phi)^2 - (\hat{\mathcal{D}}\Phi)[\zeta\nabla^2\Phi + \hat{\mathcal{D}}(\zeta\hat{\mathcal{D}}\Phi)]$$
$$- \frac{1}{2}(\nabla\Phi)^2 - \frac{1}{2}\Sigma\nabla[\nabla\zeta(\nabla\zeta)^2],$$

where $\Phi(\mathbf{x}, t)$ is the dimensionless surface velocity potential (made dimensionless by $\omega_0/k_0$) and a nonlocal operator $\hat{\mathcal{D}} = \sqrt{-\nabla^2}$ has been introduced [34]. One practical advantage of this equation is the fact that it is effectively two dimensional and hence allows an efficient numerical solution of systems of very large aspect ratio [41]. This system of equations, however, contains an uncontrolled approximation, namely that only those viscous terms that are linear in the surface variables have been retained. In spite of this shortcoming, this model still represents an improvement over the Hamiltonian approaches of Milner and Miles. In the latter, dissipative effects in the governing equations that are linear in the surface variables contribute only to linear damping terms in the amplitude equations, while cubic damping terms result entirely from nonlinear viscous terms in the boundary conditions. A more correct treatment of the free-surface boundary conditions as embodied in the quasi-potential approximation, shows that linear viscous terms alone are sufficient to produce cubic saturating terms in the standing-wave amplitude equation.

Linear analysis of the above equations indicates that the surface becomes unstable at $f_c = \gamma$ for $\gamma \ll 1$. Where $\epsilon = (f - \gamma)/\gamma \ll 1$ is used as an expansion parameter, a standard multiscale expansion [39] leads at the order of $\epsilon^{3/2}$ to a standing-wave amplitude equation:

$$\frac{\partial A_j}{\partial T} = \gamma A_j - \left[ \gamma g(1)|A_j|^2 + \gamma \sum_{i=1(i \neq j)}^{N} g(c_{ij})|A_i|^2 \right] A_j,$$

where $j = 1, 2, \ldots, N$ and $g$ is a function of $c_{ij} = \cos \theta_{ij} = \mathbf{k}_i \cdot \mathbf{k}_j$. Note that in the original unscaled variables the coefficient of the linear term is $(f - \gamma)$, whereas the coefficient of the nonlinear terms remains equal to $\gamma$. An interesting feature of the function $g(c_{ij})$ is that it incorporates the effect of triad resonant interactions through stable second-order waves. As was discussed in detail in Ref. 34, resonant excitation of these stable waves provides a dissipation channel for pairs of unstable waves that satisfy the resonance condition, resulting in large values of the function $g$ in the vicinity of that particular angle. This property of the nonlinear coefficient is seen to be determinant on pattern selection at low dissipation (where this resonance is more pronounced). Particularly interesting behavior emerges in the vicinity of $\Sigma = 1/3$ as the resonant angle approaches zero, as manifested by a sequence of quasi-periodic patterns of increasing symmetry as $\gamma$ decreases.

The quasi-potential approximation description allowed the study of two important qualitative issues: the effect of rotational flow on the cubic coefficients in the standing-wave amplitude equation and the effect of triad resonant interactions on pattern selection. Both were not properly treated in the near-Hamiltonian approach. However, the analysis neglects all terms proportional to $\gamma$ that were nonlinear in the surface variables and fails to describe the experimental phenomenology for moderate or high viscous dissipation.

More recently, Chen and Viñals [27] derived a standing-wave amplitude equation free of any approximation except for those that are inherent in a multiple-scale expansion. Although the analysis is complicated enough so as to require a substantial use of computer algebra, the results confirmed the earlier calculations of Zhang and Viñals in the region in which they were expected to be valid and also reproduced quite accurately the more recent experiments carried out in fluids of large viscosity. This calculation has confirmed the earlier observation that both irrotational and rotational flows contribute to order $\gamma$ to cubic coefficients in the amplitude equation and thus that a correct formulation of this problem requires a viscous treatment from the start. The analysis also correctly obtains stripe patterns at higher damping and gives transition frequencies between patterns of different symmetry within a few percent of the experiments [6, 19].

As usual, the field variables are expanded, e.g., the $z$ component of the velocity field $w$, as

$$w = \epsilon^{1/2} w_0 + \epsilon w_1 + \epsilon^{3/2} w_2 + \cdots,$$

with $\epsilon = (f - f_0)/f_0$ and $w_0$ combination of the linearly unstable modes with wave vectors of different orientation:

$$w_0 = \sum_m \cos(\mathbf{k}_m \cdot \mathbf{x}) B_m(T) \sum_{j=1,3,5\ldots} e^{ji\omega_0/2} \left[ \left( \frac{1}{2} ji\omega_0 + 2\nu k^2 \right) e^{kz} - 2\nu k^2 e^{q_j z} \right].$$

Here $\mathbf{x}$ is the two-dimensional vector, $\omega_0$ is the driving frequency, $\nu$ is the viscosity, $q_j^2 = k^2 + ji\omega/2\nu$, and $\omega = \omega_0/2$. The expression inside the brackets is the linear solution first obtained in a linear stability study by Kumar and Tuckerman [24]. The first term is the irrotational flow that yields a contribution proportional to $\nu$ to the linear threshold. The second term is the rotational flow that yields a contribution proportional to $\nu^{3/2}$ to the threshold.

Expansion of the equation of motion and boundary conditions at order $\epsilon$ yields equations for $w_1$ and $\zeta_1$ that need to be solved explicitly. It is at this order that triad resonant interactions appear, such that the first-order solution becomes very large when the angle between two wave vectors is close to the resonant value. This then leads to large cubic terms in the next-order equations in the vicinity of the resonant angle at low $\gamma$.

Further expansion to order $\epsilon^{3/2}$ yields a solvability condition that leads to a standing-wave amplitude equation in the usual form,

$$\frac{dB_1}{dT} = \alpha B_1 - g_0 B_1^3 - \sum_{m \neq 1} g(\theta_{m1}) B_m^2 B_1,$$

with the nonlinear coefficient $g$ explicitly given. Separation of the rotational and irrotational flow components in $g$ shows that both contribute at leading order $\nu$. Here $\theta_{m1}$ is the angle between $\mathbf{k}_m$ and $\mathbf{k}_1$. This equation is of gradient form, and the associated Lyapunov function allows the determination of the selected patterns as a function of the fluid parameters and of the frequency of the driving acceleration. The resulting predictions can be summarized as follows: (i) stripe patterns for all frequencies at high viscosity; (ii), at low viscosity, square patterns in the capillary wave region (high frequency), and hexagons in the mixed gravity-capillary wave region (low frequency), with two exceptions. First, very close to the gravity wave limit the effective dissipation is always large and the selected pattern is stripes. Second, inside the hexagonal region and in the vicinity of $\Sigma = 1/3$, a sequence of higher symmetry patterns appear, as in the quasi-potential approximation calculation [33, 34].

## Acknowledgments

The research described in this paper is motivated by earlier work in collaboration with Wenbin Zhang, to whom we are indebted. This research has been supported by the U.S. Department of Energy, contract DE-FG05-95ER14566, and also in part by the Supercomputer Computations Research Institute, which is partially funded by the U.S. Department of Energy, contract DE-FC05-85ER25000.

## References

[1] M. Faraday, Philos. Trans. R. Soc. London **121**, 319 (1831).
[2] J. Miles and D. Henderson, Ann. Rev. Fluid Mech. **22**, 143 (1990).
[3] M. Cross and P. Hohenberg, Rev. Mod. Phys. **65**, 851 (1993).
[4] B. Christiansen, P. Alstrøm, and M. Levinsen, Phys. Rev. Lett. **68**, 2157 (1992).

[5]  W. Edwards and S. Fauve, J. Fluid Mech. **278**, 123 (1994).

[6]  D. Binks and W. van de Water, Phys. Rev. Lett. **78**, 4043 (1997).

[7]  N. Tufillaro, R. Ramshankar, and J. Gollub, Phys. Rev. Lett. **62**, 422 (1989).

[8]  W. Wright, R. Budakian, and S. Putterman, Phys. Rev. Lett. **76**, 4528 (1996).

[9]  L. Daudet, V. Ego, S. Manneville, and J. Bechhoefer, Europhys. Lett. **32**, 313 (1995).

[10] R. Lang, J. Acoust. Soc. Amr. **34**, 6 (1962).

[11] A. Ezerskii, M. Rabinovich, V. Reutov, and I. Starobinets, Zh. Eksp. Teor. Fiz. **91**, 2070 (1986); [Sov. Phys. JETP **64**, 1228 (1986)].

[12] S. Ciliberto, S. Douady, and S. Fauve, Europhys. Lett. **15**, 23 (1991).

[13] H. Müller, Phys. Rev. Lett. **71**, 3287 (1993).

[14] A. Kudrolli and J. Gollub, Physica D **97**, 133 (1996).

[15] H. Lamb, *Hydrodynamics* (Cambridge U. Press, Cambridge, 1932).

[16] J. Miles, Proc. R. Soc. London A **297**, 459 (1967).

[17] S. Milner, J. Fluid Mech. **225**, 81 (1991).

[18] H. Müller, H. Wittmer, C. Wagner, J. Albers and K. Knorr, Phys. Rev. Lett. **78**, 2357 (1997).

[19] D. Binks, M.-T. Westra, and W. van de Water, Phys. Rev. Lett. **79**, 5010 (1997).

[20] Nam Hong U, Bull. Russi. Acad. Sci. Phys. Supp. **57**, 131 (1993).

[21] T. Benjamin and F. Ursell, Proc. R. Soc. London Ser. A **225**, 505 (1954).

[22] L. Landau and E. Lifshitz, *Mechanics* (Pergamon, New York, 1976).

[23] T. Lundgren and N. Mansour, J. Fluid Mech. **194**, 479 (1988).

[24] K. Kumar and L. Tuckerman, J. Fluid. Mech. **279**, 49 (1994).

[25] J. Bechhoefer, V. Ego, S. Manneville, and B. Johnson, J. Fluid Mech. **288**, 325 (1995).

[26] E. Cerda and E. Tirapegui, Phys. Rev. Lett. **78**, 859 (1997).

[27] P. Chen and J. Viñals, Phys. Rev. Lett. **79**, 2670 (1997).

[28] S. Douady, S. Fauve, and O. Thual, Europhys. Lett. **10**, 309 (1989).

[29] S. Fauve, K. Kumar, C. Laroche, D. Beysens and Y. Garrabos, Phys. Rev. Lett. **68**, 3160 (1992).

[30] S. Kiyashko, L. Korzinov, M. Rabinovich, and L. Tsimring, Phys. Rev. Lett. **54**, 5037 (1996).

[31] M. Torres, I. Pastor, I. Jiménez, and F. M. de Espinosa, Chaos Solitons Fractals **5**, 2089 (1995).

[32] K. Kumar and K. Bajaj, Phys. Rev. E **52**, R4606 (1995).

[33] W. Zhang and J. Viñals, Phys. Rev. E **53**, R4283 (1996).

[34] W. Zhang and J. Viñals, J. Fluid Mech. **336**, 301 (1997).

[35] J. Miles, J. Fluid Mech. **146**, 285 (1984).

[36] J. Miles, J. Fluid Mech. **248**, 671 (1993).

[37] V. Zakharov, Zh. Prikl. Mekh. Tekh. Fiz. **9**, 86 (1968); [J. Appl. Mech. Tech. Phys. **9**, 190 (1968)].

[38] L. Landau and E. Lifshitz, *Fluid Mechanics* (Pergamon, New York, 1959).

[39] A. Newell and J. Whitehead, J. Fluid Mech. **38**, 279 (1969).

[40] P. Hansen and P. Alstrøm, J. Fluid Mech. **351**, 301 (1997).

[41] W. Zhang and J. Viñals, Physica D **116**, 225 (1998).

# 17

# Deformation and Rupture in Confined, Thin Liquid Films Driven by Thermocapillarity

MARC K. SMITH AND DAVID R. VRANE

## 17.1. Introduction

Surface tension is a dominant force in many kinds of fluid flows, such as those in a microgravity environment or in thin liquid films and droplets. In these situations, surface-tension gradients can produce significant free-surface tangential velocities of the order of several centimeters per second or more. An instability in these kinds of thermocapillary flows or a rupture of the thin liquid film can cause defects or failures in the associated products or processes. Thus a good understanding of the behavior of thermocapillary-driven flows in thin films or droplets would be very useful in the design of such systems.

The instability of surface-tension-driven films has been reviewed by Davis (1987). He discussed how a thermocapillary film is susceptible to convective, surface-wave, and long-wave types of instabilities. Some of these modes of instability require surface deformation and some do not. Some modes are necessarily affected by the size of the container in which the film is contained and others are not. In spite of all the past work in this area, the physical mechanisms underlying the behavior of a thin liquid film driven by thermocapillarity is still being debated.

In this chapter we bring three often neglected features of the physical system into a model of a thermocapillary film flow: end effects, geometric curvature, and free-surface deformation. We examine the thermocapillary flow in a thin film of a highly conductive liquid confined to a shallow cavity of finite length cut into the outer surface of a circular cylinder. The flow is driven by a temperature gradient imposed on the free surface of the film. Using an asymptotic analysis for thin films, we obtain a highly nonlinear system of evolution equations for the behavior of the free surface. This set of equations includes the effect of surface-tension gradients, capillary pressure produced by free-surface curvature due to the cylindrical geometry and free-surface deformation, end effects, and liquid inertia. These equations are solved numerically with spectral methods for both transient and steady-state responses in the film. Of particular interest is the parameter range in which steady-state solutions are possible and when and how the film may rupture. The results of this model should give us a much better understanding of the complex interplay between the various physical effects that occur in such systems, especially as they occur in a microgravity environment.

In Section 17.2 we describe the mathematical model used in this work. We then briefly discuss the asymptotic approximations and methods used to obtain the nonlinear evolution equations that model the behavior of the free surface of the film. Next, we give the highlights of the numerical methods used to solve the equations and then present our results. Here, we also give concise physical descriptions of the flow physics in the film that lead to the behaviors seen in our computations. Finally, we present our conclusions and discuss other possible avenues of investigation.

## 17.2. Mathematical Model

A thin liquid film of average thickness $d$ is contained in a long shallow cavity of inner radius $R_i$ and length $L$ cut into the outer surface of a circular cylinder, as shown in Fig. 17.1. The cylinder insulates the liquid from the inside while the lower end of the film is held at the constant temperature $T_h$ and the upper end at the constant temperature $T_c$. We define the temperature difference $\Delta T = T_h - T_c$. The liquid film is bounded on the outside by a passive gas in which the temperature $T_g$ varies linearly along the film from the hot temperature $T_h$ to the cold temperature $T_c$. The heat-transfer coefficient from the free surface of the liquid to the bounding passive gas is given by $h_g$. The liquid has constant density $\rho$, viscosity $\mu$, thermal conductivity $k$, and thermal diffusivity $\alpha$. The surface tension of the free surface $\sigma$ is a function of temperature as $\sigma = \sigma_0 - \gamma(T - T_0)$, where $\gamma$ is the negative of the rate of change of surface tension with temperature, $T_0 = (T_h + T_c)/2$ is the average temperature in the film, and $\sigma_0$ is the surface tension at the temperature $T_0$. Gravity is ignored because the film is very thin or because it is held in a microgravity environment.

The flow in the liquid film is governed by the Navier–Stokes equations, continuity, and the conservation of energy. Standard rigid-wall boundary conditions are used together with the free-surface balances of normal and tangential stress, continuity, and heat flux. The thermal boundary conditions on the rigid surfaces bounding the film are either zero heat flux or fixed temperature, as described above. The governing equations are written in cylindrical coordinates with the origin midway between the two ends of the film and with the $x$ axis parallel to the centerline of the cylinder. To simplify later development, the zero point of the radial coordinate is shifted to the inner surface of the liquid film.

This film geometry is characterized by two relevant geometric ratios: the aspect ratio $A = d/L$ and a curvature parameter $\Phi = d/R_i$. The curvature parameter indicates the magnitude of the film curvature due to the cylindrical geometry. When $\Phi = 0$, the geometry corresponds to a two-dimensional rectangular cavity. For $\Phi \to \infty$, the geometry corresponds to a liquid bridge with no inner solid surface, as considered by Xu and Davis (1983) and several others. We consider the case in which the film is thin compared with both the cavity length $L$ and the inner radius $R_i$, but in which $L$ and $R_i$ are approximately the same size. Thus we consider a lubrication film flow in which the free-surface curvature in both the axial and the circumferential directions influence the flow.

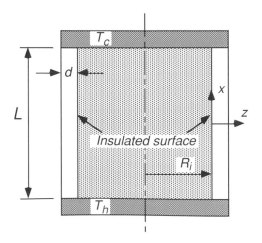

Figure 17.1. The cylindrical film-flow geometry.

To simplify the analysis, we solve the governing equations for the case of axisymmetric motion in the film. We scale the axial and radial spatial variables $x$ and $r$, the axial and radial velocities $u$ and $w$, pressure $p$, temperature $T$, and time $t$ by using standard lubrication scalings as follows:

$$x = L\hat{x}, \qquad r = R_i + d\hat{z}, \qquad u = U\hat{u}, \qquad w = AU\hat{w}, \qquad p = \hat{p}\frac{\mu U}{Ad},$$
$$T = \Delta T \hat{T} + \frac{1}{2}T_0, \qquad t = \frac{L}{U}\hat{t},$$

(1)

where carets denote dimensionless quantities. Note that we have shifted the radial coordinate to the inner solid boundary of the film and changed the coordinate in this direction to $\hat{z}$. The velocity scale $U$ is based on a thermocapillary balance at the free surface of the film, $U = \gamma\Delta T/\mu$. Using additional ordering scales on the resulting dimensionless groups from this first scaling, we obtain the following $O(1)$ dimensionless parameters:

$$R = \frac{\gamma\Delta T d}{\mu\nu}, \qquad \text{Reynolds number;}$$

(2a)

$$M = \frac{\gamma\Delta T d}{\mu\alpha}A^{-2}, \qquad \text{Marangoni number;}$$

(2b)

$$C = \frac{\gamma\Delta T}{\sigma_0}A^{-3}, \qquad \text{capillary number;}$$

(2c)

$$\Gamma = \Phi A^{-1} = L/R_i, \qquad \text{curvature number;}$$

(2d)

$$B = h_g L/k, \qquad \text{Biot number;}$$

(2e)

where $\nu = \mu/\rho$ is the kinematic viscosity. Refer to Vrane (1996) for further details of this scaling procedure. Also note that after this point we suppress the caret notation for dimensionless variables.

Our interest in this system is in the case in which $A \ll 1$, known as the lubrication limit. It has been shown by Sen and Davis (1982) and Vrane (1996) that the film flow in this limit can be divided into three distinct regions. First, there are two end regions that extend approximately one or two film thicknesses away from each end. Between these two end regions is the core-flow region. In the asymptotic limit of small $A$, the flow in the core is an almost parallel flow and the function of the end regions is to turn the flow around and complete the circulation from the core flow. One could solve for this flow explicitly and match it to the core flow. However, for our purposes it is enough to observe that the parameter scalings we have used force the free surface in the end regions to stay locally planar or flat. Thus the contact-line boundary conditions used in these regions are communicated directly to the film in the core. While Vrane (1996) considered both pinned contact lines and fixed contact angles at the ends, we consider just pinned ends.

Proceeding with the asymptotic analysis for the core-flow region, we expand all of the dependent variables in a power series in $A$, substitute these into the governing equations, and then isolate the terms of like powers in $A$. The first component of this solution is a large constant pressure in the film,

$$p_{-1} = \Gamma/CA.$$

(3)

This is the capillary pressure in the film due to its cylindrical nature. Since it is constant, it does not affect the dynamics of the rest of the flow and we may safely ignore it.

The next component of the solution is given by the following $O(1)$ core-flow system of equations:

$$R(u_t + uu_x + wu_z) = -p_x + u_{zz}, \tag{4a}$$

$$p_z = 0, \tag{4b}$$

$$u_x + w_z = 0, \tag{4c}$$

$$T_{zz} = 0; \tag{4d}$$

$$u = w = T_z = 0, \quad \text{on } z = 0; \tag{5a–5d}$$

$$\left.\begin{aligned} u_z &= -(T_x + T_z h_x) \\ p &= -C^{-1}(h_{xx} + \Gamma^2 h) \\ T_z + B(T - T_g) &= 0 \end{aligned}\right\} \quad \text{on } z = h;
\begin{aligned} &\text{(6a)} \\ &\text{(6b)} \\ &\text{(6c)} \end{aligned}$$

$$T_g = -x; \tag{7}$$

$$h_t + q_x = 0, \qquad q = \int_0^h u\,dz,$$

$$\int_{-1/2}^{1/2} h\,dx = 1, \qquad h(\pm 1/2) = 1. \tag{8a–8d}$$

The external linear temperature variation imposed on the film by the bounding passive gas and the end heating is given by Eq. (7). Equation (8a) is the integrated kinematic condition on the free surface, where the net axial volume flux in the film $q$ is defined by Eq. (8b). Finally, Eq. (8c) is a global conservation of mass condition, and Eq. (8d) is the pinned-end condition on the core flow of the film obtained by asymptotic matching with the flow in the small end regions of the film.

In these equations, we now see the effect of the scalings chosen above. The axial momentum equation (4a) balances viscous effects due to axial shear stresses with the pressure and the acceleration of the fluid in the axial direction. The fluid acceleration was brought into the problem by our choice of the scalings for the Reynolds number. The normal momentum equation (4b) shows that the pressure is constant across the depth of the film; a standard result in lubrication theory. The temperature field given by Eq. (4d) shows that we have pure conduction across the film. Convective effects have been neglected by our choice of scalings for the Marangoni number. We can also see in the normal-stress balance equation (6b) that there is a component to the capillary pressure due to axial curvature variations in the film profile as well as a component due to circumferential curvature variations that appear when the radius of the free surface of the film changes. These components appear because of our definition of the curvature parameter $\Gamma$ and our scaling of the capillary number $C$.

A partial solution of these equations for the temperature and pressure in the film yields

$$T = T_g = -x, \tag{9a}$$

$$p = -C^{-1}(h_{xx} + \Gamma^2 h). \tag{9b}$$

If we neglect inertia by setting $R = 0$, the solution of the axial momentum equation (4a) yields a parabolic axial-velocity profile. When inertia is present, we solve for the net flow rate in the film by using the Karman–Pohlhausen method. We assume an axial velocity with a parabolic form:

$$u = \frac{3}{2h^3}\left(\frac{1}{2}h^2 - q\right)(z^2 - 2hz) + z. \tag{10}$$

This velocity profile satisfies boundary conditions (5a) and (6a), and it has a net flow rate $q$. Substituting this velocity into the continuity equation (4c), we can find the corresponding normal velocity in the film. Both components of velocity are then substituted into the axial momentum equation (4a) and the result is integrated over the thickness of the film. The result is the axial momentum equation

$$R\left\{q_t + \frac{1}{40}h^2h_x - \frac{9}{20}hq_x + \frac{1}{20}qh_x + \frac{12}{5}\frac{qq_x}{h} - \frac{6}{5}\frac{q^2h_x}{h^2}\right\}$$
$$= C^{-1}(h_{xxx} + \Gamma^2 h_x)h + \frac{3}{2} - 3\frac{q}{h^2}. \tag{11}$$

This momentum equation is solved together with the film continuity equation and boundary conditions given by Eqs. (8a), (8c), and (8d).

## 17.3. Numerical Solution

The film-flow equations (11), (8a), (8c), and (8d) were solved numerically. First, the time derivatives were discretized by use of a backward Euler method. With this method, an equation such as $q_t = f(q, h)$ would be discretized as

$$q^{(i+1)} = q^{(i)} + \Delta t f\left[q^{(i+1)}, h^{(i+1)}\right], \tag{12}$$

where the superscripts refer to the discretized temporal time coordinate and $\Delta t$ is the time step. We have used a time step of 0.01 in all of the results that follow.

In the present problem, the resulting nonlinear equations in $h^{(i+1)}$ and $q^{(i+1)}$ were linearized by use of the Newton–Kantorovich method. Consider the nonlinear equation $N(q, h) = 0$. Using Frechet derivatives of the operator $N$, we produce the following equation:

$$N(q + \Delta q, h + \Delta h) = N(q, h) + N_q\Delta q + N_h\Delta h + \cdots, \tag{13}$$

where the subscripts on $N$ refer to the appropriate Frechet derivative and $\Delta$ refers to the change in the corresponding variable. Using the standard Newton approximation, we write the equation

$$N_q\Delta q + N_h\Delta h = -N(q, h) \tag{14}$$

for the change in the solution $\Delta q$ and $\Delta h$.

Equation (14) is solved with standard pseudospectral methods that use Chebyshev polynomials. We used 100 Chebyshev polynomials in all of the results that follow. Further details for all of the above techniques can be found in Vrane (1996) and Boyd (1989).

## 17.4. Results and Discussion

The steady-state behavior of a liquid film heated at its left end and cooled at its right is shown in Fig. 17.2 for $\Gamma = 0$, $R = 0$, and various values of $C$. For this parameter set, the film is two dimensional with no inertia and finite surface tension. The figure shows that the film is thicker on its right side and thinner on its left and that the free-surface deformation increases as the capillary number increases. The physics behind this behavior is a simple balance between the flows driven by two distinct forces. The first is the temperature gradient applied to the film, which produces a thermocapillary shear stress on the free surface that drives a flow toward the right. This flow is exactly balanced by a return flow to the left driven by the capillary-pressure gradient. The necessary gradient is produced by having the deformation shown in the figure in which the capillary pressure is large on the right side and small on the left. The resulting flow to the left exactly balances the themocapillary flow to the right so that the net mass flux in the film at any axial location is zero. Since the capillary pressure also depends on surface tension, decreasing the surface tension means that the deformation of the film should increase in order to produce the necessary capillary-pressure gradient to balance the thermocapillary flow. This gives us the trend we see for the effect of the capillary number. In addition, given the particular ordering we used in the analysis that set up this model, the capillary number can be increased by increasing the length of the layer. Thus, longer films will have more free-surface deformation as well.

The effect of cylindrical curvature in the film is shown in Fig. 17.3 for $C = 30$, $R = 0$, and various values of $\Gamma$ in the range $(0, 2\pi)$. Here, we see that the free-surface deformation increases as $\Gamma$ increases. In a cylindrical film, the net capillary pressure is produced by free-surface curvature in both the axial and the circumferential directions. Circumferential curvature in the film is inversely proportional to the radius of the free surface. It decreases when the film thickness increases and increases when the film thickness decreases. For the film profiles shown in Fig. 17.3, the effect of this additional curvature is a reduction in the capillary pressure on the right side of the film and an increase on the left. As a result, the

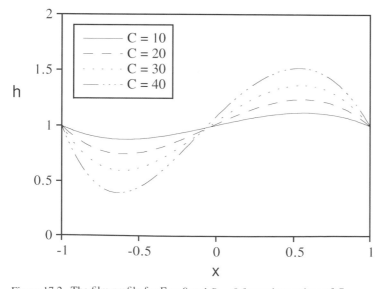

Figure 17.2. The film profile for $\Gamma = 0$ and $R = 0$ for various values of $C$.

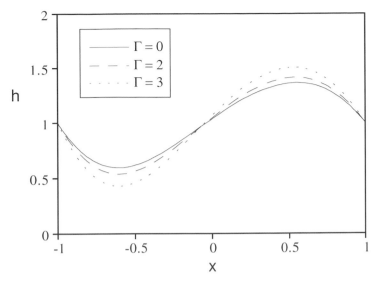

Figure 17.3. The film profile for $C = 30$ and $R = 0$ for various values of $\Gamma$ in the range where the axial curvature is dominant.

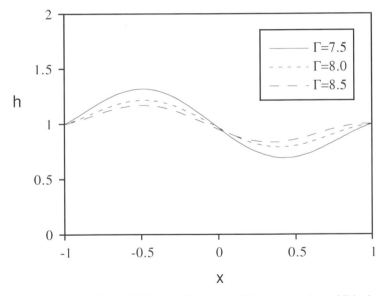

Figure 17.4. The film profile for $C = 10$ and $R = 0$ for various values of $\Gamma$ in the range where the circumferential curvature is dominant.

capillary-pressure-driven flow to the left is reduced. In order to maintain the balance with the thermocapillary-driven flow to the right, the free-surface deformation must increase. Even though both components of the curvature are affected as the free surface deforms in this way, in the parameter range $\Gamma = (0, 2\pi)$ the axial curvature is dominant and so a new flow balance can be achieved.

For $\Gamma$ in the range $(2\pi, 8.987)$, the circumferential curvature is dominant, and the effect of $\Gamma$ and the sense of the free-surface deformation are completely reversed. Figure 17.4 shows that

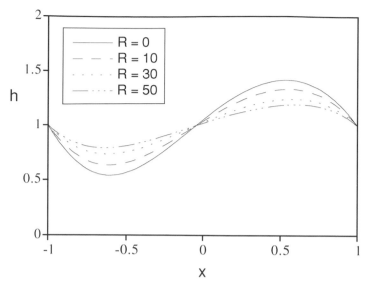

Figure 17.5. The film profile for $C = 30$ and $\Gamma = 2$ for various values of $R$.

the film is now thinner on the right and thicker on the left. The capillary-pressure-driven flow is still to the left, but now it is the circumferential curvature that is primarily responsible for the flow and the axial curvature that retards it. As $\Gamma$ is increased, the free surface becomes less deformed in order to reduce the circumferential-curvature gradient and maintain a balance with the thermocapillary-driven flow to the right. Even though these solutions are interesting, our transient computations have demonstrated that the film profiles for this range of $\Gamma$ are unstable.

Figure 17.5 shows that the effect of liquid inertia in the film is to decrease the free-surface deformation. The reason lies in the fact that the momentum of the liquid in the film near the free surface is larger than the momentum of the liquid near the rigid wall because the velocities near the free surface are larger. The flow near the free surface is primarily driven by the thermocapillary shear stress and it must accelerate from the left side of the film to the right as the thickness of the film increases. When inertia is not important, the thermocapillary stress primarily balances the viscous shear stresses in the bulk of the film and the velocity at the free surface is large. As the momentum of the liquid becomes more important, i.e., as $R$ increases, the flow driven to the right by the thermocapillary stress is reduced because this stress must now work against the viscous stresses in the bulk of the film and it must accelerate the liquid near the free surface. Since the flow to the right is smaller, the capillary-pressure gradient needed to balance the flow is reduced and so less deformation of the free surface is needed.

A more complete discussion of the behavior of a cylindrical film and other related geometries can be found in Vrane (1996). This includes a thorough explanation of the mechanism of the flow, an analysis of the stability of the steady-state film flow, a study of the multiple steady-state solutions found in the film, and a discussion of the possibility of film rupture in these systems.

Film rupture is indicated by steady-state solution curves such as the one shown in Fig. 17.6. This is a plot of the $L_2$ norm of the film deformation versus the capillary number $C$. Our interest in this graph lies in the fact that there is a well-defined limit point $C_{\lim}$ past which no steady-state solutions could be found. Preliminary transient computations discussed below, show that the film will rupture when $C > C_{\lim}$. We have tracked this limit point in the present cylindrical film with inertia by using the software AUTO94 by Doedel and Wang (1994). The

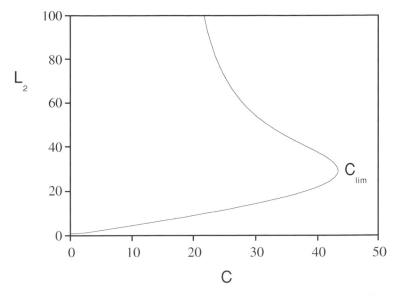

Figure 17.6. The $L_2$ norm of the film deformation versus $C$ for $R = \Gamma = 0$. The limit point $C_{\text{lim}}$ is defined by the nose of the curve.

resulting data are shown in the Fig. 17.7. Figure 17.7(a) shows the limit-point curves for the smaller values of $\Gamma$. Here, the film behaves much like a two-dimensional film, with the axial curvature being the dominant contribution to the capillary pressure. We see that for any fixed value of $R$, we can cross the limit-point curve and rupture the film by increasing the capillary number $C$ or by increasing the curvature parameter $\Gamma$. However, for a fixed value of $C$ and $\Gamma$, the effect of increasing the Reynolds number is to stabilize the film. Thus we can have a film that would be ruptured at smaller values of $R$, but that remains intact for larger values.

Figure 17.7(b) shows the limit-point curves for larger values of $\Gamma$ in which the cylindrical curvature is the dominant contribution to the capillary pressure. For all of these curves, our transient computations suggest that the associated steady-state film profiles are unstable. Thus the film always tends to rupture in this range of parameters, at least for the initial conditions that we used.

In order to demonstrate film stabilization and destabilization, transient computations were done starting from an initial condition of a uniform film at rest in the cylindrical cavity. At time $t = 0$, the temperature difference is applied to the ends of the film and the passive gas heats the free surface. Because the film is highly conductive, the temperature profile in the film becomes linear instantaneously. This drives a thermocapillary flow that starts deforming the free surface of the film. We display the temporal behavior of the film in Figs. 17.8–17.10 by plotting the minimum film thickness as a function of time.

In Fig. 17.8, we see the destabilization of the film as the capillary number increases past the limit point with $R = 0$ and $\Gamma = 2$. For the smaller values of $C$, the minimum film thickness approaches the steady-state value rapidly and in a monotonic fashion. When $C$ exceeds the limit-point value of approximately 39, the minimum film thickness continually decreases to zero. The exact point of film rupture is never reached because the relevant physical models for film rupture were not included in this work.

Starting with the parameters $C = 50$ and $\Gamma = 2$, Fig. 17.9 shows how the film can be stabilized as the Reynolds number increases. For the smaller values of $R$, the film ruptures. However, when $R$ exceeds $\sim$8.5, a steady-state profile is attained, although with liquid inertia included

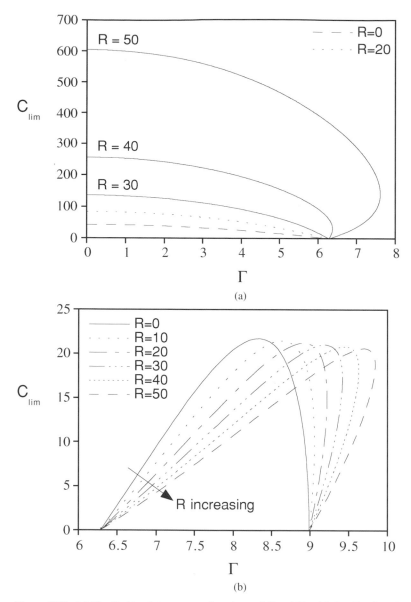

Figure 17.7. (a) The limit-point curves as functions of $R$ and $\Gamma$, with $0 < \Gamma < 2\pi$. A stable intact film is possible below and to the left of these curves. (b) The limit-point curves as functions of $R$ and $\Gamma$, with $2\pi < \Gamma < 8.987$. No stable intact-film solution could be found near these curves.

the approach to the steady-state profile occurs in an oscillatory manner. But even so, the effect of having significant liquid inertia in the flow is to stabilize the film and prevent rupture.

Finally, Fig. 17.10 shows how the film can be stabilized by decreasing the curvature of the film. We set $C = 10$ and $R = 10$ and decrease the curvature parameter from 8 to 5. For the larger values of $\Gamma$, the film ruptures. Interestingly, there are some oscillations in the minimum film thickness even as it approaches zero thickness. When $\Gamma$ is decreased below $\sim 5.7$, the film stabilizes, but again the behavior of the film during its approach to the steady-state profile is oscillatory.

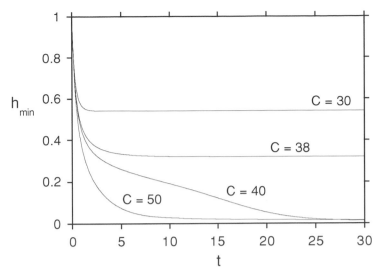

Figure 17.8. The transient behavior of the minimum film thickness for various values of $C$ with $R = 0$ and $\Gamma = 2$.

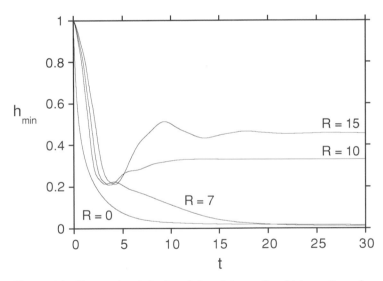

Figure 17.9. The transient behavior of the minimum film thickness for various values of $R$ with $C = 50$ and $\Gamma = 2$.

## 17.5. Conclusions

In this chapter we have examined the behavior of a thin, cylindrical liquid film contained in a shallow cavity cut into the outer surface of a circular cylinder. A flow in the film is driven by surface-tension gradients induced by heating the film along its free surface and from the ends. We have examined the situation in which the film is thin compared with its length $L$ and its inner radius $R_i$, but have required the dimensions $L$ and $R_i$ to be of the same magnitude. The film has a high thermal conductivity, and inertial effects are important. Using these approximations, we have performed an asymptotic analysis to produced a pair of evolution equations that describe the behavior of the film. These equations were solved numerically with spectral techniques.

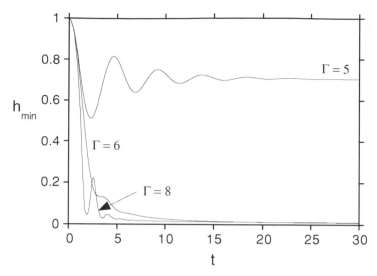

Figure 17.10. The transient behavior of the minimum film thickness for various values of $\Gamma$ with $C = 10$ and $R = 10$.

Our results show that steady intact films with an internal flow driven by thermocapillarity and capillary pressure are possible if the capillary number and the curvature number are small enough for a given Reynolds number. When this is not the case, the film will rupture.

For small curvature numbers, the film is nearly two dimensional. The flow in an intact film is a balance between a free-surface flow driven by surface-tension gradients and a return flow driven by the capillary pressure produced by the axial curvature of the free surface. When this balance cannot be maintained, the film will rupture. This occurs if the temperature gradient on the free surface is too large, the length of the film is too large, or if the surface tension is too small.

Increasing the curvature number $\Gamma$ adds an additional capillary pressure to the film because of the circumferential curvature of the free surface. This extra capillary-pressure gradient works with the thermocapillary flow against the axial capillary-pressure gradient to make the film more susceptible to rupture. Thus the limit-point value of the capillary number decreases as the curvature number increases.

Increasing the Reynolds number of the flow increases the range of capillary and curvature numbers over which an intact film is possible. This occurs because the effect of the Reynolds number is to reduce the strength of the thermocapillary-driven flow. This makes it easier for the flow to be balanced by the capillary-pressure-driven flow. In addition, inertial effects in the film allow oscillations to appear in the transient response of the film. These flow perturbations are always damped as the film evolves to either an intact or a ruptured state.

Our results also show that the limit-point capillary number goes to zero when the curvature parameter is $2\pi$. This marks the capillary-pinching limit for a circular cylindrical interface, first discussed by Plateau (1863) and Rayleigh (1879). When the curvature parameter is above this point and the Reynolds number is not too large, the capillary pressure due to circumferential curvature dominates all of the forces and causes the film to rupture. Such a rupture is shown in Fig. 17.10 for $\Gamma = 8$, $C = 10$, and $R = 10$. One interesting result of this work is that if the Reynolds number is large enough, the capillary-pinching instability can be suppressed by the inertia of the liquid flowing in the film. We see in Fig. 17.7(a) that this limit is extended to $\Gamma = \sim 7.6$ for $R = 50$, an increase of $\sim 21\%$. Russo and Steen (1989) and Dijkstra and Steen (1991) were also interested in this kind of film stabilization. However, they approached the problem by examining the competition between the capillary- and surface-wave linear instabilities in a thin annular film of infinite length. They suggested that the interaction between these disturbances was responsible for the stabilization. Our nonlinear analysis of a finite-length cylindrical film shows that a different stabilization mechanism may be at work in these films, at least in the parameter range we have investigated.

Further work on this film-flow model is needed. It would be interesting to see if any stable nonlinear flows are possible in the film when the curvature parameter is greater than the capillary-pinching limit. Although we did not find any stable intact films by using the transient calculations reported in this chapter, a more systematic search may be more productive. In particular, are there any stable shapes corresponding to the second branch of limits points of Fig. 17.7(b) that are similar to the shapes shown in Fig. 17.4? Another area of investigation concerns the type of stable intact-film profiles that are possible. We have only found steady-state profiles, although the transient approach to these states may be oscillatory. Are there any stable oscillatory film profiles for this model? This question would be very interesting to answer because of its relationship to other thermocapillary instabilities in thin liquid films and in other similar systems.

## References

Boyd, J. P., "Chebyshev and Fourier spectral methods," in *Lecture Notes in Engineering*, Brebbia, C. A. and Orszag, S. A., eds. (Spring-Verlag, New York, 1989), Vol. 49.

Davis, S. H., "Thermocapillary instabilities," *Ann. Rev. Fluid Mech.* **19**, 403–35 (1987).

Dijkstra, H. A. and Steen, P. H., "Thermocapillary stabilization of the capillary breakup of an annular film of liquid," *J. Fluid Mech.* **229**, 205–28 (1991).

Doedel, E. and Wang, X., "AUTO94 Software for continuation and bifurcation problems in ordinary differential equations," Applied Mathematics Report, California Institute of Technology, Pasadena, CA 91125 (July 1994).

Plateau, J. A. F., "Experimental and theoretical Researches on the figures of equilibrium of a liquid mass," (translated) in Annual Reports of the Smithsonian Institution (1863–1866).

Rayleigh, Lord J. W. S., "On the instability of jets," *Proc. London Math. Soc.* **10**, 4–13 (1879).

Russo, M. J. and Steen, P. H., "Shear stabilization of the capillary breakup of a cylindrical interface," *Phys. Fluids A* **1**(12), 1926–37 (1989).

Sen, A. K. and Davis, S. H., "Steady thermocapillary flows in two-dimensional slots," *J. Fluid Mech.* **121**, 163–86 (1982).

Vrane, D. R., "The stability of thermocapillary-driven flows in finite regimes," Ph.D. dissertation, Georgia Institute of Technology (1996).

Xu, J.-J. and Davis, S. H., "Liquid bridges with thermocapillarity," *Phys. Fluids* **26**, 2880–86 (1983).

# 18

# Linear and Nonlinear Waves in Flowing Water

CHIA-SHUN YIH AND WILLIAM W. SCHULTZ

## 18.1. Introduction

The subject of gravity waves in water in rectangular channels has occupied researchers' attention since the nineteenth century, in the latter part of which a particularly fruitful approach for studying long nonlinear water waves was initiated by Boussinesq (1871) and carried further by Korteweg and de Vries (1895). In comparison, far fewer studies of gravity waves in shear flows have been carried out, and all of these, so far as we are aware, have appeared after World War II.

There were first the papers[*] of Thompson (1949) and Biésel (1950). But these authors managed to obtain results only for flows with a constant velocity gradient (or a uniform shear). Later Burns (1953) obtained results for the general velocity profile assumed herein, but only for infinitely long waves, i.e., without considering the dispersion effect. If this effect is not taken into account, a Korteweg–de Vries (KdV) equation cannot be obtained, as is well known. Burns' work was followed by those of Benjamin (1962) and Yih (1972). Yih studied the stability of free-surface shear flows, investigating the variation of the wave velocity $c$ with the wave number $k$. Benjamin mainly applied the momentum principle to nonlinear waves, which involves many integrals that can be evaluated only after one knows the vertical structure of the wave motion. This structure can be supplied only by a detailed analysis of the linear waves, which was not provided in his work.

In the present work we, like Benjamin, consider a general velocity for the basic flow. However, we provide a detailed determination of the eigenvalues for $c$ (including the dispersion effect) and the eigenfunction for the stream function, thus enabling us to derive a KdV equation for nonlinear long waves going in one direction, downstream or upstream. The appearance of the wave velocity and the dispersion effect for linear waves and the curvature of the velocity of the basic flow at the free surface in the coefficients of the KdV equation shows how a less detailed derivation of that equation cannot give definite results.

For any given value of the Froude number $F$, it is shown that there is one and only one positive $c$, which is greater than $F$, corresponding to waves running downstream, and one and only one negative $c$, corresponding to waves running upstream, in agreement with the work of Burns (1953). The positive $c$ approaches $F$ and the negative $c$ approaches zero, as $F$ increases indefinitely. The effect of this is that at large $F$ the amplitude of coherent waves, such as solitary and cnoidal waves, becomes very small if they run downstream, and exceedingly so if they run

---

[*] Burns (1953) criticized these works for regarding the wavelength as finite and yet assuming the pressure distribution to the hydrostatic. This criticism is not well founded, for Thompson and Biésel merely considered the *mean* pressure to be hydrostatic, and their free-surface condition is correct even for finite wavelengths.

234

upstream (with a very small velocity). The implication of this is discussed comparatively with Benjamin and in connection with Ursell's results (1953).

For modest values of $F$, even if it is greater than 1 and as much as 2 or 3, the KdV equation derived allows solutions for solitary waves and cnoidal waves. The application of Hirota's solution (1971), as presented in Whitham (1974), determines the interaction of solitary waves going in the same direction.

In the following sections we consider a basic flow of an inviscid liquid in a horizontal rectangular channel. We measure $x$ along the direction of the flow and $y$ vertically upward from the bottom of the channel. The mean depth of water in the channel is denoted by $h$, and the gravitational acceleration is denoted by $g$. We use $h$, $(gh)^{1/2}$, and $(h/g)^{1/2}$ as the scales of length, velocity (including wave velocity), and time $t$, respectively.

For the basic flow, we denote the dimensionless velocity by $FU(y)$, where $F$ is the Froude number defined by

$$F^2 = \frac{V^2}{gh},\tag{1.1}$$

with $V$ indicating the dimensional velocity at the free surface, where $y = 1$. We assume that

$$U(0) = 0, \qquad U'(0) > 0, \qquad U'(1) = 0, \qquad U''(y) < 0.\tag{1.2}$$

Then $U$ increases monotonically from zero at the bottom to l at the free surface. Hence $F$ is the maximum dimensionless velocity and $V$ is the maximum dimensional velocity. The first and third conditions in Eqs. (1.2) are imposed with an eye on application of the results to real flows.

## 18.2. Linear Waves

We treat linear waves first. The scaling described in Section 18.1 are used throughout this paper. The stream function of the perturbation is assumed to have the form

$$\psi = \phi(y)\exp ik(x - ct),$$

where $c$ is the wave velocity and $k$ is the wave number. Then the velocity components for the perturbation denoted by $(u, v)$, are given by

$$u = \phi'(y)\exp ik(x - ct), \qquad v = -ik\phi(y)\exp ik(x - ct),\tag{2.1}$$

with the understanding that the real parts of the right-hand sides are to be used. As is well known, the equation for $\phi$ can be derived from the linearized Euler equations and is

$$(FU - c)(\phi'' - k^2\phi) - FU''\phi = 0,\tag{2.2}$$

which is called the Rayleigh equation.

The boundary condition at the bottom is

$$\phi(0) = 0.\tag{2.3}$$

The condition at the free surface is obtained when a kinematic condition is combined with the dynamic condition that the pressure vanishes there and is given by

$$(F - c)^2\phi'(1) = \phi(1),\tag{2.4}$$

which is Eq. (11) in Yih (1972, henceforth referred to as I) in dimensionless form, if we recall the scaling of the velocity adopted here, and that $U(1) = 1$ in the present notation. Note that the $U$ in I is the $FU$ here.

The system consisting of Eqs. (2.2)–(2.4) constitutes the eigenvalue problem. This problem has been studied by Burns (1953) and in I. Yih showed that the eigenvalues of $c$ are real, and both Burns and Yih found that they are outside of the range of $FU$. Burns considered only the extreme case of $k^2 = 0$. In I $k^2$ is arbitrary, but in this chapter we consider only long waves. In I the actual solution of $\phi$ for any arbitrary $U$ satisfying Eq. (1.2) was not attempted. Here we give this solution first. Below a specific $U$ of special interest is considered as an example.

We take this opportunity to clarify a point concerning the derivation of (I 23) [or Eq. (23) in I]. First, the right-hand sides of (I 22) and (I 23) should be divided by 2. Second, a referee of I noted (correctly) that the derivation of (I 23) from (I 22) depends on the existence of unstable modes in the neighborhood of the neutral one under consideration. This concern can be addressed by abandoning (I 22) and adopting Tollmien's (1935) derivation, as presented in Yih (1988, pp. 477–479). Then no existence of neighboring unstable modes need be assumed.

It is advantageous to use the substitution

$$\phi = (FU - c)f. \tag{2.5}$$

(We use $\phi$ and $f$ in place of $f$ and $F$ in I since we now use $F$ to denote the Froude number.) Then Eq. (2.2) becomes the standard Sturm–Liouville form

$$[(FU - c)^2 f']' - k^2 (FU - c)^2 f = 0, \tag{2.6}$$

and Eqs. (2.3) and (2.4) become, respectively,

$$f(0) = 0, \qquad (F - c)^2 f'(1) = f(1), \tag{2.7}$$

because $U'(1) = 0$. Since this chapter is concerned with long waves, we expand $f$ and $c - F$ in $k^2$, and write

$$f = f_0 + k^2 f_1 + k^4 f_2 + \cdots, \tag{2.8}$$

$$c - F = (c_0 - F)(1 + \alpha k^2 + \cdots). \tag{2.9}$$

Then Eq. (2.6) gives, for the lowest order in $k^2$,

$$[(FU - c_0)^2 f_0']' = 0. \tag{2.10}$$

The solution that satisfies Eqs. (2.7) is

$$f_0 = 1 + b \int_1^y (FU - c_0)^{-2} \, dy, \tag{2.11}$$

where

$$b^{-1} = \int_0^1 (FU - c_0)^{-2} \, dy = 1. \tag{2.12}$$

This gives $c_0$ for any $FU(y)$. Equations (2.10) and (2.11) show that $f_0(y)$ is positive.

We know from I that

$$c_0 > F \quad \text{or} \quad c_0 < 0 \tag{2.13}$$

for waves traveling downstream or upstream, respectively. For $F = 0$ we obtain $c_0^2 = 1$, as we should. To see how $c_0$ changes with $F$, we differentiate Eq. (2.9) with respect to $F$ and obtain

$$\frac{d(c_0 - F)}{dF} = \frac{K}{H} - 1, \tag{2.14}$$

where

$$H = \int_0^1 D^{-3}\, dy, \qquad K = \int_0^1 U D^{-3}\, dy, \qquad D = c_0 - FU. \tag{2.15}$$

Since $D$ does not change sign, inequalities (2.13) obviously show that $|H| > |K|$. For $c_0 > F$, Eq. (2.14) shows that $c_0 - F$ decreases as $F$ increases, and $c_0$ approaches $F$ as $F \to \infty$. It cannot approach $F + \delta$, however small $\delta$ is, for otherwise the integral of Eq. (2.12) would be zero. For $c_0 < 0$,

$$\frac{dc_0}{dF} = \frac{K}{H} > 0, \tag{2.16}$$

and $c_0$ must approach zero as $F$ increases indefinitely. It cannot approach a definite number $-\delta$, however small $\delta$ is, for otherwise the integral in Eq. (2.12) would be zero. For a given $F$, Eq. (2.12) indicates that there is only one positive eigenvalue for $c_0$, which is greater than $F$, for waves traveling downstream and only one negative value for $c_0$ for waves traveling upstream.

The next step is to determine $f_1$ and $\alpha$ in Eqs. (2.8) and (2.9). From Eqs. (2.6) and (2.10) we have

$$(D^2 f_1')' = -2\alpha\gamma (Df_0')' + D^2 f_0, \qquad \gamma = c_0 - F. \tag{2.17}$$

The solution of this equation is, on use of Eqs. (2.11) and (2.12),

$$f_1 = \alpha I_1(y) + I_2(y), \tag{2.18}$$

where

$$I_1 = -2\gamma \int_0^y D^{-3}\, dy, \qquad I_2 = \int_0^y \left[ D^{-2} \int_1^y D^2 f_0\, dy \right] dy. \tag{2.19}$$

The limits of integration have been arranged to satisfy

$$f_1(0) = 0, \qquad I_2'(1) = 0, \tag{2.20}$$

the first of which is necessary and the second is for convenience. It is clear that $I_1$ and $I_2$ are negative definite, whether the waves go downstream or upstream. We recall that $f_0(y)$ is positive.

To determine $\alpha$, we use Eqs. (2.7), which give, in view of Eqs. (2.8), (2.9), (2.11), (2.12), and (2.18),

$$f_1(1) = 0, \quad \text{or} \quad \alpha I_1(1) = -I_2(1) \tag{2.21}$$

after the cancellation of two terms. Hence $\alpha$ is always negative. Whether $c_0$ is positive or negative, this means the group velocity of the waves is less in magnitude than the phase velocity, as one would expect for gravity waves. For $F = 0$ we have from Eqs. (2.21) that for all waves

$$\alpha = -\frac{1}{6}. \tag{2.22}$$

This agrees with the classical theory for waves in (otherwise) still water.

For $c_0 > F \gg 1$, the main contribution to the integral in Eq. (2.12) comes from the neighborhood of $y = 1$, in which $U$ behaves like a parabola. We can therefore replace $U(y)$ with a parabola with the (constant) curvature $U''(1)$. Then a direct integration of Eq. (2.12) between zero and 1 shows that, on neglect of higher-order terms,

$$c_0 - F = (\pi^2/16F)^{1/3} \quad \text{for} \quad F \gg 1. \tag{2.23}$$

For $c_0 < 0$ and $F \gg 1$, obviously

$$c_0 - F \rightarrow -F, \tag{2.24}$$

since $c_0$ approaches zero as $F$ approaches infinity.

An interesting and realistic example is provided by

$$U(y) = 2y - y^2,$$

for this velocity distribution is that in a viscous liquid layer flowing down an incline. For $c_0 > 0$, we have

$$f_0 = \frac{1}{2q^2 F^2} \left( \frac{z}{z^2 + q^2} + \frac{1}{q} \arctan \frac{z}{q} \right), \tag{2.25}$$

in which

$$z = y - 1, \qquad Fq^2 = \gamma = c_0 - F.$$

Equation (2.10) yields

$$\frac{1}{2q^2} \left( \frac{1}{1 + q^2} + \frac{1}{q} \arctan \frac{1}{q} \right) = F^2. \tag{2.26}$$

For very large $F$, this gives

$$\gamma = \left( \frac{\pi^2}{16F} \right)^{1/3},$$

as noted above. For $c_0 < 0$, we have

$$f_0 = \frac{1}{2q^2 F^2} \left( \frac{z}{q^2 - z^2} + \frac{1}{q} \ln \frac{q + z}{q - z} \right), \tag{2.27}$$

where

$$z = y - 1, \qquad -Fq^2 = \gamma = c_0 - F, \qquad q > 1.$$

Equation (2.12) now yields

$$\frac{1}{2q^2}\left(\frac{1}{q^2-1}-\frac{1}{q}\ln\frac{q-1}{q+1}\right)=F^2. \tag{2.28}$$

For large $F$,

$$\gamma \to -F \quad \text{and} \quad c_0 \to 0,$$

as noted above. Equations (2.26)–(2.29) agree with the work of Burns (1953). The dispersion coefficient $\alpha$ in Eq. (2.18) can be determined once $f_0$ is known.

### 18.3. Nonlinear Waves in Shear Flows with a Curvilinear Velocity

Retaining the scaling specified in Section 18.2 and using $\rho g h$ as the scale of the pressure $p$, we can write the Euler equations of motion as

$$a = u_t + (FU+u)u_x + v(FU+u)_y = -p_x, \tag{3.1}$$

$$b = v_t + (FU+u)v_x + vv_y = -p_y - 1, \tag{3.2}$$

where subscripts indicate partial differentiation and $a$ and $b$ are the acceleration components. Clearly $p + y$ is the acceleration potential. For clarity it is denoted by $G(t, x, y)$. Then Eqs. (3.1) and (3.2) become

$$a = G_x, \qquad b = G_y. \tag{3.3}$$

Integrating the latter of Eqs. (3.3) from $y = 0$ to $1 + \eta$ (where $\eta$ is the surface displacement) and then differentiating the result with respect to $x$, we have

$$[G(t, x, 1+\eta) - G(t, x, 0)]_x = -[p(t, x, 1+\eta) - p(t, x, 0) + \eta]_x. \tag{3.4}$$

But

$$G_x(t, x, 0) = -p_x(t, x, 0),$$
$$p(t, x, 1+\eta) = 0$$

is the dynamic condition at the free surface. Hence Eq. (3.4) becomes

$$G_x(t, x, 1+\eta) = -\eta_x, \tag{3.5}$$

in which $\eta$ is a function of $t$ and $x$. The left-hand side of Eq. (3.5) is equal to

$$G_x(t, x, y)|_{y=1+\eta} + G_y(t, x, y)\eta_x,$$

the second term of which is $b\eta_x$ and is $O(\varepsilon^2 k^3)$, if $\varepsilon$ is the magnitude of $u$ and $k$ is the wave number (not necessarily a single wave number) for the long waves under consideration. It is therefore negligible compared with the first term. Therefore we can write Eq. (3.5) as

$$a = -\eta_x, \tag{3.6}$$

where $a$ is given by Eq. (3.1), with all quantities therein evaluated at $y = 1 + \eta$. For waves in otherwise still water, we can write the Bernoulli equation for the free surface, since the flow

can be assumed irrotational. Differentiating the Bernoulli equation with respect to $x$ then gives
Eq. (3.6) for $F = 0$. When a basic flow is present, one does not have that convenience. But we
have been able to derive Eq. (3.6) in a very efficacious way.

To reduce the expression for the acceleration in Eq. (3.1) to a more convenient form, note
that for the basic flow assumed we have

$$U(1 + \eta) = 1 - \frac{\beta}{2}\eta^2 + \cdots ,$$

$$U'(1 + \eta) = -\beta\eta + \cdots , \qquad \beta = -U''(1),$$

where primes indicate differentiation with respect to $y$. Then

$$vFU' = -vF\beta\eta = \beta F\eta\psi_x, \tag{3.7}$$

where $\psi$ is the stream function for the wave motion. As to the $u_y$ in Eq. (3.1), we recall
that for the linear case $\phi$ in Eq. (2.2) represents $\psi$ and $\phi''$ represents $u_y$. Hence at the free
surface

$$vu_y = -\frac{\beta F}{c - F}\psi\psi_x. \tag{3.8}$$

We express $\eta$ in Eq. (3.7) in terms of $\psi$ below.

The dynamical condition at the free surface is Eq. (3.6). We need an equation corresponding to Boussinesq's integrated equation of continuity. The kinematic condition at the free
surface is

$$\left[\frac{\partial}{\partial t} + (F + u)\frac{\partial}{\partial x}\right]\eta = v \tag{3.9}$$

and is exactly the equation needed. The quantity $\psi(t, x, 1 + \eta)$ is the total discharge (due to
the wave `motion`) at any cross section (i.e., for any given $x$). But at the free surface

$$v = -[\psi_x(t, x, y = 1 + \eta)] = -\psi_x(t, x, 1) - \psi_{xy}\eta, \tag{3.10}$$

So Eq. (3.9) can be written as

$$\left[\frac{\partial}{\partial t} + F\frac{\partial}{\partial x}\right]\eta + (u\eta)_x = -\psi_x, \tag{3.11}$$

where $\psi$ is evaluated at $y = 1$. Then in Eq. (3.7) we can write

$$\eta = \frac{1}{c - F}\psi + O(\varepsilon^2). \tag{3.12}$$

Note that in Eqs. (3.9) and (3.11) $F$ has been written for $FU$. Since $U'(1) = 0$ this replacement
can introduce only an error of $O(\varepsilon^3 k)$. Using Eq. (3.12) in Eq. (3.7), we see that the result
cancels exactly the right-hand side of Eq. (3.8), so that Eq. (3.6) can be written as

$$u_t + Fu_x + uu_x = -\eta_x, \tag{3.13}$$

neglecting terms of $O(\varepsilon^3 k)$. But in Eq. (3.13), the $u$ is evaluated at $y = 1 + \eta$. Using Eqs. (2.2) and (3.12), where $\phi$ represents $\psi$ and $\phi''$ represents $u_y$, we have, neglecting a term of $O(\varepsilon^2 k^2)$,

$$u(t, x, 1 + \eta) = u(t, x, 1) + u_y \eta = u(t, x, 1) + \frac{\beta F}{(c - F)^2} \psi^2. \tag{3.14}$$

We use $u(t, x, 1)$ as the $u$ in the nonlinear terms in Eqs. (3.11) and (3.13) instead of the $u$ at the free surface. This introduces an error of $O(\varepsilon^3 k)$, which is negligible.

Now we introduce a coordinate frame moving with velocity $F$ to the right and take into account the effect of dispersion. Then in the moving frame with $\gamma = c_0 - F$ and noting

$$(\partial/\partial t + F \partial/\partial x)(\eta, u) = \left( \eta_t - \alpha \gamma \eta_{txx}, u_t - \frac{\beta F}{\gamma} (\psi^2)_x - \alpha \gamma u_{tx} \right),$$

Eqs. (3.11) and (3.13) become

$$\eta_t + (u\eta)_x = -\psi_x + \alpha \psi \eta_{txx}, \tag{3.15}$$

$$u_t + u u_x - 2\beta F \gamma^{-1} \psi \psi_x = -\eta_x + \alpha \gamma u_{txx}. \tag{3.16}$$

Without the dispersive terms, in the absence of the nonlinear terms Eqs. (3.15) and (3.16) would give the wave velocity $c_0$ in the frame at rest or the wave velocity $c_0 - F$ in the moving frame, regardless of the value of the wave number.

We can obtain an equation for either $u$ or $\eta$ by combining Eqs. (3.15) and (3.16), but we choose to obtain one for $\eta$. First, note that the vertical structures of the stream function $\psi(t, x, y)$ and the velocity component $u(t, x, y)$ are assumed to be given by $\phi(y)$ and $\phi'(y)$, so that the relation between $\psi$ and $u$ at $y = 1$, given by Eq. (2.4), is

$$\psi = \gamma^2 u. \tag{3.17}$$

Then Eqs. (3.15) and (3.16) become

$$\eta_t + (u\eta)_x = -\gamma^2 u_x + \alpha \gamma \eta_{txx}, \tag{3.18}$$

$$u_t + (1 - 2\beta \gamma^3 F) u u_x = -\eta_x + \alpha \gamma u_{txx}. \tag{3.19}$$

Differentiating Eq. (3.18) with respect to $t$ and differentiating $\gamma^2$ times Eq. (3.19) with respect to $x$ and subtracting gives

$$\eta_{tt} + (u\eta)_{xt} - \gamma^2 (1 - 2\beta \gamma^3 F)(u u_x)_x = \gamma^2 \eta_{xx} \alpha \gamma \eta_{ttxx} - \alpha \gamma^3 u_{txxx}. \tag{3.20}$$

The terms of the lowest order in Eq. (3.15) give

$$\psi = \gamma \eta.$$

Then Eq. (3.17) gives

$$\eta = \gamma u, \tag{3.21}$$

which can be used in the nonlinear parts of Eq. (3.20). Exchanging temporal derivatives with spatial derivatives as before gives

$$\eta_{tt} - \gamma^2 \eta_{xx} - m(\eta \eta_x)_x + 2\alpha \gamma^3 \eta_{xxxx} = 0, \tag{3.22}$$

where

$$m = 3 - 2F\beta\gamma^3.$$

The basic flow affects the wave motion through the wave velocity $\gamma$, the velocity curvature $\beta F$ at the surface, and the dispersion coefficient $\alpha$. Below we discuss the sign of $m$ when $\gamma$ is positive (downstream-running waves) and $F$ is large. We assume $m$ to be positive. The appearance of $\beta [= -U''(0)]$ in $m$ is something that cannot be revealed by a rough analysis.

Then Eq. (3.22) is of the Boussinesq type. But we emphasize that when a value is assigned to $\gamma$, positive or negative, we are committed to that value of $\gamma$ and are then dealing with waves going either right or left, but not simultaneously in both directions. We are saved from Boussinesq's mistake by the fact that for any given $F$ the absolute value of $\gamma$ for downstream-running waves is different from that of upstream-running ones and by the fact that $\gamma^3$ appears in $m$.

For waves going in one direction we can derive a KdV equation from Eq. (3.22), which Boussinesq did not do for waves in still water. We introduce the variables

$$\xi_1 = x - \gamma t, \qquad \xi_2 = x + \gamma t,$$

and note that $\eta$ is mainly a function of $\xi_1$. (It would be a function of $\xi_1$ exactly if the nonlinear and dispersive terms were not there.) Treating those terms as functions of $\xi_1$ only, we have

$$(-m\eta\eta_x + 2\alpha\gamma^3 \eta_{xxx})_x = (-m\eta\eta x + 2\alpha\gamma^3 \eta_{xxx})_{\xi_1}. \tag{3.23}$$

Since

$$\eta_{tt} - \gamma^2 \eta_{xx} = -4\gamma^2 \eta_{\xi_2 \xi_1},$$

integration of Eq. (3.22) with respect to $\xi_1$ gives

$$-4\gamma^2 \eta_{\xi_2} - m\eta\eta_x + 2\alpha\gamma^3 \eta_{xxx} = 0$$

or

$$\eta_t + \gamma \eta_x + \frac{m}{2\gamma} \eta\eta_x - \alpha\gamma^2 \eta_{xxx} = 0, \tag{3.24}$$

since

$$x = \frac{1}{2}(\xi_1 + \xi_2), \qquad t = \frac{1}{2\gamma}(\xi_2 - \xi_1). \tag{3.25}$$

Equation (3.24) is a KdV equation. The KdV equation for flowing water has been derived by Benney (1966), Freeman and Johnson (1970), and Hwang and Chang (1987), among others. Benney's work is more general and is for finite Reynolds number. Few details are given and we cannot reduce his equations to those given here. Little interpretation is also given for the inviscid case considered in Hwang and Chang or in Oikawa et al. (1987). Most recent work relies heavily on symbolic computation as the algebra becomes tedious.

The coefficient of the crucial nonlinear term of Freeman and Johnson's KdV equation (4.15) does not contain our $\beta$, indicating that the last term in our Eq. (3.14) is neglected. The resulting error is important when $F$ is not small. Otherwise our Eq. (3.24) agrees with Freeman and Johnson's Eq. (5.15), which is for a frame of reference moving with the linear long-wave velocity.

For $F = 0$ it is easily verified that $\gamma^2 = 1$. For waves going to the right $\gamma = 1$, $m = 3$, and $\alpha = -1/6$. So Eq. (3.24) reduces to

$$\eta_t + \eta_x + \frac{3}{2}\eta\eta_x + \frac{1}{6}\eta_{xxx} = 0, \tag{3.26}$$

which is the standard form of the KdV equation for waves in otherwise still water. For waves going left, $\gamma = -1$ and an equation identical to Eq. (3.26), but with the sign of the first term (or the signs of the last three terms) changed.

Equation (3.24) can be put in the standard form by substituting [$b$ is not as in Eq. (3.3)]

$$\tilde{t} = \gamma bt, \qquad \tilde{x} = bx, \qquad \tilde{\eta} = \frac{m}{3\gamma^2}\eta, \qquad b^2 = -\frac{1}{6\alpha}. \tag{3.27}$$

The result is exactly Eq. (3.26), with a tilde on all the variables. Now suppose there is a solution for a solitary wave or for cnoidal waves, with displacement $\tilde{\eta}$ of $O(\varepsilon)$. The corresponding $\eta$, according to Eq. (3.27), is $O(F^{-2/3}\varepsilon)$ and hence exceedingly small for large $F$. For $\eta$ greater than this very small magnitude, the effect of the dispersive term cannot balance the effect of the nonlinear term, and coherent waves like solitary waves and cnoidal waves cannot exist. The wave motion is mainly governed by the shallow-water theory. We refrained from going into that theory in this paper.

For downstream-running waves at least $m$ is finite for very large values of $F$, as can be seen from Eqs. (2.23) and (3.23). For a parabolic $U(y)$, $\beta = 2$, and Eqs. (2.23) and (3.23) give

$$m = 3 - \frac{\pi^2}{4} > 0.$$

There are many other profiles for $U(y)$ with a smaller $\beta$. So we can assume $m$ to be positive without loss of cases of practical importance. For upstream-running waves, however, $\gamma$ approaches $-F$. So for large values of $F$ we have a positive $m$ of $O(F^4)$. The coefficient of $\eta$ in Eq. (3.27) is $O(F^2)$. Hence if $\tilde{\eta} = O(\varepsilon)$, the true displacement $\eta$ is of $O(F^{-2}\varepsilon)$, even smaller than $\eta$ for downstream-running waves at large $F$. What has been said of such waves can be said of upstream-running waves at large $F$.

For very large $F$, Benjamin (1962) doubted the existence of upstream-running waves and asserted that "in fact the waves will be convected downstream at an absolute velocity not much different from $\bar{U} - C_0$." (His $U$ is our $FU$, and $C_0$ is the wave speed in quiet water. The bar means average.) The present analysis, in concluding that upstream-running waves have extremely small amplitude, supports Benjamin's doubt, since a vanishing amplitude is tantamount to nonexistence, if he had in mind coherent waves, like solitary and cnoidal waves. But the same reasoning would apply to downstream-running waves, leading to the same conclusion. However, there are other waves than smooth, coherent waves. For instance, a hydraulic jump of very large amplitude can propagate upstream. As to Benjamin's assertion mentioned above, one notes that the average velocity of the basic flow does not enter the analysis at all. His assertion can be only approximately right if the difference of maximum and minimum values of the basic-flow velocity is very small, in which case the shear is weak and the average velocity is nearly the maximum velocity.

We have been discussing at length the solution of Eq. (3.24) for large $F$. Naturally Ursell (1953) comes to mind. Ursell showed that for any depth and wavelength, coherent waves (solitary or cnoidal) have limits for their amplitude. When these limits are exceeded, coherent waves cannot exist. His work resolved the long-wave paradox and the conflict between Airy and Rayleigh (Ursell, 1953). Here we see that at large $F$ the limit on the amplitude becomes very stringent.

For modest $F$, even if it is greater than 1 and as large as 3, Eq. (3.24) governs the motion of coherent waves going in the same direction. A supercritical velocity of the basic flow at the surface does not necessarily forbid coherent waves of amplitudes comparable with those of coherent waves in still water. Observable solitary and cnoidal waves can occur, and, when there is more than one solitary wave going in the same direction, their interaction can be dealt with by Hirota's elegant solution (1971), admirably presented in Whitham (1974, pp. 580–82).

## 18.4. Concluding Remarks

Roll waves are an interesting phenomenon sometimes observed in shallow water flowing in a slightly inclined canal. These have white and turbulent crests, and occur when the conditions (e.g., the roughness of the bottom, depth, the value of $F$) are just right for its occurrence. Does the present analysis account for any aspect of this phenomenon? The following suggestion is not entirely speculative. One can look at the roll waves as breaking cnoidal waves when the amplitude is too large to maintain smooth (and symmetric) cnoidal waves, the amplitude of which must have an upper limit, beyond which the Boussinesq regime does not reign. From Eq. (3.27) it is clear that this limit on $\tilde{\eta}$ is easily exceeded at large $F$. Then the wave motion will take on the feature of steepening at the wave front, well known in the shallow-water theory, which cannot be balanced by the dispersiveness of the waves, and the waves will break at the crest and become roll waves. Their analysis is beyond the pale of the present theory. Indeed Dressler (1949) noted that the condition of the crests resembles that of a hydraulic jump and built a solution for roll waves on that notion.

Finally, we mention that the present study is relevant to laminar flows of a viscous liquid down a very slight incline at large values of the Reynolds number $R$. The stability of such flows has been studied by Benjamin (1957) and Yih (1963) only for small values of the product of $R$ and wave number $k$. The stability for gravity waves at large values of $kR$ is still to be investigated.

## Acknowledgment

This manuscript was originally submitted by Chia-Shun to a journal for publication. Chia-Shun passed away before being able to respond to referee's comments. With the journal editor's and Chia-Shun's family blessing, WWS withdrew the paper and modified it to answer the reviewers concerns, add some references, and shorten the manuscript for publication in this symposium in his honor, trying to make these revisions with Chia-Shun's voice. Although this paper is written almost entirely by him, WWS would like to dedicate it to him as a departed but not forgotten friend and colleague.

## References

1. Benjamin, T. B. 1957. Wave formation in the laminar flow down an inclined plane. *J. Fluid Mech.* **2**, 554–74.

2. Benjamin, T. B. 1962. The solitary wave on a stream with an arbitrary distribution of vorticity. *J. Fluid Mech.* **12**, 97–116.

3. Benney, D. J. 1966. Long waves in liquid films. *J. Math. Phys.* **45**, 150–55.

4. Biésel, F. 1950. Êtude théorique de la houle en eau courante. *Houille Blanche* **5**, 279–85.

5. Boussinesq, J. 1871. Théorie de l'intumescence liquide appelée onde solitaire ou de translation se propageant dans un canal rectangulaire. *Comptes Rendus Académie des Sciences Paris* **72**, 755–59.

6. Burns, J. C. 1953. Long waves in running water. *Proc. Cambridge Philos. Soc.* **49**, 695.

7. Dressler, R. F. 1949. Mathematical solution of the problem of roll-waves in inclined open channels. *Comm. Pure Appl. Math.* **2**, 149–94.

8. Freeman, N. C. and Johnson, R. S. 1970. Shallow water waves on shear flows. *J. Fluid Mech.* **42**, 401–409.

9. Hirota, R. 1971. Exact solution of the Korteweg–de Vries equation for multiple collision of solitons. *Phys. Rev. Lett.* **27**, 1192–94.

10. Hwang, S. H. and Chang, H. C. 1987. Turbulent and inertial roll waves in inclined film flow. *Phys. Fluids* **30**, 1259–68.

11. Korteweg, D. J. and de Vries, G. 1895. On the change of form of long waves in a rectangular channel and on a new type of long stationary waves. *Philos. Mag.* **39**, 422–43.

12. Oikawa, M., Chow, K., and Benney, D. J. 1987. The propagation of nonlinear-wave packets in a shear-flow with a free-surface. *Stud. Appl. Math.* **76**, 69–92.

13. Thompson, P. D. 1949. The propagation of small surface disturbances through rotational flow. *Ann. NY Acad. Sci.* **51**, 463–74.

14. Tollmien, W. 1935. Ein allgemeines Kriterium der Instabilität laminarer Geschwindigkeitsverteilungen, *Nachr. Ges. Wiss. Göttingen, Math. Phys. Kl., Fachgruppe* 1, **1**, 79–114.

15. Ursell, F. 1953. The long-wave paradox in the theory of gravity waves. *Proc. Cambridge Philos. Soc.* **49**, 685–94.

16. Whitham, G. B. 1974. *Linear and Nonlinear Waves*. Wiley-Interscience, New York.

17. Yih, C.-S. 1963. Stability of liquid flow down an inclined plane. *Phys. Fluids* **6**, 321–34.

18. Yih, C.-S. 1972. Surface waves in flowing water. *J. Fluid Mech.* **51**, 209–20.

19. Yih, C.-S. 1988. *A Concise Introduction to Fluid Mechanics*, West River Press, Ann Arbor, MI.

# 19

# Pinned-Edge Faraday Waves

DIANE M. HENDERSON AND JOHN W. MILES

## 19.1. Introduction

We consider surface waves on water in circular cylinders for which the contact line is pinned and the quiescent surface is flat. Our major result is that by pinning the contact line and by cleaning the surface we are able to predict the natural frequencies and damping rates of the low modes (except the lowest, axisymmetric mode) so that the tuning of desired modes is possible. We present preliminary work in which we use this capability to observe steady, high-mode wave fields both with and without an obstacle present in the center of the cylinder.

The standing waves are excited by vertical oscillations of the fluid at a single frequency, so that their subharmonic frequencies vary from ~5 to ~50 Hz. For low-frequency oscillations, the system is tuned so that only a single mode is able to grow, while for high-frequency oscillations, many modes are available for excitation. Here we present predictions and measurements of the natural frequencies and damping rates of single-mode oscillations. Martel et al. (1998) showed the surprising result that it is necessary to include the effects of damping due to the interior fluid. Damping due to the interior is generally neglected because the contribution to damping from the interior is higher order with respect to the coefficient of viscosity than the contributions from the sidewall boundary layers and pinned contact line.

There is a large literature on experiments in which pinned-edge Faraday waves are used. The technique of pinning the contact line was first introduced by Benjamin and Scott (1979), who showed experimentally and theoretically that the pinned-edge condition significantly increases the wave frequency and the corresponding phase speed. Others use this pinned-edge condition in experiments to avoid the problems associated with contact-line effects, starting in particular, with Douady and Fauve (1988) and Douady (1990).

## 19.2. Theoretical Considerations

In this section we present predictions of mode shapes, frequencies, and damping rates of standing, gravity-capillary waves at the surface of an inviscid, irrotational fluid in a circular cylinder for which the contact line is pinned. The linear eigenvalue problem, providing a prediction of mode shapes and predictions and measurements of natural frequencies, is given by Henderson and Miles, 1994 (HM94). Predictions and measurements of damping rates are also given by HM94. These predictions are improved significantly by Martel et al. (1998), who expanded the problem in an inverse Reynolds number to include damping effects due to the bulk of the fluid. Their idea was simplified by Miles and Henderson (1998), who considered bulk effects by using Lamb's dissipation integral for an irrotational flow.

### 19.2.1. *Mode Shapes*

The expressions for the mode shapes are obtained from the linearized eigenvalue problem for inviscid, gravity-capillary waves with a pinned contact line and are given by HM94. The free-surface displacement is

$$\zeta(r, \theta, t) = [A_n R_n(r) \cos s\theta - AC(r) \cos s\theta \cos \omega t], \tag{2.1}$$

where

$$C(r) = \frac{I_s(r/l_c)}{I_s(1/\lambda)}, \qquad R_n(r) = \frac{J_s(k_n r)}{J_s(\varkappa_n)}, \qquad A_n = AC_n \left( \frac{\omega^2}{\omega^2 - \omega_n^2} \right), \tag{2.2a,b,c}$$

$$C_n = \left( \frac{2\lambda}{1 + \varkappa_n^2 \lambda^2} \right) \left[ \frac{I_s'(1/\lambda)}{I_s(1/\lambda)} \right] \left( \frac{\varkappa_n^2}{\varkappa_n^2 - s^2} \right), \tag{2.3}$$

$A$ is an arbitrary amplitude, $J_s$ is a Bessel function, $I_s$ is a modified Bessel function, $a$ is the cylinder radius, $l_c = (T/g)^{1/2}$ is the capillary length, and $\lambda = l_c/a$. The wave number $k_n$ is obtained from $J_s'(\varkappa_n) = 0$, where $\varkappa_n = k_n a$. The natural frequency of the corresponding wave when the contact line is free is

$$\omega_n \equiv [(g/a)\varkappa_n T_n(1 + \varkappa_n^2 \lambda^2)]^{1/2}, \qquad T_n \equiv \tanh k_n d. \tag{2.4}$$

The natural frequency of the wave when the contact line is pinned is $\omega$ and may be obtained from the eigenvalue equation

$$C_n \left( \frac{\omega^2}{\omega^2 - \omega_n^2} \right) = 1, \tag{2.5}$$

where, above and subsequently, repeated indices are summed over the complete, orthogonal set $\{R_n, k_n\}$ except where the index occurs once but is not repeated on one side of an equation. The $n$th mode has $s$ nodal diameters and $m$ nodal circles, where $m$ is the number of zeros of $J_s'(k_n r)$ from $0 < k_n r < \varkappa_n$.

### 19.2.2. *Damping Rates*

Damping arises from the Stokes boundary layers at the bottom and the sidewalls of the cylinders, contamination at the free surface, and viscous dissipation in the bulk of the fluid. The experiments discussed herein have a clean surface (see Section 19.3 for a definition of clean), so we do not include dissipation due to the free surface. If we consider the amplitude of the wave to decrease like $A = A_0 e^{-\gamma t}$, then the decay rate $\gamma$ is given by

$$\gamma = \gamma_b + \gamma_{sw} + \gamma_i, \tag{2.6}$$

where the subscripts $b$, $sw$, and $i$ correspond to contributions from the bottom, sidewalls, and interior of the fluid, respectively. Using a boundary-layer dissipation rate integral (HM94) and Lamb's integral for the dissipation rate for an irrotational fluid motion (Miles and Henderson, 1998) we obtain

$$\gamma_b = \frac{1}{4} \omega \frac{l_v}{a} \frac{\delta_{mn}(\varkappa_n^2 - s^2)(1 - T_n^2)\phi_m \bar{\phi}_n}{(\varkappa_l^2 - s^2)\varkappa_l^{-1} T_l \phi_l \bar{\phi}_l} \tag{2.7a}$$

for the bottom contribution, which becomes negligible as $k_n d$ becomes large,

$$
\gamma_{sw} = \frac{1}{4}\omega\frac{l_v}{a}\frac{(\varkappa_m\varkappa_n + s^2)\left(\frac{T_m+T_n}{\varkappa_m+\varkappa_n}\right) - (\varkappa_m\varkappa_n - s^2)\left(\frac{T_m-T_n}{\varkappa_m-\varkappa_n}\right)\phi_m\bar{\phi}_n}{(\varkappa_l^2 - s^2)\varkappa_l^{-1}T_l\phi_l\bar{\phi}_l}
\tag{2.7b}
$$

for the sidewall contribution, and

$$
\gamma_i = \omega\left(\frac{l_v}{a}\right)^2\left[\frac{\varkappa_n(\varkappa_n^2 - s^2)T_n|\phi_n|^2 - (\varkappa_m T_m - \varkappa_n T_n)(\varkappa_m^2 - \varkappa_n^2)^{-1}\phi_m\bar{\phi}_n}{\varkappa_l^{-1}(\varkappa_l^2 - s^2)T_l|\phi_l|^2}\right]
\tag{2.7c}
$$

for the contribution from the interior of the fluid. In Eqs. (2.7),

$$
l_v = (2v/\omega)^{1/2}
\tag{2.8a}
$$

is the viscous length scale, where $v$ is the kinematic coefficient of viscosity,

$$
\phi_n = i(\omega k_n T_n)^{-1}\omega_n^2 A_n
\tag{2.8b}
$$

is the amplitude of the velocity potential of the $n$th mode and the overbar indicates complex conjugate.

## 19.3. Experimental Apparatus

The experimental apparatus comprised acrylic circular cylinders, an electromagnetic shaker with feedback, high-performance liquid chromotography (HPLC) water, a nonintrusive wave gauge, and standard 35-mm photographs. The cylinder for the low-mode experiments had a radius $a = 2.766$ cm and depth $d = 3.80$ cm measured from the bottom to the brim of the cylinder, which was machined to have a sharp edge. The cylinders for the high-mode experiments had a radius $a = 5.080$ cm and depth $d = 1.27$ cm. Two cylinders were specially machined out of solid acrylic to have an obstacle in the center that had a uniform cross section and spanned the depth of the cylinder. The cross section for one obstacle was pie shaped with an angle of $2\pi/3$ rad. The cross section for the second obstacle was circular. Both obstacles had radii of 1.000 cm. In experiments with an obstacle present, the water was pinned both at the outer walls of the cylinder and at the obstacle in the center. We cleaned the cylinders before filling by soaking in Micro (brand) laboratory cleaner, a nontoxic, biodegradable alternative to chromic acid that cleans plastics as well as glass. After soaking the cylinder, we rinsed it in distilled water and vacuumed it dry.

The experiments discussed herein used HPLC water ($v = 0.01$ cm$^2$/s) from Sigma chemicals in an effort to obtain a clean surface. We overfilled the cylinder and then vacuumed the excess (along with surface contamination) with a micropipette attached to a vacuum pump. A simple test of pure water is provided by rising bubbles (Scott, 1979); if a bubble persists at the surface for as long as 0.5 s the water is considered to be contaminated. Bubbles on the surface of the HPLC water persisted 0.09 s. [This measurement was obtained with a high-speed digital imaging system (Kodak Ektapro) operating at 500 frames/s. The bubbles were created by an intrusive pipette.] But Kitchener and Cooper (1959) give a bubble-persistence time of "roughly 0.01 s" for a clean surface, which suggests that even with the HPLC water the surfaces may not have been adequately pure.

The surface was determined to be flat by observation of a horizontal laser beam that was flush with the cylinder's brim. The meniscus of the overfull cylinder diffracted the beam into a vertical line on a white sheet behind the cylinder; this line contracted to a point when the water was flat. We also observed the reflection of a vertical metal rod in the water surface; the reflection appeared undistorted across the contact line when the surface was flat. We measured the time series of the decaying wave field within a minute after cleaning; however, the results did not change over a 3-h period. The experimental apparatus was enclosed in an acrylic box (that was not temperature controlled).

A Bruel & Kjaer minishaker Type 4810 drove the cylinder vertically to excite waves at half the forcing frequency. A noncontacting position transducer (Kaman model KD-2310) monitored the shaker motion and provided a signal to a servocontroller to ensure proper motion. A Micro-VAX II workstation with analog-to-digital, digital-to-analog, and two independent real-time clocks provided the forcing signals. The forcing frequency was accurate to within $10^{-6}$ Hz.

A nonintrusive Wayne–Kerr capacitance probe measured the surface displacement by measuring the field between the water surface and the probe. It averaged over the surface area of the probe, which had a circular cross section with diameter of 0.25 in. The probe had to be within millimeters of the water surface to ensure linearity between surface displacement and output voltage; accordingly, the wave amplitudes were kept small. We filtered the gauge signal through a Krohn–Hite Model 3323 low-pass, analog filter, which also provided a 20-db gain and digitized it with the computer at 350 Hz. The time series was complex demodulated at half the forcing frequency to obtain the amplitude and the phase of the wave as functions of time. The phase information provides the measured natural frequencies of the waves as described in HM94.

## 19.4. Results and Discussion

In this section we discuss the experiments on low-mode and high-mode excitations. The experiments on low modes are concerned with measurements and predictions of natural frequencies and damping rates. The experiments on high modes are preliminary and concerned with mode shapes.

### 19.4.1. *Low Modes*

Table 19.1 shows the results of experiments on the lowest six modes in the circular cylinder. The natural frequencies of the pinned-edge modes, $\omega$, range from ~10% to 20% higher than those of their free counterparts, $\omega_n$. The significant result concerning natural frequencies is that when the contact line is pinned and when the surface is cleaned, the measured and the predicted frequencies agree. Typically, measurements of natural frequencies differ from predictions either because of contact-line effects or because damping is large enough to change significantly both the predicted and the measured values of frequency. Examples of the former are given by Cocciario et al. (1991), who showed that when wave amplitudes were small, the contact line affected the natural frequency, and by Henderson et al. (1992), who showed that the natural frequencies could vary depending on whether or not the fluid/container combination was hydrophobic or hydrophillic. An example of the latter is given by Henderson and Miles (1990), who obtained good agreement between the measured and the predicted frequency of the (0, 1) mode when the large damping rate was incorporated in the predictions. It is also well known that surfactants have a significant effect on the damping rates (e.g., Davies and Vose,

Table 19.1. *Measured and Predicted Frequencies and Damping Rates (Nondimensionalized so that $\Delta = 4a\gamma\omega/l_v$), the Ratios of the Interior and Boundary-Layer Damping, and the Ratios of Measured to Predicted Natural Frequencies and Damping Rates*

| | | | Measurements | | | Calculations | | | | $\Delta^{\text{meas}}/\Delta^{\text{calc}}$ | |
|---|---|---|---|---|---|---|---|---|---|---|---|
| $s$ | $m$ | $\varkappa_n$ | $\omega_n/2\pi$ (Hz) | $\omega/2\pi$ (Hz) | $\Delta$ | $\omega/2\pi$ (Hz) | $\Delta$ | $\dfrac{\gamma_i}{\gamma_b + \gamma_{\text{sw}}}$ | $\dfrac{\omega/2\pi^{\text{meas}}}{\omega/2\pi^{\text{calc}}}$ | HM | Martel et al. |
| 1 | 0 | 1.841 | 4.105 | 4.65 | 1.4 | 4.68 | 1.36 | 0.19 | 0.99 | 1.06 | 1.02 |
| 2 | 0 | 3.054 | 5.465 | 6.32 | 1.8 | 6.35 | 1.70 | 0.37 | 1.00 | 1.06 | 1.03 |
| 0 | 1 | 3.832 | 6.266 | 6.84 | 1.2 | 6.85 | 0.93 | 1.63 | 1.00 | 1.29 | 1.26 |
| 3 | 0 | 4.201 | 6.642 | 7.80 | 2.2 | 7.84 | 2.04 | 0.59 | 0.99 | 1.08 | 1.04 |
| 4 | 0 | 5.318 | 7.794 | 9.26 | 2.4 | 9.29 | 2.39 | 0.83 | 1.00 | 1.01 | 0.97 |
| 1 | 1 | 5.331 | 7.809 | 8.57 | 1.5 | 8.60 | 1.43 | 2.01 | 1.00 | 1.05 | 1.03 |

*Note:* The subscript $n$ corresponds to the $n$th mode of the linearized boundary-value problem for which the contact line is free.

1965). With the combination of a pinned-edge contact line, for which the boundary condition is known, and a clean surface, which minimizes damping and also allows the boundary condition at the surface to be known, the natural frequencies of the mode may be predicted so that tuning for the excitation of particular modes is possible.

The damping rates are also shown in Table 19.1. Martel et al. (1998) expanded the boundary-value problem for pinned-edge standing waves in an inverse Reynolds number to show that the contribution to damping from the interior of the fluid is comparable with the contribution because of boundary layers for experiments like the ones discussed herein. Miles and Henderson (1998) used Lamb's dissipation integral for an irrotational flow to include the effects of interior damping, with the results shown in column 11 of Table 19.1. The ratio of interior to boundary-layer damping, $\gamma_i/(\gamma_b + \gamma_{\text{sw}})$, shows that, indeed, interior damping is significant and in some cases more important than boundary-layer damping, even though it is higher order in viscosity than the boundary-layer contribution. The agreement between measured and predicted damping rates is somewhat better when the results of Martel et al. are used (column 12 of Table 19.1), which also include second-order boundary-layer effects that are neglected here. There is a significant discrepancy between predicted and measured damping rates for the (0, 1) mode, the lowest axisymmetric mode, for which we do not have an explanation.

### 19.4.2. *High Modes*

We have begun looking at high-mode oscillations and present some preliminary results here. Experiments in which the pie-shaped and circular obstacles in the cylinder were used were suggested by J. P. Eckmann (private communication), who was interested in the scattering problem caused by obstacles with various shapes.

Figure 19.1 shows photographs of three experiments in which the cylinder with the pie-shaped obstacle was used. The wave field in Fig. 19.1(a) was unsteady with a frequency of 27.35 Hz, while those in Figs. 19.1(b) and 19.1(c) were steady. Figure 19.2 shows a steady wave field in the presence of the circular obstacle. Figure 19.3 shows two steady wave fields in the cylinder with no obstacle. The frequencies $f$ of the steady wave fields are given in Table 19.2 along with the azimuthal wave number $s$ obtained from the photographs. A possible description

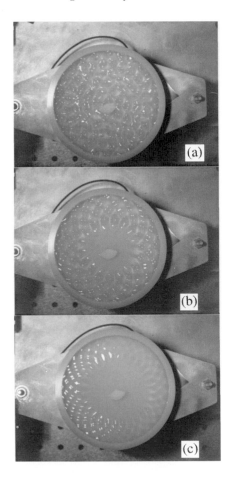

Figure 19.1. Photographs of (a) 27.35-Hz, (b) 27.17-Hz, (c) 41.40-Hz wave fields in the cylinder with the pie-shaped obstacle.

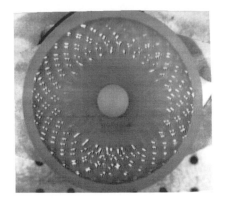

Figure 19.2. Photograph of a 44.44-Hz wave field in the cylinder with the circular obstacle.

of the wave fields is given by Eq. (2.1), which assumes no obstacle is present. We consider this description, since the $J_s$ Bessel function has values close to zero for a significant interval of $r$ when $s$ is large. Table 19.2 shows values of the wave numbers of the $(s, m)$ modes that correspond closest to the photographs, as well as corresponding pinned-edge frequencies from Eq. (2.5). The natural frequencies differ from the wave-field frequencies by 2.0% and 2.1% for

Table 19.2. *Experimental Parameters (s and f) of the Photographs in Figs. 19.1–19.3 and Corresponding Wave Numbers and Frequencies of the Boundary-Value Problem with No Obstacle from Section 19.2*

| Figure | $s$ | $m$ | $k_{sm}$ (rad/cm) | $f_{sm}$(cycle/s) | $f$ (cycle/s) |
|---|---|---|---|---|---|
| 19.1(b) | 18 | 4 | 6.60 | 27.70 | 27.17 |
| 19.1(c) | 29 | 4 | 9.12 | 42.26 | 41.40 |
| 19.2 | 28 | 5 | 9.66 | 45.25 | 44.44 |
| 19.3(a) | 11 | 1 | 2.52 | 10.87 | 10.81 |
| 19.3(b) | 30 | 4 | 10.12 | 48.40 | 47.84 |

(a)

Figure 19.3. Photographs of (a) 10.81-Hz, (b) 47.84-Hz wave fields in the cylinder with no obstacle.

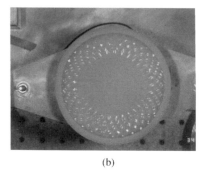

(b)

the waves in the cylinder with the pie-shaped obstacle, 1.8% for the waves in the cylinder with the circular obstacle, and 0.5% and 1.0% for the waves in the cylinder with no obstacle.

We make two observations of the wave fields shown here. First, we can obtain steady wave fields in the presence of obstacles. Second, Figs. 19.1–19.3 show wave fields that appear to have a modal structure that curves into the cylinder walls, instead of intersecting them perpendicularly, as expected from the theory of Section 19.2. Experiments on high-mode Faraday waves in pinned-edge fluid-cylinder configurations by Edwards and Fauve (1994) and Kudrolli and Gollub (1996) do show wave fields with large $s$ that have straight nodal diameters. We are still investigating the apparent differences between their results and ours.

## Acknowledgment

This work was supported in part by the U.S. National Science Foundation grants DMS92-57456 and OCE95-01508, U.S. Office of Naval Research grant N00014-92-J-1171, a David & Lucile Packard Foundation Fellowship (DMH), and an Alfred P. Sloan Fellowship (DMH).

## References

Benjamin, T. B. and Scott, J. C. 1979. Gravity-capillary waves with edge constraints. *J. Fluid Mech.* **92**, 241–267.

Cocciaro, B., Faetti, S., and Nobili, M. 1991. Capillarity effects on surface gravity waves: wetting boundary conditions. *J. Fluid Mech.* **231**, 325.

Davies, J. T. and Vose, R. W. 1965. On the damping of capillary waves by surface films. *Proc. R. Soc. London A* **260**, 218–233.

Douady, S. and Fauve, S. 1988. *Europhys. Lett.* **6**, 221.

Douady, S. 1990. Experimental study of the Faraday instability. *J. Fluid Mech.* **221**, 383–409.

Edwards, W. S. and Fauve, S. 1994. Patterns and quasi-patterns in the Faraday experiment. *J. Fluid Mech.* **278**, 123–148.

Henderson, D. M. and Miles, J. W. 1994. Surface-wave damping in a circular cylinder with a fixed contact line. *J. Fluid Mech.* **275**, 285–299.

Henderson, D. M. and Miles, J. W. 1990. Single-mode Faraday waves in small cylinders. *J. Fluid Mech.* **213**, 95–109.

Henderson, D. M., Hammack, J., Kumar, P., and Shah, D. 1992. The effects of static contact angles on standing waves. *Phys. Fluids A* **4**, 2320–2322.

Kitchener, J. A. and Cooper, C. F. 1959. Current concepts in the theory of foaming. *Q. Rev. Chem. Soc.* **13**, 71–97.

Kudrolli, A. and Gollub, J. P. 1996. Localized spatiotemporal chaos in surface waves. *Phys. Rev. E* **54**, R1052–R1055.

Martel, C., Nicolas, J. A., and Vega, J. M. 1998. Surface-wave damping in a brimful circular cylinder. *J. Fluid Mech.*, **360**, 213–228.

Miles, J. W. and Henderson, D. M. 1998. A note on interior vs boundary-layer damping of surface waves in a circular cylinder. *J. Fluid Mech.* **364**, 319–323.

Scott, J. C. 1979. The preparation of clean water surfaces for fluid mechanics in surface contamination: Genesis, Detection and Control, Vol. 1. (ed., K. L. Mittal), 477–497, Plenum, New York.

# 20

# Interfacial Shapes in the Steady Flow of a Highly Viscous Dispersed Phase

DANIEL D. JOSEPH AND RUNYUAN BAI

## 20.1. Introduction

In treating the flow of two immiscible liquids with greatly different viscosities, like bitumen and water, certain simplifications arise when the more viscous liquid is dispersed and not attached to rigid boundaries. In this case the dispersed phase may move nearly as a rigid body since the forces that arise from the motion of the continuous phase are not great enough to drive large secondary motions in the dispersed phase. The water will move bitumen dispersed in water more or less as a rigid body, provided that the bitumen is not anchored at some wall.

Here and henceforward we call the dispersed phase oil and the continuous phase water. We search for simplified mathematical descriptions as a perturbation of a rigid motion in the limit in which the ratio of the water viscosity $\mu_w$ to the oil viscosity $\mu_o$ is

$$\varepsilon = \frac{\mu_w}{\mu_o} \to 0. \tag{1.1}$$

In this chapter we confine our attention to the cases in which interfacial rheology and Maragnoni effects are neglected. These effects are greatly diminished by the high bulk viscosity of the dispersed phase and in a later work we will look to describe exactly how diminished these effects are. Generally speaking, our work here is motivated by the needs of the heavy oil industry.

## 20.2. Governing Equations

To keep the description simple, we consider the case in which the oil is free to move in the water as in the case of sedimentation of a single drop of heavier-than-water oil or in the core-annular flow studied by Bai et al. (1996).

In steady flow the oil-water interface is given by

$$F[\chi(\varepsilon), \varepsilon] = 0, \tag{2.1}$$

where $\mathbf{x} = \chi(\varepsilon)$ is the position of points on $F = 0$. The unknowns in our problem are

$\mathbf{u}(\mathbf{x}, \varepsilon), \psi(\mathbf{x}, \varepsilon)$    in the oil,

$v(\mathbf{x}, \varepsilon), \phi(\mathbf{x}, \varepsilon)$    in the water,

$\chi(\varepsilon),$                             (2.2)

where $\mathbf{u}$ and $v$ are velocities and

$$\psi = p_o + \rho_o \boldsymbol{\lambda} \cdot \mathbf{x},$$
$$\phi = p_w + \rho_w \boldsymbol{\lambda} \cdot \mathbf{x}, \tag{2.3}$$

are dynamic pressures, $p$ is pressure, and $\boldsymbol{\lambda}$ is a constant vector ($\boldsymbol{\lambda} = \mathbf{g}$ in sedimentation problems; $\boldsymbol{\lambda} = \mathbf{e}_x \beta$ for the constant part of the pressure gradient that balances the pressure drop in core-annular flow).

The equations of motion in the oil and water are

$$\rho_o \mathbf{u} \cdot \nabla \mathbf{u} = -\nabla \psi + \frac{\mu_w}{\varepsilon} \nabla^2 \mathbf{u}, \quad \text{div } \mathbf{u} = 0, \tag{2.4}$$

$$\rho_w v \cdot \nabla v = -\nabla \phi + \mu_w \nabla^2 v, \quad \text{div } v = 0. \tag{2.5}$$

At the interface, the velocity is continuous,

$$\mathbf{u}(\chi) = v(\chi), \tag{2.6}$$

and the normal component vanishes,

$$\mathbf{u}(\chi) \cdot \mathbf{n} = v(\chi) \cdot \mathbf{n} = 0, \tag{2.7}$$

where $\mathbf{n}$ is the normal from oil to water. The shear stress is continuous,

$$\boldsymbol{\tau} \cdot \mathbf{D}[\mathbf{u}(\chi) - \varepsilon v(\chi)] \cdot \mathbf{n} = 0 \tag{2.8}$$

where $\mathbf{D}[\mathbf{u}]$, the rate of strain, is the symmetric part of $\nabla \mathbf{u}$ and $\boldsymbol{\tau}$ is a unit tangent vector in the interface, $\boldsymbol{\tau} \cdot \mathbf{n} = 0$. The balance of normal stresses can be expressed as

$$-\phi(\chi) + \psi(\chi) - (\rho_w - \rho_o)\boldsymbol{\lambda} \cdot \chi + 2\mu_w \mathbf{n} \cdot \mathbf{D}[v - \mathbf{u}/\varepsilon] \cdot \mathbf{n} = 2H(\chi)\sigma \tag{2.9}$$

where $H(\chi)$ is the mean curvature and $\sigma$ is interfacial tension.

The boundary conditions apply only to water, since oil is assumed not to touch the boundary. For steady flow the velocity of the boundary at $\mathbf{x} = \mathbf{x}_b$ is

$$v(\mathbf{x}_b) = \mathbf{V}. \tag{2.10}$$

$\mathbf{V}$ is the velocity of solid walls in a coordinate system centered on the falling drop or in a coordinate system moving with the average velocity of the core in annular flow.

## 20.3. Equations when $\varepsilon \to 0$

Assuming now that all fuctions listed in Eqs. (2.2) are bounded as $\varepsilon \to 0$, we find that

$$\mathbf{u}_o(0) = 0,$$
$$\text{div } \mathbf{u}_o = 0,$$
$$\nabla^2 \mathbf{u}_o = 0, \tag{3.1}$$
$$\mathbf{u}_o(\chi_o) \cdot \mathbf{n}_o = 0,$$

$$\tau_o \cdot \mathbf{D}[\mathbf{u}_o(\chi_o)] \cdot \mathbf{n}_o = 0,$$

$$\mathbf{n}_o \cdot \mathbf{D}[\mathbf{u}_o(\chi_o)] \cdot \mathbf{n}_o = 0.$$

The function

$$\mathbf{u}_o(\mathbf{x}) \equiv 0$$

satisfies Eq. (2.10). Then, in the water we have

$$\rho_w v_o \cdot \nabla v_o = -\nabla \phi_o + \mu_w \nabla^2 v_o, \qquad \mathrm{div}\ v_o = 0$$

$$v_o(\chi_o) = 0, \tag{3.2}$$

$$v(\chi_b) = \mathbf{V}.$$

Equations (3.2) are a Dirichlet problem for $v_o(\mathbf{x})$ and $\phi_o(\mathbf{x})$ that can be solved when the interface $\chi_o$ is given. No condition on $v(\mathbf{x})$ arises from the shear stress balance of Eq. (2.8), and shear stress arising from Eqs. (3.1) is acceptable. The idea is to iterate $\chi_o$ by using the $\chi_o$ that will reduce Eq. (2.9) to an identity. For doing this iteration, more work is required.

## 20.4. Perturbation Equations at Lowest Order

Now we develop a solution in powers of $\varepsilon$, to the lowest order:

$$\mathbf{u}(\mathbf{x}, \varepsilon) = \varepsilon \mathbf{u}_1(\mathbf{x}),$$

$$\psi(\mathbf{x}, \varepsilon) = \psi_o(\mathbf{x}) + \varepsilon \psi_1(\mathbf{x}),$$

$$v(\mathbf{x}, \varepsilon) = v_o(\mathbf{x}) + \varepsilon v_1(\mathbf{x}), \tag{4.1}$$

$$\phi(\mathbf{x}, \varepsilon) = \phi_o(\mathbf{x}) + \varepsilon \phi_1(\mathbf{x}),$$

$$\chi(\varepsilon) = \chi_o + \varepsilon \chi_1.$$

At the interface, we have

$$v[\chi(\varepsilon), \varepsilon] = v_o(\chi_o) + \varepsilon v_1(\chi_o) + \varepsilon \chi_1 \cdot \nabla v_o(\chi_o), \tag{4.2}$$

$$\phi[\chi(\varepsilon), \varepsilon] = \phi_o(\chi_o) + \varepsilon \phi_1(\chi_o) + \varepsilon \chi_1 \cdot \nabla \phi_o(\chi_o). \tag{4.3}$$

Since $\mathbf{u}_o(\mathbf{x}) \equiv 0$ in the oil

$$\mathbf{u}[\chi(\varepsilon), \varepsilon] = \varepsilon \mathbf{u}_1(\chi_o) + \varepsilon^2 \mathbf{u}_2(\chi_o) + \varepsilon^2 \chi_1 \cdot \nabla \mathbf{u}_1(\chi_o), \tag{4.4}$$

but

$$\psi[\chi(\varepsilon), \varepsilon] = \psi_o(\chi_o) + \varepsilon \psi_1(\chi_o) + \varepsilon \chi_1 \cdot \nabla \psi_o(\mathbf{x}_o). \tag{4.5}$$

Moreover, since the shape of drop changes with

$$\mathbf{n}(\chi) = \mathbf{n}_o + \varepsilon \mathbf{n}_1,$$

$$\tau(\chi) = \tau_o + \varepsilon \tau_1. \tag{4.6}$$

After inserting Eqs. (4.1)–(4.6) into the basic equations (2.4)–(2.9) we find first that

$$\nabla \psi_o = \mu_w \nabla^2 \mathbf{u}_1, \qquad \text{div } \mathbf{u}_1 = 0,$$

$$\mathbf{u}_1(\chi_o) \cdot \mathbf{n}_o = 0, \tag{4.7}$$

$$\tau_o \cdot \mathbf{D}[\mathbf{u}_1(\chi_o) - v_o(\chi_o)] \cdot \mathbf{n}_o = 0.$$

This problem may be solved for $\mathbf{u}_1(x)$ and $\psi_o(\mathbf{x})$ when $\chi_o$ is given. The slow motion in the oil is driven by the shear rate in the water:

$$\tau_o \cdot \mathbf{D}[v_o] \cdot \mathbf{n}_o = \partial v_\tau(\chi_o)/\partial y_n \stackrel{\text{def}}{=} \overset{\circ}{\gamma}(\chi_o), \tag{4.8}$$

where $v_\tau(\chi_o)$ is the velocity component tangent to the interface and $y_n$ is normal at $\mathbf{x} = \chi_o$.
    The normal stress balance of Eq. (2.9) now becomes

$$-\phi_1(\chi_o) + \psi_o(\chi_o) - (\rho_w - \rho_o)\mathbf{g} \cdot \chi_o - 2\mu_w \mathbf{n}_o \cdot \mathbf{D}[\mathbf{u}_1(\chi_o)] \cdot \mathbf{n}_o = 2H(\chi_o)\sigma. \tag{4.9}$$

We may write

$$\mathbf{n}_o \cdot \mathbf{D}[\mathbf{u}_1(\chi_o)] \cdot \mathbf{n}_o = \partial u_{1n}/\partial y_n,$$

where $u_{1n}$ is the normal component of $\mathbf{u}_1$ at the interface point $\mathbf{x} = \chi_o$. In deriving Eq. (4.9) we used an easily proved result that says that

$$\mathbf{n}_o \cdot \mathbf{D}[v(\chi_o)] \cdot \mathbf{n}_o = 0$$

when $v_o(\chi_o)$ is the fluid velocity at the boundary of a rigid body. Equation (4.9) selects $\chi_o$, which until now was arbitrary.

## 20.5. Perturbation Equations at Higher Order

Continuing now to higher orders, we find that

$$\rho_w[v_o \cdot \nabla v_1 + v_1 \cdot \nabla v_o] = -\nabla\phi_1 + \mu_w \nabla^2 v_1, \qquad \text{div } v_1 = 0, \tag{5.1}$$

$$v_1(\chi_o) = \mathbf{u}_1(\chi_o) - \chi_1 \cdot \nabla v_o(\chi_o), \qquad v_1(\chi_b) = 0. \tag{5.2}$$

Equations (5.1) and (5.2) may be solved for $v_1(\mathbf{x})$ and $\phi_1(\mathbf{x})$ when $\mathbf{x}_1$ is given.
    To get $\chi_1$, we must go to order $\varepsilon^2$ in our expansion. From Eq. (2.4) we get

$$0 = -\nabla\psi_1 + \mu_w \nabla^2 \mathbf{u}_2, \qquad \text{div } \mathbf{u}_2 = 0. \tag{5.3}$$

Equation (2.7) gives rise to

$$\mathbf{n}_o \cdot [\mathbf{u}_2(\chi_o) + \chi_1 \cdot \nabla \mathbf{u}_1(\chi_o)] + \mathbf{n}_1 \cdot \mathbf{u}_1(\chi_o) = 0, \tag{5.4}$$

and Eq. (2.8) leads to

$$\tau_o \cdot \mathbf{D}[\mathbf{u}_2 - v_1 + \chi_1 \cdot \nabla(\mathbf{u}_1 - v_o)] \cdot \mathbf{n}_o + \tau_1 \cdot \mathbf{D}[\mathbf{u}_1 - v_o] \cdot \mathbf{n}_o + \tau_o \mathbf{D}[\mathbf{u}_1 - v_o] \cdot \mathbf{n}_1 = 0. \tag{5.5}$$

Equations (5.3)–(5.5) can be solved for $\mathbf{u}_2$ and $\psi_1$ when $\chi_1$ is given. The normal stress condition of Eq. (2.9) at order $\varepsilon^2$ gives rise to

$$-\phi_1(\chi_o) + \psi_1(\chi_o) - \chi_1 \cdot \nabla(\phi_o - \psi_o) - (\rho_w - \rho_o)\frac{\lambda}{h} \cdot \chi_1$$

$$+ 2\mu_w \mathbf{n}_o \cdot \mathbf{D}[v_1 - \mathbf{u}_2 + \psi_1 \cdot \nabla(v_o - \mathbf{u}_1)] \cdot \mathbf{n}_o = 2\frac{\mathrm{d}H}{\mathrm{d}\chi}(\chi_o) \cdot \chi_1 \sigma. \qquad (5.6)$$

Expression (5.6) selects the correct boundary perturbation.

Equations (2.5) and (2.6) at order $\varepsilon^2$ give rise to a perturbation problem for $v_2$ and $\phi_2$, depending on $\chi_2$ and so on:

$$\chi = \mathbf{e}_x x + \mathbf{e}_z \zeta(x, \varepsilon), \qquad \chi = \mathbf{e}_z \zeta_1(x), \qquad \zeta_1 = \frac{\partial \zeta}{\partial \varepsilon}.$$

So $\chi_1$ is just one scalar function.

## 20.6. Core-Annular Flow

Here we revisit the problem of waves on core-annular flow considered by Bai et al. (1996). They treated a steady flow in which the holdup ratio $c_o/c_w$ of average velocities $c_o = Q_o/\pi R_1^2$ and $c_w = Q_w/\pi(R_2^2 - R_1^2)$ is prescribed. Here $Q_o$ and $Q_w$ are the volume flux of oil and water, respectively, $R_2$ is the outer radius of the pipe, and $R_1$ is the mean radius of the core. In the approximation carried out by them, the core is rigid. The analysis of the steady flow of water is carried out in a coordinate system in which the core is stationary; secondary motions in the core were not treated. The shape of the interface was computed with the normal stress condition under the assumption that the pressure in the core is uniform, apart from a constant pressure gradient $\beta$ along the pipe axis $z$ (see Fig. 20.1).

The problem of core-annular flow may be treated within the framework of the perturbation theory described in Sections 20.3 and 20.4 with $\lambda \cdot \mathbf{x}$ in Eqs. (2.3) equal to $-\beta z$, where $\beta$ is a constant pressure gradient. The governing equations at zeroth order are essentially Eqs. (3.2).

$$\rho_w v \cdot \nabla v = \beta \mathbf{e}_z - \nabla p_w + \mu_w \nabla_v^2 = 0,$$

$$v = 0 \text{ on } r = f(z), \qquad (6.1)$$

$$v = -c\mathbf{e}_z \text{ on } r = R_2,$$

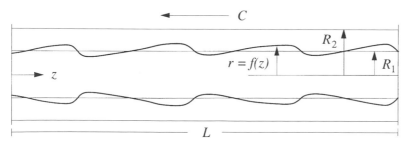

Figure 20.1. Core-annular flow. The flow is periodic with period $L$. The mean radius is $R_1$, where $R_1^2 = \frac{1}{L}\int_0^L f^2(z)\,dz$. The core moves forward with velocity $c$ and the wall is stationary; here the core has been put to rest. $\Omega_w$ is the domain occupied by water $0 \le z \le L$, $f(z) \le r \le R_2$, and $\Omega_o$ is the domain occupied by oil $0 \le z \le L$, $0 \le r \le f(z)$.

where $r = f(z)$ gives the shape of the interface and $f(z)$ was determined by Bai et al. (1996), who used the normal stress condition

$$\frac{\sigma}{f(1+f'^2)^{\frac{1}{2}}} - \frac{\sigma f''}{(1+f'^2)^{\frac{3}{2}}} = C - p_w. \tag{6.2}$$

The ratio of the average oil to water velocity $h = c/c_w$ is given by

$$h = \frac{Q_o/Q_w}{R_1^2/(R_2^2 - R_1^2)} = \frac{\pi c R_1^2}{\pi c [f^2 - R_1^2] + 2\pi \int_f^{R_2} r v \, dr} \frac{R^2 - R_1^2}{R_1^2}. \tag{6.3}$$

Although $f$ depends on $z$, $h$ is a constant, independent of $z$; $h = 2$ for perfect core flow without waves and $h = 1$ when the water is trapped between wave crests touching the pipe wall. For a wavy flow $1 < h < 2$; $h = 1.4$ occurs frequently in experiments; the selection mechanism is related to stability and is not understood. Bai et al. (1996) prescribed $h = 1.4$, ensuring waves.

Going further now than Bai et al. (1996) we consider now problem (4.7) for the flow $\mathbf{u} = \mathbf{u}_1$ in the oil core:

$$-\beta \mathbf{e}_z + \nabla p_o = \mu_w \nabla^2 \mathbf{u}, \qquad \text{div } \mathbf{u} = 0, \tag{6.4}$$

where, on $r = f(z)$, we have

$$\mathbf{u}(r, z) \cdot \mathbf{n} = 0, \tag{6.5}$$

$$\tau \cdot \mathbf{D}[\mathbf{u}] \cdot \mathbf{n} = \overset{\circ}{\gamma}(r, z), \tag{6.6}$$

where the shear rate

$$\overset{\circ}{\gamma}[f(z), z] = \tau \cdot \mathbf{D}[v] \cdot \mathbf{n} \tag{6.7}$$

is evaluated on the solution $v$ of Eqs. (6.1)–(6.3). The constant $\beta$ and function $\overset{\circ}{\gamma}(r, z)$ are prescribed.

After computing $v$ and $p_w$ from problem (6.1) and $\mathbf{u}$ and $p_o$ from problems (6.4)–(6.7), we may complete the perturbation cycle by forming the normal stress balance corresponding to Eq. (4.9). This balance replaces Eq. (6.2) with

$$\frac{\sigma}{f(1-f'^2)^{\frac{1}{2}}} - \frac{\sigma f''}{(1+f'^2)^{\frac{3}{2}}} - 2\mu_w n \cdot \mathbf{D}[\mathbf{u}] \cdot \mathbf{n} = p_o - p_w. \tag{6.8}$$

Equation (6.8) cannot be satisfied for arbitrarily selected functions $r = f(z)$ and wavelengths $L$. These parameters are iterated at each perturbation cycle until Eq. (6.8) balances and holdup ratio (6.3) is met, giving rise to converged values of $f(z)$ and $L$.

Preliminary calculations following the perturbation method just presented have been carried out with the methods of Bai et al. (1996). The additional terms in the normal stress balance (6.8) are small (see Fig. 20.2) and the difference between the rigid approximation of Bai et al. (1996) and the present calculation are also small (see Fig. 20.3). The wavelength and pressure gradient versus Reynolds number are shown in Fig. 20.4. The wavelength of perturbation is slightly larger than the wavelength of the rigid core, but the pressure gradients are the same.

A more complete study of the perturbation solution will be presented in a future calculation.

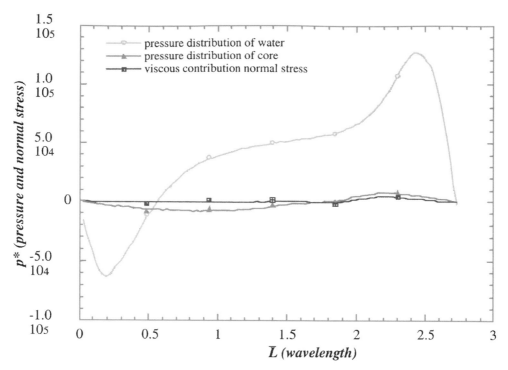

Figure 20.2. The pressure distributions and viscous contribution normal stress along the wave interface when $[\eta, h, \mathbb{R}, J] = [0.8, 1.4, 600, 13 \times 10^4]$.

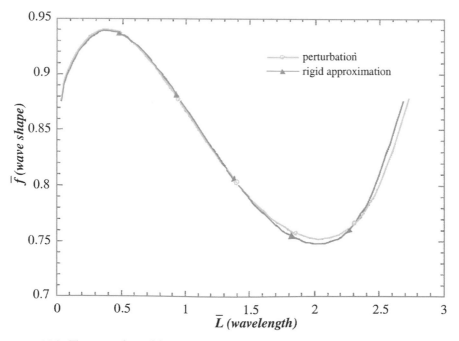

Figure 20.3. The comparison of the wave shapes approached from perturbation and rigid approximation when $[\eta, h, \mathbb{R}, J] = [0.8, 1.4, 600, 13 \times 10^4]$.

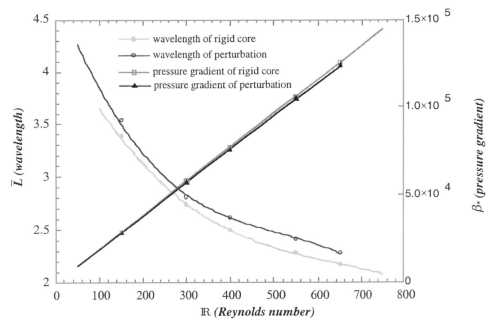

Figure 20.4. The comparison of dimensionless wavelength and pressure gradient versus Reynolds number $\mathbb{R}$ for $[\eta, h, J] = [0.8, 1.4, 13 \times 10^4]$.

## Acknowledgments

This work was supported by the U.S. Department of Energy, the U.S. National Science Foundation, and Intevep, S.A. The authors thank Clara Mata for helpful discussions.

## Reference

Bai, R., Kelkar, K., and Joseph, D. D. 1996. Direct simulation of interfacial waves in a high viscosity ratio and axisymmetric core-annular flow. *J. Fluid Mech.* **240**, 97–142.

# PART FOUR

# Multiphase Systems

# 21

# Interaction between Fluid Flows and Flexible Structures

WEI SHYY, HENG-CHUAN KAN, H. S. UDAYKUMAR, AND ROGER
TRAN-SON-TAY

## 21.1. Introduction

Many problems of practical interest involve the interaction of fluid flows and flexible structures (Shyy et al., 1996). The problems considered in this chapter are such that the interface shape and conditions cannot be simplified by means of linearization or small-perturbation techniques and must be addressed as part of a solution procedure. For such cases, although the governing laws and computational procedures for the fluid and solid domains are well developed, treatment of the whole physical system including the moving boundary remains challenging. With the presence of a flexible structure, both normal and shear forces, involving pressure, viscosity, and velocity gradients, play important roles. Together with surface tension and elasticity, the interface shape is determined. Within both fluid and solid phases, the field equations need to be solved with the location of their boundary being determined simultaneously. From a continuum mechanics viewpoint, the phase interface is a discontinuity in the continuum field. Within the context of finite grid resolution, this discontinuity needs to be accurately tracked in time and in space. Examples relevant to the present work include physiological flows (Dong et al., 1991; Kan et al., 1998), free-surface flows (Osher and Sethian, 1988; Kothe and Mjolsness, 1992; Sussman et al., 1994), solidification fronts (Kessler et al., 1988; Juric and Tryggvason, 1996; Shyy et al., 1996; Udaykumar et al., 1996), the breakup of liquid drops (Stone, 1994; Sheth and Pozrikidis, 1995; Sadhal et al., 1997) and jets (Vassalo and Ashgriz, 1991; Richards et al., 1994), flows over marine sails (Smith and Shyy, 1996) and deformable airfoils (Shyy and Smith, 1997). A number of numerical techniques with different strengths and limitations have been proposed in the literature to treat moving boundary problems (Peskin, 1977; Sethian, 1996; Shyy et al., 1996).

In this chapter we deal with the issues related to the interaction between viscous flows and flexible structures. Both pure Lagrangian and combined Eulerian–Lagrangian approaches are adopted to study two types of problems. Specifically, the Lagrangian method is applied to study the aerodynamics of flexible airfoils at low Reynolds numbers ($10^4$–$10^5$) under unsteady flow conditions. The mixed Eulerian–Lagrangian method with the immersed boundary technique (IBT) (Peskin, 1977; Fauci and Peskin, 1988; Udaykumar et al., 1997) is used to investigate the deforming and recovering dynamics of single and compound viscous drops as well as meniscus characteristics in a convective two-fluid system. In the Lagrangian method, the interfacial condition is applied explicitly and individual phases treated distinctively, while in the mixed Eulerian–Lagrangian method, moving boundaries are treated by distribution of the momentum sources at the interface onto the grid points surrounding the phase boundary. A number of parameters have been investigated, including those addressing relative effects between surface

265

tension and elasticity, convection, viscosity, capillary number, Reynolds number, and also the time scales between different domains.

## 21.2. Aerodynamics of Low Reynolds Number Flexible Airfoils

Many flying machines from self-inflating parawings to birds use lifting surfaces that deform significantly during flight. With the right choice of materials, pretension, and unstrained shape, an aerodynamically effective equilibrium configuration can often be achieved to improve the flight performance. The interest of the present work is to study the aerodynamic performance of a low Reynolds number ($10^4$–$10^5$) airfoil for microair vehicle applications (Shyy and Smith, 1997). The well-known CLARK-Y airfoil is adopted when a portion of the top surface (from 10% to 100% chord) is considered to be a massless membrane that will instantaneously adjust its curvature according to the membrane equilibrium equation,

$$\kappa = -\frac{1}{\Pi_1^3} \frac{\Delta p}{\sigma},$$ (1)

where $\Pi_1$ is a nondimensional parameter, $\Delta p$ is the pressure difference across the membrane, and $\sigma$ is the membrane tension. For situations in which the membrane tension is dominated by elastic strain, $\Pi_1$ is given by the elasticity number

$$\Pi_1 = \left[\frac{Eh}{q_\infty c}\right]^{1/3},$$ (2)

where $E$ is the elastic modulus, $h$ is the membrane thickness, $c$ is the airfoil chord, and $q_\infty$ is the free-stream stagnation pressure. In the present case, the pressure on the outside of the airfoil is obtained from the flow solution, and the pressure on the inside of the airfoil is set equal to the stagnation pressure.

Based on the pressure difference, the airfoil can adjust its shape to accommodate an oscillating flow caused by wind gusts, for example. The membrane shape is updated by enforcing the Young–Laplace equation, Eq. (1), by use of the shooting method. The procedure converges when the membrane's trailing-edge $y$ coordinate coincides with the rigid trailing-edge $y$ coordinate of the bottom surface. A new pressure distribution is then calculated, and another membrane shape is determined corresponding to the new pressure difference. The free-stream velocity oscillates at a single frequency. The computation is conducted at every time instant to obtain a new equilibrium shape corresponding to the transient flow conditions. A number of computations are done to assess the relative merits of a flexible over a rigid airfoil. A coupled inviscid and thin-layer viscous flow computational strategy (Drela and Giles, 1987; Drela, 1989), with the aid of a mixing-length turbulence closure and a simplified instability growth model to track the laminar-turbulence transition, is adopted. A CLARK-Y profile serves as the base shape, upon which a membrane top was attached. Computations for a Reynolds number oscillating around $8.0 \times 10^4$ were made for both a rigid and a flexible case to compare the influence of the membrane. An example of the movements of the membrane top for a CLARK-Y profile during a full period oscillation of the free-stream velocity for $Re_{mean} = 8.0 \times 10^4$ with $\Pi_1 = 8.8$ (indicating that the time-scale ratio between the fluid and the membrane is ~26) is displayed in Fig. 21.1. The time-dependent lift coefficient $C_L$ and power index $C_L^{3/2}/C_D$ are displayed in Fig. 21.2.

For both $\alpha = 0°$ and $\alpha = 3°$, the lift coefficient of the rigid profile noticeably degrades as the free-stream velocity approaches a minimum. However, the flexible airfoil gives overall more favorable aerodynamic characteristics by adjusting the membrane shape in accordance with

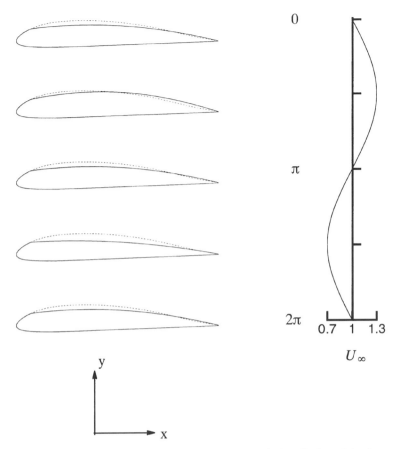

Figure 21.1. Membrane deformations during a full period oscillation of the free-stream velocity, shown on the right side, with $Re_{mean} = 8.0 \times 10^4$ and $\Pi_1 = 8.8$. Dashed and solid curves are original and flexible profiles corresponding to the instantaneous velocity magnitude, respectively.

the instantaneous pressure difference. Qualitatively, the airfoil thickness increases as the flight speed increases and decreases as the flight speed decreases. This trend is consistent with the general expectation that a thinner airfoil performs better with a reduced Reynolds number. The wide performance gap can best be seen for the zero-angle-of-attack case in Fig. 21.2. Similar conclusions have been drawn in work by Shyy and Smith (1997), in which a thin membrane airfoil was compared with a similar rigid airfoil.

## 21.3. Dynamics of Single and Compound Drops

In the Lagrangian method, in which boundary conforming curvilinear grids are used, if the shape of the domain is changed, the grid is regenerated to describe the new geometry and the flow is recomputed. If the geometry is evolving as the flow solution proceeds, as in moving boundary problems, the grid has to adapt to this moving boundary. In problems in which the boundary deformation is large, it is advantageous to perform computations on a fixed grid, while the boundary itself deforms arbitrarily and moves through the grid. The moving boundary can be evolved by means of a fixed variable (such as volume of fluid, level set or phase field methods) or as curves/surfaces in space. When the interface is explicitly defined as a curve/surface and

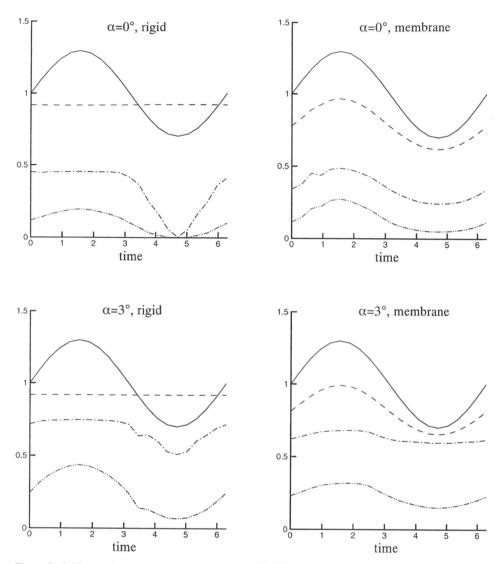

Figure 21.2. Comparison between rigid and flexible CLARK-Y profiles at $Re = 8.0 \times 10^4$. ———, free-stream velocity; _·—·_, $C_L$; − − −, $y$ coord $\times 10$ at $x = c/2$; ·······, $C_L^{3/2}/C_D \times 10^{-2}$.

tracked over a fixed grid, the approach is called mixed Eulerian–Lagrangian. Examples in this category include the immersed boundary method (Peskin, 1977; Unverdi and Tryggvason, 1992; Udaykumar et al., 1997) and the immersed interface method (LeVeque and Li, 1994).

The deformation and the recovery of a viscous drop subjected to different flow conditions are investigated by a mixed Eulerian–Lagrangian method with the IBT. A three-layer incompressible Newtonian fluid system of density $\rho_i$ and viscosity $\mu_i$, which occupies the regions $\Omega_i$ ($i = 1, 2, 3$), is adopted to define the problems under investigations. In the case of a compound drop, $\Omega_1$, $\Omega_2$, and $\Omega_3$ represent the suspending fluid, shell (or cytoplasm), and core (or nucleus), respectively. The surface tensions $\sigma_{12}$ and $\sigma_{23}$ at the two interfaces $\Gamma_{12}$ and $\Gamma_{23}$ are assumed to be constant. The lengths $R_o$ and $R_i$ are the undeformed radii of the drop (or cell) and core (or nucleus), respectively.

The choices of characteristic length, velocity, viscosity, density, time, and pressure are as follows:

$$l_c = R_o, \qquad u_c = Gl_c, \qquad \mu_c = \mu_1, \qquad \rho_c = \rho_1, \qquad t_c = \frac{1}{G}, \qquad p_c = \frac{\mu_c u_c}{l_c},$$

where $\mu_1$ and $\rho_1$ are the viscosity and the density of the suspending fluid, respectively. $G$ is the strain rate of the imposed flow, $\sigma_c = \sigma_{12}$ is the surface tension of the outer drop and suspending fluid interface. Then the dimensionless governing equations for Navier–Stokes flows are

(i) continuity equations in regions $\Omega_i$ ($i = 1, 2, 3$):

$$\nabla \cdot \mathbf{V}_i = 0. \tag{3}$$

(ii) Momentum equations are, for the suspending fluid (region $\Omega_1$),

$$\beta_1 \left[ \frac{\partial \mathbf{V}_1}{\partial t} + \nabla \cdot (\mathbf{V}_1 \mathbf{V}_1) \right] = -\nabla p_1 + \frac{\lambda_1}{\text{Re}} \nabla^2 \mathbf{V}_1 + \frac{1}{\text{We}} \mathbf{f}_{12}, \tag{4}$$

for the enclosed fluids, namely the shell (or cytoplasmic) fluid ($\Omega_2$),

$$\beta_2 \left[ \frac{\partial \mathbf{V}_2}{\partial t} + \nabla \cdot (\mathbf{V}_2 \mathbf{V}_2) \right] = -\nabla p_2 + \frac{\lambda_2}{\text{Re}} \nabla^2 \mathbf{V}_2 + \frac{1}{\text{We}} (\mathbf{f}_{12} + \gamma \mathbf{f}_{23}), \tag{5}$$

and for the core (or nucleus) in a compound drop model ($\Omega_3$),

$$\beta_3 \left[ \frac{\partial \mathbf{V}_3}{\partial t} + \nabla \cdot (\mathbf{V}_3 \mathbf{V}_3) \right] = -\nabla p_3 + \frac{\lambda_3}{\text{Re}} \nabla^2 \mathbf{V}_3 + \frac{\gamma}{\text{We}} \mathbf{f}_{23} \tag{6}$$

In the above equations, $\beta_i$ and $\lambda_i$ are the nondimensional density and viscosity of the suspending fluid ($i = 1$), shell fluid ($i = 2$), and core fluid ($i = 3$), respectively. $\mathbf{V}_i$ and $p_i$ represent the velocity and the pressure, respectively, in region $\Omega_i$. $\gamma$ is the nondimensional surface-tension ratio between inner and outer drops. The relevant parameters that arise from the nondimensionalization are the Reynolds number [$\text{Re} = 2\rho_c (Gl_c) l_c / \mu_c$] and Weber number [$\text{We} = 2\rho_c (Gl_c)^2 l_c / \sigma_c$]. The former provides a measure of the relative dominance of inertia to viscous effects, while the latter is the ratio of inertia to capillary forces. The forces $\mathbf{f}_{12}$ and $\mathbf{f}_{23}$ denote the Eulerian contributions of the normal stress discontinuity due to the capillary forces evaluated on the interfaces $\Gamma_{12}$ and $\Gamma_{23}$. These forces are transmitted from the interfaces to the fluid by use of the IBT (Peskin, 1977; Unverdi and Tryggvason, 1992).

For creeping flows ($\text{Re} \approx 0$), the Navier–Stokes equations reduce to Stokes equations when the unsteady and convective terms are excluded from the momentum equations. Therefore Eqs. (4)–(6) become

$$0 = -\nabla p_1 + \lambda_1 \nabla^2 \mathbf{V}_1 + \frac{1}{\text{Ca}} \mathbf{f}_{12}, \tag{7}$$

$$0 = -\nabla p_2 + \lambda_2 \nabla^2 \mathbf{V}_2 + \frac{1}{\text{Ca}} (\mathbf{f}_{12} + \gamma \mathbf{f}_{23}), \tag{8}$$

$$0 = -\nabla p_3 + \lambda_3 \nabla^2 \mathbf{V}_3 + \frac{\gamma}{\text{Ca}} \mathbf{f}_{23}, \tag{9}$$

where Ca denotes the capillary number that is defined as ($\mu_c Gl_c / \sigma_c$). The capillary number represents the strength of the viscous forces over surface tensions. The pressure variable in Eqs. (7)–(9) differs from that in Eqs. (4)–(6) by a factor of Weber number.

The nondimensional velocity condition for the imposed axisymmetric extensional flow is

$$
\mathbf{V}_\infty = \mathbf{E} \cdot \mathbf{R} = \frac{1}{2}
\begin{bmatrix} 2 & 0 & 0 \\ 0 & -1 & 0 \\ 0 & 0 & -1 \end{bmatrix}
\cdot
\begin{bmatrix} \mathbf{z} \\ \theta \\ \mathbf{r} \end{bmatrix},
\tag{10}
$$

where $\mathbf{E}$ is the strain-rate tensor and $\mathbf{R}$ is the position vector.

The boundary conditions to be satisfied on each interface for cases involving no mass exchange across it are the continuity conditions and the stress balances:

$$
(\mathbf{V})_{\Gamma_{12}} = (\mathbf{V})_1 = (\mathbf{V})_2,
\tag{11}
$$

$$
(\mathbf{V})_{\Gamma_{23}} = (\mathbf{V})_2 = (\mathbf{V})_3,
\tag{12}
$$

$$
\mathbf{n}_2 \cdot \tilde{\tau}_1 - \lambda_2 \mathbf{n}_2 \cdot \tilde{\tau}_2 = \mathbf{n}_2 \left( \nabla_{\Gamma_{12}} \cdot \mathbf{n}_2 \right),
\tag{13}
$$

$$
\mathbf{n}_3 \cdot \tilde{\tau}_2 - \lambda_3 \mathbf{n}_3 \cdot \tilde{\tau}_3 = \gamma \mathbf{n}_3 \left( \nabla_{\Gamma_{23}} \cdot \mathbf{n}_3 \right).
\tag{14}
$$

Solutions of several cases are presented to highlight the role of the capillary number, Reynolds number, time scales, and arbitrary geometries. All computations are performed on uniform Cartesian grids. The boundaries (moving as well as stationary) are tracked curves that lie over the fixed grid.

### 21.3.1. *Dynamics of Drops in Constricted Tubes*

First we study the deformation of drops in viscosity- and inertia-dominated flows through a constricted tube. The interaction of drops with the geometry in which they are constrained to traverse has received some attention from experimental (Olbricht and Kung, 1992; Olbricht and Leal, 1993) and numerical viewpoints (Tsai and Miksis, 1994; Manga, 1996a, 1996b). However, those works deal with creeping flow regimes only. In Fig. 21.3(a), we show the results of the simulation of drop motion through a constricted tube at Re = 50. The constriction in the tube is a sine wave with amplitude $0.6R$ and wavelength $2R$, where $R$ is the radius of the tube, Ca = 0.1, the viscosity ratio $\lambda_2 = 0.01$, and the ratio of the undeformed drop radius to the tube radius $a = 0.75$. For the area ratio and Reynolds number considered, there appears to be an unsteady motion in the region following the constriction. A region of recirculating flow is formed and travels downstream of the throat. This region of the flow has a profound effect on the interface shape. The interface gets stretched into an elongated shape downstream of the constriction because of the higher flow speeds generated by a rapid area change and takes a mushroomlike shape. For comparison, the continuous deformed drop shapes for the case of Re = 0 and $a = 0.9$ are shown in Fig. 21.3(b), where the drop deforms essentially in conformity with the shape of the constriction. At entry to the constriction the front of the drop is sharpened, and at exit the rear of the drop is flattened by the acceleration of the flow at the constriction.

### 21.3.2. *Deformation of a Compound Drop in a Uniaxial Extensional Flow*

Next, we examine the deformation characteristics of a compound drop ($\lambda_2 = \lambda_3 = 1$, $\gamma_2 = \gamma_3 = 1$, and radius $R_i = 0.5$) under an elongational flow with and without inertia effects. Figures 21.4(a) and 21.4(b) show the results for creeping (Ca = 0.1, Re $\approx 0$) and noncreeping flow conditions

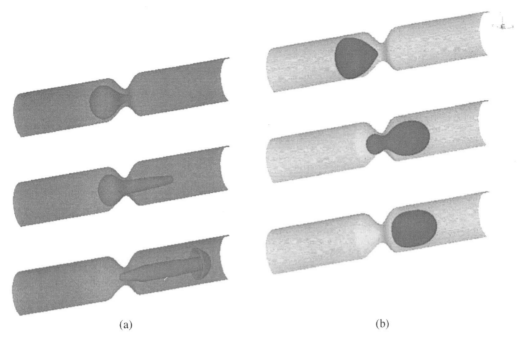

(a)                                                      (b)

Figure 21.3. Deformation of a liquid drop ($\lambda_2 = 0.01$) into a constricted tube at Ca = 0.1. (a) For Re = 50, & $a = 0.75$ drop shapes at nondimensional times $t = 0.5$, 0.625, and 1.12; (b) for Re = 0 & $a = 0.9$, drop shapes at nondimensional times $t = 0.25$, 1.25, and 1.5.

(Re = 100, We = 10), respectively. It is noted that the core deforms into an oblate spheroid, while the shell becomes prolate for the creeping flow condition. The recirculating flows within the annular region cause the inner drop to deform toward the outer drop interface when the compound drop reaches an equilibrium state. On the other hand, on the inclusion of inertia forces, the compound drop undergoes unsteady deformation. The convective effects not only significantly enhance the degree of the deformation of the compound drop, but also force the inner drop into a highly stretched cylindrical configuration and to stretch to large extents that cannot be achieved in the creeping flow regime.

### 21.3.3. *Effects of the Core Deformation and Time Scales on Compound Drop Recovery*

To highlight the essential elements of the dynamics of compound drops, we consider the recovery of compound drops with two different sets of ratios ($\gamma_2/\lambda_2$) and ($\gamma_3/\lambda_3$). These ratios are related to the time scales of the shell layer and the core, respectively. These two cases are (a) $\gamma_2 = 1$, $\lambda_2 = 40$, $\gamma_3 = 1$, and $\lambda_3 = 4000$, which corresponds to a highly viscous core, and (b) $\gamma_2 = 1$, $\lambda_2 = 40$, $\gamma_3 = 1$, and $\lambda_3 = 400$. The shape of the drop is taken to be that obtained after an extensional flow deformation to $L_2 = 2.1$ and $L_3 = 1$ under Ca = 0.14. The deformed core is assumed to be a cylinder with two hemispherical caps.

Figure 21.5(a) shows the instantaneous drop shapes during the recovery process of case (a), in which the time scales between the shell and the core are disparate. Clearly, in the recovery stage, the response of the compound drop is largely dominated by the shell, while the core itself serves mainly as a solid-like obstruction to the flow field. These characteristics of the flow fields lead to the rapid recoil response of the compound drop and provide a non-Newtonian-like behavior.

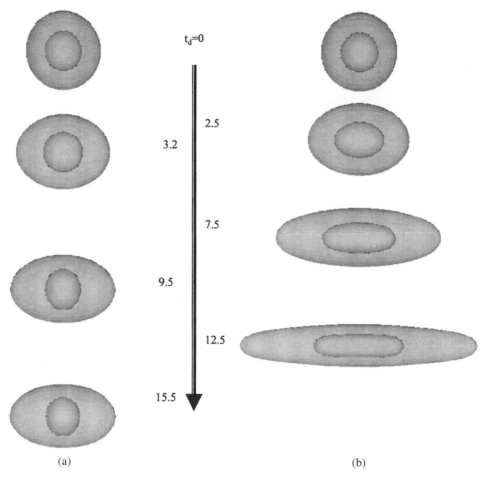

Figure 21.4. Effect of the inertia on the deformation ($L_2 = 3$) of the compound drop ($\gamma_2 = 1$, $\lambda_2 = 1$, $\gamma_3 = 1$, $\lambda_3 = 1$, and $R_i = 0.5$) under extensional flow. (a) For creeping flow conditions (Re $= 0$, Ca $= 0.1$), (b) for noncreeping flow conditions (Re $= 100$, We $= 10$). Scaled deformation times $t_d$ ($t_d = 1/G$, where $G$ is the shear rate of the imposed flow) are shown next to the corresponding shapes.

Figure 21.5(b) shows the recovery shapes of the compound drop that has compatible time scales between the shell, $\tau_{\text{shell}} = \lambda_2/\gamma_2$, and core, $\tau_{\text{core}} = \lambda_3/\gamma_3$. Although the shell is deformed to the same extent and the resulting capillary-induced flow from the outer interface is of the same magnitude, the recovery dynamics changes drastically from the first case.

### 21.3.4. *Recovery Dynamics of a Single Drop after Large Deformation*

In addition to the study of the deformation and recovery dynamics of the compound drop, we also investigate the recovery characteristics of a single drop ($\lambda_2 = 1$ and $\gamma_2 = 1$) after it undergoes large deformation in an elongational flow. The drop is initially deformed to a specified extent ($L/D = 5.4$, where $L$ is the elongated length of the drop and $D$ is the undeformed drop diameter), then recovers after the flow field is switched off. Both the initial and the sequential drop shapes during the relaxation process are presented in Fig. 21.6. The results demonstrate that the relaxation process leads to an impending breakup phenomenon in a later recovery stage for this highly elongated initial configuration. It is clearly seen in Fig. 21.6 that the breakup

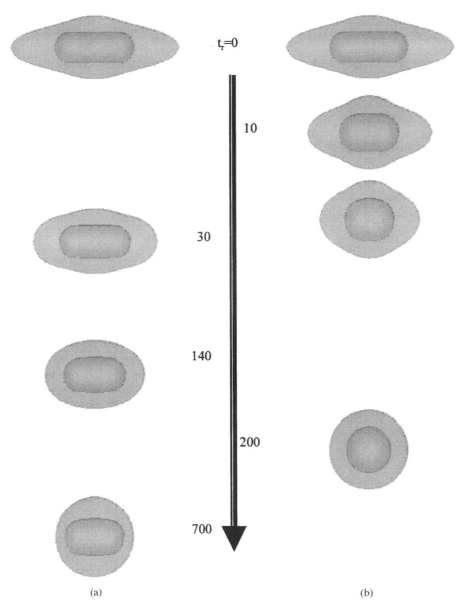

(a)                                                    (b)

Figure 21.5. Effect of time scales of the shell $(\lambda_2/\gamma_2)$ and core $(\lambda_3/\gamma_3)$ on recovery characteristics of two compound drops with the same initial elongated lengths $(L_2 = 2.1, \ L_3 = 1)$. Scaled recovery times $t_r$ $(t_r = R_o \mu_s/\sigma_{12})$ are indicated in each case. (a) Disparate time scales between layers $(\gamma_2 = 1, \lambda_2 = 40, \gamma_3 = 1,$ and $\lambda_3 = 4000)$, (b) compatible time scales between layers $(\gamma_2 = 1, \lambda_2 = 40, \gamma_3 = 10,$ and $\lambda_3 = 400)$.

would lead to the formation of a small satellite drop along with two larger drops in the final recovery process.

## 21.4. Meniscus Characteristics in a Convective Two-Fluid System

We now present an example dealing with the meniscus formation in convective two-phase flows. A schematic of the computational domain and boundary conditions is shown in Fig. 21.7,

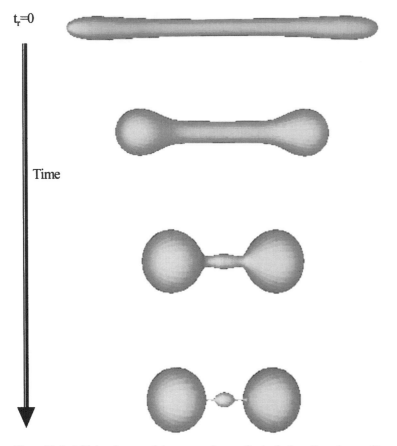

Figure 21.6.  Initial and sequential recovery shapes of a single drop ($\lambda_2 = 1$, $\gamma_2 = 1$) after it undergoes large deformation in a straining flow.

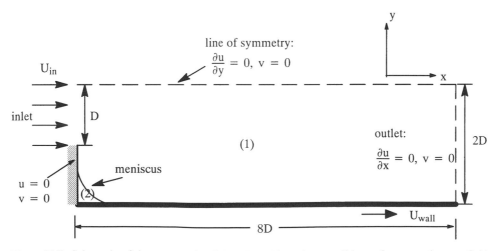

Figure 21.7.  Schematic of the computational domain and boundary conditions of a convective two-fluid system.

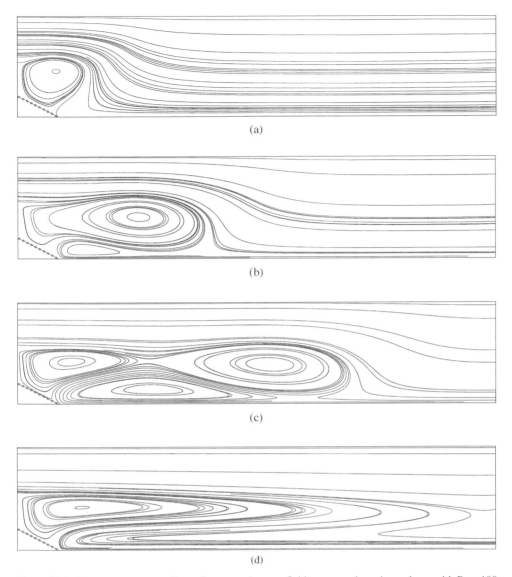

Figure 21.8. Instantaneous streamlines of a convective two-fluid system and meniscus shape with Re $=190$ and We $= 0.09$ (a) at nondimensional time $t = 3$, (b) $t = 8$, (c) $t = 15$, (d) $t = 73$. Meniscus shapes are represented by open circles.

which is two-dimensional planar, and consists of open inlet/outlet and a moving solid wall at the bottom boundary. Equations (3)–(5) are used with the following characteristic scales:

$$l_c = D, \qquad u_c = U_{\text{in}}, \qquad \mu_c = \mu_s, \qquad \rho_c = \rho_s, \qquad \sigma_c = \gamma, \qquad t_c = \frac{D}{U_{\text{in}}}, \qquad p_c = \rho_s U_{\text{in}}^2,$$

where $D$ is the height of the inlet, $U_{\text{in}}$ denotes the peak velocity of the inlet flow, $\gamma$ is the meniscus surface tension, and $\mu_s$ and $\rho_s$ are the viscosity and the density of the incoming fluid, respectively.

The top computational boundary is taken to be the line of symmetry, and the bottom wall velocity is equal to the peak inlet flow velocity. The inlet flow is assigned to be fully developed with a fixed flow rate. The inlet Reynolds number is 190, and the density and viscosity ratios between fluids (2) and (1) are 0.02 and 0.1, respectively. The Weber number is 0.09, the contact location at left wall is fixed at 40% height, and the contact angle at the bottom wall is 30°. This case is motivated by our interest in studying the casting process of aluminum.

The governing equations are solved with the IBT to determine the flow field and the resulting shape of the meniscus at steady state. Figure 21.8 shows streamlines of the convective flow field at different time instants. Since the surface tension is large, the meniscus is only slightly deformed, but the flow field displays time-dependent behavior. At the meniscus, a large separation bubble periodically sheds. Thus the meniscus remains steady while the flow becomes unsteady.

## 21.5. Concluding Remarks

In this study, we investigate the dynamics characteristics of viscous flows interacting with flexible structures. Computations and physical aspects in treating such problems are summarized with the aid of two examples, namely, low Reynolds number flexible airfoil aerodynamics and multiphase drop and meniscus dynamics. Because of its soft upper surface, a flexible airfoil provides a more desirable aerodynamic performance when compared with the corresponding rigid profile in a low Reynolds number, oscillating free stream. For a deforming drop pushed though a constricted tube, the interplay among convection, shear, pressure, and surface tension results in distinctively different shapes and flow fields. For the deformation of a compound drop in uniaxial flows, it is demonstrated that the inertia effects can significantly affect the degree of deformation of the compound drop in all layers. Furthermore, a compound drop behaves like a homogeneous, simple liquid drop if the core is sufficiently deformed and the time scale of the core, related to the combination of its viscosity and capillarity, is compatible with that of the shell layer. Disparate time scales between the core and shell layer result in a rapid initial recoil of the drop during which the shell fluid is the primary participant in the hydrodynamics, followed by a slower relaxation period during which the core and shell layer interact with each other. Finally, the meniscus characteristics corresponding to given fluid and flow parameters have major implications in materials processing applications.

The cases presented demonstrate the complicated characteristics and potential insight one can gain from studying the interaction between fluid flows and flexible structures. Such knowledge can be valuable in many engineering and biomedical applications.

## Acknowledgment

The present work has been partially supported by the U.S. Air Force Office of Scientific Research, Eglin Air Force Base, the Alcoa Foundation, and the U.S. National Institutes of Health.

## References

Dong, C., Skalak, R., and Sung, K.-L. P., 1991, "Cytoplasmic rheology of passive neutrophils," *Biorheology* **28**, 557–567.

Drela, M., 1989, "XFOIL: an analysis and design system for low Reynolds number airfoils," in *Low Reynolds Number Aerodynamics*, Mueller, T. J., ed., pp. 1–12, Springer-Verlag, New York.

Drela, M. and Giles, M. B., 1987, "Viscous-inviscid analysis of transonic and low Reynolds number airfoils," *AIAA J.* **25**, 1347–1355.

Fauci, L. J. and Peskin, C. S., 1988, "A computational model of aquatic animal locomotion," *J. Comput. Phys.* **77**, 85–108.

Juric, D. and Tryggvason, G., 1996, "A front tracking method for dendritic solidification," *J. Comput. Phys.* **123**, 127–148.

Kan, H.-C., Udaykumar, H. S., Shyy, W., and Tran-Son-Tay, R., 1998, "Hydrodynamics of a compound drop with application to leukocyte modeling," *Phys. Fluids* **10**, 760–774.

Kessler, D. A., Koplik, J., and Levine, H., 1988, "Pattern selection in fingered growth phenomena," *Adv. Phys.* **37**, 255–339.

Kothe, D. B. and Mjolsness, R. C., 1992, "RIPPLE: a new method for incompressible flows with free surfaces," *AIAA J.* **30**, 2694–2700.

LeVeque, R. J. and Li, Z., 1994, "The immersed interface method for elliptic equations with discontinuous coefficients and singular sources," *SIAM J. Num. Anal.* **31**, 1019–1044.

Manga, M., 1996a, "Dynamics of droplets in cavity flows: aggregation of high viscosity ratio droplets," *Phys. Fluids* **8**, 1732–1737.

Manga, M., 1996b, "Dynamics of droplets in branched tubes," *J. Fluid Mech.* **315**, 105–117.

Olbricht, W. L. and Kung, D. M., 1992, "The deformation and breakup of liquid droplets in low Reynolds number flow through a capillary," *Phys. Fluids* **4**, 1347–1354.

Olbricht, W. L. and Leal, L. G., 1993, "The creeping motion of immiscible droplets through a converging/diverging nozzle," *J. Fluid Mech.* **134**, 329–355.

Osher, S. and Sethian, J. A., 1988, "Fronts propagating with curvature dependent speed: algorithms based in Hamilton–Jacobi formulations," *J. Comput. Phys.* **79**, 12–49.

Peskin, C. S., 1977, "Numerical analysis of blood flow in the heart," *J. Comput. Phys.* **25**, 220–252.

Richards, J. R., Lenhoff, A. M., and Beris, A. N., 1994, "Dynamic breakup of liquid-liquid jets," *Phys. Fluids* **6**, 2640–2655.

Sadhal, S. S., Ayyaswamy, P. S., and Chung, J. N., 1997, *Transport Phenomena with Drops and Bubbles*, Springer-Verlag, New York.

Sethian, J. A., 1996, *Level Set Methods: Evolving Interfaces in Geometry, Fluid Mechanics, Computer Vision, and Materials Science*, Cambridge U. Press, New York.

Sheth, K. S. and Pozrikidis, C., 1995, "Effects of inertia on the deformation of liquid drops in simple shear flow," *Comput. Fluids* **24**, 101–119.

Shyy, W. and Smith, R., 1997, "A study of flexible airfoil aerodynamics with application to micro aerial vehicles," AIAA paper 97-1933, AIAA, Washington, D.C.

Shyy, W., Udaykumar, H. S., Rao, M. M., and Smith, R. W., 1996, *Computational Fluid Dynamics with Moving Boundaries*, Taylor & Francis, Washington, D.C.

Smith, R. and Shyy, W., 1996, "Computation of aerodynamic coefficients for a flexible membrane airfoil in turbulent flow: a comparison with classical theory," *Phys. Fluids* **8**, 3346–3353.

Stone, H. A., 1994, "Dynamics of drop deformation and breakup in viscous fluids," *Ann. Rev. Fluid Mech.* **26**, 65–102.

Sussman, M., Smereka, P., and Osher, S., 1994, "A level set approach for computing solutions to incompressible two-phase flow," *J. Comput. Phys.* **114**, 146–159.

Tsai, T. M. and Miksis, M. J., 1994, "Dynamics of a droplet in a constricted capillary tube," *J. Fluid Mech.* **274**, 197–217.

Udaykumar, H. S., Rao, M. M., and Shyy, W., 1996, "ELAFINT – a mixed Eulerian–Lagrangian method for fluid flows with complex and moving boundaries," *Int. J. Num. Methods Fluids* **22**, 691–704.

Udaykumar. H. S., Kan, H.-C., Shyy, W., and Tran-Son-Tay, R., 1997, "Multiphase dynamics in arbitrary geometries on fixed Cartesian grids," *J. Comput. Phys.* **137**, 366–405.

Unverdi, S. O. and Tryggvason, G., 1992, "A front tracking method for viscous, incompressible, multi-fluid flows," *J. Comput. Phys.* **100**, 25–37.

Vassalo, P. and Ashgriz, N., 1991, "Satellite formation and merging in liquid jet breakup," *Proc. R. Soc. London*, **433**, 269–286.

# 22

# Numerical Treatment of Moving Interfaces in Phase-Change Processes

SURESH V. GARIMELLA AND JAMES E. SIMPSON

## 22.1. Introduction

In problems involving moving interfaces between dissimilar phases, the interface plays a key role in the solution to the problem; the interface simultaneously influences, and deforms in response to, the thermal, compositional, and flow fields throughout the domain. As a result, in the numerical computation of such problems there is a need to resolve accurately the shape and the position of the interface, so that the phenomenological effects at the interface may be adequately described. In general, this interface may be very complicated and characterized by large deformations, multiple interfaces, merging and fragmenting, and drastic jumps in property values across the interface, thus severely complicating numerical investigations. These problems are common to a host of diverse applications including melting and solidification, evaporation and condensation, bubble growth dynamics, and flame fronts in combustion.

Moving interfaces have been treated by interface-capturing and interface-tracking methods (Floryan and Rasmussen, 1989). In interface-capturing methods, a minimal effort is made to resolve the details of the structure of the interface. The interface is modeled only to the extent that it is known to be someplace in the vicinity of a determining (usually temperature) gradient. Properties are changed either discretely at the interface or continuously over a range near the interface. The enthalpy method (Alexiades and Solomon, 1993; Shamsundar and Sparrow, 1975) and the apparent heat capacity method (Dantzig, 1989; Yao et al., 1997) are examples of this one-domain approach. Interface-capturing methods are acceptable for problems for which the matching conditions at the interface can be handled by a change in material properties at the interface and the constitutive equations are solved throughout the entire domain. For more complicated problems, such as those involving a free surface, these methods have limited utility since the position and the shape of the interface determine the size and the shape of the computational domain and so the interface must be determined accurately.

Interface-tracking methods include volume-tracking and front-tracking approaches. Volume-tracking methods are suitable for specifying complex interfaces and for treating three-dimensional interactions such as merging and fragmenting. The earliest technique for simulating free-surface problems by volume tracking was the marker and cell (MAC) method (Welch et al., 1965). In this method, massless marker points are advected by use of the local velocity field. The disadvantages of this method are that (i) computational overhead incurred in storing and updating the location of each particle is high, (ii) particles may accumulate in one portion of the interface, resulting in other portions being underrepresented or false void regions being created, and (iii) boundary conditions are difficult to specify. More recent volume-tracking methods solve an advection equation for a color function (Rudman, 1997) that represents the volume fraction of a mesh cell that is filled with a fluid of a particular type. A system of $m$ fluids requires $m - 1$

of these color functions. Standard advection techniques may be applied to advect this indicator function; however, such schemes introduce either artificial diffusion or instability into the solution scheme. Numerical diffusion results in the sharpness of the advecting phase front being smeared over several cells while instability results in unphysical oscillations in the interface shape. While many techniques have been proposed to limit both artificial diffusion and instability, these remain the main obstacles to effective simulation with volume-tracking methods (Labonia et al., 1998). Volume-tracking methods do not store a representation of the interface but reconstruct it from the color function whenever necessary. The most well-known methods for reconstructing the front are the volume-of-fluid (VOF) method of Hirt and Nichols (1981) and the simple line interface calculation (SLIC) method of Noh and Woodward (1976). A more thorough review of volume-tracking methods is available in the work of Chen et al. (in press); the performance of various volume-tracking methods was recently evaluated by Rudman (1997).

Front-tracking methods explicitly specify the location of the interface through an ordered set of marker points located on the interface. The grid is deformed so that mesh points lie on the interface itself. The advantages of front-tracking methods are their ability to resolve the features of the interface (even to the subgrid level) and the ease of applying boundary conditions. In addition, the information regarding the location and the orientation of the interface is available throughout the entire solution procedure. However, complex phenomena such as merging and fragmenting cannot readily be simulated; computational requirements are also large. In general, few extensions to three dimensions have been demonstrated. The simplest of these methods uses a height function that specifies the location of the interface above a fixed reference line. This method breaks down if the interface is multivalued and so can be applied in only the simplest cases. More advanced methods include Lagrangian approaches such as adaptive grid generation (Zhang and Moallemi, 1995) and boundary-fitted grids (Glimm et al., 1981), Eulerian approaches such as level-set (Sethian, 1996) and phase-field (Murray et al., 1995) methods, and mixed Eulerian–Lagrangian methods (Unverdi and Tryggvason, 1992; Udaykumar et al., 1996; Shyy et al., 1996).

The aim of this chapter is to present brief summaries of methodologies used for four moving-interface problems. It is seen that the selection of an algorithm for advecting and reconstructing the interface depends on the nature of the problem to be solved; each method has its own characteristics that make it suitable for the solution of a certain class of problems. The rise and the development of a gas bubble in two and three dimensions are explored by use of a modified VOF method (Chen et al. 1997) to identify the front. The propagation of a solidification front past an array of reinforcing fibers [as found in metal-matrix composites (MMCs)] will be examined by use of the single-domain enthalpy method (Simpson et al., 1998b). Directional solidification (Simpson and Garimella, 1998) and microgravity crystal growth (Simpson et al., 1998a) are examined with a hybrid of the single-domain enthalpy method and Hirt and Nichols (1981) front reconstruction. Finally, the performance and the utility of a new algorithm for explicit front tracking on a fixed Eulerian grid (Labonia et al., 1998) are demonstrated through the results of some test problems.

## 22.2. Methodologies for Treatment of the Interface in Different Applications

### 22.2.1. *Analysis of Bubble Rise and Deformation with a Modified VOF Method*

Complex bubble interactions, including the rise of single and mutiple bubbles, bubble–wall interactions, and bubble merging/breakup in two and three dimensions have been studied with

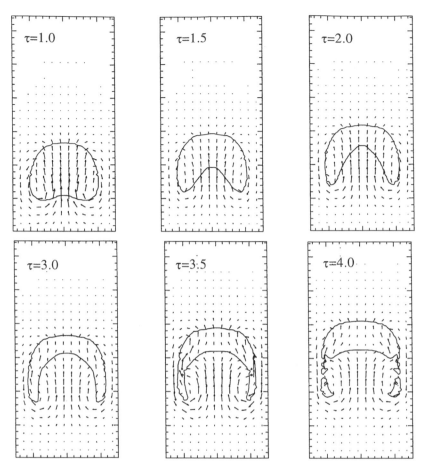

Figure 22.1. Influence of container wall on the shape of a single rising bubble. The shape development and velocity vectors are shown. The ratio of cylinder to bubble radius is 1.5.

a modified VOF method (Chen et al., 1997) in cylindrical coordinates. Previously, such effects could not be studied with VOF because of excessive numerical diffusion at the interface. The new modified VOF scheme described here is able to overcome both the complexities of formulating an accurate VOF scheme in cylindrical coordinates and the problem of numerical diffusion.

In traditional VOF methods (Floryan and Rasmussen, 1989), a donor–acceptor scheme is used to calculate the advective fluxes. This involves using a combination of both first-order upwind and downwind convective fluxes on a staggered Cartesian grid. The use of either the upwind scheme or the downwind scheme is dependent on the orientation of the interface. The upwind convective flux is diffusive while the downwind flux is unstable; a suitable solution procedure seeks to use the two in a systematic way that reduces both artificial diffusion and instability. In traditional VOF approaches, an approximate height function is used to determine the orientation of the interface and then whether upwind or downwind fluxes should be used. This gives reasonable accuracy on a two-dimensional Cartesian grid [although extensions to this scheme have been proposed that use additional flux correction as in the work of Rudman (1997)] but it is very cumbersome and inaccurate to implement in three dimensions, particularly for cylindrical coordinates.

The modified VOF scheme developed by Chen et al. (1997) is more accurate for three-dimensional cylindrical coordinates. The selection of fluxes is as follows. The surface normal is determined with the gradient of the volume fraction. The orientation of the interface within a given control volume is then taken to be parallel to the largest component of this surface normal. When the interface is parallel to the flow, an upwind scheme is used. When the interface is normal to the flow, a downwind scheme is used. The downwind scheme contains an additional flux term that prevents the oscillatory downwind scheme from advecting too little or too much flux out of the cell. The constitutive equations for mass conservation and momentum transport are solved with the SIMPLE (Patankar 1980) algorithm. Appropriate boundary conditions involving surface tension and phase change at the liquid vapor interface are used. It should also be noted that the solution scheme uses a nonstaggered grid, which simplifies the application of the pressure boundary condition at the interface.

A selection of results obtained with this modified VOF approach (Chen et al., 1997) are presented here. Figure 22.1 illustrates the influence of the container wall on the shape of an isolated bubble as it rises in a container whose diameter is comparable with that of the bubble. The walls are seen to exert a definite influence on the bubble shape. The bubble departs from having a spherical shape to having that of a shell that is concave from below. The trailing edges of the bubble eventually break off into a series of small gas bubbles. The algorithm is clearly able to resolve the interface under the condition of bubble breakup. It is precisely this sort of phenomenon that volume-tracking algorithms excel in resolving.

Results for merging bubbles are shown in Fig. 22.2. The motion of the upper bubble induces

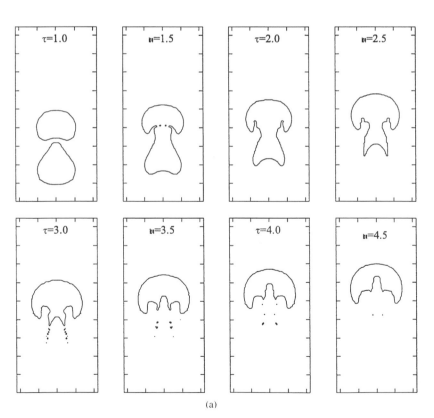

(a)

Figure 22.2. Merging of two bubbles as they rise and the subsequent deformation of the complex merged bubble. The ratio of the cylinder to bubble radius is 2.16, while the initial separation of the bubbles is 0.24 times the bubble diameter: (a) Re = 10, (b) Re = 20.

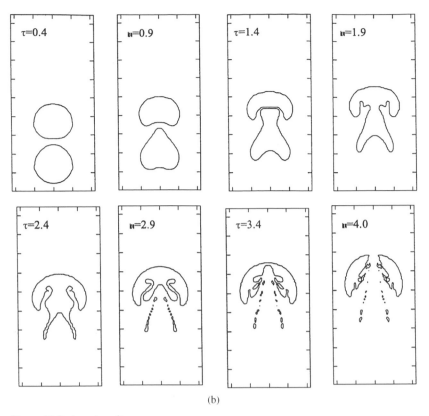

(b)

Figure 22.2. *(continued)*.

a deformation in the lower bubble, imparting it a pearlike shape. With time, the lower bubble merges into the upper bubble, resulting in a taillike structure that eventually breaks up into a trail of gas bubbles. The process of merger and breakup is much more vigorous at the higher Reynolds number of 20 considered in Fig. 22.2(b). Again, the key point to note about this figure is the capacity of the algorithm to resolve adequately the front under the complex processes of bubble merger and breakup.

### 22.2.2. *Solidification Processing of Metal-Matrix Composites*

As a simulation of the processing of MMCs, the solidification of a pure or alloy matrix material in the presence of aligned reinforcing fibers has been investigated (Simpson et al., 1998b) by use of the single-domain enthalpy method approach to account for the change of phase. Because of the small length scales present in the problem, convection effects were ignored. Both a pure material and a binary alloy were used for the matrix. Accounting for the presence of the fibers, which take part in the simulation process by means of heat conduction but do not undergo phase change themselves, required the development of a modified enthalpy method. This solution scheme can be generalized to account for the inclusion of $N$ phases that do not undergo phase change. Previous studies of this problem have used a more cumbersome Lagrangian-based method (Khan and Rohatgi, 1994) that involved the use of boundary-fitted grids. This method did not yield the temperature distributions throughout the fibers nor could it be extended to simulate the solidification of a dilute binary alloy instead of a pure material.

The enthalpy method (Alexiades and Solomon, 1993; Shamsundar and Sparrow, 1975) is arguably the most robust and widely used means of numerically describing phase change at an interface. The phase front is not tracked explicitly, and the constitutive equations are solved throughout the entire solution domain. The absorption of latent heat as new solid forms is handled by means of a specially formulated source term: this is the crux of the method. Heat fluxes through the faces of each cell due to temperature gradients are calculated, and the temperature of each cell is updated. Should a cell reach the melting temperature of the material, it is maintained at the melting temperature until the latent heat of solidification is absorbed by that cell. Thermophysical properties in the solid and the liquid phases may be different. Temperature–enthalpy relations provide the logic that determines if a cell is fully liquid, fully solid, or is to undergo phase change.

The generalization of this standard enthalpy method to account for the presence of the fiber phase consists of forming density, specific heat, and thermal conductivity matrices that are weighted according to the presence of the volume of solid and liquid matrix material and fiber present in each cell. Additionally, some alterations in the temperature–enthalpy relations that determine the disposition (solid, liquid, or phase change) of the matrix material in a given cell are required. These property matrices replace the constant values used in the standard enthalpy method. There is surprisingly little additional CPU burden imposed, although there is a definite increase in the amount of dynamic memory required. The solution scheme is rounded out by the inclusion of an algorithm for solving the concentration equation and accounting for solute segregation at the interface; full details are supplied in the work of Simpson et al. (1998b).

Solidification in the domain shown in Fig. 22.3 has been investigated by Simpson et al. (1998b). Both alumina (low-conductivity) and copper (high-conductivity) fibers were used. The

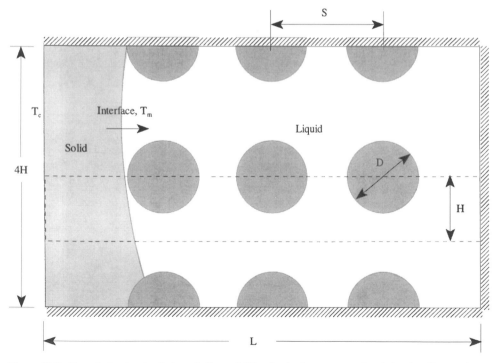

Figure 22.3. Domain for the simulation of alloy solidification in the presence of a reinforcing fibrous phase. The computational domain is indicated by the dashed outline.

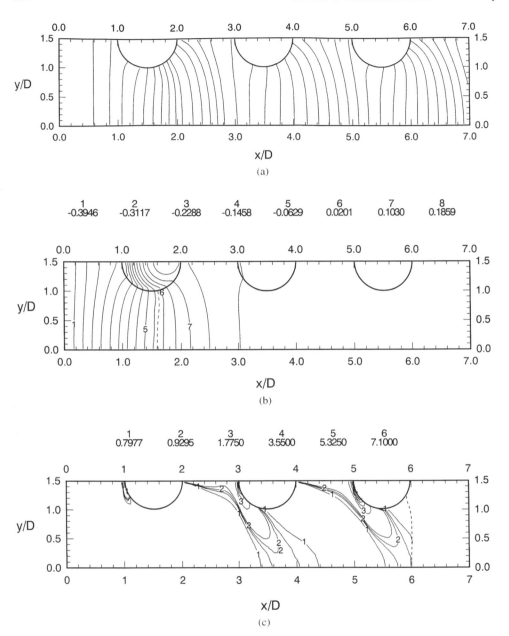

Figure 22.4. (a) Front locations, (b) temperature contours (pure Al matrix), (c) solute concentrations (Al–1.0-wt.% Cu matrix) for solidification in the presence of a fibrous phase. In (a), the first 10 fronts are at time intervals of 50 time steps, the next 5 at 100 time-step intervals, and the final fronts at 200 time-step intervals. Temperature contours in (b) are shown after 300 time steps; the dashed curve shows the interface position.

matrix material was either pure Al or Al–1.0-wt.% Cu alloy. The domain was discretized by a uniform $168 \times 36$ mesh. The diameter of the fibers used was $10\,\mu$m, with a domain size $L \times H$ of $70 \times 15\,\mu$m and a fiber pitch of $S = 2D$. Material properties and further details may be found in the original reference. Simulations required approximately an hour on a workstation.

Representative results for the solidification of the pure material and alloy in the presence of alumina fibers are presented here. Figure 22.4(a) is a plot of the front locations at various time intervals. The front locations were found as a postprocessing operation by a Hirt and Nichols (1981) front-reconstruction technique. But it is emphasized that this reconstruction was performed simply for plotting purposes; the solution scheme itself is purely single domain and never requires the explicit location or orientation of the front (in contrast to alloy solidification results in Subsection 22.2.3). The results clearly indicate the strong influence of the fibers on the shape of the propagating front. Had the fibers not been present in the domain, a one-dimensional problem would result and the interface would remain planar throughout the simulation. At 50 time steps, the front is still planar since it is remote from the influence of the fiber. As the front approaches the fiber, it arcs toward the fiber. This is because of the much lower thermal conductivity of the fiber compared with the melt; the melt between the fiber and the approaching front is lower in temperature than the matrix material lower in the domain. As the front passes the fiber, the fiber acts to delay the propagation of the front since it is at a higher temperature. Once past the first fiber, the front is distorted from the normal. As it approaches the second fiber, the front remains distorted. As the front passes the second fiber, its curvature is influenced once again. The third fiber causes further distortion. Note that the front does not restore to the vertical between successive fibers; the fiber pitch is smaller than the critical value required for complete relaxation of the front to vertical between fibers.

Figure 22.4(b) is a plot of the temperature distribution after 300 time steps, when the front has passed approximately halfway around the first fiber. This figure illustrates the impact of the fiber on the thermal field and hence the driving force behind the front distortion. The single-domain approach is able to resolve the temperatures in both the phase-change material and the fiber despite the high gradients and large variations in thermophysical properties present. Figure 22.4(c) is a plot of solid solute concentration contours for the solidification of an Al–1.0-wt.% Cu matrix, at the time when the front has almost passed beyond the third fiber. The maxima in the solute concentration levels occur in the matrix preceding each fiber. Other solute-rich regions are also present at the top of the domain between the second and the third fibers; these extend diagonally below the second and the third fibers. These solute-rich bands are formed as parts of the front accelerate because of the influence of the fibers on the thermal field.

### 22.2.3. *Directional Solidification and Microgravity Crystal Growth*

The directional solidification of pure Sn, as well as of a Sn–0.5-wt.% Bi alloy under terrestrial conditions were simulated by Simpson and Garimella (1998). The solution scheme includes the effects of thermosolutal convection. The presence of the solidification front is handled by a hybrid of the single-domain traditional enthalpy method with a two-domain approach for the solution of the vorticity-vector potential representation of the momentum and continuity equations. This solution scheme has subsequently been extended to simulate microgravity crystal growth of pure Bi and a Bi–Sn alloy with Sn concentrations of 0.1 and 1.0 at.% (Simpson et al., 1998a).

The primary disadvantage of the single-domain enthalpy method is that no explicit phase front is identified; thus it becomes impossible to specify the requisite boundary conditions for velocity when solving the momentum equations in the melt. This limits the utility of the enthalpy method to conduction-only cases such as the one discussed in Subsection 22.2.2 above.

However, the location of the front can be determined with a Hirt and Nichols (1981) style front reconstruction. The enthalpy values are used to determine a cell liquid fraction, which

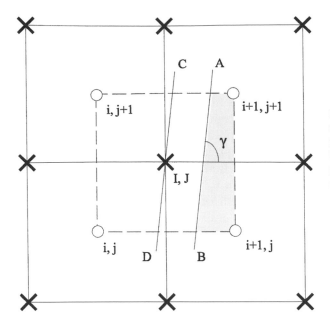

Figure 22.5. Details of front reconstruction. The interface (line AB) is considered planar in each cell. For simplicity, vorticity boundary conditions are applied at the mesh point I, J (line CD) irrespective of where the interface actually lies.

is used as the color function for the reconstruction. Note that this is not the same as the VOF method: here, the enthalpy method supplants the advection equation for the volume fraction. Once the front location and orientation are determined, this information can be used to apply appropriate velocity boundary conditions at the interface. In this way, the solution procedure becomes a hybrid between the single-domain enthalpy method approach for the solution of the energy equation (and for the solution of the concentration equation if a binary alloy is being considered) and a two-domain approach for the solution of the momentum and mass conservation equations.

Details for the enthalpy method itself have already been discussed above. The modification that generalized the approach to include the presence of the fiber regions can be used to account for other effects such as conduction through domain walls (Simpson et al., 1998a). In solving for the velocity field with the vorticity-vector potential representation, the interfacial condition for the vector potential is that it vanishes. The boundary condition on vorticity is more difficult and depends on the orientation of the front within the cell. A further complication is that since this is a fixed-grid solution scheme, the boundary condition may be applied at only a fixed point while the true location of the interface may lie away from a mesh point. Such an arrangement is shown in Fig. 22.5. The interface may actually be along line AB but is taken to be along line CD. The angle $\gamma$ is needed to apply the boundary condition.

The performance of the solution scheme was evaluated by computation of the directional solidification of pure Sn in a rectangular cavity and comparison of the results with those of Zhang et al. (1996) and Raw and Lee (1991) and with the experimental data of Wolff and Viskanta (1988). Reasonable agreement was obtained (to within 10% in most cases) as described in the work of Simpson and Garimella (1998). Zhang et al. (1996) were able to track more accurately the progression of the interface at later stages of solidification. Their simulation used an adaptive grid scheme and required 45 h of CPU time on a workstation while the scheme presented here required less than 9 h on a comparable machine.

Directional solidification of alloy Sn–0.5-wt.% Bi in a rectangular cavity was investigated next. Both thermal and solutal convections were considered. Figure 22.6(a) shows velocity vectors and isotherms at time $t = 0.621$ h. Note the curvature of the interface and the complex convection pattern that results. There is a primary convective cell that is driven by thermal gradients and a weaker secondary convective cell below it. This secondary cell rotates in the same direction as the primary convective cell. Similar flow patterns have been observed during the solidification of pure Ga (Dantzig, 1989) for a comparable melt apect ratio.

Figure 22.6(b) is a plot of solute concentrations at the same time of $t = 0.621$ h. In the melt region, solute rejection at the interface and the influence of the convection cells on the solute distribution are readily apparent. In the solid region, the complex solute distribution is a result of the propagation of the solidification front, convection in the melt, and solute rejection in the interface. The most significant detail is the solute-deficient region in the upper portion of the solid while the lower portion is, overall, solute rich in comparison.

The solution algorithm was next used to simulate the space-borne crystal growth of a Bi–1.0-at.% Sn alloy (Simpson et al., 1998a). The Bridgman crystal growth process was simulated. Previously, studies of this process were limited to considering only pure Bi since the low partition coefficient for the Bi–Sn alloy system caused numerical difficulties (Yao et al., 1997). Velocity vectors and isotherms are shown in Fig. 22.7. Conduction in the ampoule walls is included in the solution scheme. The key features of this plot are the curvature evident in the solid/liquid interface, which is a result of different thermal properties in the solid and liquid phases, the dominant clockwise rotating primary convective cell, which is driven by thermal gradients, the secondary counterclockwise rotating convective cell that develops with time as the interface rejects more and more solute into the melt, and the influence of conduction in the ampoule wall on the thermal field. This result highlights the utility and robustness of the solution scheme.

### 22.2.4. *Explicit Three-Dimensional Front Tracking on a Fixed Grid*

A new front-tracking algorithm has been developed by Labonia et al. (1998) for a fixed grid in three dimensions. A fixed-grid approach enables straightforward computations for the constitutive equations and is able to handle large deformations and multiple surfaces. A disadvantage of earlier fixed-grid front-tracking methods was their inability to track the location of the interface precisely. The new method discussed here is able to overcome this shortcoming.

Three distinct processes are involved: interface tracking, calculation of normal velocities, and the solution of the governing equations over the entire domain. At time step $t$, the domain is as shown in Fig. 22.8(a). The location of a set of marker points that lie on the intersection of the interface and the regular Cartesian grid is known, as is the interface normal velocity at each of these marker points. During time interval $\Delta t$, these marker points propagate with the interface normal velocity to new locations, known as advected points. In general, these points will not lie on the grid [Fig. 22.8(b)]. New marker points at time $t + \Delta t$ must be recovered by use of the location of these advected points. For maximum accuracy, a local coordinate transformation plus bilinear interpolation is used to deduce the location of the new marker points. Full details of this transformation are provided in the original reference.

Now that the marker points at time $t + \Delta t$ are known, the new interface normal velocities must be calculated so that the front may be propagated over the next time step. The first step in this procedure is the calculation of surface normals at each marker point by means of cubic spline fits. Next, a set of intersection points is determined. Intersection points are the locations

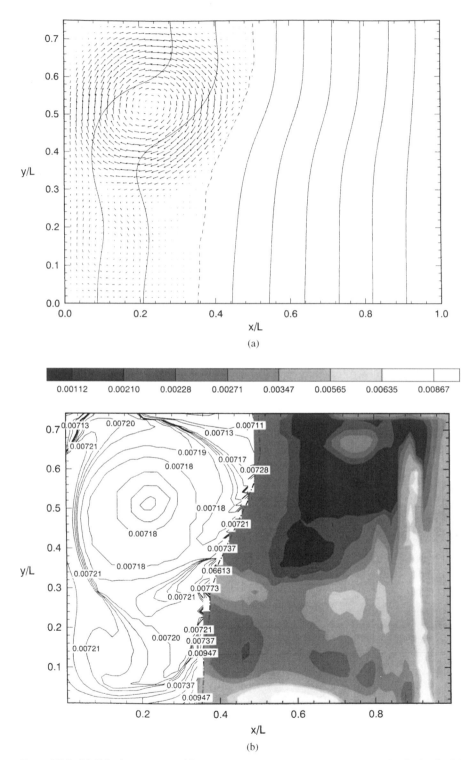

Figure 22.6. (a) Velocity vectors and isotherms, (b) contours of solute concentration in the liquid and solid for Sn–0.5% Bi at $t = 0.621$ h. Isotherms are at intervals of $\Delta\theta = 0.1$; dashed curve is the front location.

| Level | 1 | 2 | 3 | 4 | 5 | 6 | 7 | 8 |
|---|---|---|---|---|---|---|---|---|
| $\lambda$ : | -0.4671 | -0.2570 | -0.0468 | 0.1633 | 0.3735 | 0.5836 | 0.7938 | 1.0039 |

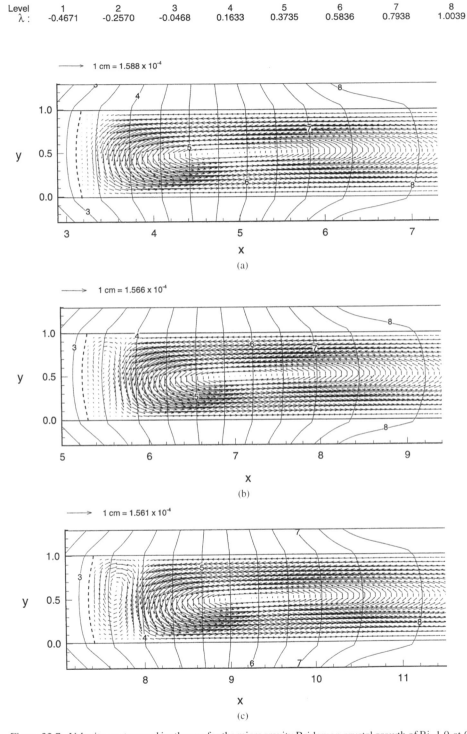

Figure 22.7. Velocity vectors and isotherms for the microgravity Bridgman crystal growth of Bi–1.0-at.% Sn at (a) 1.10 h, (b) 2.15 h, (c) 3.19 h. The heavy dashed curve indicates the solid/liquid interface. A primary counterclockwise convective cell, driven by thermal gradients, is evident in all three plots; a secondary clockwise cell, driven by solutal gradients, develops with time.

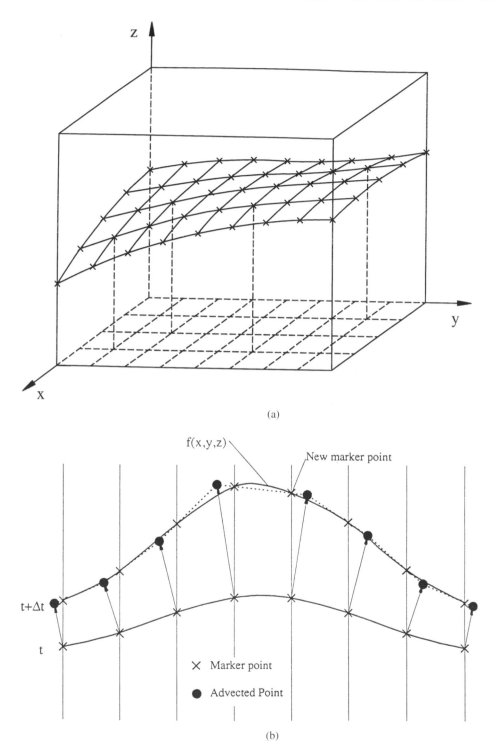

(a)

(b)

Figure 22.8. (a) Physical domain showing the interface, Cartesian grids, and marker points (×) for the explicit interface tracking approach; (b) determination of new marker points at time step $t + \Delta t$ from the advected points (●). Bilinear interpolation in a coordinate-transformed plane is used for three-dimensional calculations; a two-dimensional case is shown here for clarity.

where the interface cuts the horizontal grid lines. The number of intersection points is variable and depends on the shape of the interface. These intersection points and surface normals are then used to determine (for a solidification problem) the normal temperature gradients. As for the calculation of the marker points, a coordinate transformation is used when these gradients are calculated for each grid point. The surface normal velocities may then be recovered from the temperature gradients.

As the final step, the constitutive equations for energy, species, mass, and momentum transport may be solved over the entire domain. The interface shape, position, and normal velocity are known and so matching conditions at the interface are fully defined.

This new approach is best appreciated by examination of the results from a few test cases. Further results are given in the work of Labonia et al. (1998). Front tracking in the solidification of a pure material is the sample problem selected. For simplicity, no constitutive equations are solved; rather a three-dimensional temperature profile is imposed throughout the domain and the front moves in response to the gradients in this imposed profile. The front propagates as a result of different thermal conductivities imposed for the solid and the liquid phases. Results are shown in Fig. 22.9 for two imposed temperature functions – one parabolic in $x$ and $y$ and the other sinusoidal in $x$ and $y$ and nonlinear in $z$ (equations shown in figure caption). For the case shown in Fig. 22.9(a), a $50 \times 50 \times 20$ regular mesh was used while a finer $50 \times 50 \times 50$ mesh was required for the case shown in Fig. 22.9(b) (there are larger temperature gradients in the $z$ direction). A time step of 0.0001 was used. The results show that the front propagates as expected from the functional form of the imposed temperature profile. Note the ability of the algorithm to track the front even as it becomes highly distorted in Fig. 22.9(b).

## 22.3. Summary

Four different methodologies for solving moving boundary problems have been discussed. For complex bubble growth dynamics in three-dimensional cylindrical coordinates, the modified VOF method (Chen et al., 1997) was shown to be suitable. For the simulation of alloy solidification in the presence of aligned fibers, the modified enthalpy method (Simpson et al., 1998b) was presented and discussed. This method offers inherent efficiency, and a single-domain approach is well suited since the front need not be known explicitly for this conduction-dominated problem. For the terrestrial and microgravity simulation of convection–conduction-driven alloy solidification, use of the hybrid solution scheme of Simpson and Garimella (1998) was illustrated. This fixed-grid scheme incurs much reduced computational time while retaining accuracy comparable with more complicated methods. However, if less dilute alloys were involved, a mushy zone would need to be considered, with an associated dendritic interface. In the present work, the interface is considered smooth and hence is limited to dilute alloys. Finally, preliminary results for a new front-tracking algorithm have been presented. This method overcomes accuracy obstacles that have hitherto hindered the development of such a scheme for a fixed grid (Labonia et al., 1998). Selective use of localized coordinate transformations along with careful interpolation and curve fitting have resulted in a robust algorithm. The test studies indicate that the interface is successfully resolved even as it undergoes large distortions. This new methodology has appeal for solidification and other moving boundary problems and is being further developed in ongoing work.

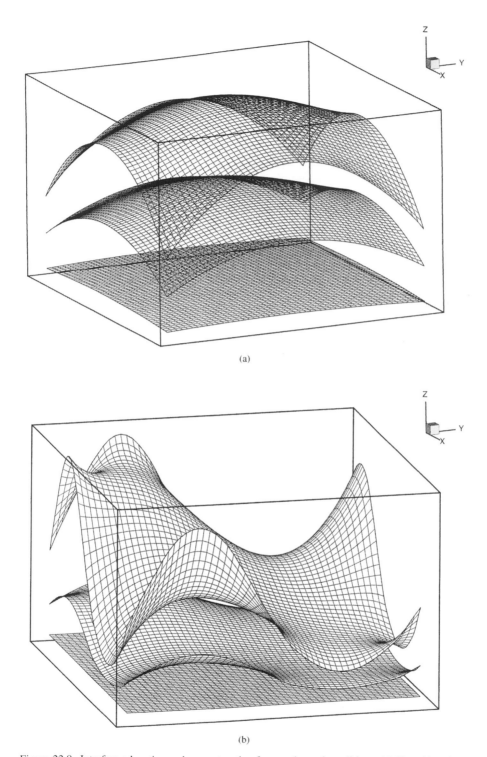

Figure 22.9. Interface advection and reconstruction for two thermal conditions: (a) $T = z[1 + 4x(1 - x) + 4y(1 - y)]$, the interface is shown after 50, 500, and 1000 time steps and the solid to liquid thermal conductivity ratio was 1.25; (b) $T = z^2[2 + \sin(2\pi x)][2 + \sin(2\pi y)]$, the interface is shown after 1, 600, and 1200 time steps and the conductivity ratio was 2.25, and the front propagation in $z$ is no longer linear since the temperature varies as $z^2$.

# References

Alexiades, V. and Solomon, A. D. 1993. *Mathematical Modeling of Melting and Freezing Processes*. Hemisphere, Washington, D.C.

Chen, L., Garimella, S. V., Reizes, J. A., and Leonardi, E. 1997. Motion of interacting gas bubbles in a viscous liquid including wall effects and evaporation. *Num. Heat Transfer* **31**, 629–654.

Chen, L., Garimella, S. V., Reizes, J. A., and Leonardi, E. The development of a bubble rising in a viscous liquid. *J. Fluid Mech.* (in press).

Dantzig, J. A. 1989. Modeling liquid–solid phase changes with melt convection. *Int. J. Num. Methods Eng.* **28**, 1769–1785.

Floryan, J. M. and Rasmussen, H. 1989. Numerical methods for viscous flows with moving boundaries. *Appl. Mech. Rev.* **39**, 323–341.

Glimm, J., Marchesin, D., and McBryan, O. 1981. A numerical method for two phase flow with an unstable interface. *J. Comput. Phys.* **39**, 179–200.

Hirt, C. W. and Nichols, B. D. 1981. Volume of fluid (VOF) method for the dynamics of free boundaries. *J. Comput. Phys.* **39**, 201–225.

Khan, M. A. and Rohatgi, P. K. 1994. Numerical solution to a moving boundary problem in a composite medium. *Num. Heat Transfer* **25A**, 209–222.

Labonia, G., Timchenko. V., Simpson, J. E., Garimella, S. V., Leonardi, E., and de Vahl Davis, G. 1998. Reconstruction and advection of a moving interface in three dimensions on a fixed grid. *Num. Heat Transfer*, **34A**, 121–138.

Murray, B. T., Wheeler, A. A., and Glicksman, M. E. 1995. Simulations of experimentally observed dendritic growth behavior using a phase-field model. *J. Crystal Growth* **154**, 386–400.

Noh, W. F. and Woodward, P. 1976. SLIC (simple line interface calculation). In *Lecture Notes in Physics*. A. I. van de Vooren and P. J. Zandbergen, eds. Springer-Verlag, New York, Vol. 59, pp. 330–340.

Patankar, S. V. 1980. *Numerical Heat Transfer and Fluid Flow*. Hemisphere, Washington, D.C.

Raw, W. Y. and Lee, S. L. 1991. Application of weighting function scheme on convection-conduction phase change problems. *Int. J. Heat Mass Transfer* **34**, 1503–1513.

Rudman, M. 1997. Volume-tracking methods for interfacial flow calculations. *Int. J. Num. Methods Fluids* **24**, 671–691.

Sethian, J. A. 1996. *Level Set Methods*. Cambridge U. Press, New York.

Shamsundar, N. and Sparrow, E. M. 1975. Analysis of multidimensional conduction phase change via the enthalpy model. *J. Heat Transfer*, **97**, 333–340.

Shyy, W., Udaykumar, H. S., Rao, M. M., and Smith, R. W. 1996. *Computational Fluid Dynamics with Moving Boundaries*. Taylor & Francis, Washington, D.C.

Simpson, J. E. and Garimella, S. V. 1998. An investigation of the solutal, thermal and flow fields in unidirectional alloy solidification. *Int. J. Heat Mass Transfer*, **41**, 2485–2502.

Simpson, J. E., Garimella, S. V., and de Groh III, H. C. 1998a. Melt convection effects in the Bridgman crystal growth of an alloy under microgravity conditions. In *Proceedings of the 7th AIAA/ASME Joint Thermophysics and Heat Transfer Conference*, Albuquerque, NM, ASME HTD-Vol. 357–4, 123–132.

Simpson, J. E., Garimella, S. V., and Guslick, M. M. 1998b. Interface propagation in the presence of a fibrous phase in alloy solidification. In *Heat Transfer* 1998, Vol. 7, pp. 235–240, 1998.

Udaykumar, H. S., Shyy, W., and Rao, M. M. 1996. ELAFINT: a mixed Eulerian–Lagrangian method for fluid flows with complex and moving boundaries. *Int. J. Num. Methods Fluids* **22**, 691–712.

Unverdi, S. O. and Tryggvason, G. 1992. A front-tracking method for viscous, incompressible, multi-fluid flows. *J. Comput. Phys.* **100**, 25–37.

Welch, J. E., Harlow, F. H., Shannon, J. P., and Daly, B. J. 1965. The MAC method. A computing technique for solving viscous, incompressible, transient fluid-flow problems involving free surfaces. LASL Rep. LA-3425, Los Alamos, NM.

Wolff, F. and Viskanta, R. 1988. Solidification of a pure metal at a vertical wall in the presence of liquid superheat. *Int. J. Heat Mass Transfer* **31**, 1735–1744.

Yao, M., de Groh III, H. C., and Abbaschian, R. 1997. Numerical modeling of solidification in space with MEPHISTO-4 (Part 1). AIAA paper 97-0449. AIAA, Washington, D.C.

Zhang, H. and Moallemi, M. K. 1995. A multizone adaptive grid generation technique for simulation of moving and free boundary problems. *Num. Heat Transfer* **27B**, 255–276.

Zhang, H., Prasad, V., and Moallemi, M. K. 1996. Numerical algorithm using multizone adaptive grid generation for multiphase transport processes with moving and free boundaries. *Num. Heat Transfer* **29B**, 399–421.

# 23

# Accuracy and Convergence of Continuum Surface-Tension Models

M. W. WILLIAMS, D. B. KOTHE, AND E. G. PUCKETT

## 23.1. Introduction

The basic idea underlying all continuum surface tension (CST) models is the representation of surface tension as a continuous force per unit volume that acts in a neighborhood of the interface. This representation is quite different from previous models in which surface tension is applied as a discrete boundary condition. In the limit as the transition region becomes infinitely thin, the surface-tension boundary condition should be recovered from the CST force. Similar approaches to modeling a force acting on a fluid boundary date back to the work of Peskin (1997) and are still in wide use (e.g., see Unverdi and Tryggvason, 1992). CST methods provide a robust mechanism for modeling flows with topologically complex interfaces that experience merging and breakup. Our focus, however, is to develop robust CST models that remain high-order accurate under such difficult flow conditions.

In CST models, surface tension at the fluid interface (denoted by $\Gamma$), is replaced by a smoothly varying force that is zero outside a local neighborhood of $\Gamma$. Surface tension is expressed as a force per unit volume, $\mathbf{F}_s$, concentrated on the interface $\Gamma$ (see Brackbill, Kothe, and Zemach, 1992):

$$\mathbf{F}_s(\mathbf{x}) = \int_\Gamma \mathbf{f}_s(\mathbf{x})\delta(\mathbf{x} - \mathbf{x}_s)\,\mathrm{d}S, \tag{1}$$

where $\mathbf{x}_s$ are points on $\Gamma$, $\delta$ is the Dirac delta function, and $\mathrm{d}S$ is an arc length in two dimensions and a surface area in three dimensions. The surface-tension force per unit interfacial area, $\mathbf{f}_s$, is given by

$$\mathbf{f}_s = \sigma\kappa\mathbf{n} + \nabla_s\sigma, \tag{2}$$

where $\sigma$ is the surface-tension coefficient, $\nabla_s$ is the gradient taken along the interface $\Gamma$, $\mathbf{n}$ is the unit normal to the interface $\Gamma$, and $\kappa$ is the mean curvature of the interface (Weatherburn, 1927):

$$\kappa = -\nabla \cdot \mathbf{n}. \tag{3}$$

The first term in Eq. (2) is a force that acts normal to the interface and is proportional to the curvature $\kappa$ and the surface-tension coefficient $\sigma$. The second term in Eq. (2) is a force that acts tangent to the interface in the direction of larger values of $\sigma$. The normal force tends to smooth the interface by damping regions of high curvature, whereas the tangential force tends to drive fluid along the interface toward regions of larger $\sigma$. The surface-tension coefficient $\sigma$ can be dependent on the temperature and concentration of impurities (e.g., surfactants). We assume here that $\sigma$ is constant and therefore neglect the tangential force $\nabla_s\sigma$ in Eq. (2).

In many Eulerian methods for modeling flows with fluid interfaces, the presence or absence of a given fluid at a point **x** is represented by an auxiliary scalar function. This function is sometimes referred to as the color function, which represents an approximation to the characteristic function $\chi$ of one of the fluids:

$$\chi(\mathbf{x}) = \begin{cases} 1 & \text{if } \mathbf{x} \text{ lies in fluid 1} \\ 0 & \text{if } \mathbf{x} \text{ lies in fluid 2} \end{cases}. \tag{4}$$

In this chapter we take the color function to be the (discrete) volume fraction function, denoted by $f$ and defined as

$$f_i = \frac{\int_{\Omega_i} \chi(\mathbf{x}) \, d\Omega_i}{|\Omega_i|} = \text{volume of fluid 1 in the } i\text{th cell}, \tag{5}$$

where $\Omega_i$ denotes the $i$th cell and $|\Omega_i|$ denotes the volume of the $i$th cell. These volume fractions transition abruptly across the interface, causing problems with some CST formulations. Approximations to first- and second-order spatial derivatives of $f$ (required for estimating **n** and $\kappa$) can therefore be prone to noise and inaccuracies (Brackbill, Kothe, and Zemach, 1992).

A solution to this problem proposed in the original CSF (Continuum surface force) model (Brackbill, Kothe, and Zemach, 1992) (hereafter referred to as the BKZ model) is first to convolve $f$ with a smooth kernel **K** to construct a smoothed or mollified function $\tilde{f}$. Given $\tilde{f}$, **n** follows from

$$\mathbf{n} = \frac{\nabla \tilde{f}}{|\nabla \tilde{f}|}, \tag{6}$$

with the use of any standard form for the discrete gradient operator ($\nabla$). The curvature $\kappa$ is then computed with Eq. (3) with a finite-difference approximation to the discrete divergence operator.

In the BKZ model, the kernel **K** is a quadratic B-spline, but this choice for **K** is by no means unique. A variety of kernels have been introduced by others seeking more accurate approximations to **n** and $\kappa$. Some examples include a Gaussian kernel (Monaghan, 1992), cubic and quintic B-splines (Morris, 1997; Morton, 1997; Rudman, 1998), the Nordmark kernel (Aleinov and Puckett, 1995), and the Peskin kernel (Bussman, Mostaghimi, and Chandra, 1997; Bussman et al., 1998). Many of these same kernels have also been used to discretize the surface integral in Eq. (1), where they represent smooth, finite-width approximations to the Dirac delta function $\delta$. One kernel of note is the cosine kernel of Peskin (Peskin, 1997). This kernel and its subsequent modified versions are seen in a wide variety of methods, including both front-tracking (Juric and Tryggvason, 1997) and level-set (Sussman and Smereka, 1997) methods.

Section 23.2 begins with a brief review of the role of kernels in CST models. We then introduce a new kernel that has been specifically designed to improve the accuracy of estimates to the curvature $\kappa$. Next, in Section 23.3, we use this kernel in a combination of convolution and finite differences to construct a high-order accurate approximation to the curvature of a simple circular interface. Finally, in Section 23.4, we test several CST models on a spherical drop in static equilibrium (Brackbill, Kothe, and Zemach, 1992; Kothe et al., 1996), which is known to produce the so-called parasitic currents (Lafaurie et al., 1994) that can often compromise solution quality. In Section 23.5 we conclude with suggestions for future work.

## 23.2. Kernels in CST Models

### 23.2.1. *Continuum Surface-Tension Models*

In a CST model, a smooth color function results from convolving a discrete form of the characteristic function $\chi$ in Eq. (4) with a smooth kernel. This convolution smoothes the discontinuity in $\chi$. Here, we use the volume fraction function $f$ defined in Eq. (5) as the discrete $\chi$, although the signed distance function in a level-set method (e.g., Sussman and Smereka, 1997), the phase-field function in a phase-field method (e.g., Jacqmin, 1998), or the indicator function in a front-tracking method (e.g., Unverdi and Tryggvason, 1992) could just as easily replace $\chi$. A smooth color function, denoted by $\tilde{f}(\mathbf{x})$, is given by

$$\tilde{f}(\mathbf{x}) = \mathbf{K} * f(\mathbf{x}) = \int_{\Omega_K} f(\mathbf{x}')\mathbf{K}(\mathbf{x}' - \mathbf{x})\,d\mathbf{x}', \qquad (7)$$

where $\Omega_K$ denotes the support of the kernel $\mathbf{K}$ [i.e., those points $\mathbf{x}$ for which $\mathbf{K}(\mathbf{x}) \neq 0$], which is typically compact (i.e., of finite extent).

The effect of convolving the volume fraction function $f$ with a smooth kernel $\mathbf{K}$ is illustrated in the following two-dimensional example. Consider a circular drop centered in a $4 \times 4$ domain where the characteristic function $\chi$ is 1 inside the circle and 0 outside the circle. Since in this example the drop is static, $f$ is identical to $\chi$ in all cells that lie entirely inside or outside of the drop, i.e., $f_i = \chi(\mathbf{x}_i)$, where $\mathbf{x}_i$ denotes the center of the $i$th cell.

Figure 23.1(a) shows that near the interface the change in $f$ is too abrupt for standard finite-difference discretizations of $\mathbf{n}$ and $\kappa$ to be reliable. On the other hand, it is apparent from Fig. 23.1(b) that accurate approximations to $\mathbf{n}$ and $\kappa$ are possible by taking finite differences of $\tilde{f}$.

Consider now a few useful definitions and observations. For mollified functions $\tilde{f}$ we define the transition region $\tilde{\Gamma}$ to be those $\mathbf{x}$ where $\tilde{f}(\mathbf{x})$ lies between zero and one:

$$\tilde{\Gamma} = \{\mathbf{x} : 0 < \tilde{f}(\mathbf{x}) < 1\}. \qquad (8)$$

Note that if $\mathbf{K}$ has continuous partial derivatives of all orders $n \leq k$, then $\tilde{f}$ also has continuous partial derivatives of all orders $n \leq k$. In other words, if $\mathbf{K}$ can be differentiated $n$ times, then $\tilde{f} = \mathbf{K} * f$ can also be differentiated $n$ times, even if $f$ is a discontinuous function.

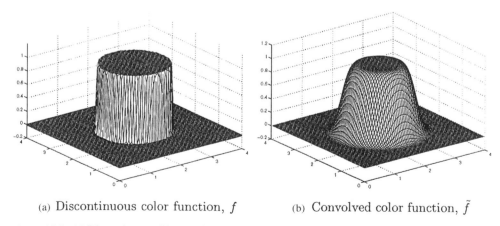

(a) Discontinuous color function, $f$       (b) Convolved color function, $\tilde{f}$

Figure 23.1. (a) Discontinuous, (b) convolved volume fractions for a unit circle on a $4 \times 4$ domain.

In practice, there are a variety of ways to approximate the integral in Eq. (7). We currently use the midpoint rule to perform the quadrature:

$$\mathbf{K} * f(\mathbf{x}) = \int_{\Omega_K} \mathbf{K}(\mathbf{x}' - \mathbf{x}) f(\mathbf{x}') \, d\mathbf{x}' \approx \sum_i \mathbf{K}(\mathbf{x}'_i - \mathbf{x}) f(\mathbf{x}'_i) \Delta \mathbf{x}'_i. \tag{9}$$

A useful property of Eq. (7) is that the derivatives of $\tilde{f} = \mathbf{K} * f$ can be evaluated by convolving the function $f$ with the derivatives of the kernel $\mathbf{K}$. For example, the first derivative of $\tilde{f} = \mathbf{K} * f$ with respect to $x$ is given by

$$\frac{\partial \tilde{f}(\mathbf{x})}{\partial x} = \frac{\partial (\mathbf{K} * f)}{\partial x}(\mathbf{x}) = \left( \frac{\partial \mathbf{K}}{\partial x} * f \right)(\mathbf{x}) = \int_{\Omega_K} \frac{\partial \mathbf{K}}{\partial x}(\mathbf{x}' - \mathbf{x}) f(\mathbf{x}') \, d\mathbf{x}'. \tag{10}$$

We use this property below to approximate the interface unit normal $\mathbf{n}$.

### 23.2.2. *Choosing a Kernel*

The choice of $\mathbf{K}$ is critical for the design of an accurate CST model. Many types of kernels have been used in the past, such as Gaussians, splines, and polynomials. Some of these kernels are radially symmetric while others are products of one-dimensional functions, i.e., $\mathbf{K}(\mathbf{x}) = q(x) \cdot q(y) \cdot q(z)$, where $q$ is an approximation to the one-dimensional Dirac delta function. We now present some guidelines that have proven useful in our kernel selection. We begin with some basic requirements, then introduce a new kernel that satisfies these requirements.

Recall that the portion of the domain over which the kernel $\mathbf{K}$ is nonzero is called the support of the kernel and is denoted by $\Omega_K$. We let $\epsilon$ represent the size of the support of the kernel. In particular, for a radially symmetric kernel, $\epsilon$ is the radius of the region over which $\mathbf{K}$ is nonzero. Since the kernel is also a function of $\epsilon$, we can write $\mathbf{K}(\mathbf{x}, \epsilon)$. The kernel $\mathbf{K}(\mathbf{x}, \epsilon)$ should

1. have compact support,
2. be monotonically decreasing with respect to $r = |\mathbf{x}|$,
3. be radially symmetric: $\mathbf{K}(\mathbf{x}) = \mathbf{K}(r)$,
4. be sufficiently smooth; i.e., for some $k \geq 3$, $\mathbf{K}$ should be $k$ times continuously differentiable: $\mathbf{K} \in C^k$,
5. have a normality property, i.e., $\int_{\Omega_K} \mathbf{K}(\mathbf{x}, \epsilon) \, d\mathbf{x} = 1$,
6. approach the Dirac delta function $\delta(\mathbf{x})$ in the limit $|\Omega_K| \to 0$.

Properties 1–3 are desirable, but not necessary. For example, a kernel with the entire domain $\Omega$ for its support (e.g., a Gaussian) is possible, but this violates the locality associated with surface-tension forces. Kernels that are nonmonotonic, i.e., lacking property 2, tend to develop highly singular oscillations as $|\Omega_K| \to 0$ (see Subsection, 23.2.3). In our experiences, kernels that are radially symmetric tend to produce more uniformly accurate results. The smoothness property allows the derivatives of $\tilde{f}$ to be computed by analytically differentiating $\mathbf{K}$, as shown in Eq. (10). Property 6 is a necessary condition for the numerical method to converge to the exact solution of the underlying problem as $h \to 0$, where $h$ is a characteristic cell length. A detailed investigation of the requirements that guarantee property 6 and whether or not this property is truly a necessary condition for the convergence of the numerical method is beyond the scope of this paper.

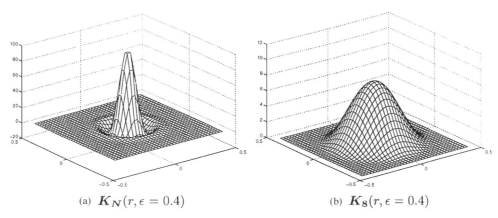

(a) $K_N(r, \epsilon = 0.4)$            (b) $K_8(r, \epsilon = 0.4)$

Figure 23.2. (a) Nordmark, (b) eighth-degree kernels on a unit domain.

Two example kernels are illustrated in Fig. 23.2, namely the Nordmark kernel $K_N$ (Nordmark, 1991) and a new eighth-degree polynomial kernel, denoted $K_8$, given by

$$K_8(r, \epsilon) = \begin{cases} A[1 - (r/\epsilon)^2]^4 & \text{if } r < \epsilon \\ 0 & \text{otherwise} \end{cases}, \tag{11}$$

where $A$ is a normalization constant that ensures $\int_{\Omega_K} K_8(r, \epsilon) \, dr = 1$. A plot of this kernel (for $\epsilon = 0.4$) is shown in Fig. 23.2(b). The kernel $K_8$ satisfies all of the six properties listed above, while the Nordmark kernel $K_N$ satisfies all but the monotonicity requirement, as is evident in Fig. 23.2(a). The simple form of $K_8$ results in a trivial and computationally efficient implementation, which is important since numerical approximation of the convolution in Eq. (7) can require many kernel evaluations.

### 23.2.3. *Avoiding Singularities in the Kernel Derivatives*

Figure 23.3 demonstrates another important aspect of kernels. As $|\Omega_K| \to 0$ or, equivalently, as $\epsilon \to 0$, the derivatives of the kernel become very large in magnitude, since they become singular in the limit $\epsilon \to 0$. Resolving these extrema in the kernel and its derivatives may require an excessive number of discrete points within the kernel's support. Thus, for any given kernel, it is important to understand and quantify the relationship between the limits $\epsilon \to 0$ and $h \to 0$. For example, in Fig. 23.3 it is apparent that the second derivative extrema are growing at a much faster rate than those for the first derivative (note the different plot scales). Growth of these singularities is even more severe for nonmonotonic kernels. Any numerical approximation to the convolution of second-order and higher derivatives, therefore, will require many discrete points within $|\mathbf{x}| < \epsilon$ to approximate to the convolution accurately. Table 23.1 provides evidence for this tendency, in which the error between the exact convolution integral of $\partial^2 K_8/\partial x^2$ with $f$ and $\partial^2 K_8/\partial y^2$ with $f$ and our numerical approximations to these quantities is quite large, even with many discrete points within $\epsilon$.

### 23.3. Determining the Interface Topology

It is difficult to obtain high-order accurate approximations to $\mathbf{n}$ and $\kappa$ when the interface is represented by a color function with a steep transition region. This is the case when the characteristic function associated with the interface is the discrete volume fraction function

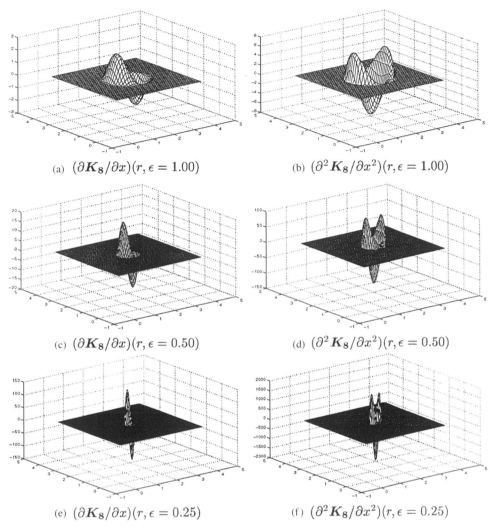

(a) $(\partial K_8/\partial x)(r, \epsilon = 1.00)$

(b) $(\partial^2 K_8/\partial x^2)(r, \epsilon = 1.00)$

(c) $(\partial K_8/\partial x)(r, \epsilon = 0.50)$

(d) $(\partial^2 K_8/\partial x^2)(r, \epsilon = 0.50)$

(e) $(\partial K_8/\partial x)(r, \epsilon = 0.25)$

(f) $(\partial^2 K_8/\partial x^2)(r, \epsilon = 0.25)$

Figure 23.3. First and second derivatives of the $K_8$ kernel for different values of the support radius $\epsilon$. A $4 \times 4$ domain is used with $\Delta x = \Delta y = \epsilon/10$.

$f$. As discussed above, the original BKZ approach used standard finite-difference approximations to the first and the second derivatives of the mollified volume fractions $\tilde{f}$. Although this algorithm is straightforward to implement (Kothe et al., 1996), it is difficult (and sometimes not possible) to demonstrate convergence under mesh refinement. We have therefore developed a hybrid method for estimating $\mathbf{n}$ and $\kappa$, which involves both convolution and standard finite-difference techniques. Similar techniques have recently been developed by others (Morris, 1997; Morton, 1997; Rudman, 1998; Bussman, Mostaghimi, and Chandra, 1997; Bussman et al., 1998).

In our hybrid model, high-order accurate approximations to $\kappa$ are possible with a centered, finite-difference approximation to the divergence in Eq. (3), computed at cell centers, while $\mathbf{n}$ is found by convolving $f$ with the first derivatives of kernel as shown in Eq. (10). The divergence in Eq. (3) can also be approximated by convolving $f$ with the second derivatives of the kernel as was done in (Aleinov and Puckett, 1995). However, as shown in Table 23.1, the

Table 23.1. *Discretization Errors in the Convolution Integral, Taken at the Point of Maximum Error in the Curvature Approximation along a Unit Circle* ($\epsilon = 0.2$)

| $\epsilon/\Delta x$ | Value of Convolution | $(\partial \mathbf{K}_8/\partial x) * f$ | $(\partial \mathbf{K}_8/\partial y) * f$ | $(\partial^2 \mathbf{K}_8/\partial x^2) * f$ | $(\partial^2 \mathbf{K}_8/\partial y^2) * f$ |
|---|---|---|---|---|---|
| 4 | Analytic | −4.75514 | −2.86529 | 16.5622 | 2.00181 |
|   | Numeric | −4.69157 | −2.80470 | 12.4195 | −0.97652 |
|   | % Error | 1.337 | 2.114 | 25.013 | 148.78 |
| 8 | Analytic | −1.96967 | −5.47108 | −5.16573 | −1.24848 |
|   | Numeric | −1.96178 | −5.44712 | −5.84872 | −1.83220 |
|   | % Error | 0.4006 | 0.4379 | 13.222 | 46.75 |
| 16 | Analytic | 4.03704 | 4.10382 | 2.29008 | 1.97206 |
|   | Numeric | 4.03189 | 4.10329 | 2.24012 | 2.52696 |
|   | % Error | 0.1276 | 0.0129 | 2.182 | 28.14 |

relative error associated with discretizing the convolution of the kernel second derivatives with $f$ can be an order of magnitude larger than the relative error associated with discretizing the convolution of the kernel first derivatives with $f$. Furthermore, at least for the case shown in Table 23.1, the support radius $\epsilon$ must be an order of magnitude greater than $\Delta x$ to reduce the relative error associated with the second derivatives of the kernel $\mathbf{K}$ to reasonable levels. For example, even with $\epsilon/\Delta x = 16$, the second derivative convolutions are not well resolved, while $\epsilon/\Delta x = 4$ provides a reasonably accurate approximation to the first derivative convolutions. This evidence suggests the need to pursue alternate strategies for computing the second derivatives of $\tilde{f}$. The hybrid approach we have developed has proven effective in predicting curvatures ranging over several orders of magnitude while only $\epsilon/\Delta x = 4$ is used.

In our hybrid method, the convolved interface normals are computed with the kernel $\mathbf{K}_8$ rather than the Nordmark kernel $\mathbf{K}_N$, as was proposed by Aleinov and Puckett (Aleinov and Puckett, 1995). This is principally because $\mathbf{K}_N$ is nonmonotonic and highly peaked, therefore requiring larger values of $\epsilon/\Delta x$ to resolve accurately the convolution of $\mathbf{K}_N$ and its derivatives with $f$. In addition to satisfying all of conditions 1–6 in Subsection 23.2.2, the kernel $\mathbf{K}_8$ also has a larger effective smoothing radius than $\mathbf{K}_N$, since it approaches zero as $r \to \epsilon$ from below more slowly than $\mathbf{K}_N$ (see Fig. 23.2).

Tables 23.2 and 23.3 show curvature errors that result from applying our hybrid method to a unit circle on a $4 \times 4$ domain. These results indicate that greater than second-order accuracy in the $L^\infty$ norm (i.e., pointwise) is possible. Furthermore, $L^\infty$ errors are reduced by more than an order of magnitude with $\mathbf{K}_8$ instead of $\mathbf{K}_N$.

## 23.4. A Hybrid Continuum Surface-Tension Model

### 23.4.1. *The Surface-Tension Force in a CST Model*

In this section we propose various discretizations for the CST force $\mathbf{F}_s$ in Eq. (1). This requires a discrete approximation to the integral along the interface $\Gamma$ that yields a function $\tilde{\mathbf{F}}_s$ varying smoothly through the transition region $\tilde{\Gamma}$. In the BKZ model (Brackbill, Kothe, and Zemach,

Table 23.2. *Unit Circle Curvature Errors when the Hybrid Method is used, in which Curvatures are Estimated by Discretizing Normals that are Convolved with* $\mathbf{K}_8(r, \epsilon = 0.4)$

| $\Delta x$ | $\epsilon/\Delta x$ | $L^\infty$ Errors($\times 10^{-2}$) | Convergence Rate |
|---|---|---|---|
| 0.05000 | 8 | 0.897640 | |
| 0.02500 | 16 | 0.160431 | 2.484 |
| 0.01250 | 32 | 0.021017 | 2.932 |
| 0.00625 | 64 | 0.003208 | 2.712 |

Table 23.3. *Unit Circle Curvature Errors when the Hybrid Method is used, in which Curvatures are Estimated by Discretizing Normals that are Convolved with* $\mathbf{K}_N(r, \epsilon = 0.4)$

| $\Delta x$ | $\epsilon/\Delta x$ | $L^\infty$ Errors($\times 10^{-1}$) | Convergence Rate |
|---|---|---|---|
| 0.05000 | 8 | 4.92363 | |
| 0.02500 | 16 | 0.83809 | 2.555 |
| 0.01250 | 32 | 0.06907 | 3.601 |
| 0.00625 | 64 | 0.00735 | 3.232 |

1992), the authors approximated $\mathbf{F}_s$ as

$$\mathbf{F}_s(\mathbf{x}) = \sigma\kappa(\mathbf{x})\nabla\tilde{f}(\mathbf{x}) = \sigma\kappa(\mathbf{x})\,\mathbf{n}(\mathbf{x})|\nabla\tilde{f}(\mathbf{x})|. \tag{12}$$

Consistent with the BKZ notation, $\mathbf{n}(\mathbf{x})$ and $\kappa(\mathbf{x})$ denote approximations to the unit normal and curvature of the interface that have been extended in some way to points $\mathbf{x} \in \tilde{\Gamma}$. The authors denote these terms as $\tilde{\mathbf{n}}$ and $\tilde{\kappa}$, respectively, since our tilde quantities typically refer to functions that have a support of $\tilde{\Gamma}$. The force in Eq. (12) can also be expressed as

$$\mathbf{F}_s = \sigma\kappa(\mathbf{x})\mathbf{n}(\mathbf{x})\tilde{\delta}_\Gamma = \mathbf{f}_s\tilde{\delta}_\Gamma, \tag{13}$$

where $\mathbf{f}_s$ is given by Eq. (2) with $\sigma$ constant and $\tilde{\delta}_\Gamma = |\nabla\tilde{f}|$ represents a discrete (one-dimensional) Dirac delta function $\delta_\Gamma$ for the interface $\Gamma$. The function $\delta_\Gamma$ is constructed by differentiating the characteristic function $\chi$ along paths normal to the interface.

For a CST model, the force in Eq. (13) can be expressed in general form as

$$\tilde{\mathbf{F}}_s(\mathbf{x}) = \sigma\tilde{\kappa}(\mathbf{x})\tilde{\mathbf{n}}(\mathbf{x})\mathcal{I}_{\tilde{\Gamma}}(\mathbf{x}), \tag{14}$$

where $\mathcal{I}_{\tilde{\Gamma}}$ is an interface indicator function and $\tilde{\mathbf{n}}(\mathbf{x})$ and $\tilde{\kappa}(\mathbf{x})$ are numerical approximations to $\mathbf{n}$ and $\kappa$ on $\Gamma$ and their extension to $\tilde{\Gamma}$. As with $\tilde{\mathbf{n}}$ and $\tilde{\kappa}$, $\mathcal{I}_{\tilde{\Gamma}}$ typically has its support coincident with $\tilde{\Gamma}$. If the form for $\mathcal{I}_{\tilde{\Gamma}}$ exhibits some density dependence, then $\mathbf{F}_s$ becomes a body force, as discussed in (Brackbill, Kothe, and Zemach, 1992). Variants of CST models can be devised if different $\mathcal{I}_{\tilde{\Gamma}}$ are chosen and different techniques are used to extend $\mathbf{n}$ and $\kappa$ to $\tilde{\mathbf{n}}(\mathbf{x})$ and $\tilde{\kappa}(\mathbf{x})$, respectively. We now discuss three ways of determining $\tilde{\kappa}$ and their combinations with various $\mathcal{I}_{\tilde{\Gamma}}$ to approximate the surface-tension force throughout $\tilde{\Gamma}$.

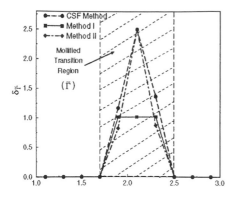

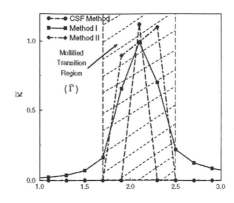

(a) Interface indicator function, $I_{\widetilde{\Gamma}}$.          (b) Smoothed curvatures $\tilde{\kappa}$.

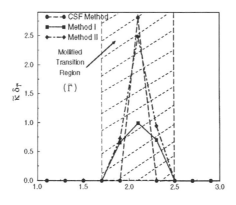

(c) The product $\tilde{\kappa}\,\delta_{\widetilde{\Gamma}}$.

Figure 23.4. Three models for the variation of the interface indicator function $I_{\tilde{\Gamma}}$, smoothed curvature $\tilde{\kappa}$, and their products $\delta_{\tilde{\Gamma}}\tilde{\kappa}$ through the mollified interface transition region $\tilde{\Gamma}$.

**The BKZ Method.** For $\tilde{\mathbf{F}}_s$ in Eq. (14), the original BKZ method used

$$\tilde{\kappa} = -\nabla \cdot \tilde{\mathbf{n}},$$
$$\mathcal{I}_{\tilde{\Gamma}} = |\nabla \tilde{f}|, \tag{15}$$

where $\tilde{\mathbf{n}}$ results from a finite-difference approximation to the right-hand side of Eq. (6) and $\tilde{\kappa}$ follows from a finite-difference approximation to the divergence of $\tilde{\mathbf{n}}$. The discretization stencil used to determine $\mathcal{I}_{\tilde{\Gamma}} = |\nabla \tilde{f}|$ results in a smooth function centered in the transition region, as seen in Fig. 23.4(a).

**Method I.** The curvature $\kappa$ on $\Gamma$ can be extended over any portion of the domain by convolving $\kappa$ with some kernel to determine $\tilde{\kappa}$ (Rider et al., 1998):

$$\tilde{\kappa}(\mathbf{x}) = \int_{\Omega_K} \kappa(\mathbf{y}) \mathbf{K}_{\kappa}(\mathbf{x} - \mathbf{y}) \, d\mathbf{y},$$
$$\mathbf{K}_{\kappa}(\mathbf{x}) = 1/10\Delta x r^4. \tag{16}$$

Here, a simple step function is used for the indicator function:

$$\mathcal{I}_{\tilde{\Gamma}}(\mathbf{x}) = \begin{cases} 1 & \text{if } \mathbf{x} \in \tilde{\Gamma} \\ 0 & \text{otherwise} \end{cases}. \tag{17}$$

The curvatures $\kappa$ are first found only on the interface $\Gamma$ by use of the hybrid method described in Section 23.3. Here the interface resides in those cells for which the unmollified volume fraction function $f$ satisfies $0 < f_i < 1$. We then extend $\kappa$ off the interface by convolving $\kappa$ with the kernel $\mathbf{K}_{\kappa}$ above to obtain curvatures $\tilde{\kappa}(\mathbf{x})$ defined throughout the transition region $\tilde{\Gamma}$. This kernel $\mathbf{K}_{\kappa}$ is not the same as $\mathbf{K}_8$, but rather is a singular kernel having infinite support. This convolution yields a function $\tilde{\kappa}$ that varies smoothly from near zero at the edges of $\tilde{\Gamma}$ to a value equal to the original value of $\kappa$, as shown in Fig. 23.4(b). The indicator function $\mathcal{I}_{\tilde{\Gamma}}$ is simply the characteristic function associated with the transition region; it is equal to unity inside $\tilde{\Gamma}$ and zero elsewhere, as shown in Fig. 23.4(a).

**Method II.** Another approach is to construct $\tilde{\kappa}$ by applying our hybrid method everywhere within $\tilde{\Gamma}$:

$$\tilde{\kappa} = -\nabla \cdot \tilde{\mathbf{n}}. \tag{18}$$

In this method we have chosen to use a parabolic indicator function,

$$\mathcal{I}_{\tilde{\Gamma}}(\mathbf{x}) = 4\tilde{f}(1 - \tilde{f}). \tag{19}$$

When the hybrid method is used to estimate $\tilde{\kappa}$, the radius of support $\epsilon$ of the kernel $\mathbf{K}$ should extend several cells beyond those inside of $\tilde{\Gamma}$. For this method, $\tilde{\kappa}$ correctly exhibits the inverse radius dependence expected for a spherical interface, as seen in Fig. 23.4(b). The indicator function $\mathcal{I}_{\tilde{\Gamma}}$ varies quadratically with $\tilde{f}$, attaining a maximum value at $\Gamma$, as shown in Fig. 23.4(a).

For any CST model, our experience has been that one should use caution when constructing the force $\tilde{\mathbf{F}}_s$, since too much smoothing can violate the locality of surface tension and lead to erroneous surface pressures induced by the force $\tilde{\mathbf{F}}_s$.

Plots of the interface indicator function, $\mathcal{I}_{\tilde{\Gamma}}$, for each of the three methods are shown in Fig. 23.4(a). The extended curvatures $\tilde{\kappa}$ are shown in Fig. 23.4(b). Given the forms chosen for the indicator function $\mathcal{I}_{\tilde{\Gamma}}$ and curvature $\tilde{\kappa}$, the product $\tilde{\kappa}\mathcal{I}_{\tilde{\Gamma}}$ for each of the three methods is shown in Fig. 23.4(c).

### 23.4.2. *A Spherical Drop In Static Equilibrium*

Modeling a spherical drop in static equilibrium serves as an ideal test for comparing different CST formulations. The net surface force should be zero, since at each point on the drop surface the tension force is counteracted by an equal and opposite force at a diametrically opposed point. The correct solution is a zero velocity field and a pressure field that rises from a constant value of $p_0$ outside the drop to a value of $p_0 + 2\sigma/R$ inside the drop. The degree to which a CST model maintains a zero velocity field is a good measure of its validity, similar to the notion of a numerical method being free-stream preserving.

The computational domain, $8^3$ in size and partitioned with $40^3$ cells, is occupied by two inviscid, incompressible fluids: a drop fluid and a background fluid. The drop fluid resides inside a spherical drop of radius $R = 2$ centered in the domain. The background fluid, having one-tenth the drop density, occupies the remainder of the domain. For all simulations a value of

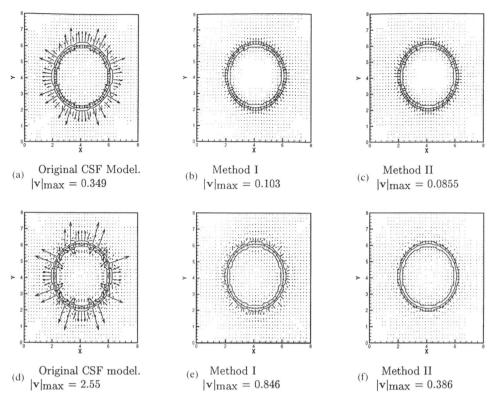

(a)    Original CSF Model.
$|\mathbf{v}|\text{max} = 0.349$

(b)    Method I
$|\mathbf{v}|\text{max} = 0.103$

(c)    Method II
$|\mathbf{v}|\text{max} = 0.0855$

(d)    Original CSF model.
$|\mathbf{v}|\text{max} = 2.55$

(e)    Method I
$|\mathbf{v}|\text{max} = 0.846$

(f)    Method II
$|\mathbf{v}|\text{max} = 0.386$

Figure 23.5. Velocity distributions in an $xy$ plane through the drop center. Plots (a)–(c) are after one time step, $t = 0.001$, while plots (d)–(f) are after 50 time steps, $t \approx 0.05$.

$\sigma = 73.0$ and a time step $\delta t \approx 0.001$ are used. In Fig. 23.5 we present computed velocity fields for the three different CST models described in Subsection 23.4.1. The velocities are shown for the $xy$ plane passing through the drop center at 1 and 50 time steps.

It is evident from Fig. 23.5 that both methods I and II reduce the magnitude of the parasitic currents relative to the BKZ model. In this example, the reductions for methods I and II are a factor of 3 to 5. The increased accuracy in our approximation to the curvature as well as a more smoothly varying force $\tilde{\mathbf{F}}_s$ is largely responsible for these improved results. Additional measures of the accuracy, such as the error in approximating the pressure jump across the interface and the growth of kinetic energy of the system, will guide further CST model development.

## 23.5. Conclusions and Summary

Despite the widespread use of CST methods to model surface-tension effects along topologically complex interfaces, inaccurate approximations to interface topology are commonplace. However, high-order accurate approximations are possible by use of kernels that satisfy our proposed design requirements. We have introduced a new polynomial kernel $\mathbf{K}_8$ that satisfies these requirements and is easy to implement. A hybrid technique, which combines convolutions with standard finite differences, is capable of approximating high-order accurate curvatures for circular and spherical interfaces. These techniques can be also be used to approximate surface-tension forces. Improved estimates of the topology and force distribution help to reduce the parasitic currents induced by computations of an equilibrium static drop.

However, much algorithm development is needed. In particular, approximations to the surface-tension force that results from quadratures of the interface integral in Eq. (1) will be investigated. Reduced-pressure forms (Sussman and Smereka, 1997; Rider et al., 1998), which result from energy considerations in which forces are estimated from the spatial gradients of surface free energies (Jacqmin, 1998), will also be studied. Our hybrid discretization technique will be tested on more realistic interfaces having curvatures that vary rapidly along the interface. Finally, a systematic assessment of the impact of grid shape and mesh/interface orientation on the accuracy of these models will be conducted.

## References

Aleinov, I. and Puckett, E. G. Computing surface tension with high-order kernels. In Dwyer, H. A. ed., *Proceedings of the Sixth International Symposium on Computational Fluid Dynamics*, pp. 13–18, Lake Tahoe, NV, 1995.

Bussman, M., Aziz, S. D., Mostaghimi, J., and Chandra, S. Modeling the fingering of impacting droplets. Technical report, University of Toronto, Toronto, Canada, 1998. Submitted to *Phys. Fluids*.

Brackbill, J. U., Kothe, D. B., and Zemach, C. A continuum method for modeling surface tension. *J. Comput. Phys.* **100**:335–354, 1992.

Bussman, M., Mostaghimi, J., and Chandra, S. On a three-dimensional volume tracking model of droplet impact. Technical report, University of Toronto, Toronto, Canada, 1997. Submitted to *Phys. Fluids*.

Jacqmin, D. A variational approach to deriving smeared interface surface tension models. In Venkatakrishnan, V., Salas, M. D., and Chakravarthy, S. R. eds., *Workshop on Barriers and Challenges in Computational Fluid Dynamics*, pp. 231–240, Boston, MA, 1998, Kluwer Academic.

Juric, D. and Tryggvason, G. Computations of boiling flows. Technical report LA-UR-97-1145, Los Alamos National Laboratory, 1997. Accepted for publication in the *Int. J. Multiphase Flow*.

Kothe, D. B., Rider, W. J., Mosso, S. J., Brock, J. S., and Hochstein, J. I. Volume tracking of interfaces having surface tension in two and three dimensions. Technical report AIAA 96-0859, AIAA, 1996. Presented at the 34rd Aerospace Sciences Meeting and Exhibit. Available at http://www.lanl.gov/home/Telluride/Text/publications.html.

Lafaurie, B. Nardone, C., Scardovelli, R., Zaleski, S., and Zanetti, G. Modelling merging and fragmentation in multiphase flows with SURFER. *J. Comput. Phys.* **113**:134–147, 1994.

Monaghan, J. J. Smoothed particle hydrodynamics. *Ann. Rev. Astron. Astrophys.* **30**:543–574, 1992.

Morris, J. P. Simulating surface tension with smoothed particle hydrodynamics. *Int. J. Num. Methods Fluids*, 1997. Submitted.

Morton, D. E. Numerical Simulation of an Impacting Drop. Ph.D. thesis, The University of Melbourne, Australia, 1997.

Nordmark, H. O. Rezoning for higher order vortex methods. *J. Comput. Phys.* **97**:366–397, 1991.

Peskin, C. S. Numerical analysis of blood flow in the heart. *J. Comput. Phys.* **25**:220–252, 1977.

Rider, W. J., Kothe, D. B., Puckett, E. G., and Aleinov, I. D. Accurate and robust methods for variable density incompressible flows with discontinuities. In Venkatakrishnan, V., Salas, M. D., and Chakravarthy, S. R. eds., *Workshop on Barriers and Challenges in Computational Fluid Dynamics*, pp. 213–230, Boston, MA, 1998, Kluwer Academic. Available at http://www.lanl.gov/home/Telluride/Text/publications.html.

Rudman, M. A volume-tracking method for incompressible multifluid flows with large density variations. *Int. J. Num. Methods Fluids*, 1998. In press.

Sussman, M. and Smereka, P. Axisymmetric free boundary problems. *J. Fluid Mech.* **341**:269–294, 1997.

Unverdi, S. O. and Tryggvason, G. A front-tracking method for viscous, incompressible, multifluid flows. *J. Comput. Phys.* **100**:25–37, 1992.

Weatherburn, C. E. On differential invariants in geometry of surfaces, with some applications to mathematical physics. *Q. J. Math.* **50**:230–269, 1927.

# 24

# Interaction of Convection and Solidification at Fluid–Solid Interfaces

L. BÜHLER, P. EHRHARD, AND U. MÜLLER

## 24.1. Introduction

Phase change from the liquid to the solid state and the reverse can be commonly observed in nature, for example in the freezing of water in rivers and lakes and during volcanic activities of the Earth in terms of lava flows exposed to heat losses. There are many technical processes in metallurgy, such as casting, welding of materials, and crystal growth from melts, in which the control of the liquid–solid phase change is of utmost importance for the quality of the fabricated product. For the safety of nuclear power plants severe core melt accidents are being analyzed and designed devices are being tested. There is the requirement to collect core melts and control them at any time by adequate cooling systems within the reactor containment. The spreading of the core melt under the effect of partial solidification or even melt attack of structural material is currently the subject of extensive technological investigations.

In a single-component material a phase transition from liquid to solid is generally associated with a heat removal from the liquid interface. The heat flows through the solid to a heat sink by conduction only. In the liquid the heat transport to the interface can be significantly enhanced by convection. Thus the solidification process at the interface is controlled by heat conduction in the solid and by conduction and convection in the liquid. The convection in the liquid phase may be induced by external forcing, volume shrink by solidification, and/or buoyancy. The release of latent heat will change the temperature gradient, and segregation in multicomponent alloys will change the concentration near the interface and thus influence thermal and solutal convection. In any case, convection will alter the thermal and solutal properties near the interface with strong implications for the solidification process.

Solidification problems in engineering and applied science can roughly be divided into closed-domain and open-domain problems. Simple analytical or semianalytical solutions have been obtained only for the one-dimensional freezing or melting process in a semispace. The principal features of conductive and convective heat transport in the solidification process are briefly outlined by comparison of the freezing of a semi-infinite space filled by liquid at rest and at intensive mixing motion, respectively. The heat is extracted at the horizontal upper plane. The problem is sketched in Fig. 24.1.

If all liquid is at solidification temperature $T_m$ the solidification is controlled only by conduction in the solid phase and there is no convection. The following simplified initial boundary-value problems have to be solved:

$$\partial_t T = \kappa \partial_{zz} T, \qquad \text{in } 0 < z < Z(t); \tag{1a}$$

$$T(z, t) = T_B, \qquad z = 0, \qquad t \geq 0; \tag{1b}$$

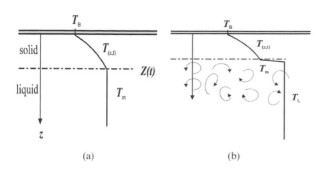

Figure 24.1. Principle sketch of freezing a liquid layer from above a liquid layer (a) at rest, (b) intensively mixed, e.g., buoyancy driven.

$$T(z, t) = T_m = T_L, \qquad z = Z(t); \tag{1c}$$

$$\rho L d_t Z = k \partial_z T, \qquad z = Z(t); \tag{1d}$$

where $T$ is the temperature in the solid, $\kappa$ and $k$ are the thermal diffusivity and the heat conductivity, respectively, of the solid, $Z(t)$ is the position of the interface, $L$ is the latent heat of fusion, $T_B$ is the temperature of the cooling plate, $T_m$ is the solidification temperature of the single-component liquid, and $T_L$ is the temperature of the bulk liquid. There is a self-similar solution to this problem in the form $T(z, t) = T_B + (T_m - T_B)\,\mathrm{erf}\,[z/(2\sqrt{\kappa t})]/\mathrm{erf}\,\lambda$. The propagation of the interface is given by the relationship $Z(t) = 2\lambda(\kappa t)^{1/2}$, where $\lambda$ is the solution of the transcendental algebraic equation $\pi^{1/2}\lambda \exp(\lambda^2) \cdot \mathrm{erf}(\lambda) = S^{-1}$, with $S = L/[c_P(T_m - T_B)]$, the dimensionless Stefan number (Carslaw & Jaeger, 1959). Obviously the melt front propagates for all times as $Z(t) \sim \sqrt{\kappa t}$. The described problem is commonly denoted as the Stefan problem.

The problem has to be modified if there is an imposed heat flux $q_L$ across the interface from the liquid side to the solid side. It is assumed – as sketched in Fig. 24.1(b) – that this heat flux is caused by a laminar or turbulent convective heat transfer from the fluid to the interface. When Newton's cooling law is used, this heat flux can be given by the general form $q_L = \alpha(T_L - T_m)^\gamma$, where $\gamma$ depends on the mode of heat transfer, which may be $\gamma = 5/4$ for laminar and $\gamma = 4/3$ for turbulent natural convection (Worster, 1996).

The problem can then be formulated as

$$\partial_t T = \kappa \partial_{zz} T, \qquad \text{in } 0 < z < Z(t); \tag{2a}$$

$$T(z, t) = T_B, \qquad z = 0, \quad t \geq 0; \tag{2b}$$

$$T(z, t) = T_m, \qquad z = Z(t); \tag{2c}$$

$$k \partial_z T = [\rho L + \rho c_p (T_L - T_m)]\, d_t Z(t) + q_L, \qquad z = Z(t); \tag{2d}$$

$$\rho c_p [H - Z(t)]\, d_t T_L = -q_L, \qquad z > Z(t). \tag{2e}$$

Here $H$ is the liquid bulk extension. Equations (2d) and (2f) describe the heat balance at the interface and for the liquid bulk. It can be shown that for very short times and $Z(t)/H \ll 1$ the problem described by Eqs. (2) reduces to the purely conduction-controlled solidification process similar to that described by Eqs. (1). While in Eqs. (1) the conductive heat flux removes the latent heat only, now in addition the sensible heat $\rho c_p (T_L - T_m)$ has to be removed. This leads to a slight modification in the equation determining the value of $\lambda$. Instead of $S$ on the right-hand side one should use now $S + (T_L - T_m)/(T_m - T_B)$. Since more heat is released in total, the initial speed of the interface is smaller than in the pure Stefan problem. In an intermediate time interval the progression of the melt front is governed by a balance of the convective heat transfer in the liquid and a conductive transfer on the solid side. The latent heat

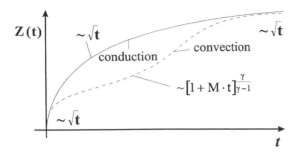

Figure 24.2. Melt front propagation: ——— liquid at rest, - - - - convecting liquid, $M = (\gamma - 1)q_0/[\rho c_p(T_{L0} - T_m)]$.

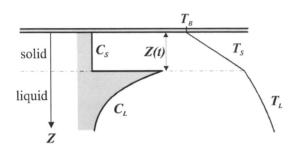

Figure 24.3. Principal sketch of the concentration and the temperature for one-dimensional solidification fronts in binary mixtures.

has a minor influence. The melt front propagation is then

$$Z \sim \left[1 + (\gamma - 1)\frac{q_0}{\rho c_p(T_{L0} - T_m)}t\right]^{\gamma/(\gamma-1)},$$

where the subscript 0 denotes initial conditions at an intermediate time scale. During the intermediate state the melt loses its temperature until $T_L = T_m$ and one recovers the original Stefan problem.

A qualitative solution of the melt front propagation is depicted in Fig. 24.2. If the liquid consists of more then one component, say two, the solidification temperature depends on the concentration $c$ of the solute, i.e., $T_m = T_m(c)$. Generally the equilibrium concentration in the solid $c_S$ is different from that in the liquid $c_L$, and a segregation of the solute is observed at the solidification front. There are good reasons to argue that mass diffusion occurs in the liquid phase only. The solidification problem defined by Eqs. (1) has to be modified by the addition of a mass diffusion equation in the liquid and a mass conservation relation at the interface in the forms

$$\partial_t c_L = D\partial_{zz}c_L, \qquad \text{in } 0 < z < Z(t), \tag{3a}$$

$$(c_L - c_S)d_t Z = +D\partial_z c_L \qquad \text{at } z = Z(t), \tag{3b}$$

where $D$ is the mass diffusion coefficient in the liquid. For liquids in general $D \ll \kappa$ holds, that is, the concentration boundary layers are much thinner than the temperature boundary layers in the liquid. Since the temperature and concentration equations are decoupled it can be shown that the propagation of the solidification front is mainly controlled by mass diffusion. A self-similar solution gives $Z(t) \sim \sqrt{Dt}$. A sketch of the situation is given in Fig. 24.3.

Naturally the concentration in the liquid phase may be modified by enhanced convective mixing in the liquid. However, since the mass diffusion boundary layer is much thinner than any thermal or viscous boundary layer, there is predominantly mass diffusion control.

If the propagation of the solidification front is forced by an external speed, for example, by a moving fixed-temperature gradient – known as directional solidification – planar interfaces may become corrugated because of morphological instabilities. Flow-induced convective mass transport may affect these instabilities (Bühler and Davis, 1998; see section 3).

## 24.2. Bénard Convection in Liquid Layers Solidifying from Above

### 24.2.1. *The Single-Component System*

If a horizontal layer is heated from below and cooled from above so that the enclosed single-component liquid is frozen in the upper part, the heat may be transferred in the remaining liquid layer by conduction only or by conduction and convection, depending on whether the temperature gradient in the liquid zone is subcritical or supercritical. For a subcritical conductive situation the liquid–solid interface is flat; for supercritical conditions it becomes corrugated under the influence of the convection pattern in the liquid layer. The situation is sketched in Fig. 24.4. In Bénard convection generally the flow occurs in cellular form. Roll patterns are observed for perfectly conducting rigid boundaries up to Rayleigh numbers $\mathrm{Ra} \sim 2 \times 10^4$ and very high Prandtl numbers (Busse, 1967). The Rayleigh number is defined as $\mathrm{Ra} = \beta g \Delta T_L h_L^3 / (\kappa v)$, where $\beta$ is the coefficient of thermal expansion of the liquid, $g$ is the acceleration of gravity, $\Delta T_L$ is the temperature difference across the liquid layer, $h_L$ is the height of the liquid layer, $\kappa$ is the thermal diffusion coefficient, and $v$ is the kinematic viscosity. The Prandtl number is defined as $\mathrm{Pr} = v/\kappa$. The question is: In what way does the convection pattern interact with the solid–liquid interface? Experiments have been conducted to clarify this. The tests were performed in two geometries, an extended slab container of aspect ratio height:length:depth $= 1:60:60$ and a slender channel-type cavity of aspect ratio height:length:depth $= 1:20:2$. A transparent liquid, cyclohexane, was used as the test fluid. Its physical properties show normal behavior near the freezing point and their thermal variations are negligible. For identifying the corrugations in the frozen layer of cyclohexane, the upper plate was removed when a steady state had been reached for a particular temperature difference across the layer and photographs of the deformed ice surface were taken. Figure 24.5 shows three typical photographs. Depending on the mean temperature in the test section, thin and thick frozen layers of cyclohexane can be established at the upper boundary. If the ice layer is very thin, line patterns are observed, as shown in Fig. 24.5(a), indicating roll-like convection patterns in the liquid layer underneath the ice. A more or less regular hexagonal pattern is observed for ice layers whose thicknesses are comparable with the depth of the liquid layer. This is shown in Fig. 24.5(c). In an intermediate range both polygonal and line patterns coexist, as displayed by Fig. 24.5(b). The range of occurrence of the different ice patterns – to be considered as a print of the convective pattern into the frozen

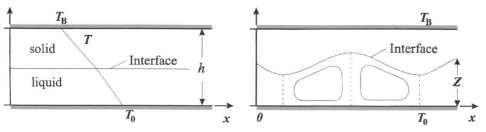

Figure 24.4. Schematic drawing of a partially solidified liquid layer.

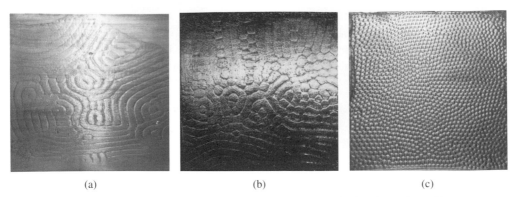

Figure 24.5. Corrugated solid–liquid interfaces: (a) roll pattern, (b) mixed polygonal – roll pattern, (c) hexagonal pattern.

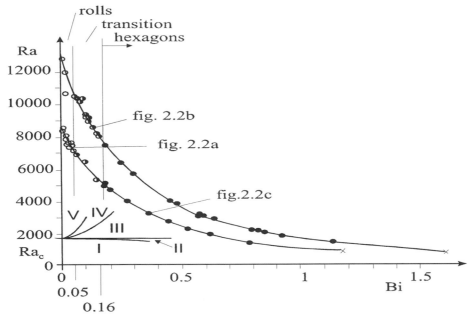

Figure 24.6. Convection pattern regime chart in the Ra–Bi plane. Theory: Labels I–V refer to different stable convective states according to theory: I, conductive regime; II, subcritical hexagons; III, supercritical hexagons; IV, mixed-mode regime; V, rolls. Experimental findings: o, d rolls; o, rolls and polygons; •, hexagons; ×, state at rest.

cyclohexane – has been determined experimentally. It turns out that a dimensionless parameter, the Biot number Bi, characterizing the thermal conductivities of the liquid and solid phase and depending on the ratio of the liquid and the solid layer heights, fixes the ranges where different patterns exist.

The deflection of the liquid–solid interface can lead to a sufficient degree of vertical asymmetry so that hexagonal convection can be generated (Davis and Segel, 1968). The experimental observations indicate that this is the case for large Biot numbers and thick ice. For small Biot numbers and thin ice, line patterns are observed and roll convection exists. By using a weakly nonlinear analysis Davis et al. (1984) have theoretically explained the observations. The experimental and theoretical results are summarized in a pattern regime chart in Fig. 24.6.

Although the experiments were performed partly at highly supercritical Rayleigh numbers while the theory was limited to slightly supercritical conditions, the qualitative agreement of the dependence of the different patterns on the Biot number is striking.

The experimental observations of the interfacial corrugation in the channel cavity fit into the overall picture of patterns observed in the above-discussed slab layer. For height ratios $h_S/h_L < 1$ a regular line pattern with troughs and crests aligned with the shorter side of the cavity are seen, indicating regular convection rolls of a fixed wavelength with axes parallel to the shorter side. If the temperature gradient across the liquid layer is several times the critical value, bimodal corrugations of the liquid–solid interface occur. If the solidified portion of the layer is thick, i.e., $h_S/h_L \gg 1$, more or less regular three-dimensional interfacial deformations of polygonal shape are formed, which, once established, become stationary. The crests and troughs are ordered by two distinct wavelengths. The typical interfacial patterns are shown in the photographs of Fig. 24.7.

A graphical interpretation of the photographs is shown in Fig. 24.8.

There is a plausible explanation for the bimodal patterns. At higher supercritical Rayleigh numbers two counterrotating secondary line vortices form in the vicinity of flow separation

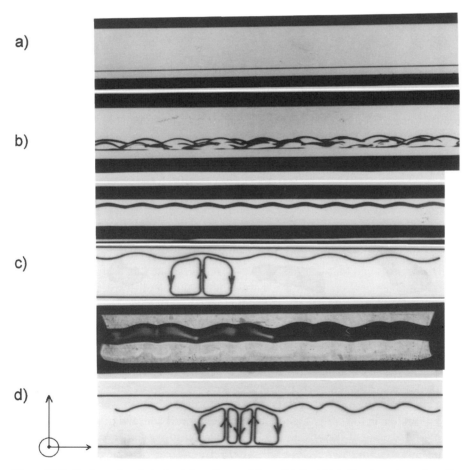

Figure 24.7. Photographs of steady-state patterns of the solid–liquid interface in a channel cavity, (a) near-critical conditions, $\mathrm{Ra} \sim 1450$, flat interface $\mathrm{Ra} < \mathrm{Ra}_c$, $h_S/h_L < 1$; (b) three-dimensional polygonal form, $\mathrm{Ra} < \mathrm{Ra}_c$, $h_S/h_L > 1$; (c) two-dimensional single-mode form, $\mathrm{Ra} < \mathrm{Ra}_c$, $h_S/h_L < 1$; (d) two-dimensional bimodal form, $\mathrm{Ra} \gg \mathrm{Ra}_c$, $h_S/h_L < 1$.

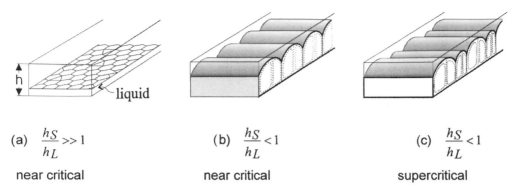

(a) $\dfrac{h_S}{h_L} \gg 1$                    (b) $\dfrac{h_S}{h_L} < 1$                    (c) $\dfrac{h_S}{h_L} < 1$

near critical                            near critical                            supercritical

Figure 24.8. Schematic drawing of the interfacial corrugation patterns in the channel cavity.

lines at rigid boundaries. If the boundary is compliant, as in the case of the ice layer, the two smaller vortices imprint their length scale into the ice layer and simultaneously extend in the vertical direction to the opposite rigid boundary and then become arrested between two larger original convection rolls. This transient process has been observed by use of interferometry.

The onset and suppression of convection in the liquid layer beneath a thick ice layer [see Fig. 24.8(b)] shows hysteretic behavior, as can be expected under symmetry breaking conditions.

We consider next the onset of convection at a slightly supercritical temperature difference across the liquid layer. The liquid height increases in a jump transition to approximately double its value while simultaneously a polygonal corrugation pattern forms. The reverse transition to a flat interface and no convection in the liquid phase occurs at a significantly smaller overall temperature difference across the layer. The melt front progression and its reverse, the retreat of the solidification front, exhibit under the influence of convection the features displayed in the graph of Fig. 24.2. A measured transient of the mean liquid layer height from the state of conduction to three-dimensional convection is seen in Fig. 24.9. More details of the experimental investigations have been reported by Dietsche and Müller (1985).

### 24.2.2. The Two-Component System with Thermal Mass Diffusion

Binary mixtures subject to temperature gradients can generate concentration gradients as well. The Soret effect (DeGroot and Mazur, 1969) derives from the solute flux as a function of both gradients. An imposed temperature gradient can thus induce double-diffusive convection. The thermally induced solute flux may either support or diminish thermal buoyancy. In the latter case the opposing gradients may generate oscillatory convection. In the Bénard case with a desta-bilizing temperature gradient a stabilizing thermal mass diffusion effect occurs if the lighter component diffuses toward colder regions. In this case the thermal mass diffusion coefficient (Soret coefficient) is negative and the convection starts as a traveling roll pattern. Experiments have been performed with ethanol–water mixtures in a cavity of aspect ratio 3:20:200, heated at its lower side and cooled from above such that ice may form at the upper wall. (Zimmermann et al., 1992). With no ice in the cavity a permanent traveling roll convection pattern has been identified by interferograms. If there is an ice layer attached to the upper wall, the onset of convection also starts in form of a traveling roll pattern. But instead of achieving a permanent state of traveling waves, the convection undergoes a transition to a steady three-dimensional convection with corresponding steady quasi-polygonal corrugations of the interface. The ob-served critical values for the onset of traveling roll convection beneath an ice layer, the Rayleigh

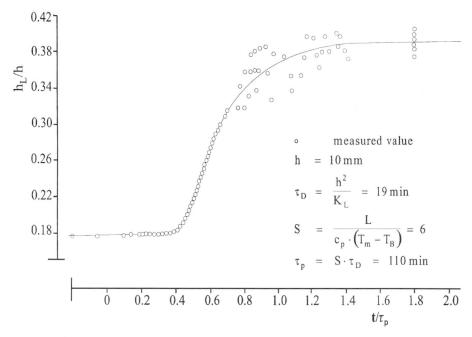

Figure 24.9. Transient development of the layer height after onset of three-dimensional convection in a partly solidified layer.

number, wave number, and the frequency, have been found to be in good agreement with predictions by a linear stability theory (Karcher and Müller, 1994). However, the dependency of the Soret coefficient and of the thermal and the solutal expansion coefficient on the concentration of the water–ethanol mixtures has to be taken into account.

## 24.3. Directional Solidification of Dilute Binary Alloys with Long-Scale External Flow

In directional solidification of dilute binary alloys a planar interface is pulled through a fixed-temperature gradient. Solute rejection and diffusion in a motionless melt build up a solute profile at the front that leads to morphological instabilities with wavelength $2\pi/\beta$ if the morphological number $M$ exceeds a critical value. $M$ represents the ratio of the concentration and temperature gradients near a planar interface.

We consider now a solid–liquid interface in the presence of a steady, two-dimensional cellular flow pattern in the melt. Such flows are typically created by thermal or solutal convection in unstably stratified liquid layers. If the wavelength $2\pi/\alpha$ of the flow is much larger than $2\pi/\beta$, it can be shown asymptotically (Bühler and Davis, 1998) that the front feels a shear imposed by a quasi-parallel remote flow, periodic along the interface. If the wavelength of the flow is large compared with the thickness of the concentration layer, the component of the flow normal to the interface is negligible at leading order.

The concentration $c$ of the solute in the melt, in a frame moving with the front, is governed by the nondimensional convection diffusion equation,

$$\partial_t c + [-P(\eta)f(z)\hat{x} - \hat{z}] \cdot \nabla c = \nabla^2 c,$$

where the remote flow has an amplitude $P(\eta)$ slowly varying on the long scale $\eta = \alpha x$ and normal structure $f$, $f(z) = e^{-z/s}$. Here $s$ measures the thickness of the viscous layer at the front. The unit vectors $\hat{x}$ and $\hat{z}$ are chosen tangential and normal to the undisturbed front. The analysis is performed with an exponentially compressed coordinate normal to the interface such that $\zeta = \exp(H - z)$, which transfers the front position $z = H(x, t)$ to $\zeta = 1$ and $z \to \infty$ to $\zeta \to 0$. In this notation, the basic state for $\alpha = 0$ consists of a planar interface $H_0 = 0$ and an exponential concentration profile $c_0 = \zeta$.

The planar interface is subjected to small disturbances of amplitude $\varepsilon$. Linear stability is governed by the $O(\varepsilon)$ terms, $[H_1(x), c_1(x, \zeta)] \exp(\sigma t)$, where $\sigma$ represents the complex growth rate. $c_1$ is expressed as a power series in $\zeta$:

$$c_1 = \zeta H_1 + \sum_{\nu=0}^{\infty} C_\nu \zeta^{\mu+\nu/s},$$

where $H_1$ and $C_\nu$ depend on $x$ and $\eta$. Collecting powers of $\zeta$, one finds at lowest order that

$$\partial_{xx} C_0 + P(\eta)\partial_x C_0 + \beta_0^2 C_0 = 0,$$

where $\beta_0^2 = \mu^2 - \mu - \sigma$ defines $\mu$.

If there were no convection, $P = 0$, then $\beta_0$ would represent the wave number of the cellular solidification at marginal stability. For $\alpha \to 0$ one uses a WKB approximation by writing $C_0 = \exp[P(\eta)/\alpha]$ and retaining only the leading order term in $\alpha$. For the case of $P(\eta) = P_0 \sin \eta$, one finds finally up to an arbitrary factor

$$C_0 = \exp\left[\frac{P_0}{2\alpha}(\cos \eta - 1)\right] \cos\left[E\left(\eta, \frac{P_0}{2\beta_0}\right)\frac{\beta_0}{\alpha}\right],$$

where $E$ is the elliptic integral of the second kind with modulus $\frac{1}{2}P_0/\beta_0$. In the limit $\alpha \to 0$, $C_0$ asymptotes to the expression

$$C_0 = \exp\left(-\frac{1}{4}\frac{P_0}{\alpha}\eta^2\right)\cos \beta_0 x,$$

which is proportional to the interface deflection $H_1$. A typical example is shown in Fig. 24.10. The instability is highly localized on the scale $\eta = \alpha x$ of the flow but is still wide enough on the morphological scale $x$ to exhibit a large number of modes near the inward surface stagnation points.

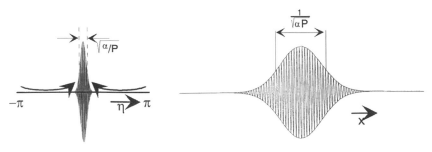

Figure 24.10. Morphological interface pattern influenced by a long-scale periodic remote flow in the liquid phase.

## 24.4. Spreading Flows with Basal Solidification

### 24.4.1. *A Spreading Model for* $Pr \gg 1$

The spreading of oxidic melts with low heat conductivities is observed in lava flows and is considered in the safety analysis of nuclear reactor accidents. We consider here the plane spreading of a poor heat conducting melt on a horizontal substrate of temperature $T_w < T_{\text{sol.}}$. We expect solidification of the melt at the substrate. This problem features two free interfaces, namely a liquid–gas interface between the melt and the ambient gas and a liquid–solid interface due to solidification. For all that follows we restrict ourselves to the viscous-gravitational regime, i.e., we have

$$\text{Re} = \frac{h_0 u_0}{i} = 0(1) \quad \text{and} \quad \text{Fr} = \frac{u_0^2}{h_0 g} = 0(\varepsilon).$$

Here $h_0$ and $u_0$ denote inflow height and average inflow velocity. Moreover, we consider poorly conducting melts only, such that $\text{Pr} = i/\hat{e} \gg 1$ holds.

We introduce separate scales for a slender melt region of horizontal extent $\sim l_0$ and vertical extent $\sim h_0$ and the dimensionless variables

$$(X, Z) = \left[\frac{x}{l_0}, \frac{z}{(\varepsilon l_0)}\right], \qquad (U, W) = \left[\frac{u}{u_0}, \frac{w}{(\varepsilon u_0)}\right], \qquad P = \frac{p}{\left(\mu_0 l_0 u_0/h_0^2\right)},$$

$$\Theta = \frac{(T - T_\infty)}{(T_0 - T_\infty)}, \qquad \tau = \frac{t}{(l_0/u_0)},$$

where $T_0$ is the initial melt temperature and $T$ is the ambient temperature. There is furthermore the small parameter $\varepsilon = h_0/l_0 \ll 1$. Then, to the leading order for nonisothermal thin-film flow, the following equations hold:

$$\partial_X U + \partial_Z W = 0,$$

$$\partial_X P = \partial_{ZZ} U,$$

$$\partial_Z P = -\frac{\varepsilon \text{Re}}{\text{Fr}},$$

$$\varepsilon \text{Re} \, \text{Pr}\{\partial_\tau \Theta + U \partial_X \Theta + W \partial_Z \Theta\} = \partial_{ZZ} \Theta.$$

The boundary conditions are formulated on the substrate $Z = 0$, the liquid–gas interface $Z = H(X, \tau)$, the liquid–solid interface $Z = S(X, \tau)$ and the inlet $X = 0$. They read as

$$U = 0, \qquad W = 0, \qquad \Theta = 0, \qquad Z = 0;$$
$$\partial_Z U = 0, \qquad W = \partial_X H U + \partial_\tau H, \qquad \partial_Z \Theta = 0, \qquad Z = H;$$
$$U = 0, \qquad W = 0, \qquad \Theta = \Theta_{\text{sol.}}, \qquad Z = S;$$
$$\Theta = 1, \qquad X = 0.$$

Within this approximation we take the liquid–gas interface as adiabatic and neglect capillary forces. At the liquid–solid interface, latent heat release is negligible to leading order and the melt is at solidification temperature $\Theta_{\text{sol.}}$.

Bunk and Ehrhard (1998) deduced an approximate solution to the above problem in three steps:

1. From the work of Huppert (1982), the isothermal flow field is inferred. Integrating the momentum equations subject to the kinematic boundary conditions at both the substrate and the free interface gives

$$U = \frac{\varepsilon \text{Re}}{\text{Fr}} \partial_X H \left( \frac{Z^2}{2} - HZ \right),$$

$$W = \frac{\varepsilon \text{Re}}{6\text{Fr}} [3(\partial_X H)^2 Z^2 + 3H \partial_{XX} H Z^2 - \partial_{XX} H Z^3]$$

for the velocity field. For the layer height $H(X, \tau)$ the evolution equation

$$\partial_\tau H - \frac{\varepsilon \text{Re}}{3\text{Fr}} \partial_X (H^3 \partial_X H) = 0$$

holds, subject to conditions

$$V(\tau) = \int_0^{A(\tau)} H(X, \tau) \, dX = C_V \tau^\alpha, \qquad H[A(\tau), \tau] = 0. \tag{4}$$

Here the volume $V(\tau)$ is given in form of a power law in time and $A(\tau)$ is the position of the spreading melt front. The solution to this problem for a given $\alpha$ can be found by a similarity transformation and subsequent numerical integration (see Huppert, 1982).

2. Based on the previously computed flow field a solution to the temperature field is developed. Because of the assumption that $\text{Pe} = \text{Re} \cdot \text{Pr} \gg 1$ there is a thin thermal boundary layer developing at the substrate, while the rest of the melt remains at inlet temperature $\Theta_0 = 1$. This calls for a matched asymptotic solution. The vertical coordinate is rescaled for an inner solution in the thermal boundary layer by

$$(\hat{Z}, \hat{H}) = (Z, H)(\varepsilon \text{Pe})^{1/2}.$$

To leading order the quasi-steady heat transport equation

$$\partial_{\hat{Z}\hat{Z}} \Theta = \frac{\varepsilon \text{Re}}{\text{Fr}} (\varepsilon \text{Pe})^{-3/2} \hat{Z} \left[ \frac{1}{2} \partial_X (\hat{H} \partial_X \hat{H}) \hat{Z} \partial_{\hat{Z}} \grave{E} - (\hat{H} \partial_X \hat{H}) \partial_X \grave{E} \right]$$

has to be solved subjected to the above boundary condition at the substrate and the matching conditions toward the bulk flow, e.g.,

$$\grave{E} \to 1 \quad \text{for} \quad \hat{Z} \to \infty.$$

An analytical solution to the above thermal problem is obtained by means of a similarity transformation. A typical result for the thermal field is given in Fig. 24.11. In addition to isotherms, the liquid–gas interface $H(X, \tau)$ and the liquid–solid interface $S(X, \tau)$ are plotted. The liquid–solid interface coincides with the solidification isotherm at temperature $\Theta_{\text{sol.}}$.

3. Given the solidified region we solve the free interface flow problem once more on top of the crust. By enforcing the no-slip condition on $S(X, \tau)$, we end with a modified evolution

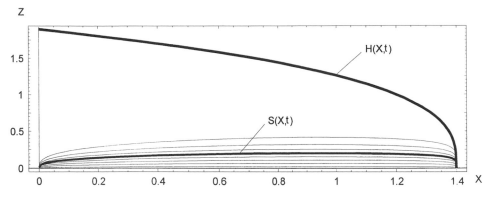

Figure 24.11. Interfaces and temperature field for $Pr \gg 1$.

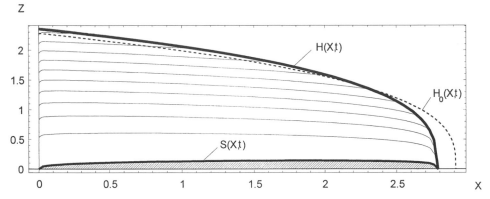

Figure 24.12. Flow field and interfaces in the presence of the solidified basal crust.

equation for the free interface $H(X, \tau)$, in the form

$$\partial_\tau H - \frac{\varepsilon Re}{Fr} \partial_X [(H - S)^3 \partial_X H] = 0.$$

This evolution equation has to be solved subjected to conditions (4). A typical result is presented in Fig. 24.12. The actual interfaces $H(X, \tau)$, $S(X, \tau)$ are depicted as bold solid curves, while the liquid–gas interface without solidification, $H_0(X, \tau)$, obtained in the first step of approximation, is given as a dashed curve for comparison. In addition to the interfaces, instantaneous streamlines are displayed as thin curves in Fig. 24.12.

We recognize from Fig. 24.12 a spreading melt flow tangential to the (quasi-steady) basal crust. The liquid–gas interface $H(X, \tau)$ is at a different position if compared with the interface in absence of the basal crust $H_0(X, \tau)$. The difference indicates that the front propagation is slower and the height of the melt at the inlet has increased.

### 24.4.2. *Experiments with Canauba Wax*

To validate the theoretical model, we have conducted small-scale experiments by using a high Prandtl number melt. The wax ($Pr = 560$) is poured at a controlled temperature $T_0$ and at a constant volumetric flux $\dot{V}$ into a thermally insulated cavity. As soon as the cavity is filled, spreading occurs across a chrome-coated copper plate, which is controlled at temperature $T_w$.

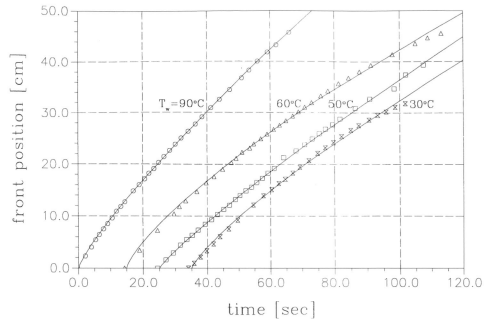

Figure 24.13. Front propagation for different substrate temperatures.

Aside from all relevant temperatures (measured by thermocouples), we measure optically both the interface shape $h(x, t_0)$ at several fixed times $t_0$ and the position of the front $a(t)$. We vary the Reynolds number in the range $0.5 \leq \mathrm{Re} \leq 5.2$ and the substrate temperature in the range $20°\mathrm{C} \leq T_w \leq 90°\mathrm{C}$, such that either solidification ($T_{\mathrm{sol.}} = 85°\mathrm{C}$) or pure liquid spreading occurs.

A set of results for the front propagation at different substrate temperatures is presented in Fig. 24.13. We have chosen a low Reynolds number of $\mathrm{Re} = 0.56$ in these experiments such that the effect of solidification at the substrate is sufficiently strong. The different curves in Fig. 24.13 are artificially shifted on the time axis to clarify the overall picture. We find for a plate temperature of $T_w = 90°\mathrm{C}$ pure liquid spreading, since the initial melt temperature in the experiments is chosen to be $T_0 = 90°\mathrm{C}$ and all temperatures are well above solidification temperature of the melt, $T_{\mathrm{sol.}} = 85°\mathrm{C}$. As soon as we reduce the substrate temperatures to $T_w < T_{\mathrm{sol.}}$, solidification of the melt occurs at the substrate and all experimental curves for $T_w = 60, 50$, and $30°\mathrm{C}$ show a different behavior for the front propagation. We find for substrate temperatures $T_w < T_{\mathrm{sol.}}$ a slower progression of the melt front. This becomes obvious from the reduced slope of the corresponding curves $a(t)$ in Fig. 24.13. This experimental finding of slower melt spreading due to basal solidification confirms the theoretical results. The theoretical predictions likewise give slower spreading rates as soon as a crust is formed at the substrate.

## 24.5. Summary

Convection in many ways affects solidification at free solid–liquid interfaces. There is a rich set of complex phenomena associated with this interaction. Its experimental and theoretical investigation is a challenge to engineers, physicists, and applied mathematicians. Any progress in the understanding and modeling of complex solidification processes is of significant importance for several fields of engineering and science.

## Acknowledgement

The authors gratefully acknowledge the long-lasting fruitful collaboration with S. H. Davis in this particular interdisciplinary field of thermofluid dynamics. They are grateful to their collaborators M. Bunk and U. Siegel for carrying out significant parts of the analytical and experimental work. They thank particularly D. Köhler for her speedy, but nevertheless careful typing and organizing of the manuscript.

## References

Bühler, L. and Davis, S. H. 1998. Flow-induced changes of the morphological stability in directional solidification. *J. Crystal Growth* **186**, 629–647.

Bunk, M. and Ehrhard, P. 1998. Theoretische Beschreibung der Ausbreitung und Erstarrung viskoser Schmelzen. *Z. Angew. Math. Mech.*, in press.

Busse, F. 1967. On the stability of two-dimensional convection in layer heated from below. *J. Math. Phys.* **46**, 140–149.

Carslaw, H. S. and Jaeger, J. C. 1959. *Conduction of Heat in Solids*. Oxford: Clarendon.

Davis, S. H., Müller, U., and Dietsche, C. 1984. Pattern selection in single-component systems coupling Bénard convection and solidification. *J. Fluid Mech.* **144**, 133–151.

Davis, S. H. and Segel, L. A. 1968. Effects of surface curvature and property variation on cellular convection. *J. Fluid Mech.* **11**, 470–476.

DeGroot, S. R. and Mazur, P. 1969. *Non-Equilibrium Thermodynamics*. Amsterdam: North-Holland.

Dietsche, C. and Müller, U. 1985. Influence of Bénard convection on solid–liquid interfaces. *J. Fluid Mech.* **161**, 249–268.

Huppert, H. 1982. The propagation of two-dimensional and axisymmetric viscous gravity currents offer a rigid horizontal surface. *J. Fluid Mech.* **121**, 43–58.

Karcher, C. and Müller, U. 1994. Onset of oscillatory convection in binary mixtures with Sorét effects and solidification. *Int. J. Heat Mass Transfer* **37**, 2517–2523.

Worster, M. G. 1996. Private communication.

Zimmermann, G., Müller, U., and Davis, S. H. 1992. Bénard convection in binary mixtures with Soret effects and solidification. *J. Fluid Mech.* **238**, 657–682.

# 25

# Interfacial Motion of a Molten Layer Subject to Plasma Heating

P. S. AYYASWAMY, S. S. SRIPADA, AND I. M. COHEN

## 25.1. Introduction

The transient and quasi-steady-state energy transport from plasmas associated with low-current ($\sim$50-mA) low-energy arcs between two dissimilar electrodes has many industrial applications. Here we are using the term arc in the same sense as Thompson did (1962). In particular, microelectronics packaging technology, which uses the wire-bonding process, has significantly benefited from analyses of this regime of plasma science. Although there are a number of detailed studies of high-energy, high-current ($\sim$100-A) arcs (typically in welding technology) and the associated transport, the mechanisms governing transport at low currents are different and warrant special investigation. With low-current plasmas, electrical dissipation and radiation from the plasma are unimportant while conduction (particularly for a wire electrode) and power input by electrons entering the electrode surface are dominant. The average velocity of particles is much smaller than the random thermal velocity, and convective effects are of lesser importance. These are opposite characteristics from those of high-current plasmas. With the low-current arc, the voltage between electrodes is high. The formulation for estimating the effect of electric field in this case must include Gauss' law for the electric-field intensity together with the conservation equations for the low-current problem. High-current arc studies have usually involved a point-plane geometry, and the heat transfer to the plate has been of primary concern. With the intended important application (wire bonding in packaging), there is a need to investigate cylinder-(wire-) to-plane discharges, and heat transfer to the cylinder and subsequent melting of the cylinder are of major concern.

The study and analysis of the transient heating and melting of a wire electrode by energy transport from a low-energy plasma may be divided into four distinct substudies. These are the analysis of

1. the breakdown of the fluid medium in the gap between the two electrodes, starting from an initial electric field, followed by initiation of an arc,
2. the self-sustained arc consisting of a plasma (a mixture of neutrals, positively charged ions, and electrons), and the energy transport to the wire electrode,
3. the melting of the wire electrode due to the heat transfer from the arc along with radiation and convection from the wire to the ambient, and motion of the molten layer,
4. the resolidification of the molten layer after termination of the arc.

We now describe our approach to the investigation of each of these problems.

## 25.2. Breakdown and Discharge

A low-energy arc, like all arcs, is initiated by a breakdown of the insulating medium between the electrodes. The breakdown is achieved by applying a voltage across the electrodes that is high enough to cause an ionization avalanche. During the breakdown phase (which has a duration of a microsecond), the medium between the electrodes becomes conductive, and a temporally increasing current begins to flow in the circuit. Simultaneously, the potential difference between the electrodes begins decreasing. At the end of the breakdown phase, a steady current and voltage are established across the gap between the electrodes. The fluid medium in the arc is composed of neutrals, ions, and electrons and is termed a plasma to refer to a fluid consisting of charged particles in sufficient quantity to be a good electrical conductor but yet overall is electrically neutral.

The initial breakdown in a nonuniform gap between a wire and a plane has been analyzed by Ramakrishna et al. (1989), Donovan and Cohen (1991), and Jog et al. (1991). The wire electrode is taken there to be a hyperboloid of revolution that facilitates the use of a prolate spheroidal coordinate system. The studies concluded that (i) the breakdown occurs at a faster rate when the wire is the positive electrode, (ii) with the wire as anode, the maxima in particle density and electron temperature occur on the discharge axis, and (iii) a steep rise in electron temperature precedes the rapid increase in electron density followed by a gradual decrease in electron temperature as the ionization progresses. We have also numerically simulated the arc between two dissimilar electrodes, a cylindrical anode and a planar cathode, by using a nonorthogonal body-fitted coordinate system based on elliptic grid generation (Qin, 1997).

Low-energy plasmas at atmospheric pressure may be considered collision dominated and are adequately modeled by continuum conservation equations when Kn $\ll$ 1. Here, Kn is the Knudsen number defined as the ratio of the representative mean free path in the plasma to the macroscopic length scale of the problem (e.g., anode diameter). We consider two-temperature plasmas consisting of three species: electrons, positive ions, and neutrals. It is assumed that each specie is in equilibrium with its like particles and the background gas is homogeneous and at rest. Since good collision coupling exists between the heavy particles, positive ion and neutral particle temperatures are taken to be the same. The governing conservation equations for the analysis of the breakdown are (see Wilkins and Gyftopoulos, 1966):

$$\frac{\partial N_{e,i}}{\partial t} + \nabla \cdot \mathbf{j}_{e,i} = P_i(N_e) + P_t(N_e) - R_t(N_e), \tag{1}$$

where the electron and ion number fluxes are given by

$$\mathbf{j}_{e,i} = -\frac{\mu_{e,i}}{e} \nabla p_{e,i} \mp \mu_{e,i} N_{e,i} \mathbf{E}, \quad \text{where} \quad p_{e,i} = N_{e,i} k T_{e,i}. \tag{2}$$

The electron energy equation is

$$\frac{\partial}{\partial t} \left( \frac{3}{2} N_e k T_e \right) + \nabla \cdot \mathbf{q}_e = -e\mathbf{j}_e \cdot \mathbf{E} - Q_{ei}, \quad \text{with} \quad \mathbf{q}_e = \frac{5}{2} k T_e \mathbf{j}_e - \kappa_e \nabla T_e. \tag{3}$$

$Q_{ei}$ is the rate of transfer of energy during inelastic collisions considering that successive excitation collisions result in ionization. The self-consistent electric field is given by Gauss' law,

$$\nabla \cdot \mathbf{E} = \frac{e}{\epsilon_0}(N_i - N_e), \quad \text{where} \quad \mathbf{E} = -\nabla V. \tag{4}$$

The assumption that the electric-field intensity $\mathbf{E}$ is expressible by the gradient of a potential has been justified by Ramakrishna (1989) for the range of parameters considered here. There it is shown that the contribution to the electric field due to magnetic induction is orders of magnitude smaller than the applied electric field. Here, $N$ is the charged particle number density, the subscript $e$ refers to electrons, $i$ refers to ions, $p$ is pressure, $\mu$ is the mobility of the charged specie, $k$ is the Boltzmann constant, $\epsilon_0$ is permittivity of free space, $e$ is the magnitude of electrical charge on a single electron, $\mathbf{E}$ is the electric-field intensity, $V$ is the electric potential, $\kappa_e$ is the electron thermal conductivity which is a function of $N_e$ and $T_e$ (Wilkins and Gyftopoulos, 1966). $P_t(N_e)$ is the thermal ionization rate, and $R_t(N_e)$ is the three-body recombination rate. We use the Saha equation (Mitchner and Kruger Jr., 1973) to obtain the net ionization due to thermal ionization and recombination:

$$P_t(N_e) - R_t(N_e) = \gamma \left[ \frac{2g_i N_n}{g_n} \left( \frac{2\pi m_e k T_e}{h^2} \right)^{\frac{3}{2}} \exp\left( -\frac{eV_i}{kT_e} \right) - N_e N_i \right], \tag{5}$$

where the three-body recombination coefficient $\gamma = 1.09 \times 10^{-20} T_e^{-\frac{9}{2}} N_e$ m³/s is a strong function of electron temperature (see Hinnov and Hirschberg, 1962). In the above, the $g$'s are statistical weights, $N_n$ is the number density of neutrals, $V_i$ is the ionization potential, $m_e$ is the electron mass, and $h$ is Planck's constant. The inelastic collision energy transfer rate is $Q_{ei} = eV_i[P_t(N_e)]$.

The initial conditions are small charged-particle densities $N_0$, equal to those in the ambient because of background radiation, uniform temperature that is the same as the ambient, and the electric field given by Laplace's equation for the applied voltage boundary conditions at the electrodes.

The boundary conditions are

At the cathode plate: $N_{e,i} = N_0$, $\quad V = V_{ap} - I_w R_c$, $\quad T_e = T_\infty$.
At the anode (wire tip and surface): $N_{e,i} = N_0$, $\quad V = 0$, $\quad T_e = T_s$.
At the outer boundary: $N_{e,i} = N_0$, $\quad \nabla V \cdot \mathbf{n} = 0$, $\quad \nabla T_e \cdot \mathbf{n} = 0$.

Here, $N_0$ is the small ambient electron and ion density due to omnipresent background radiation. The electrodes are considered to be nearly perfect absorbing surfaces for the ionic species. $N_0$ is assigned to the electrodes as well to avoid the singular behavior that may result because of a zero-density condition. This also accounts for the weak reflection and emission from the electrode surfaces. If the arc is decoupled from the heat transfer in the anode interior, then an anode surface temperature profile $T_s$ is assumed. During the short duration of the discharge breakdown, this temperature is taken to be equal to that of the ambient. In a coupled formulation, the coupling interface conditions provide the anode boundary conditions for the arc. $V_{ap}$ is the applied external voltage between the electrodes. However, as current flows through the circuit, the actual voltage appearing at the cathode is less than the applied voltage because of the potential drop across an external circuit resistance $R_c$. This controls the magnitude of the final steady-state current flowing through the circuit. We use this boundary condition. Here, $I_w$ is the current collected at the wire (anode) and is expressed as

$$I_w = e \int_{\text{wire}} (\mathbf{j}_i - \mathbf{j}_e) \cdot d\mathbf{A}_w,$$

where the integral is taken over the entire wire surface and $d\mathbf{A}_w$ is the elemental area vector of the anode surface. The Dirichlet condition at the outer boundary for the number densities is justified by the observation that the densities must fall to $N_0$ if the plasma is assumed to be confined within the domain. The Neumann conditions on the electric potential and temperature are due to the definition of the outer boundary. As may be observed from the expression for the number fluxes, the potential gradient and thermal gradient drive the ionic species' diffusive fluxes. If these fluxes are to be zero at the outer boundary, the gradients must be zero.

The above vector equations are written out algebraically depending on the domain of interest and the corresponding coordinate system. For an irregular domain a body-fitted coordinate system (Knupp and Steinberg, 1993) may be generated to map the domain to a computational rectangle (in two dimensions). The conservation equations are then solved in this computational domain, and results are mapped back to the physical domain. For example, we have obtained orthogonal grids (Sripada et al., 1997) on which the conservation equations have been solved to simulate low-energy plasmas.

During the breakdown, current $I_w$ is zero initially, but slowly starts to rise as charged-particle number densities begin to rise in the gap and electrons drift toward the anode and are collected there. As this current rises, the voltage drop across the external circuit resistance causes the potential across the gap to steadily decrease. Because of the decreased potential, the rate of increase of current starts decreasing until a steady state is reached when the potential is just sufficient to support a steady current. This steady state is reached rapidly (in 1–2 $\mu$s) and a self-sustained plasma is established.

## 25.3. Steady Discharge and Energy Transport to Wire

Once the interelectrode gap has undergone breakdown, the steady arc established across the electrodes provides heat input to the wire tip. The model for a continuum, collision-dominated plasma presented in Section 25.2 is adequate for the analysis of the steady arc. Since the arc is steady, only the time-dependent terms from the breakdown model need be dropped. We have earlier analyzed electrode heating due to a wire-to-plane arc (Jog et. al., 1992) by using a prolate spheroidal coordinate system. During the transient breakdown phase, as the charged-particle densities become significant, the effect of space charge modifies the initial harmonic potential into that given by Gauss' law for the self-consistent electric field. However, because of the nearly perfectly absorbing electrode boundary conditions, the charged-particle densities drop steeply, resulting in the formation of electrode sheaths that are regions of high gradients of potential and temperature. A quasi-neutral region is established between the sheaths. The gradients in the quasi-neutral region are finite while the gradients in the sheaths are large. These arguments have been used by Jog et al. (1992) as the basis for a two-region model of the arc. One region consists of sheaths and the other consists of the quasi-neutral region. Results indicate that when the wire is the anode, the maximum heat flux occurs at the wire tip. The cathode voltage drop is higher than the anode drop and the maximum temperature location is in the interior of the arc approximately one-third of the gap length from the wire tip.

For the types of discharges considered, the anode begins to melt and deform into a spherical segment because of surface tension. During this deformation, the boundary conditions for the discharge begin to change because of the evolving anode shape. Therefore a need was felt to be able to model anode shapes that were not simple coordinate surfaces. For a significant fraction of the arc duration, the anode has a spherical pendent molten tip at the end of a long cylinder. We have modeled and simulated the steady-state arc established with such an anode and computed the heat transfer to the anode tip. The calculation of the heat transfer is next summarized.

Our past work has shown that the heat flux to the wire is higher when the wire is the anode than when it is the cathode (Huang et al., 1991). This heat flux from the plasma to the wire has two major components, the work function energy flux $q_n$ and the conductive energy flux $q_c$. In comparison with $q_n$ and $q_c$, all other modes of energy transport are negligible (Jog et al., 1992). Incoming electrons at the anode tip release energy to the electrode surface equivalent to the work function $\phi$. The net surface normal energy flux $q_n$ and electron thermal conductive heat flux $q_c$ are expressed as

$$q_n = a_e e \phi \mathbf{j}_e \cdot \mathbf{n}, \qquad q_c = -\kappa_e \nabla T_e \cdot \mathbf{n}, \tag{6}$$

where $a_e = 0.9$ is the electron energy accommodation coefficient for the gold anode surface (Wiedmann and Trumpler, 1946) and $\mathbf{n}$ is the surface normal vector. In the energy flux expressions, the electron flux $\mathbf{j}_e$ and temperature field $T_e$ are obtained from a simulation of the steady-state arc.

The charged-particle number conservation equations along with the boundary conditions presented in Section 25.2 constitute a set of highly nonlinear coupled conservation equations to be solved in the interelectrode region in which the discharge occurs. This region is that between a slender, spherical-tipped anode oriented normal to a circular planar cathode at a finite distance from the spherical wire tip. In a coupled formulation, in which the evolving anode shape due to anode melting must be included, this distance increases as the anode melts. Because of the symmetries in the domain, an axisymmetric section in a cylindrical coordinate $(r, z)$ system is considered. The center of the circular planar cathode is located at $z = 0$, and the cathode radius is $r_p$. The spherical tip of the wire is represented by a spherical segment attached to a long slender wire of length $l_w$ oriented along the $z$ axis. The tip is at a distance $l_g$, the gap length, from the center of the cathode plate. All the dimensions are input parameters in the simulation.

We choose to use an orthogonal grid system to minimize grid-based errors and errors due to approximation of boundary orthogonal fluxes such as heat flux to the wire tip. Additionally, the transformed equations in the computational domain also retain the same structure as the original conservation equations and cross-derivative terms are absent, as in the physical domain. Now we briefly describe the method to obtain an orthogonal grid system.

### 25.3.1. *Orthogonal Grid Generation*

It is well known that not all domains with prescribed boundary point distributions can be orthogonally mapped into a rectangular computational domain (Eca, 1996; Duraiswami and Prosperetti, 1992). However, an iterative method suggested by Eca (1996) resulted in satisfactory grids for domain shapes considered in the wire–plane discharge problem. We summarize the method below.

A solution of the nonlinear coupled covariant Laplace equations (Ryskin and Leal, 1983; Eca, 1996) with the prescribed domain coordinates as boundary conditions provides the required mapping of the physical domain to a rectangular computational domain:

$$\frac{\partial}{\partial \xi}\left(f\frac{\partial r}{\partial \xi}\right) + \frac{\partial}{\partial \eta}\left(\frac{1}{f}\frac{\partial r}{\partial \eta}\right) = 0, \qquad \frac{\partial}{\partial \xi}\left(f\frac{\partial z}{\partial \xi}\right) + \frac{\partial}{\partial \eta}\left(\frac{1}{f}\frac{\partial z}{\partial \eta}\right) = 0,$$

$$\mathbf{R} = \mathbf{R}_b, \quad \text{at} \quad \eta = 0, \quad \xi \in [0, 1], \qquad \mathbf{R} = \mathbf{R}_t, \quad \text{at} \quad \eta = 1, \; \xi \in [0, 1],$$

$$\mathbf{R} = \mathbf{R}_l, \quad \text{at} \quad \xi = 0, \quad \eta \in [0, 1], \qquad \mathbf{R} = \mathbf{R}_r, \quad \text{at} \quad \xi = 1, \; \eta \in [0, 1],$$

where $\mathbf{R}(r, z)$ is the position vector in the $(r, z)$ coordinate system, and the subscripts $b, t, l$, and $r$ denote the bottom, top, left, and right domain boundaries, respectively. $f$ is the distortion function of the scaling factors, defined as the ratio $h_\eta/h_\xi$ in the $(\xi, \eta)$ coordinate system. Here,

$$h_\xi = \sqrt{g_{11}} = \sqrt{\left(\frac{\partial r}{\partial \xi}\right)^2 + \left(\frac{\partial z}{\partial \xi}\right)^2}, \qquad h_\eta = \sqrt{g_{22}} = \sqrt{\left(\frac{\partial r}{\partial \eta}\right)^2 + \left(\frac{\partial z}{\partial \eta}\right)^2},$$

where $g_{11}$ and $g_{22}$ are the diagonal components of the covariant metric tensor. As the distortion function $f$ involves derivatives of both the unknown mapping functions $[r(\xi, \eta), z(\xi, \eta)]$, the grid generation equations are highly nonlinear and coupled. The finite-volume method was used to discretize the above equations, and, because of the strong nonlinearities, an iterative algorithm is used to obtain the solution. At a given iteration level, the system is nominally linear, i.e., $f$ is known in terms of the solution at the previous iteration level. At a given iteration level, a low relative error tolerance of $10^{-6}$ was used. Solvers based on the multigrid principle (Wesseling, 1992) significantly improve the rate of convergence of basic iterative methods of solution of large systems of linear or nonlinear algebraic equations. Therefore a robust multigrid solver, MGD9V (de Zeeuw, 1990, 1997), is used to solve the nominally linear equations. MGD9V uses incomplete Line LU decomposition as the relaxation procedure and matrix-dependent grid transfer operators. The grid error is monitored by means of the maximum deviation from orthogonality (MDO), defined as

$$\text{MDO} = \max\left(|90^\circ - \theta_{i,j}|\right), \quad \text{where} \quad \theta = \cos^{-1}\left(\frac{g_{12}}{h_\eta h_\xi}\right) \quad \text{and} \quad g_{12} = \frac{\partial r}{\partial \xi}\frac{\partial r}{\partial \eta} + \frac{\partial z}{\partial \xi}\frac{\partial z}{\partial \eta}.$$

Here, $g_{12}$ is the off-diagonal component of the covariant metric tensor and $\theta_{i,j}$ is the angle between the $\xi$ and the $\eta$ grid lines at a grid location $\mathbf{R}(r_i, z_j)$.

The numerical accuracy of the computer programs were tested by grid generation for simple geometries like an annular sector in a polar coordinate system. In all cases the grid obtained gave an MDO of $\leq 0.5^\circ$.

Figure 25.1 illustrates a $41 \times 41$ grid generated with this method. For the grid presented, MDO $= 0.85^\circ$, which represents an orthogonality error of $\pm 0.94\%$ apart from the standard discretization error resulting from the use of central differences to represent first-order derivatives in the partial differential equations. This error is $O(\Delta \xi^2) \approx 10^{-4}$.

### 25.3.2. *Numerical Solution of Plasma Equations*

The governing equations for the plasma are first written in the $(r, z)$ coordinate system and then transformed to the computational $(\xi, \eta)$ system. This preserves the axisymmetric nature of the original problem. A finite-volume method was used in conjunction with the power-law scheme (Patankar, 1980) to discretize the partial differential equations.

In order to use the power-law scheme, a novel interpretation for the grid Peclet number for the discretized plasma equations is necessary. In the equations, the electric field $\mathbf{E}$ plays the

**41x41 ORTHOGONAL GRID**

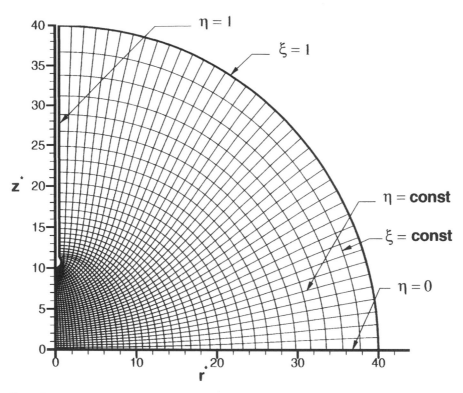

Figure 25.1. Grid system: all lengths are normalized by anode diameter.

role of a convection velocity in the fluid dynamics sense. Therefore the ratio $e\mathbf{E} \cdot \mathbf{t}_\zeta \Delta\zeta / kT_e$, the electronic Peclet number, is used in lieu of the standard grid Peclet number. Here $\mathbf{t}_\zeta$ is the local tangent vector to the $\zeta$ coordinate line, where $\zeta$ may be either of $(\xi, \eta)$, as appropriate. All dependent variables are colocated at the grid points. Since the potential is defined at each grid point, the gradient is defined naturally at the control volume interfaces. The grid generation process also provides the metrics and the Jacobian of the transformation.

The discretized plasma equations are strongly nonlinear and coupled and thus must be solved iteratively. The strong coupling necessitates severe underrelaxation of the iterative procedures. The system of discretized equations is solved iteratively with the multigrid solver MGD9V (de Zeeuw, 1990, 1997).

The simulations were performed on a CRAY C90 at the Pittsburgh Supercomputer Center. Results from a simulation of the arc are presented here. All the results presented here correspond to the time instant when the anode has a spherical tip of diameter $1.5d$. The parameters are an applied voltage of $-2500$ V, an anode (wire) diameter $d = 1$ mil $= 25.4$ $\mu$m, a discharge gap $l_g = 10d$, an anode length $l_{an} = 30d$ from the anode tip, the cathode plate radius $r_p = 40d$, and $R_c = 100$ k$\Omega$. The melting temperature of the anode material was prescribed to be $T_m = 1336$ K (this corresponds to a gold anode, which is often used in microelectronic applications). The ambient reference temperature is $T_\infty = 300$ K, and the ambient pressure is 1 atm.

The equipotential contours for a normalized applied voltage of $V^* = eV/kT_\infty = -96618$ (which corresponds to $V_{ap} = -2500$ V) are shown in Fig. 25.2. This is the potential field

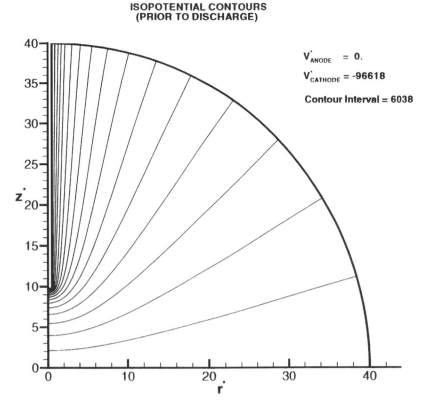

**ISOPOTENTIAL CONTOURS
(PRIOR TO DISCHARGE)**

$V_{ANODE}^{\cdot} = 0.$

$V_{CATHODE}^{\cdot} = -96618$

**Contour Interval = 6038**

Figure 25.2. Equipotential contours: nondimensional potential.

before the breakdown and discharge when the anode still has a cylindrical tip. Because of the relatively sharp features of the anode tip, strong potential gradients exist in the anode vicinity. The equipotential contours are orthogonal to the left and the right boundaries because of the Neumann boundary conditions at the symmetry axis and at the outer boundary. A small number of ambient electrons present in the medium are accelerated by the potential field to initiate the breakdown process until enough ions and electrons have been generated to reach a steady state.

The medium is a self-sustaining plasma, at steady state, with a flux of electrons absorbed at the anode and a flux of ions absorbed at the cathode. As may be anticipated, the fluxes of electrons at the cathode and of positive ions at the anode are very small. A steady-state current of 21.54 mA is sustained by the discharge for the assumed outer circuit resistance $R_c = 100$ k$\Omega$. At steady state, this results in a cathode voltage of $-346$ V. The arc is mostly quasi-neutral (nearly equal ion and electron number densities) except in the electrode sheaths, where strong density gradients exist to support the intense charged-particle fluxes appropriate to the electrodes. The variation of the nondimensional electron number density $N_e^* = N_e/N_{ref}$ along the discharge axis is shown in Fig. 25.3. $N_{ref} = kT_\infty \epsilon_0/e^2 d^2 = 2.22 \times 10^{15}$ m$^{-3}$. The cathode is at $z^* = 0$, and the anode tip is at $z^* = 10$. Impact ionization becomes a strong ionization source as electrons accelerate to large drift velocities near the anode because of the strong local potential gradients. Thus there is a peak in the electron density near the anode. However, the strongly absorbing electrode boundary condition causes a severe gradient in the density at the anode. A similar trend is seen for ions since ions are generated at the same locations as electrons.

**Electron Density & Temperature Variation**

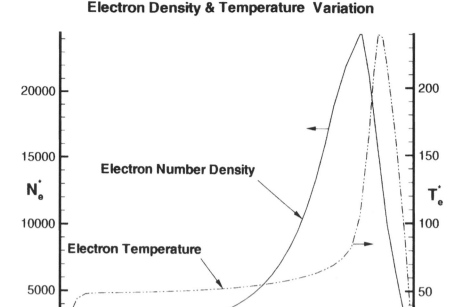

Figure 25.3. Variation of nondimensional electron density and temperature along discharge axis.

The variation of the electron temperature along the discharge axis is also shown in Fig. 25.3. Because of the strong anode sheath gradients, electrons undergo much stronger acceleration and assume large kinetic energies as reflected in the temperature. The anode surface is a strong energy-absorbing sink, and this explains the temperature peak in the anode vicinity. The magnitude of the energy flux on the anode surface decreases away from the tip. This is shown in Fig. 25.4, in which the variation of $q^*$, the dimensionless energy flux, with distance along the anode surface from the tip is shown. The heat flux is normalized by

$$q_{ref} = kT_\infty \frac{\mu_e N_{ref} V_{ref}}{d},$$

which is $3.26 \times 10^{-4}$ W/m$^2$ for the parameters considered. The minimum in the heat flux variation appears at the neck of the anode, where the melt is attached to the solid cylindrical portion of the anode. Beyond the spherical tip, there is enough ionization in the sheath to provide an energy flux to the anode surface, albeit orders of magnitude lower than that in the vicinity of tip. For the chosen set of parameters, 0.097 W of thermal power is transferred to the anode from the plasma. The minimum in the heat flux is explained by the shielding effect of the bulge of the melt layer against all energy-driving gradients. This indicates the crucial role played by electrode geometry in the determination of the heat flux from an arc plasma.

### 25.4. Melting of the Wire Electrode

The energy deposited by the condensation of electrons at the tip of the wire electrode provides sufficient heat to raise the temperature of the tip beyond its melting point (for a gold or copper

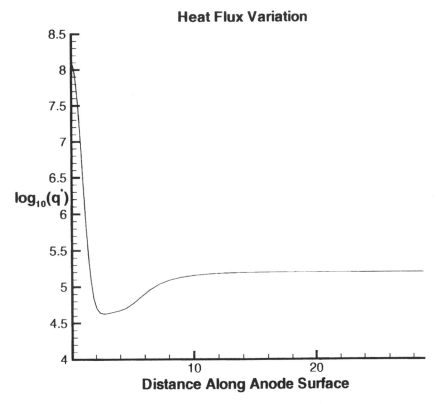

Figure 25.4. Variation of nondimensional heat flux along anode surface.

anode). As the tip melts, the molten layer moves under the influence of surface tension and gravity. In our past studies (Huang et al., 1995; Cohen et al., 1995), we have described both model and experiments for the transient heating of a thin wire causing the tip to melt, rollup of the melt, and subsequent solidification on termination of the arc. The shape of the melt is analytically/numerically determined assuming equilibrium based on minimum-energy principles. A constant heat flux from the arc is assumed and is an input parameter for the simulation of the melting process. The fluid in the melt is assumed static, and the temperature field in the melt is determined by the solution of the transient energy equation in a body-fitted coordinate system accounting for phase change. The domain is remeshed at each time step. Comparisons with scaled-up experimental results are presented. The experimental results included high-speed motion pictures of the melting process for large length-scale and time-scale experiments. (It is very difficult to conduct experiments at the true length scale that also yield transient temperature data.) Results for equilibrium shape calculations of the free surface indicate a very small influence of gravity and Marangoni, i.e, surface-tension gradient effects.

To assess the relative influence of surface tension and gravity on the motion of the melt we calculate the Eötvös number defined as Eo $= \Delta\rho g d^2/\sigma$, where $\Delta\rho$, $g$, $d$, and $\sigma$ are the density difference between the two media, gravitational acceleration, wire diameter (the appropriate length scale), and surface tension, respectively. For applications of interest in microelectronic packaging, $d = 25.4$ $\mu$m, and with gold wire in air, Eo $\approx 9 \times 10^{-5}$. This indicates that for the length scale of interest, gravity has a negligible effect compared with that of surface tension. Therefore surface tension is taken to be the prime cause for motion of the molten layer. The velocity field is then determined purely by the interaction between the two material properties:

surface tension ($\sigma$) and viscosity ($\hat{\mu}$) of the fluid. The appropriate velocity scale is then the ratio $\sigma/\hat{\mu}$. This naturally leads to a time scale, $d\hat{\mu}/\sigma$, and pressure scale, $\rho\sigma^2/\hat{\mu}^2$, with the choice of anode diameter $d$ as the length scale. With this choice for the various scales, the nondimensional continuity and Navier–Stokes equations governing the incompressible motion of the melt become

$$\nabla \cdot \mathbf{u} = 0, \qquad \frac{\partial \mathbf{u}}{\partial t} + \mathbf{u} \cdot \nabla \mathbf{u} = -\nabla p + \text{Oh}^2 \nabla^2 \mathbf{u}. \tag{7}$$

Here $\mathbf{u}$ and $p$ are the velocity field and the pressure in the melt, respectively. The Ohnesorge number, $\text{Oh} = \hat{\mu}/(\rho\sigma d)^{1/2}$, represents the ratio of the viscous forces to those due to surface tension (Sadhal et al., 1997). The interface between the melt and the arc may be considered a hydrodynamic free surface as the ambient gas (air) is otherwise considered to be at rest in the arc problem. Then the flow field in the melt has to satisfy the stress-free boundary conditions at the free surface:

$$\tau : \mathbf{nn} + (p - p_0) + \sigma \left( \frac{1}{R_1} + \frac{1}{R_2} \right) = 0, \qquad \tau : \mathbf{nt} = 0. \tag{8}$$

Here, $\tau$ is the viscous stress tensor in the melt, $p - p_0$ is the static pressure in the melt in excess of the ambient pressure, $(R_1, R_2)$ are the principal radii of curvature of the free surface, $\mathbf{n}$ is the local surface normal vector, and $\mathbf{t}$ is the local surface tangent vector. The free surface is also subject to the kinematic boundary condition,

$$\frac{D\mathbf{R}_s}{Dt} = \mathbf{u}|_{\mathbf{R}=\mathbf{R}_s}, \tag{9}$$

where $D/Dt$ is the material derivative and $\mathbf{R}_s$ is the position vector of any point on the free surface. At the interface between the melt and the solid regions, the flow is subject to the no-slip boundary condition, $\mathbf{u} \cdot \mathbf{t} = 0$ and the normal velocity condition for a melt–solid interface given by Equation (13) below.

The fluid dynamics of the melt is inherently coupled to the heat transfer because the fluid volume depends on the heat flux from the arc at the free surface. This volume increases in time as more of the anode becomes molten. As the free surface moves, i.e., the anode deforms, the gap between the electrodes is modified, which affects the arc. Thus the arc problem becomes coupled to the fluid dynamics and transport of the melt. Thus the free surface represents an interface that involves a complex interaction between the arc on one side and the fluid dynamics and heat transfer on the melt side.

The heat transfer in the melt is governed by the unsteady energy equation,

$$(\rho C)_l \left( \frac{\partial T}{\partial t} + \mathbf{u} \cdot \nabla T \right) = \nabla \cdot (k_l \nabla T), \tag{10}$$

where $T$ is the temperature, and $C$ and $k$ are the specific heat and the thermal conductivity respectively. The subscript $l$ refers to the liquid phase. The energy equation is subject to a heat flux boundary condition at the free surface:

$$-k\nabla T \cdot \mathbf{n} = q_c + q_n - q_L, \tag{11}$$

where, on the right-hand side, $q_c + q_n$ is the energy flux from the arc given by Eqs. (6). Here, $q_L$ is the thermal loss by radiation and natural convection from the free surface. The heat transfer in the melt is also coupled to a similar energy equation in the solid phase through the conditions at the solid–liquid interface:

$$T_s = T_l = T_m, \qquad -k_s \nabla T_s \cdot \mathbf{n} + k_l \nabla T_l \cdot \mathbf{n} = \rho_l L(\mathbf{u}_l - \mathbf{u}_i) \cdot \mathbf{n}, \tag{12}$$

$$\rho_l(\mathbf{u}_l - \mathbf{u}_i) \cdot \mathbf{n} = \rho_s(\mathbf{u}_s - \mathbf{u}_i) \cdot \mathbf{n}, \tag{13}$$

where subscript $s$ refers to the solid phase, $m$ refers to melting, $\mathbf{n}$ is the local surface normal vector at the phase-change interface, $L$ is the latent heat of fusion, and $\mathbf{u}_i$ is the velocity of the moving phase-change interface. Equation (13) expresses mass conservation at the interface (see Sadhal et al., 1997) and provides an expression for $\mathbf{u}_i$. The energy equations are also governed by the initial and boundary conditions of the phase-change problem. The boundary conditions for the solid region include losses due to radiation and natural convection to the ambient from the lateral surfaces of the anode. There is also a temperature boundary condition ($T = T_\infty$) at the far end of the slender anode.

Thus a simulation of the moving anode surface involves a solution of the nonlinear coupled governing equations for the plasma, for the flow and heat transfer in the melt, and the energy equation in the solid region of the wire. These equations are all subject to their boundary conditions and to the interfacial conditions at the two interfaces, one between the plasma and the melt, and the other between the solid and the melt regions of the anode.

Next, we summarize a procedure that we used to obtain a solution to this problem by using modern computational techniques.

### 25.4.1. *Numerical Methodology for Anode Evolution*

The anode tip melts and rolls up because of surface tension as more and more energy is deposited at the anode surface by the plasma. Analysis of problems involving moving interfaces and evolving domains have been of much interest in the literature (see Tsai and Yue, 1996; Shyy et al., 1996). Numerical methods for moving boundary problems may be classified as Lagrangian or Eulerian methods. In Lagrangian methods, the moving boundary surface is explicitly tracked and the grid (discretized domain) conforms to the shape of the moving interface. In Eulerian methods, a fixed grid is used while the interface is not explicitly tracked but reconstructed from the field variables such as temperature or volume fraction. As the Lagrangian methods involve movement of grids with the moving interface, these are also referred to as moving-grid methods. The grid may be structured or unstructured. The choice of the method is strongly dependent on the type of problem being solved and the nature of information to be obtained from the simulation. Techniques that combine the advantages of Lagrangian and Eulerian methods have been developed that include the arbitrary Lagrangian–Eulerian (ALE) method based on finite elements (see Shyy et al., 1996; Ramaswamy et al., 1996; Chippada et al., 1996). The ALE method is the method of choice when severe deformation of the interface is expected as, for example, in wave motion and fluid instabilities.

For the problem of anode melting and motion of the free surface, we have chosen a Lagrangian method since explicit tracking of the moving interface between the arc and melt is desired. On the other hand, a Eulerian method is used to simulate the moving phase-change interface between the melt and the solid regions.

The orthogonal grid generation method (see Subsection 25.3.1) is used to discretize the evolving anode domain at each instant of time. The Navier–Stokes equation and the energy equation are transformed from the axisymmetric physical domain to a rectangular computational domain. A finite-volume method and a power-law scheme (Patankar, 1980) are used to discretize the governing equations. The iterative Semi-Implicit Pressure Linked Equation Revised (SIMPLER) algorithm of Patankar (1980) is used to solve the momentum equation [see also the extension to orthogonal coordinates by Raithby et al. (1986) and to general curvilinear coordinates by Karki and Patankar (1988); Shyy (1994)]. The procedure uses a staggered grid in which all the dependent variables except velocities are located at centers of the finite volumes (discretization units). The velocities are located at grid points that lie on the control volume faces. This staggered arrangement prevents pressure oscillations in the numerical solution. The algorithm couples the continuity equation with the momentum equation and iteratively obtains the correct velocity and pressure fields such that mass conservation is strictly satisfied.

With regard to the motion of the interface, the steady-state free-surface tracking problem in an orthogonal coordinate system has been analyzed with a finite-difference solution of the stream function–vorticity formulation for the Navier–Stokes equation by Ryskin and Leal (1984). The unsteady version of this method has been presented by Kang and Leal (1987). Our formulation fundamentally differs from these methods in that primitive variables are used in the solution of the momentum equation. An explicit implementation of the kinematic boundary condition of Eq. (9) must then be used to advance the free surface at each time step. The free-surface normal velocity required in the kinematic boundary condition is obtained from the stress boundary conditions. The curvature information required for evaluating the surface-tension forces is obtained from a quadratic fit of the grid points on the free surface (see also Shyy et al., 1996). Since the shape of the physical domain is time dependent, care must be taken to transform the time derivatives from the physical domain $(r, z, t)$ to the computational domain $(\xi, \eta, t)$.

With regard to the phase-change interface, the equivalent heat capacity method of Hsiao (1985) is used (see Section 25.5). The choice of this Eulerian method is based on our experience that an explicit tracking of two interfaces (one for the free surface and one for the phase change) leads to grid instabilities. These require exhaustive numerical experimentation for the iterative relaxation parameters in order to obtain satisfactorily convergent solutions. Moreover, since the exact tracking of the phase-change interface is not of fundamental interest in the application, an Eulerian method that reduces an otherwise two-region problem to a one-region problem with a strong nonlinearity in a material property is preferred.

## 25.5. Solidification of the Melt

Once the arc is extinguished, the anode loses heat by conduction along its axis and by radiation and natural convection from its surfaces. Huang et al. (1995) have studied the melting and solidification of thin wires as a class of phase-change problems with interfacial motion based on a nonorthogonal body-fitted coordinate system based on elliptic grid generation (Knupp and Steinberg, 1993). They considered phase-change of copper wires (1.6-mm diameter). The use of a nonorthogonal grid system, however, introduces cross-derivative terms that considerably complicate the numerical solution with attendant convergence difficulties.

In the melt and the solid wire, the temperature field is unsteady and is governed by the unsteady energy conservation equations:

$$(\rho C)_s \frac{\partial T}{\partial t} = \nabla \cdot (k_s \nabla T), \qquad (\rho C)_l \frac{\partial T}{\partial t} = \nabla \cdot (k_l \nabla T), \tag{14}$$

where $T$ is the temperature and $\rho$, $C$, and $k$ are the density, specific heat, and thermal conductivity, respectively. During the solidification, the fluid is assumed to be at rest. The subscripts $s$ and $l$ refer to the solid and the liquid phases, respectively. The two energy equations are coupled through conditions at the solid–liquid interface:

$$T_s = T_l = T_m, \qquad -k_s \nabla T_s \cdot \mathbf{n} + k_l \nabla T_l \cdot \mathbf{n} = \rho_l L (\mathbf{u}_l - \mathbf{u}_i) \cdot \mathbf{n}, \tag{15}$$

$$\rho_l (\mathbf{u}_l - \mathbf{u}_i) \cdot \mathbf{n} = \rho_s (\mathbf{u}_s - \mathbf{u}_i) \cdot \mathbf{n}, \tag{16}$$

where the subscript $m$ refers to melting, $\mathbf{n}$ is the local surface normal vector at the phase-change interface, $L$ is the latent heat of fusion, and $\mathbf{u}_i$ is the velocity of the moving phase-change interface. The equations are also subject to the initial and boundary conditions of the phase-change problem.

This is a two-region problem and is nonlinear, in general, because of the presence of the unknown location of the moving interface in the interfacial conditions. Both finite-difference methods (Bonacina et al., 1973; Huang et al., 1995) and finite-element methods (Yoo and Rubinsky, 1983) have been applied to obtain solutions of phase-change problems with a moving interface. Either a time-variant mesh or a fixed-mesh approach have been used. In the time-variant mesh method (Yoo and Rubinsky, 1983), the mesh is regenerated at each time step to account for the moving phase-change interface. In the fixed-mesh method (Hsiao, 1985), the latent heat of fusion is usually absorbed into the material's specific heat (equivalent specific heat model) or enthalpy (enthalpy model), but this introduces a severe nonlinearity in the material property. The enthalpy method treats the enthalpy as an additional variable (Shamsundar and Sparrow, 1975). The equivalent specific heat model (Bonacina et al., 1973; Hsiao, 1985) superimposes the latent heat effect on the specific heat of the system over a small temperature interval $2\Delta T$ that acts as a range of temperature over which the phase transition takes place. This introduces a severe nonlinearity in the specific heat, and the choice of $\Delta T$ becomes crucial for convergence and prevention of oscillatory solutions. An effective algorithm that is insensitive to the choice of $\Delta T$ has been presented by Hsiao (1985), which is the method chosen for this study. This is described next.

The two-region solidification is reduced to a one-region problem with a jump in the specific heat over a temperature interval $2\Delta T$ around the melting point for the material by use of the equivalent heat capacity method and the stable algorithm of Hsiao (1985). The algorithm is summarized next.

It has been shown (Bonacina et al., 1973) that if the heat capacity per unit volume ($\rho C$) is defined as

$$\rho C = \begin{cases} (\rho C)_s, & \text{if } T < T_m, \\ \rho_s L \delta (T - T_m), & \text{if } T = T_m, \\ (\rho C)_l, & \text{if } T > T_m, \end{cases} \tag{17}$$

and

$$k = \begin{cases} k_s, & \text{if } T < T_m, \\ k_l, & \text{if } T > T_m, \end{cases}$$

then, Eqs. (14) along with the interfacial coupling conditions in Eqs. (15) and (16) may be reduced to

$$\rho C \frac{\partial T}{\partial t} = \nabla \cdot (k \nabla T). \tag{18}$$

In Eq. (17), $\delta$ is the Dirac delta function. This reduction of the system to one energy equation with a highly nonlinear specific heat capacity allows for the use of a fixed-grid approach, and a two-region problem can be treated as an equivalent one-region problem through the entire phase-change process.

In a numerical solution, the Dirac delta function in Eq. (17) is replaced by a finite delta function with a large specific heat value over a small temperature interval $2\Delta T$ across the fusion temperature. Thus Eq. (17) is replaced by

$$\rho C = \begin{cases} (\rho C)_s, & \text{if } T < T_m - \Delta T, \\ \dfrac{\rho_s L}{2\Delta T} + \dfrac{(\rho C)_s + (\rho C)_l}{2}, & \text{if } T_m - \Delta T \le T \le T_m + \Delta T, \\ (\rho C)_l, & \text{if } T > T_m + \Delta T. \end{cases} \tag{19}$$

As mentioned above, the choice of $\Delta T$ significantly affects the course of the solution process (convergent or oscillatory). A small $\Delta T$ is preferable for phase change of a pure material with a sharp melting/freezing point. On the other hand, a large $\Delta T$ is desirable to reduce the nonlinearity of the latent heat term in the expression for the specific heat. In the algorithm proposed by Hsiao (1985), this difficulty is eliminated by use of more nodal temperature values to calculate the property at any given location. Instead of using the temperature at a particular grid location to calculate the specific heat according to Eq. (19), an effective specific heat is calculated as an average of the influence of all the neighboring nodal temperatures. This eliminates the dependence of the convergence process on the choice of $\Delta T$. Also, the phase-change interface is the isotherm $T = T_m$. The band $\Delta T$ around $T = T_m$ is the so-called mushy zone. For alloys this would be the range of temperatures for phase transition.

The reduced energy equation is numerically solved in a geometrically irregular domain consisting of a molten ball at the end of a long cylinder of diameter $d$. The molten tip of the wire is represented by a spherical segment attached to a long slender wire of length $l_w$ oriented along the $z$ axis. All the dimensions are input parameters in the simulation.

The boundary conditions are

Along the end face of the cylinder ($0 \le r \le d/2$, $z = l_w$): $T = T_\infty$.
At the ball surface and curved surface of the cylinder ($\mathbf{r} = \mathbf{r}_s$): $-k\nabla T \cdot \mathbf{n} = h(T - T_\infty)$ $+ \varsigma\epsilon(T^4 - T_\infty^4)$.
At the axis of symmetry ($r = 0$, $0 \le z \le l_w$): $\nabla T \cdot \mathbf{n} = 0$.

Here, $T_\infty$ is the ambient temperature, $h$ is the convective heat transfer coefficient, $\epsilon$ is the emissivity of the material surface, and $\varsigma$ is the Stefan–Boltzmann constant. The lateral surfaces are taken to lose heat by radiation and convection, while the end face of the cylinder is considered to be at ambient temperature. In the convective boundary condition, the position vector $\mathbf{r}_s$ is used to denote the anode surface wherein lies the geometry difficulty. The boundary condition cannot be prescribed in a simple fashion in the axisymmetric cylindrical coordinate system. This necessitates the generation of an orthogonal boundary-conforming grid system (described in Subsection 25.3.1).

The initial conditions are those that exist at the instant the arc is extinguished and the anode ceases melting.

The governing equations are non-dimensionalized by the wire diameter $d$ as the length scale, $T_\infty$ as the temperature scale, and

$$\frac{d^2}{\alpha} = d^2 \frac{(\rho C)_\infty}{k_\infty}$$

as the time scale. The variable properties $(\rho C)$ and $k$ are nondimensionalized by $(\rho C)_\infty$ and $k_\infty$, respectively.

A finite-volume method (Patankar, 1980) is used to discretize the unsteady energy equation. The system of discretized equations at each time step is solved iteratively with the multigrid solver MGD9V. At the end of each iteration, the temperature is used to update the properties. The Neumann boundary conditions are implemented by artificial Dirichlet conditions in the numerical procedure that are updated after each iteration to simulate Neumann conditions. In a converged solution, this results in a correct implementation of the Neumann conditions.

The overall simulations were made on a 64-Mbyte RAM IBM RISC/6000 workstation and took ~1.5 h for a typical run.

Multigrid-based simulations were made for solidification of a molten gold ball pendent at the end of a long slender gold cylinder (wire). Parameters used for the simulation were wire diameter $d = 1$ mil $= 25.4$ $\mu$m, a wire length $l_w = 30d$, and molten ball radius $= 0.75d$. The melting temperature of gold is $T_m = 1336$ K. The temperature band $\Delta T$ in the approximation of the Dirac delta function is taken to be 1 K. Even with the choice of 2 K for $\Delta T$, the results were found to be insensitive to $\Delta T$. All physical properties were from interpolations of handbook data for gold (Miller, 1952; Ho et al., 1974). The ambient reference temperature is $T_\infty = 300$ K.

The isotherms in the vicinity of the anode tip region after 400 nondimensional time steps of $\delta t = 0.5$ are shown in Fig. 25.5. Each nondimensional time step is equivalent to 24.6 $\mu$s. For the choice of parameters, it takes 36 ms for the domain to attain steady state, which corresponds to a uniform ambient temperature.

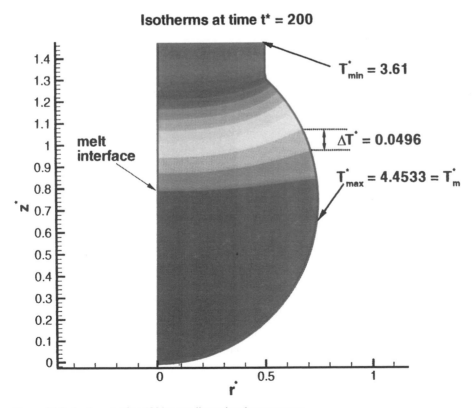

Figure 25.5. Isotherms after $400\delta t$: nondimensional temperature.

**Phase-Change Interface Location**

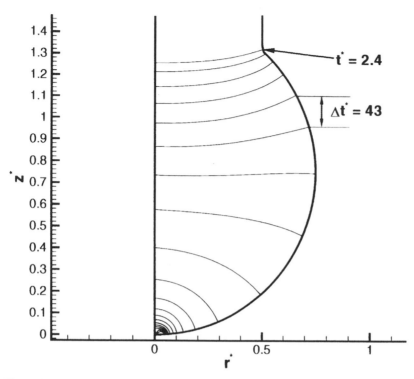

Figure 25.6. Position of phase-change interface at different times.

The position of the phase-change interface at different times during the simulation is shown in Fig. 25.6. The solidification progresses from the neck (joint between the melt to the solid region), downward and toward the axis of symmetry. This is explained by the fact that conduction up along the axis of the cylinder is the primary mode of heat loss from the molten ball. The interface moves in a direction opposite to that of the peak heat flux direction. Owing to the axisymmetric nature of the problem and the fact that there is convective heat loss from the ball surface, the solidification progresses toward the axis. These observations are also supported by our large-scale experimental studies (Sripada et al., 1996) for this configuration in which heat conduction is found to be the primary mode of heat dissipation from the molten ball.

### 25.6. Conclusions

The theory of low-current arcs and a method of calculation of the energy transfer to the anode has been described. The plasma in the arc is taken to consist of ions, neutrals, and electrons in the continuum regime (Knudsen number ≪ 1). The interface between the bottom of an initially solid metallic cylinder and a weakly ionized plasma (dominated by charge-neutral collisions) has been considered. The analysis of the breakdown and steady discharge has been presented. Because of the heat transfer from the arc to the anode, it melts and rolls up because of surface tension into a pendent ball at the tip of the anode. The theory for the melting and for the motion of the free surface of the melt has been presented. A discussion of current numerical techniques and the numerical methodology for a simulation of the melting and the free-surface motion have been presented. The theory and numerical methodology for an analysis of the solidification process (once the arc is extinguished) have also been presented. All of the numerical methodology is based on finite-volume discretization of the governing equations and a multigrid iterative solver. Results for simulations of a sample application in microelectronic packaging, the electronic flame-off discharge used in wire bonding, have been presented.

## Acknowledgments

The authors gratefully acknowledge support of this work by the U.S. National Science Foundation through NSF grant CTS-94-21598. The support of the Pittsburgh Supercomputer Center through a grant from the U.S. National Science Foundation is gratefully acknowledged. The support of the University of Pennsylvania Research Foundation through the grant of an IBM RS/6000 workstation is also acknowledged.

## References

Bonacina, C., Comini, G., Fasans, A., and Primiceris, M. 1973. Numerical solution of phase change problems. *Int. J. Heat Mass Transfer* **16**, 1825–1832.

Chippada, S., Jue, T. C., Joo, S. W., Wheeler, M. F., and Ramaswamy, B. 1996. Numerical simulation of free-boundary problems. *Int. J. Comput. Fluid Dyn.* **7**, 91.

Cohen , I. M., Huang, L. J., and Ayyaswamy, P. S. 1995. Melting and solidification of thin wires: a class of phase-change problems with a mobile interface-II. Experimental confirmation. *Int. J. Heat Mass Transfer* **38**, 1647–1659.

De Zeeuw, P. M. 1990. Matrix-dependent prolongations and restrictions in a blackbox multigrid solver. *J. Comput. Appl. Math.* **33**, 1–27.

De Zeeuw, P. M. 1997. Acceleration of iterative methods by coarse grid corrections. PhD thesis, University of Amsterdam.

Donovan, K. G. and Cohen, I. M. 1991. Two-dimensional analysis of electrical breakdown between a wire and a plane-extended. *J. Appl. Phys.* **70**, 4132–4138.

Duraiswami, R. and Prosperetti, A. 1992. Orthogonal mapping in two dimensions. *J. Comput. Phys.* **98**, 254–268.

Eca, L. 1996. 2D orthogonal grid generation with boundary point distribution control. *J. Comput. Phys.* **125**, 440–453.

Hinnov, E. and Hirschberg, J. G. 1962. Electron–ion recombination in dense plasmas. *Phys. Rev.* **125**, 795–801.

Ho, C. Y., Powell, R. W., and Liley, P. E. 1974. Thermal conductivity of the elements: a comprehensive review. *J. Phys. Chem. Ref. Data* **3** (Suppl. 1).

Hsiao, J. S. 1985. An efficient algorithm for finite-difference analysis of heat transfer with melting and solidification. *Num. Heat Transfer* **8**, 653–666.

Huang, L. J., Ayyaswamy, P. S., and Cohen, I. M. 1995. Melting and solidification of thin wires: a class of phase-change problems with a mobile interface-I. Analysis. *Int. J. Heat Mass Transfer* **38**, 1637–1645.

Huang, L. J., Jog, M. A., Cohen, I. M., and Ayyaswamy, P. S. 1991. Effect of polarity on heat transfer in ball formation process. *ASME J. Electron. Packag.* **113**, 33–39.

Jog, M. A., Cohen, I. M., and Ayyaswamy, P. S. 1991. Breakdown of a wire-to-plane discharge. *Phys. Fluids B* **3**, 3532–3536.

Jog, M. A., Cohen, I. M., and Ayyaswamy, P. S. 1992. Electrode heating in a wire-to-plane arc. *Phys. Fluids B* **4**, 465–472.

Kang, I. S. and Leal, L. G. 1987. Numerical solution of axisymmetric, unsteady free-boundary problems at finite Reynolds number. I. Finite-difference scheme and its application to the deformation of a bubble in a uniaxial straining flow. *Phys. Fluids* **30**, 1929–1940.

Karki, K. C. and Patankar, S. V. 1988. Calculation procedure for viscous incompressible flows in complex geometries. *Num. Heat Transfer* **14**, 295–307.

Knupp, P. and Steinberg, S. 1993. *Fundamentals of Grid Generation*. Boca Raton, FL: CRC Press.

Miller, R. R. 1952. Physical properties of liquid metals. In *Liquid-Metals Handbook*, 2nd ed., R. N. Lyon et al., eds., Chap. 2, pp. 38–102. Washington, DC: Office of Naval Research, Dept. of Navy.

Mitchner, M. and Kruger Jr., C. H. 1973. *Partially Ionized Gases*. New York: Wiley.

Patankar, S. V. 1980. *Numerical Heat Transfer and Fluid Flow*. Washington, DC: Hemisphere.

Qin, W. 1997. Numerical and experimental studies of heat transfer phenomena in microelectronic packaging. PhD dissertation, University of Pennsylvania, Philadelphia.

Raithby, G. D., Galpin, P. F., and Van Doormaal, J. P. 1986. Prediction of heat and fluid flow in complex geometries using general orthogonal coordinates. *Num. Heat Transfer* **9**, 125–142.

Ramakrishna, K. 1989. Study of electrical breakdown phenomena and wire heating for ball formation in ball bonding. PhD dissertation, University of Pennsylvania, Philadelphia.

Ramakrishna, K., Cohen, I. M., and Ayyaswamy, P. S. 1989. Two-dimensional analysis of electrical breakdown in a nonuniform gap between a wire and a plane. *J. Appl. Phys.* **65**, 41–50.

Ramaswamy, B., Chippada, S., and Joo, S. W. 1996. A full-scale numerical study of interfacial instabilities in thin-film flows. *J. Fluid Mech.* **325**, 163–194.

Ryskin, G. and Leal, L. G. 1983. Orthogonal mapping. *J. Comput. Phys.* **50**, 71–100.

Ryskin, G. and Leal, L. G. 1984. Numerical solution of free-boundary problems in fluid mechanics. Part 1. The finite-difference technique. *J. Fluid Mech.* **148**, 1–17.

Sadhal, S. S., Ayyaswamy, P. S., and Chung, J. N. 1997. *Transport Phenomena with Bubbles and Drops*. New York: Springer.

Shamsundar, N. and Sparrow, E. 1975. Analysis of multidimensional phase change via the enthalpy model. *J. Heat Transfer, Trans. ASME* **19**, 333–340.

Shyy, W. 1994. *Computational Modeling for Fluid Flow and Interfacial Transport*. Amsterdam: Elsevier.

Shyy, W., Udaykumar, H. S., Rao, M. M., and Smith, R. W. 1996. *Computational Fluid Dynamics with Moving Boundaries*. Washington, DC: Taylor & Francis.

Sripada, S. S., Ayyaswamy, P. S., and Cohen, I. M. 1997. Numerical computation of the heat transfer to a spherical-tip anode during an electronic flame off process. In *Proceedings of the ASME Heat Transfer Division, Proceedings of the International Mechanical Engineering Congress & Expo*, Vol. HTD-351 (1), pp. 55–61. New York: ASME.

Sripada, S. S., Cohen, I. M., and Ayyaswamy, P. S. 1996. A study of the electronic flame off discharge process used for ball bonding in microelectronic packaging. In *Transport Phenomena in Materials Processing and Manufacturing, Proceedings of the International Mechanical Engineering Congress & Expo*, Vol. HTD-336, pp. 129–136. New York: ASME.

Thompson, W. B. 1962. *An Introduction of Plasma Physics*, Chap. 3. Oxford: Pergamon.

Tsai, W. T. and Yue, D. K. P. 1996. Computation of nonlinear free-surface flows. *Ann. Rev. Fluid Mech.* **28**, 249–278.

Von Engel, A. 1965. *Ionized Gases*, 2nd ed. London: Oxford U. Press.

Wesseling, P. 1992. *An Introduction to Multigrid Methods*. W. Sussex, UK: Wiley.

Wiedmann, M. L. and Trumpler, P. R. 1946. Thermal accommodation coefficients. *Trans. ASME* **68**, 57–64.

Wilkins, D. R. and Gyftopoulos, E. P. 1966. Transport phenomena in low energy plasmas. *J. Appl. Phys.* **37**, 3533–3540.

Yoo, J. and Rubinsky, B. 1983. Numerical computation using finite elements for the moving interface in heat transfer problems with phase change transformation. *Num. Heat Transfer* **6**, 209–222.

# 26

# The Fluid Mechanics of Premelted Liquid Films

M. G. WORSTER AND J. S. WETTLAUFER

## 26.1. Introduction

It is a well-known fluid-mechanical phenomenon that a droplet of one fluid immersed in another migrates up an imposed temperature gradient (Levich, 1962). Such thermophoresis is effected by thermocapillarity – the surface tension between two fluids is a decreasing function of temperature, so thermal gradients give rise to tangential stresses at a fluid–fluid interface. Thermophoresis of solid particles immersed in a fluid can also occur, driven, for example, by gradients in the strength of a surrounding electrical double layer (Anderson, 1989). The converse, a fluid droplet migrating through a solid under the influence of an ambient temperature gradient, can occur by a process of melting and resolidification. Sometimes known as temperature gradient zone migration (TGZM), this was the subject of a talk by Yih at the GI Taylor Memorial Symposium in 1987 that one of us (MGW) was privileged to hear (Yih, 1986, 1987). This chapter describes the fundamental mechanisms underlying a related process, thermal regulation, by which a solid particle immersed in another solid can migrate up an imposed temperature gradient. Thermal regulation is similar to TGZM in that it involves melting and resolidification of the surrounding solid. However, whereas in TGZM the melting (really dissolution) is caused by solute gradients within the liquid inclusion, in thermal regulation the melting is caused by intermolecular forces (such as van der Waals forces) altering the equilibrium states of matter near interfaces. Specifically, although a material held in thermodynamic equilibrium below its bulk freezing temperature is solid in the main, it may nevertheless be liquid in a thin layer adjacent to its contact with a foreign substrate.

That such (premelted) liquid films might exist was suggested by Faraday (1860) and can be predicted from interfacial thermodynamics (e.g., Lipowsky, 1982; 1984). More recently, direct evidence of their existence has come from a number of different experiments (see Dash, et al., 1995, for a review). Such experiments reveal the breakdown of long-range periodic order at the surface of a crystalline solid, where it makes contact either with its own vapor or with a foreign substrate. That this disordered state, which has been variously referred to as quasi-liquid or as being liquidlike (e.g., Löwen et al., 1989), is fluid is implicit in any understanding of thermal regulation and other related phenomena. There is now, however, explicit experimental evidence that premelted liquid flows in response to an applied temperature gradient (Wilen and Dash, 1995). Further, fluid-mechanical calculations based on an assumption that the premelted liquid behaves as a Newtonian fluid with the same viscosity as the bulk molten material are adequate to predict the results of the experiments.

In this chapter we describe the thermodynamic arguments used to predict premelting and to calculate the relationship between pressure and temperature in premelted liquid films. We review some of the calculations made of flows involving premelted films, including thermal

339

regelation, and introduce a new model, based on these and additional premelting effects, of the growth of an ice lens in a water-saturated soil – a process fundamental to the phenomenon of frost heave.

Premelting can be induced by a variety of intermolecular forces. For uniformity of presentation we discuss only the case of van der Waals forces. The fluid-mechanical processes we describe and the associated free-boundary problems are similar in the other cases.

## 26.2. Premelting due to van der Waals Forces

We present here an outline of the physical principles underlying the phenomenon of premelting sufficient to determine the fluid pressure driving the flows examined in the remainder of the chapter. For a more precise and detailed treatment, see Wettlaufer and Worster (1995).

Consider a semi-infinite solid ($s$) separated by its own melt ($m$) from a semi-infinite substrate ($w$). The van der Waals forces between all the molecules in the three layers give rise to a force per unit area between the solid and the substrate (the thermomolecular pressure):

$$p_T = \frac{A}{6\pi d^3},\tag{1}$$

where $d$ is the thickness of the liquid film. The effective Hamaker constant $A$ depends on the dielectric properties of all three materials in the layered system and can have either sign (Tabor, 1991). The pressure represented by Eq. (1) is familiar in the context of thin-film fluid mechanics and can, for example, lead to rupturing of a liquid film on a solid substrate when $A$ is negative and the interfacial interaction is therefore attractive.

We are interested in cases in which $A$ is positive so that there is a force of repulsion (disjoining pressure) between the media bounding the liquid film. The external pressure applied to the substrate $p_w$, equal to that applied to the solid $p_s$, balances the thermomolecular pressure $p_T$ and the hydrodynamic pressure $p_l$. Thus

$$p_w = p_s = p_T + p_l.\tag{2}$$

Finally, the Gibbs–Duhem relationship can be applied on each side of a solid–melt interface to show in general that in equilibrium

$$p_s - p_l = \frac{\rho_s q (T_m - T)}{T_m},\tag{3}$$

where $\rho_s$ is the density of the solid, $q$ is the latent heat of solidification per unit mass, $T_m$ is the bulk melting temperature of the solid, and $T$ is the temperature of the system.

Combining Eqs. (1)–(3) shows that the equilibrium thickness of the liquid film is

$$d = \lambda \left( \frac{T_m - T}{T_m} \right)^{-1/3},\tag{4}$$

where $\lambda^3 = A/6\pi \rho_s q$. Equation (4), which shows that a premelted liquid film can exist (when $A > 0$) at temperatures below the bulk freezing point and gives the relationship between its thickness and the temperature, has received experimental confirmation for many different materials (Dash et al., 1995).

## 26.3. Flow of Premelted Liquid

Equation (4) shows that the thickness of a premelted liquid film decreases as the temperature decreases. Correspondingly, the thermomolecular pressure increases. Thus, if the external pressure is held constant, for example, then the pressure in the liquid decreases and there is a tendency for premelted liquid to flow from warmer to cooler regions.

An experiment to investigate premelting by measuring such a flow was designed by Wilen and Dash (1995). In it, water was sandwiched in a cylindrical layer between a lower glass slide and a flexible upper membrane, from the center to the outside of which was imposed a steady temperature gradient (Fig. 26.1). The axis of the cylindrical layer was held at a temperature below the bulk freezing temperature (0°C), causing a disk of ice to grow radially outward until its edge coincided with the 0°C isotherm. Without premelting, this would have been the ultimate steady configuration of the system. However, the thermomolecular pressure gradients in the premelted liquid film between the ice and the flexible membrane drew water radially inward, which resulted in the membrane being deflected laterally. In order to maintain the equilibrium thickness of the premelted film, the additional water froze onto the upper surface of the disk of ice.

This process can be modeled simply, with lubrication theory to analyze the flow in the premelted liquid film (Wettlaufer et al., 1996). The membrane is significantly displaced only

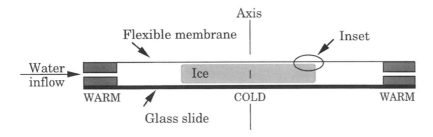

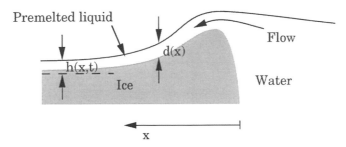

Figure 26.1. Schematic diagram of the experiment of Wilen and Dash (1995), showing an axial cross section through a cylindrical apparatus. A disk of ice grew radially outward between a lower glass slide and an upper flexible membrane until its outer edge coincided with the 0°C isotherm. The pressure gradient in the premelted liquid between the ice and the membrane drew water radially inward, as shown in the inset, which has a greatly exaggerated vertical scale.

near the rim of the disk of ice, so a two-dimensional analysis of a radial cross section is sufficient to model the experiments.

The temperature field is steady and given by

$$T = T_m - Gx,$$

where $G$ is the constant imposed temperature gradient and $x$ is distance from the leading edge of the ice toward the center of the disk. The volumetric flow rate $q$ per unit length transverse to the flow in the premelted liquid film is given by lubrication theory to be

$$q = -\frac{d^3}{12\mu}\frac{\partial p_l}{\partial x} = -\frac{\lambda^3 T_m}{12\mu Gx}\frac{\partial p_l}{\partial x}. \tag{5}$$

For small wall displacement $h(x, t)$, the pressure exerted by the elastic membrane is

$$p_w \approx -\gamma h_{xx}, \tag{6}$$

where $\gamma$ is the tension in the membrane. So from Eqs. (2) and (3), the liquid pressure is

$$p_l = -\gamma h_{xx} - \frac{\rho_s q}{T_m}Gx. \tag{7}$$

Combining Eqs. (5) and (7) into the conservation equation

$$h_t + q_x = 0 \tag{8}$$

leads to the evolution equation for the wall displacement

$$h_t + D[x^{-1}(h_{xxx} + \beta)]_x = 0, \tag{9}$$

where

$$D = \frac{1}{12}\frac{\lambda^3 T_m}{G}\frac{\gamma}{\mu}, \qquad \beta = \frac{\rho_s q G}{\gamma T_m}.$$

Equation (9) admits a similarity solution,

$$h = \beta(Dt)^{3/5}f(\eta), \qquad \text{with} \quad \eta = \frac{x}{(Dt)^{1/5}},$$

where the dimensionless wall displacement satisfies

$$f'''' - \frac{f''' + 1}{\eta} - \frac{1}{5}\eta^2 f' + \frac{3}{5}\eta f = 0 \tag{10}$$

and boundary conditions

$$f' = f'' = 0 \quad (\eta = 0), \qquad f \to 0 \quad (\eta \to \infty). \tag{11}$$

The solution of this ordinary-differential system is displayed in Fig. 26.2.

All the physical parameters in this theory are known independently of the experiments of Wilen and Dash (1995), with the exception of $\lambda$. Additionally, it has often been speculated that, although premelted liquid behaves as a Newtonian fluid (Mantovani et al., 1980), its viscosity may be significantly different from that of bulk liquid. Since $\lambda$ and $\mu$ appear only in ratio, in the parameter $D$, it is not possible to use these experiments to determine values for $\lambda$ and

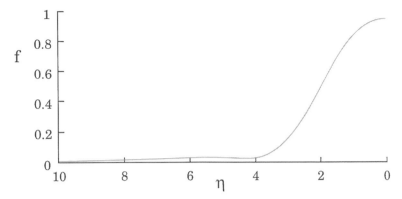

Figure 26.2. The similarity solution for the deflection $f$ of the flexible membrane shown in Fig. 26.1, where $\eta$ is a scaled $x$ coordinate. The deformation is largest at high temperatures at which the gradient of volume flux is largest. Typical physical parameter values give a horizontal-to-vertical aspect ratio of a few hundred.

$\mu$ separately. Wettlaufer et al. (1996) presented a family of similarity solutions for different power-law interactions (including the case of van der Waals) and found good agreement with the experimental results when $p_T \propto d^{-2}$, corresponding to long-range electrostatic interactions. Assuming that the viscosity of the premelted liquid is the same as that of bulk liquid led them to a prediction for the multiplicative constant (the equivalent of $\lambda$) consistent with independent calculations (Wilen et al., 1995) suggesting that the bulk value of viscosity is appropriate. A similar conclusion was drawn by Gilpin (1980) from wire-regelation experiments, which are discussed in Section 26.4 below.

## 26.4. Thermal Regelation

A solid particle encased in ice, for example, migrates toward warmer regions of the ice. This phenomenon was analyzed by Gilpin (1979) alongside the closely related problem of pressure-induced regelation, whereby a weighted wire is pulled through a block of ice.[*] Although the latter can be effected by pressure melting (the decrease in bulk freezing temperature of ice under pressure), it was shown by Telford and Turner (1963) that regelation still occurs at temperatures and pressures too low for pressure melting to operate. Both pressure-induced regelation and thermal regelation can be explained in terms of premelting. The description given here follows closely that of Gilpin (1979), although expressed in terms of van der Waals interactions rather than a spatially varying chemical potential.

Consider a solid sphere of radius $a$ surrounded by ice, as shown in Fig. 26.3. We assume that the thermal properties of the sphere, ice, and premelted water are identical and that the latent heat released and absorbed as the sphere migrates is negligible. The temperature is then given everywhere by

$$T = T_0 + Gx = T_0 - Ga\cos\theta, \tag{12}$$

where $T_0$ is the temperature at the center of the sphere, $x$ is measured upward from the center

---

[*] The term regelation was used by Tyndall (1858) and Faraday (1860) to describe the sintering of ice crystals by the freezing of (premelted) liquid layers on their surfaces.

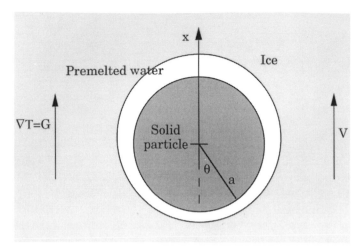

Figure 26.3. A definition sketch for the migration of a solid particle through
a block of ice by the process of thermal regelation.

of the sphere, and $\theta$ is the polar angle measured from the downward vertical. Therefore, if the
dominant intermolecular interactions are van der Waals, this temperature field gives rise to a
premelted liquid film around the sphere of thickness:

$$d = \lambda T_m^{1/3}(T_m - T_0 + Ga\cos\theta)^{-1/3}. \tag{13}$$

The thermomolecular pressure within this film,

$$p_T = \frac{\rho_s q}{T_m}(T_m - T_0 + Ga\cos\theta),$$

provides a net upward force on the sphere:

$$F_T = \frac{4\pi a^3}{3}\frac{\rho_s q}{T_m}G. \tag{14}$$

Mass conservation during steady migration of the sphere at speed $V$ gives the local volumetric
flow rate per unit length transverse to the flow in the thin film to be

$$q = -\frac{1}{2}a\sin\theta V = -\frac{d^3}{12\mu}\frac{1}{a}\frac{\partial p_l}{\partial \theta}. \tag{15}$$

The latter equation, derived from lubrication theory, can be readily integrated to give

$$p_l = p_0 - \frac{6\mu a^2 V}{\lambda^3 T_m}\left(T_m - T_0 + \frac{1}{2}Ga\cos\theta\right)\cos\theta,$$

whence the net upward lubrication force on the sphere can be calculated to be

$$F_\mu = -\frac{4\pi a^3}{3}\frac{6\mu a V}{\lambda^3}\left(\frac{T_m - T_0}{T_m}\right). \tag{16}$$

If there is an external force on the particle, for example if the sphere is heavier than water, then there is an overburden force,

$$F_O = -\frac{4\pi a^3}{3} \Delta \rho g, \tag{17}$$

where $\Delta \rho$ is the density difference between particle and water and $g$ is the acceleration due to gravity. The total force $F_T + F_\mu + F_O$ is zero, so the migration velocity can be determined to be

$$V = \frac{\lambda^3}{6\mu a} \frac{T_m}{T_m - T_0} \left[ \frac{\rho_s q}{T_m} G - \Delta \rho g \right]. \tag{18}$$

This is perhaps the simplest expression embodying the processes of thermal regelation ($g = 0$) and pressure-induced regelation ($G = 0$) and shows the dominant physical parameters controlling these phenomena. In order to make quantitative comparisons with experiments, however, a more detailed analysis is required that accounts for the different thermal properties of the different phases, the release of latent heat, pressure melting, and the effects of any dissolved solutes (e.g., Nye, 1967).

## 26.5. Particle Rejection

A similar analysis to that presented above can be used to explain how solid particles can be pushed ahead of a solidification front. It is known from experiments that particles suspended in a melt are pushed ahead of the front if the solidification rate is sufficiently slow but become engulfed when the solidification rate exceeds a critical rate $V_c$ (e.g., Azouni et al., 1990; Lipp et al., 1990; Lipp and Körber 1993).

   In many analyses of this phenomenon it is assumed that the solidification front is planar. It is straightforward then to determine the van der Waals repulsion and to calculate the lubrication pressure from an analysis of the squeeze film between the ice and particle (assumed spherical). However, given this too-simple picture, it is not possible to determine the distance between the particle and the solidification front. Therefore the critical velocity $V_c$ cannot be determined either, although one can place a rather extreme bound on it by taking the distance of separation to be equal to the molecular cutoff distance for the van der Waals force. As the particle nears the solidification front, however, interfacial premelting causes the front to become concave toward the liquid and thus to begin to conform to the particle (Fig. 26.4). The lubrication force then increases markedly to a point where it can no longer be balanced by the van der Waals repulsion and the particle becomes engulfed. This effect of a nonplanar interface is widely recognized and has been incorporated in many analyses, although often the deflection of the interface has been attributed to dissolved solutes or to the difference in the thermal conductivities of the particle and the ice. Note that the latter effect can cause the interface to deflect either way, depending on the relative conductivities, and it retards engulfment if the particle is less conductive than the ice (Chernov, et al., 1977). Recent experiments (Azouni et al., 1997) indicate that particle trapping depends dominantly on the nature of the surface of the particle, which suggests that bulk thermal and solutal effects are not the primary mechanism controlling engulfment.

   An analysis of particle rejection based purely on the effects of premelting can be distilled from the work of Chernov et al. (1977). Equations (4) and (12) applied to the solidification front combine to yield

$$\frac{l^4}{d^3} = (a + d) \cos \theta - H, \tag{19}$$

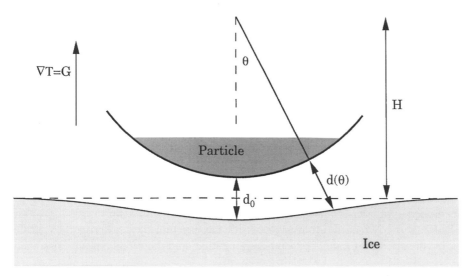

Figure 26.4. A definition sketch for the rejection of a solid particle from an advancing ice–water interface. Note that, in steady state, with the interface advancing at a fixed rate $V$, the particle rides higher than the undisturbed location of the interface. Premelting causes the interface to be deformed beneath the particle, which increases the lubrication pressure there and ultimately results in the particle being engulfed.

where $H = (T_0 - T_m)/G$ is the vertical distance between the center of the sphere and the undisturbed location of the solidification front (which is at the bulk melting isotherm), $d$ is the radial distance between the particle and the solidification front, and $l$ is a length scale given by

$$l^4 = \frac{\lambda^3 T_m}{G}. \tag{20}$$

The van der Waals force and the lubrication force are both locally dominated by the neiborhood of $\theta = 0$. Equation (19) can be differentiated twice to show that

$$d \approx d_0 + \frac{\frac{1}{2} a \theta^2}{1 + \frac{3l^4}{d_0^4}} \qquad \text{for} \quad \theta \ll 1, \tag{21}$$

where $d_0$ is the minimum thickness of the liquid film. With this approximate expression for $d$, the van der Waals and lubrication forces are readily calculated to be

$$F_T \approx \frac{Aa}{6d_0^2} \left( 1 + \frac{3l^4}{d_0^4} \right), \tag{22}$$

$$F_\mu \approx -\frac{6\pi \mu a^2 V}{d_0} \left( 1 + \frac{3l^4}{d_0^4} \right)^2. \tag{23}$$

The net force on the particle is zero, so approximations (22) and (23) show that

$$V = \frac{A}{36\pi \mu a} \frac{d_0^3}{d_0^4 + 3l^4}. \tag{24}$$

Given a solidification rate $V$, this equation can be used to determine the distance $d_0$ at which a steady equilibrium is achieved between the van der Waals and the hydrodynamic forces, which are augmented by the conformity of the phase boundary to the particle induced by premelting. If $V$ is too large, Eq. (24) has no solution; no equilibrium is then possible and the particle is engulfed.

Maximizing $V$, as given by Eq. (24), with respect to $d_0$ gives the critical solidification rate:

$$V_c = \frac{\sqrt{3}}{144\pi} \frac{A}{\mu al} = \frac{\sqrt{3}}{24} \frac{\rho_s q \lambda^{9/4}}{\mu T_m^{1/4}} \frac{G^{1/4}}{a}. \tag{25}$$

There is significant experimental support for the relationship $V_c \propto a^{-1}$ and some support for the relationship $V_c \propto G^{1/4}$, although there is also evidence in favor of other exponents in the latter expression (Pötschke and Rogge, 1989; Lipp and Körber, 1993). Note that, at the critical solidification rate, $d_0 = \sqrt{3}l$ whereas the deflection of the interface is only $l/3\sqrt{3}$, so the particle sits above the undisturbed position of the interface, as shown in Fig. 26.4.

Note that the critical solidification rate provides a different relationship between viscosity and the Hamaker constant, which, in principle, can be used in combination with the earlier result to determine values for these quantities separately.

## 26.6. Ice Lenses

The fundamental physical mechanisms discussed above underlie a unique environmental problem: frost heave. When a saturated soil is cooled below the freezing temperature of the interstitial water, it can undergo large-scale deformations. The total deformation of the partially frozen soil is accounted for by a sequence of ice lenses (layers of ice from which all the soil particles have been expelled) that form perpendicularly to the temperature gradient. Ice lenses typically form periodically in a frozen soil but it is possible, by cooling very slowly, to form a single, continuous ice lens, as illustrated in Fig. 26.5(a) (T. Ishizaki, private communication). This situation is similar to the particle rejection just discussed but differs in some important respects. Equation (19) is an approximate form of the more general interface condition

$$T_m - T = G[H + (a + d)\cos\theta] = T_m \left(\frac{\lambda}{d}\right)^3 + \frac{T_m \gamma_{sl}}{\rho_s q} \mathcal{K}, \tag{26}$$

written here in dimensional form and with $H$ redefined to symbolize the distance of the center of the particle (soil grain) behind the $0\,^\circ$C isotherm. The solid–melt interface has surface energy $\gamma_{sl}$ and curvature $\mathcal{K}$, measured positive if the interface is convex toward the water. The final term in Eq. (26) represents the Gibbs–Thomson depression of the bulk freezing temperature at a curved interface. The analysis of Section 26.5 was developed under the assumption that this latter term is negligible, which is appropriate if the isolated, rejected particle is sufficiently large (Chernov et al., 1977). Within a porous medium (soil), this term plays an important role in determining the location of the ice–water interface.

It is necessary to consider the interaction within and between two regions: an interfacial region near the base of a soil grain and the free region between grains [Fig. 26.5(b)]. We proceed from here with the assumption that $a \ll H$, i.e., that the front of the ice lens lags behind the $0\,^\circ$C isotherm by a distance that is much greater than the size of a soil grain. In the interfacial region, the film thickness $d$ is small, the interface conforms closely to the soil grain, and Eq. (26) can

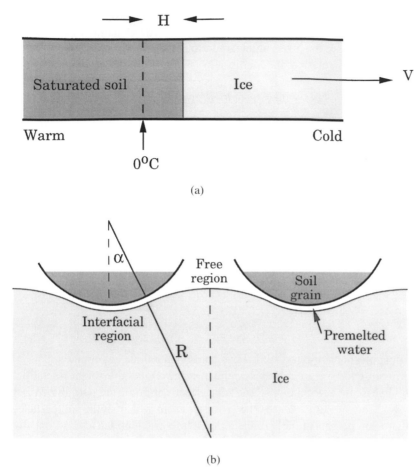

Figure 26.5. (a) Plan view of an experiment (T. Ishizaki, private communication) in which water-saturated soil, placed between two microscope slides, is pulled at constant speed $V$ through a temperature gradient. If the pulling speed is sufficiently slow, then a steady state can be acheived in which a single ice lens rejects all the soil in advance of it. The ice-lens front lags behind the $0^\circ$C isotherm. (b) A magnified sketch of the ice-lens front showing soil grains being pushed ahead of the ice by the intermolecular forces in the interfacial region. The ice–water meniscus adopts a constant curvature in the free regions between soil grains.

be approximated to

$$\left(\frac{\lambda}{d}\right)^3 = \frac{GH}{T_m} + \frac{2\gamma_{sl}}{\rho_s qa}. \tag{27}$$

Equation (1) is modified at a curved interface but Eq. (3) still holds, whence

$$p_T = \frac{\rho_s qG}{T_m}H. \tag{28}$$

The integral of the pressure is not dominated by the neigborhood of $\theta = 0$, as it was for the isolated particle, and cannot therefore be approximated asymptotically in the same way. Rather, the upper limit of integration is set to $\theta = \alpha$, where the angle $\alpha$ is determined by matching with the free region. The total thermomolecular force is then calculated to be

$$F_T = \int_0^\alpha (p_T \cos\theta) 2\pi a^2 \sin\theta \, d\theta = \frac{\pi a^2 \rho_s q G}{T_m} H(1 - \cos^2\alpha)$$

$$\sim \frac{\pi a^2 \rho_s q G}{T_m} H\alpha^2 \quad (\alpha \ll 1). \tag{29}$$

The lubrication pressure is obtained from Eqs. (15), with Eq. (27) for $d$, to be

$$p_l - p_f = -\frac{6\mu a^2 V G}{\lambda^3 T_m}(H + \Gamma)(\cos\theta - \cos\alpha),$$

where $\Gamma = 2\gamma_{sl} T_m / \rho_s q G a$, whence the net lubrication force is found to be

$$F_\mu = -\frac{\pi \mu a^4 V G}{\lambda^3 T_m}(H + \Gamma)(4 - 6\cos\alpha + 2\cos^3\alpha)$$

$$\sim -\frac{3\pi \mu a^4 V G}{2\lambda^3 T_m}(H + \Gamma)\alpha^4 \quad (\alpha \ll 1). \tag{30}$$

Within the free region, the van der Waals interaction is negligible and Eq. (26) is approximated by

$$H = \Gamma \frac{a}{R}, \tag{31}$$

where $R$ is the constant radius of curvature of the ice–water interface. There is a complicated geometrical problem within the free region to determine the shape of the interface, which makes zero contact angle with the soil grains. However, simple considerations of geometry and scaling lead to the approximate relationship $\alpha \sim a/R$, whence

$$\alpha \sim \frac{H}{\Gamma}. \tag{32}$$

The thermomolecular and lubrication forces on a soil grain must balance the overburden force $F_O$ applied to it, which for illustration here we take to be constant. Further, we consider only the case in which $\Gamma \gg H$. In dimensional terms, this inequality is $2\gamma_{sl} T_m / \rho_s qa \gg GH$, which indicates that the undercooling of the ice lens $\Delta = GH$ is much less than the Gibbs–Thomson undercooling arising from the curvature of a soil grain. Equivalently, the inequality means that the radius of curvature of the ice–water interface in the free region is much larger than the radius of a soil grain. In this case, the balance of forces leads to the equation

$$V = \frac{A}{9\pi \mu a^2} \left(\frac{2\gamma_{sl}}{a}\right) \left(\frac{T_m}{\rho_s q}\right) \frac{\Delta^3 - \Delta_c^3}{\Delta^4}, \tag{33}$$

where

$$\Delta_c = \left(\frac{F_O}{\pi a^2}\right)^{1/3} \left(\frac{2\gamma_{sl}}{a}\right)^{2/3} \frac{T_m}{\rho_s q}. \tag{34}$$

Figure 26.6 shows the possible steady growth rates $V$ for a single ice lens as a function of the undercooling $\Delta$. We see that a minimum undercooling $\Delta_c$ is required in order for the

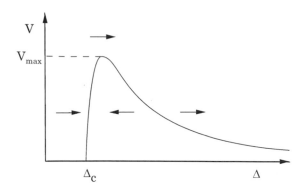

Figure 26.6. A graph of the possible steady states of single-ice-lens growth showing the undercooling of the ice-lens front $\Delta = GH$ corresponding to a given pulling speed $V$. The arrows show the directions in which the front migrates, which shows that the left-hand branch of solutions is stable, while the right-hand branch is unstable. If the pulling speed is too great then no steady state exists and the planar front recedes to positions of greater undercooling.

thermomolecular pressure to overcome the overburden. We also see that there are no steady solutions with a growth rate greater than

$$V_{\max} = \frac{A}{3.4^{4/3}\pi\mu a^2} \left(\frac{2\gamma_{sl}}{a}\right)^{1/3} \left(\frac{\pi a^2}{F_O}\right)^{1/3}. \tag{35}$$

A time-dependent analysis shows that the system evolves, as shown by the arrows in Fig. 26.6, which indicates that the left-hand branch of steady solutions is stable, the right-hand branch is unstable, and that the front of the ice lens recedes to colder temperatures if $V > V_{\max}$. It is anticipated that the planar ice-lens front will become unstable once its undercooling becomes too large, but proof of this awaits further study.

## 26.7. Epilog

The phenomenon of premelting gives rise to interesting fluid mechanical problems in which intermolecular interactions dictate both the pressure field driving a flow and the geometry of the flow domain. Confining our attention to van der Waals interactions, we have illustrated a few fundamental flows driven by premelting. The scale of premelted liquid films is of the order of nanometers but the pressures within them can be many tens of atmospheres. As a result, these microscopic flows can have enormous environmental consequences, shaping the landscape in cold regions of the Earth and having devastating effects on manmade structures built in such regions. The fluid-mechanical ideas presented here provide a basis for the development of macroscopic models of the deformation of partially frozen soils that can be used to predict these effects.

## Acknowledgments

We are very grateful to Greg Dash and to Alan Rempel for their critical reading of the manuscript and to the Natural Environment Research Council of the UK and the National Science Foundation of the US for their support of our research.

## References

Anderson, J. L. 1989. Colloid transport by interfacial forces. *Ann. Rev. Fluid Mech.* **21**, 61–99.
Azouni, M. A., Casses, P., and Sergiani, B. 1997. Capture or repulsion of treated nylon particles by an ice–water interface. *Colloids Surf. A* **122**, 199–205.

Azouni, M. A., Kalita, W., and Yemmou, M. 1990. On the particle behaviour in front of advancing liquid–ice interface. *J. Cryst. Growth* **99**, 201–205.

Chernov, A. A., Temkin, D. E., and Mel'nikova, A. M. 1977. The influence of the thermal conductivity of a macroparticle on its capture by a crystal growing from a melt. *Sov. Phys. Crystallogr.* **22**, 656–658.

Dash, J. G., Fu, H-Y., and Wettlaufer, J. S. 1995. The premelting of ice and its environmental consequences. *Rep. Prog. Phys.* **58**, 115–167.

Faraday, M. 1860. Note on regelation. *Proc. R. Soc. London* **10**, 440–450.

Gilpin, R. R. 1979. A model of the "liquid-like" layer between ice and a substrate with applications to wire regelation and particle migration. *J. Colloid Interface Sci.* **68**, 235–251.

Gilpin, R. R. 1980. Wire regelation at low temperatures. *J. Colloid Interface Sci.* **72**, 435–448.

Levich, V. G. 1962. *Physicochemical Hydrodynamics* (Prentice-Hall, Englewood Cliffs, NJ), p. 384.

Lipowsky, R. 1982. Critical surface phenomena at first-order bulk transitions. *Phys. Rev. Lett.* **49**, 1575–1578.

Lipowsky, R. 1984. Upper critical dimension for wetting in systems with long-range forces. *Phys. Rev. Lett.* **52**, 1429–1432.

Lipp, G. and Körber, Ch. 1993. On the engulfment of spherical particles by a moving ice–liquid interface. *J. Cryst. Growth* **130**, 475–489.

Lipp, G., Köber, Ch., and Rau, G. 1990. Critical growth rates of advancing ice–water interfaces for particle rejection. *J. Cryst. Growth* **99**, 206–210.

Löwen, H., Beier, T., and Wagner, H. 1989. Van der Waals theory of surface melting. *Europhys. Lett.* **9**, 791–796.

Mantovani, S., Valeri, S., Loria, A., and del-Pennino, U. 1980. Viscosity of the ice surface layer. *J. Chem. Phys.* **72**, 1077–1083.

Nye, J. F. 1967. Theory of regelation. *Philos. Mag.* **16**, 1249–1266.

Pötschke, J. and Rogge, V. 1989. On the behaviour of foreign particles at an advancing solid–liquid interface. *J. Cryst. Growth* **94**, 726–738.

Tabor, D. 1991. *Gases, Liquids and Solids*. (Cambridge U. Press, Cambridge).

Telford, J. W. and Turner, J. S. 1963. The motion of a wire through ice. *Philos. Mag.* **8**, 527–531.

Tyndall, J. 1858. On some physical properties of ice. *Proc. R. Soc. London* **9**, 76–80.

Wettlaufer, J. S. and Worster, M. G. 1995. Dynamics of premelted films: frost heave in a capillary. *Phys. Rev. E* **51**, 4679–4689.

Wettlaufer, J. S., Worster, M. G., Wilen, L. A., and Dash, J. G. 1996. A theory of premelting dynamics for all power law forces. *Phys. Rev. Lett.* **76**, 3602–3605.

Wilen, L. A. and Dash, J. G. 1995. Frost heave dynamics at a single crystal interface. *Phys. Rev. Lett.* **74**, 5076–5079.

Wilen, L. A., Wettlaufer, J. S., Elbaum, M., and Schick, M. 1995. Dispersion force effects in interfacial premelting of ice. *Phys. Rev. B* **52**, 426–433.

Yih, C-S. 1986. Movement of liquid inclusions in soluble solids: an inverse Stokes' law. *Phys. Fluids* **29**, 2785–2787.

Yih, C-S. 1987. Convective instability of a spherical fluid inclusion. *Phys. Fluids* **30**, 36–44.

# 27

# Recent Advances in Lattice Boltzmann Methods

SHIYI CHEN, GARY D. DOOLEN, XIAOYI HE, XIAOBO NIE, AND
RAOYANG ZHANG

## 27.1. Introduction

In recent years, the lattice Boltzmann method (LBM) (Chen and Doolen, 1998) has emerged as an alternative and promising numerical scheme for simulating fluid flows and modeling physics in fluids. Unlike conventional numerical schemes based on discretizations of macroscopic continuum equations such as the Navier–Stokes (NS) equations, the LBM method is based on mesoscopic kinetic equations and the particle distribution function. The fundamental idea in the LBM is to construct simplified kinetic models that incorporate the essential mesoscopic physics so that the macroscopic averaged properties obey the desired macroscopic equations. By using a simplified version of the kinetic equation, one avoids solving complicated kinetic equations such as the full Boltzmann equation, and one avoids following each particle as in molecular dynamics simulations.

From the point of view of computational fluid dynamics (CFD) the kinetic nature of the LBM provides three important advantages. First, the convection operator in the LBM is linear, allowing the incorporation of upwind algorithms and avoiding the use of nonlinear Riemann solvers. Simple convection combined with a relaxation process (or collision operator) allows the recovery of the nonlinear macroscopic advection through multiscale expansions. Second, the incompressible NS equations are obtained in the nearly incompressible limit of the LBM. This eliminates the necessity of solving the Poisson equation for the pressure, which is the normal procedure for solving the incompressible NS equations. Third, in contrast to traditional kinetic theory such as the continuum Boltzmann equation approach, the LBM seeks the minimum set of velocities in phase space. Only a very few velocities are used in the LBM, and the transformation relating the mesoscopic distribution function and macroscopic quantities, including mass, momentum, and energy, is greatly simplified, consisting of simple arithmetic calculations.

The lattice Boltzmann equation for the discrete particle distribution function, similar to the kinetic equation in the lattice gas method (Frisch et al., 1986), can be written as

$$f_i(\mathbf{x} + \mathbf{e}_i \delta x, t + \delta t) - f_i(\mathbf{x}, t) = \Omega_i[f(\mathbf{x}, t)], \qquad (i = 0, 1, \ldots, M), \tag{1}$$

where $f_i$ is the particle velocity distribution function along the $i$th direction, $M$ is the number of particle speed, and $\Omega_i = \Omega_i[f(\mathbf{x}, t)]$ is the collision operator that determines the rate of change of $f_i$ due to collision. The most straightforward choice of collision operator is the linearized collision operator with a single relaxation time $\tau$ or the Bhatnagar-Gross-Krook (BGK) approximation (Chen et al., 1991; Qian et al., 1992):

$$\Omega_i = -\frac{f_i - f_i^{(eq)}}{\tau}, \tag{2}$$

where $f_i^{(eq)}$ is the local equilibrium distribution function. $\Omega_i$ is required for satisfying conservation of total mass and total momentum at each lattice site in order to recover the NS equations:

$$\sum_i \Omega_i = 0, \qquad \sum_i \Omega_i \mathbf{e}_i = 0, \tag{3}$$

where $\sum_i \equiv \sum_{i=1}^{M}$.

The macroscopic quantities, such as density $\rho$ and momentum density $\rho\mathbf{u}$, are defined as the particle velocity moments of the distribution function $f_i$:

$$\rho = \sum_i f_i, \qquad \rho\mathbf{u} = \sum_i f_i \mathbf{e}_i, \tag{4}$$

It is assumed that the lattice spacing $\Delta x$ and the time increment $\Delta t$ in Eq. (1) can be treated as small parameters of the same order, $\varepsilon$. Performing a Taylor expansion in time and space, we obtain the following continuum form of the kinetic equation accurate to second order in $\varepsilon$:

$$\frac{\partial f_i}{\partial t} + \mathbf{e}_i \cdot \nabla f_i + \varepsilon \left( \frac{1}{2} \mathbf{e}_i \mathbf{e}_i : \nabla\nabla f_i + \mathbf{e}_i \cdot \nabla \frac{\partial f_i}{\partial t} + \frac{1}{2} \frac{\partial^2 f_i}{\partial t^2} \right) = \frac{\Omega_i}{\varepsilon}. \tag{5}$$

To derive the macroscopic hydrodynamic equation, we utilize the Chapman–Enskog expansion by assuming that

$$\frac{\partial}{\partial t} = \varepsilon \frac{\partial}{\partial t_1} + \varepsilon^2 \frac{\partial}{\partial t_2}, \qquad \frac{\partial}{\partial x} = \varepsilon \frac{\partial}{\partial x_1}.$$

Likewise, the one-particle distribution function $f_i$ can be expanded formally about the local equilibrium distribution function $f_i^{eq}$:

$$f_i = f_i^{eq} + \varepsilon f_i^{(neq)}. \tag{6}$$

Inserting the above formula into Eq. (5), we obtain, by using simple algebra, the following mass and momentum equations:

$$\frac{\partial \rho}{\partial t} + \nabla \cdot \rho\mathbf{u} = 0, \tag{7}$$

$$\frac{\partial \rho\mathbf{u}}{\partial t} + \nabla \cdot \Pi = 0, \tag{8}$$

which are accurate to second order in $\varepsilon$. Here the momentum flux tensor $\Pi$ has the form

$$\Pi_{\alpha\beta} = \sum_i (\mathbf{e}_i)_\alpha (\mathbf{e}_i)_\beta \left[ f_i^{eq} + \left( 1 - \frac{1}{2\tau} \right) f_i^{(1)} \right]. \tag{9}$$

$(\mathbf{e}_i)_\alpha$ is the component of the velocity vector $\mathbf{e}_i$ in the $\alpha$-coordinate direction.

To obtain the detailed form of $\Pi$, in the two-dimensional system we use a square lattice with nine speeds: $\mathbf{e}_i = \{\cos[\pi/4(i-1)], \sin[\pi/4(i-1)]\}$ for $i = 1, 3, 5, 7$; $\mathbf{e}_i = \sqrt{2}\{\cos[\pi/4(i-1)], \sin[\pi/4(i-1)]\}$ for $i = 2, 4, 6, 8$; $\mathbf{e}_0 = 0$ corresponds to a zero speed velocity. We also assume that the equilibrium distribution has the following form (Qian et al., 1992):

$$f_i^{eq} = \rho w_i \left[ 1 + 3\mathbf{e}_i \cdot \mathbf{u} + \frac{9}{2}(\mathbf{e}_i \cdot \mathbf{u})^2 - \frac{3}{2} u^2 \right], \tag{10}$$

with $w_0 = 4/9$, $w_1 = w_3 = w_5 = w_7 = 1/9$, and $w_2 = w_4 = w_6 = w_8 = 1/36$.

Inserting the above formula into Eq. (9), we obtain the resulting momentum equation,

$$\rho\left(\frac{\partial \mathbf{u}_\alpha}{\partial t} + \nabla_\beta \cdot \mathbf{u}_\alpha \mathbf{u}_\beta\right) = -\nabla_\alpha p + \nu \nabla_\beta \cdot (\nabla_\alpha \rho \mathbf{u}_\beta + \nabla_\beta \rho \mathbf{u}_\alpha), \tag{11}$$

which is exactly as the same as the NS equation if the density variation $\delta\rho$ is small.

## 27.2. LBM Simulation of Flows in Microelectromechanical Systems

The LBM has been widely used for simulating fluid flows, including complex flows, fluid turbulence, suspension flows, and reaction diffusion systems (see the review by Chen and Doolen, 1998). Most of the applications are related to flows in the incompressible limit whose dynamics can be described by the macroscopic NS equations. Since the LBM is intrinsically kinetic, it is more general and can be used to simulate fluid flows with mean-free-path effects associated with high Knudsen numbers, such as fluid flows in microelectromechanical systems (MEMSs) (Ho and Tai, 1998). In this section we present LBM simulation results for flows in a microchannel (Nie et al., 1998).

For flows in a microchannel, the mean free path of fluid molecules could be of the same order as the typical geometric length of the device or larger. The continuum hypothesis that is fundamental for the NS equation breaks down. An important feature in these flows is the emergence of a slip velocity at the flow boundary, which strongly affects the mass and heat transfer in the system. In our LBM simulation, the microchannel consists of two parallel plates separated by a distance $h$ and the fluid flow is driven by the pressure difference between the inlet pressure $P_i$ and exit pressure $P_e$. The channel length in the longitudinal direction is $L$. The bounce-back boundary condition (Chen and Doolen, 1998) is used for the particle distribution functions at the plates (i.e., when a particle distribution hits a wall node, the particle distribution scatters back to the fluid nodes in a direction opposite to its incoming direction). The Knudsen number is defined as $K_n = l/h$, where $l$ is the mean free path of the fluid molecular.

The slip velocity $V_s$ at the exit of the microchannel is determined with the following formula:

$$u(y) = u_0(Y - Y^2 + V_s), \tag{12}$$

where $u(y)$ is the velocity along the flow direction at the exit and $Y = y/h$. $u_0$ and $V_s$ can be obtained by fitting numerical results with the least-squares method.

In Fig. 27.1, we plot the slip velocity $V_s$ and the normalized mass flow rate $M_f = M/M_0$ as functions of Knudsen number when the pressure ratio $\eta = P_i/P_e = 2$. The normalization factor,

$$M_0 = \frac{h^3 P_e}{24\nu L}(\eta^2 - 1),$$

is the mass flow rate when the velocity slip is zero. Using a least-squares fit, from Fig. 27.1 we obtain

$$V_s = 8.7 K_n^2. \tag{13}$$

If we assume that the NS equation is valid for the microflow except that the slip boundary condition $V_s$ in Eq. (13) is used to replace the traditional nonslip condition on walls [a similar procedure has been used in the engineering model (Beskok et al., 1996)], Eqs. (11) and (13)

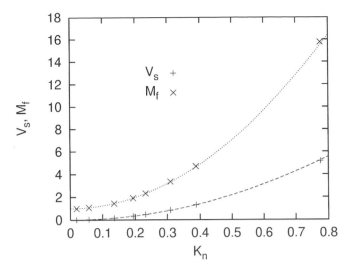

Figure 27.1. The slip velocity and the normalized mass flow rate at the exit of a microchannel flow as functions of Knudsen number for $\eta = 2$. The + and × indicate LBM numerical results. The dashed curve is for Eq. (13) and the dotted curve for Eq. (14).

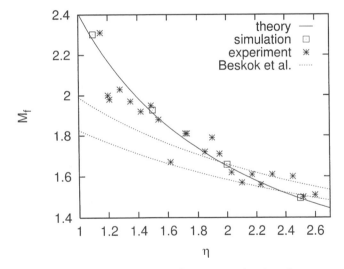

Figure 27.2. The normalized mass flow rate as a function of pressure ratio for $K_n = 0.165$. The solid curve is for Eq. (14).

give the analytical mass flow rate:

$$M_f = 1 + 12V_s(K_n)\frac{\ln(\eta)}{\eta^2 - 1}. \tag{14}$$

For $\eta = 2$, Eq. (14) becomes $M_f = 1 + 24.1 K_n^2$, which agrees with numerical results in Fig. 27.1. In Fig. 27.2 the mass flow rates as functions of pressure ratio $\eta$ for $K_n = 0.165$ are shown for our theory, the experimental work (Arkilic et al., 1995), the engineering model (Beskok et al., 1996), and the LBM simulation. Our theory and the LBM simulation agree well with the experimental measurements. It is noted that for large pressure ratios ($\eta \geq 1.8$), the LBM agrees reasonably well with that of Beskok et al. (1996). But for smaller pressure ratios, the difference increases.

## 27.3. Lattice Boltzmann Simulation of Multiphase Flows

Since its first being proposed as an useful alternative to classical CFD techniques, the LBM has proven to be useful for simulations of multiphase flow. Instead of solving macroscopic equations such as traditional CFD approaches, the LBM simulates fluid flow based on microscopic models and mesoscopic kinetic equations. Consequently, the essential microscopic processes, such as intermolecular interactions, can be easily incorporated. Phase segregation and interfacial dynamics, which are essential in multiphase flow and are difficult to handle in traditional approaches, can be accurately simulated in the LBM.

There have been several LBM models proposed for the simulation of multiphase flows. The first such model was proposed by Gunstensen et al. (1991), which was based on a two-component lattice gas model. In this model, red and blue particle distribution functions were introduced to represent two different fluids. To maintain interfaces and to separate the different phases, an extra step was introduced to force the colored fluids to move toward fluids with the same color. The second LBM model proposed by Shan and Chen (1993) used the concept of microscopic interactions between particles. An interparticle potential was introduced to model the phase segregation and surface tension. A third LBM model proposed by Swift et al. (1995) used a free-energy approach. In this model, the equilibrium distribution was modified so that the pressure tensor was consistent with that derived from a free-energy function for nonuniform fluids.

Although each of the LBM multiphase models was developed based on different physical pictures and each appears to be different, a recent study (He et al., 1998) showed that all of them have an origin in the kinetic theory. In other words, these models can be derived by discretizing the continuous Boltzmann equation by use of different approximations. An improved LBM model for multiphase flows can be obtained by systematically discretizing the continuous Boltzmann equation. In this section, we report the latest development of the multiphase model proposed by He et al. (1999) and present a benchmark study of the method applied to the Rayleigh–Taylor instability.

In this new formulation of the LBM for multiphase fluid flows (He et al., 1999), we use an index function $f$ and a pressure distribution function $g$ for tracking the evolution of the multiphase fluid. The reason for using these two distributions is that, for incompressible fluids, the density is constant in each phase and only the pressure varies. In addition, if we do not follow the density field, an index function must be used to track the two phases. The evolution equations for $f$ and $g$ are

$$\frac{\partial f_i}{\partial t} + \mathbf{e}_i \cdot \nabla f_i = -\frac{f_i - f_i^{\mathrm{eq}}}{\tau} + \frac{(\mathbf{e}_i - \mathbf{u}) \cdot \mathbf{F}_r}{RT} f_i^{\mathrm{eq}}, \tag{15}$$

and

$$\frac{\partial g_i}{\partial t} + \mathbf{e}_i \cdot \nabla g_i = -\frac{g_i - g_i^{\mathrm{eq}}}{\tau} + \frac{(\mathbf{e}_i - \mathbf{u}) \cdot (\mathbf{F}_d + \mathbf{F}_s + \mathbf{G})}{RT} g_i^{\mathrm{eq}}, \tag{16}$$

where

$$\mathbf{F}_r = \nabla \Psi(\psi), \qquad \mathbf{F}_d = \frac{1}{\rho} \nabla(\rho RT - p), \qquad \mathbf{F}_s = \kappa \rho \nabla \nabla^2 \rho, \tag{17}$$

and **G** is the gravitational force. The index function $\psi$, the pressure, and the velocity are calculated with

$$\psi = \sum f_i, \qquad p = \sum g_i, \qquad p\mathbf{u} = \sum g_i \mathbf{e}_i. \tag{18}$$

$\Psi$ controls the rate of segregation of the different phases and $\kappa$ controls the surface tension. The above governing equations are solved with a second-order finite-differencing scheme. The equilibrium distribution functions have the Maxwellian form:

$$f_i^{\text{eq}} = \frac{\psi}{(2\pi RT)^{D/2}} \exp\left[-\frac{(\mathbf{e}_i - \mathbf{u})^2}{2RT}\right], \tag{19}$$

$$g_i^{\text{eq}} = \frac{p}{(2\pi RT)^{D/2}} \exp\left[-\frac{(\mathbf{e}_i - \mathbf{u})^2}{2RT}\right]. \tag{20}$$

For two-dimensional simulation, the discrete velocities are chosen to be the nine velocities discussed above; for three-dimensional simulations, a 15-velocity model can be used (Qian et al., 1992).

## 27.4. Rayleigh–Taylor Instability

When a heavy fluid is placed on top of a light fluid in a gravitational field with gravity pointing downward, the initial planar interface is unstable. Any disturbance will grow and produce spikes of heavy fluids moving downward and bubbles of light fluids moving upward. This is the so-called Rayleigh–Taylor instability (for a review, see Sharp, 1984).

In this study, we focus on the Rayleigh–Taylor instability for a single-mode disturbance. The amplitude of the initial perturbation is chosen to be 10% of the channel width. The computational domain is a two-dimensional square or a three-dimensional box. Nonslip boundary conditions are applied at the top and the bottom walls. Periodic boundary conditions are applied at the sides. The density ratio of heavy fluid to light fluid is 3. This corresponds to an Atwood number of 0.5 [$A = (\rho_h - \rho_l)/(\rho_h + \rho_l)$ where $\rho_h$ and $\rho_l$ are densities of heavy and light fluids, respectively]. The kinetic viscosity was assumed to be the same for both heavy and light fluids. Surface tension was neglected in the simulation.

We present our results in terms of nondimensional variables. We took the channel width $W$ as the length scale and $T = \sqrt{W/g}$ as the time scale, where g is the gravitational acceleration. The characteristic speed is then defined as $\sqrt{AgW}$.

### 27.4.1. *Two-Dimensional Simulation*

A two-dimensional simulation was carried out on a $256 \times 1024$ grid. The Reynolds number, $\text{Re} = \sqrt{gW}W/\nu$ in this simulation, was 2048. Figure 27.3 shows the time evolution of the interface. The interface was represented by 19 equal-spaced isodensity contours. As expected, the heavy fluid forms downward spikes while the light fluid rises to form bubbles. The interface remains symmetric during the early stages of growth ($t < 1.0$). After that, the Kelvin–Helmholtz instability becomes important and the spike of the heavy fluid begins to roll up at the edges. The rollups become quite evident at $t = 2.5$ and $t = 3.0$. Although many of the previous Rayleigh–Taylor studies stop at this time, we continue our simulation to much later times. As

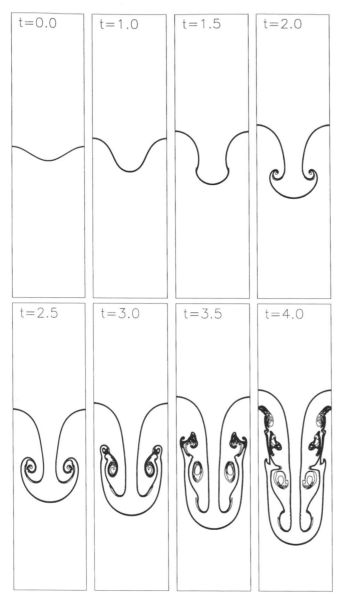

Figure 27.3. Time evolution of the interface for the two-dimensional Rayleigh–Taylor instability. The Atwood number is 0.5, the Reynolds number is 2048, and the viscosity ratio is 1. Surface tension is neglected.

shown, the bubble and the front of the spike remain rather smooth and the interfaces along the edge of the spike are stretched and folded into very complicated shapes. The mixing of heavy and light fluids is considerable.

Previous theoretical and numerical studies showed that when the amplitude becomes larger than 0.4 times the wavelength, the bubble speed approaches a constant. In our study, the final bubble speed, $V_B/\sqrt{AgW}$, was found to be 0.270. This result compares well with the value of 0.265 obtained by Tryggvason (1988) with a front-tracking approach.

### 27.4.2. *Three-Dimensional Simulation*

A three-dimensional simulation was carried out on a $128 \times 128 \times 512$ grid with a Reynolds number of 1024. Figure 27.4 shows pictures of the interface at two different time steps. The left panel is a view of the interface from the heavy fluid side, and the right panel is a view of the interface from the light fluid side. As expected, the heavy and the light fluids penetrate into each other as time increases. The spike grows at approximately the same speed as the bubble during the early stages ($t < 1.0$), but gradually the spike becomes faster than the bubble ($t > 1.0$).

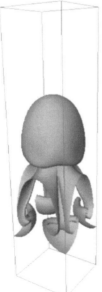 

Figure 27.4. Interface shapes at two times in three-dimensional Rayleigh–Taylor instability. The Atwood number is 0.5, the Reynolds number is 1024, and the viscosity ratio is 1. Surface tension is neglected. The left panel is a view of the interface from the heavy fluid side, and the right panel is a view of the interface from the light fluid side. The upper row is at $t = 1.0$ and the lower row is at $t = 3.0$.

The rollups of the spike and the bubble become obvious at the large-amplitude stage ($t = 3.0$). The shapes of the bubble and the spike are similar to the shapes observed by Li et al. (1996), who used a level-set approach.

## 27.5. Two-Dimensional Isotropic Multiphase Turbulence

Since there is no vortex stretching in two-dimensional NS, both the kinetic energy and the enstrophy (vorticity squared) are conserved. Therefore there are two cascade processes: the direct enstrophy cascade and the inverse energy cascade. A dimensional analysis can be performed for both scalings, which shows that $E(k) \sim \beta^{2/3} k^{-3}$ for the direct enstrophy cascade range and $E(k) \sim \varepsilon^{2/3} k^{-5/3}$ for the inverse cascade range. Here $\beta$ and $\varepsilon$ are the enstrophy flux and the energy flux, respectively.

While single-phase two-dimensional turbulence has been studied extensively (see, for example, the review article by Kraichnan and Montgomery, 1980), two-phase two-dimensional immiscible turbulence is a relatively undeveloped field. Although a combined approach in which both analytical theories and direct numerical simulations are used has been very effective for the two-dimensional single-phase turbulence, these methodologies cannot be extended to a two-phase immiscible fluid in a straightforward manner. Conventional numerical simulations have limitations that do not permit a clear physical understanding of the interface dynamics.

In order to achieve a statistical steady state, a large-scale forcing is applied in physical space: $F_i = \mathbf{e}_i \cdot \sum_{k_{1n}^2 + k_{2n}^2 \leq N^2} \{k_{2n}(-A_n \sin\phi + B_n \cos\phi)\vec{i}_x + K_{1n}(A_n \sin\phi - B_n \cos\phi)\vec{i}_y\}$ with $\phi = k_{1n} \cdot x + k_{2n} \cdot y$ and $\vec{k}_n = (k_{1n}, k_{2n})$. In this way, the energy is injected into the first three modes, $N \leq 3$, in physical space. Here $\vec{k}$ is the wave vector, $c$ is a parameter that controls the force, and $A_n$ and $B_n$ are the random numbers in the range [0, 1]. This forcing is periodic and satisfies the incompressibility constraint. The boundary conditions are periodic in both directions.

To compare single-phase and two-phase turbulence in the enstrophy cascade range, we present in Fig. 27.5(a) the energy spectrum for a stationary state from a single-phase LBM simulation. This simulation was carried out with a $1024 \times 1024$ grid. The Taylor microscale Reynolds number, $Re_\lambda$, is ~80. $Re_\lambda$ is defined as $u\lambda/\nu$, where $u$ and $\lambda$ are the root mean square of the velocity fluctuations and the Taylor microscale. Since the flow is forced at the largest scales, only the enstrophy cascade range is well resolved. The power law of the energy spectrum is seen for this LBM simulation to scale approximately as $k^{-3}$ in the enstrophy cascade range. This compares well with the theoretical prediction (Kraichnan and Montgomery, 1980).

Next, we present results from our simulations of two-dimensional two-phase immiscible fluid turbulence subject to the same forcing at large scales. The simulation was carried out with a scheme proposed by Gunstensen et al. (1993) that colors the two components red and blue. The viscosity and the density of the two components are identical. Initially we assume that the two fluids mix completely, i.e., the density at each point is assigned to be identical. After the large-scale stirring force is introduced, each component begins to move, merge, and segregate because of the competition between the external forces and surface tension. Depending on the strength of the surface tension, different flow patterns appear. After the energy spectrum reaches a stationary state, the flow properties are measured. Figure 27.6 illustrates the density and vorticity distribution of the two components at a late time. It is seen that, under the influence of the large-scale force and the surface tension, the two components are stretched and separated. Both large-scale and small-scale vortices are observed.

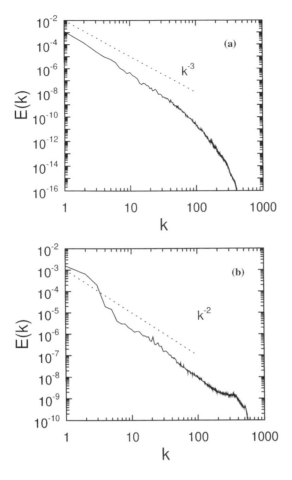

Figure 27.5. Energy spectrum for the two-dimensional forced (a) single-phase, (b) two-phase turbulence.

In Fig. 27.5(b) we present the energy spectrum of the two-phase flow. It is seen that the power law of energy spectrum is $\sim k^{-2}$ in the inertial range. This result is different from the $k^{-3}$ behavior of single-phase turbulence and the results of Esmaeeli and Tryggvason (1996). Their energy spectrum from a low Reynolds number simulation of bubbly flows gives a larger energy decay rate. The new scaling of $k^{-2}$ is similar to the typical scaling of the Burgers equation in which the shock structure dominates the dynamics of the velocity field. We argue that in two-phase fluid turbulence, in particular when the surface phenomenon dominates, the small scales are completely characterized by the surface velocity $V$ and the surface thickness $1/k$, resulting in $E(k) \sim k^{-2}$ from dimensional arguments.

## 27.6. Concluding Remarks

In this chapter we have briefly presented the basic principles of the lattice Boltzmann method and described several applications, including the LBM simulation of the microchannel, multiphase mixing, and two-dimensional multiphase turbulence. Simulation results from the LBM agree well with results from existing experimental and other numerical simulations.

We would like to emphasize that the LBM is a useful mesoscopic dynamical description of physical phenomena. The scheme is most suitable for fluid problems for which macroscopic

Figure 27.6. The density distribution of the two-dimensional forced immiscible two-phase turbulence.

hydrodynamics and mesoscopic statistics are both important. Even though the LBM originates from particle dynamics and uses the particle distribution function, the scheme describes the averaged macroscopic dynamics. In most cases the LBM has been treated as a numerical scheme rather than as a mesoscopic physical model. However, the utilization of the particle description or the kinetic equation provides the advantages of particle dynamics and kinetic theory, including clear physical understanding, easy boundary implementation and efficient parallel algorithms.

## References

Arkilic, E., Schmidt, M. A., and Breuer, K. S. 1995. Gaseous slip flow in long micro-channels. *J. MEMS.* **6**, 167.

Beskok, A., Karniadakis, G. E., and Trimmer, W. 1996. Rarefaction and compressibility effects in gas micro-flows. *J. Fluids Eng.* **118**, 448.

Chen, S., Chen, H. D., Martinez, D., and Matthaeus, W. 1991. Lattice Boltzmann model for simulation of magnetohydro-dynamics. *Phys. Rev. Lett.* **67**, 3776.

Chen, S. and Doolen, G. D. 1998. Lattice Boltzmann method for fluid flows. *Ann. Rev. Fluid Mech.* **30**, 3294.

Esmaeeli, A. and Tryggvason, G. 1996. An inverse energy cascade in two-dimensional low Reynolds number bubbly flows. *J. Fluid Mech.* **314**, 315.

Frisch, U., Hasslacher, B., and Pomeau, Y. 1986. Lattice gas automata for the Navier–Stokes equations. *Phys. Rev. Lett.* **56**, 1505.

Gunstensen, A. K., Rothman, D. H., Zaleski, S., and Zanetti, G. 1991. Lattice Boltzmann model of immiscible fluids. *Phys. Rev. A* **43**, 4320.

He, X., Shan, X., and Doolen, G. D. 1998. A discrete Boltzmann equation model for nonideal gases. *Phys. Rev. E* **57**, R13.

He, X., Zhang, R., and Chen, S. 1999. *J. Comput. Phys,* (in press).

Ho, C. M. and Tai, Y. C. 1998. Micro-electro-mechanical-systems (MEMS) and fluid flows. *Ann. Rev. Fluid Mech.* **30**, 579.

Kraichnan, R. H. and Montgomery, D. 1980. Two dimensional turbulence. *Rep. Prog. Phys.* **43**, 1385.

Li, X. L., Jin, B. X., and Glimm, J. 1996. Numerical study for the three-dimensional Rayleigh–Taylor instability through the TVD/AC scheme and parallel computation. *J. Comput. Phys.* **126**, 343.

Nie, X., Doolen, G. D., and Chen, S. 1998. Lattice Boltzmann simulation of fluid flows in MEMS, submitted to *Phys. Fluids.*

Qian, Y. H., d'Humières, D., and Lallemand, P. 1992. Lattice BGK models for Navier–Stokes equation. *Europhys. Lett.* **17**, 479.

Shan, X. and Chen, H. 1993. Lattice Boltzmann model for simulating flows with multiple phase and components. *Phys. Rev. E* **47**, 1815.

Sharp, D. H. 1984. An overview of Rayleigh–Taylor instability. *Physica D* **12**, 3.

Swift, M. R., Osborn, W. R., and Yeomans, J. M. 1995. Lattice Boltzmann simulation of nonideal fluids. *Phys. Rev. Lett.* **75**, 830.

Tryggvason, G. 1988. Numerical simulation of the Rayleigh–Taylor instability. *J. Comput. Phys.* **75**, 253.

# 28

# Bubble Dynamics in Heterogeneous Boiling Heat Transfer

RENWEI MEI, JAMES F. KLAUSNER, AND GLEN THORNCROFT

## 28.1. Introduction

Studies on nucleate boiling phenomena intrinsically deal with the nucleation, growth, and detachment of bubbles from a heater surface. The discrete bubble portion of the nucleate boiling regime is characterized by rather weak thermal and hydrodynamic interactions among individual bubbles. Nucleate flow boiling heat transfer results from the bulk two-phase turbulent convection and the phase change associated with individual bubbles. Because of the disparate length scales associated with each mechanism, the former is referred to as the macroconvective component while the latter is referred to as the microconvective component. It has been well established that microconvection contributes significantly to the overall heat transfer.

The rate of energy removal by a bubble from the heating surface due to evaporation is directly proportional to the rate of bubble growth. In a pool boiling system with an upward-facing heating surface, the bubble will lift off its nucleation site because of the increasing buoyancy force with increasing bubble size. In a flow boiling system with an upward-facing heating surface, the bubble can directly lift off the nucleation site provided the flow velocity is relatively small. If the liquid flow velocity is large, the bubble can be forced to depart from the nucleation site while continuing to slide along the heating surface. During the sliding process, the bubble withdraws energy from the heater surface and continues to grow. The bubble lifts off from the heater surface at a later stage when the buoyancy force or the shear lift force becomes sufficiently large. If the heating surface is oriented vertically, the departure and lift-off of the bubble are further complicated by the direction of the flow. Hence the microconvection heat transfer strongly depends on the rate of bubble growth, the size of the bubble at the point of its departure from the nucleation site, and the size of the bubble lifting off the heating surface.

Because of the important role of bubble detachment diameters in the boiling heat transfer process, many investigators (e.g., Fritz and Ende, 1936; Chang, 1963; Levy, 1967; Koumoutsos et al., 1968; Cole and Rohsenow, 1969; and Cooper et al., 1983) have attempted to predict the detachment diameter of bubbles in pool and flow boiling systems by considering the balance of various forces on the bubble. In general, there are a number of forces acting on the bubble in a flow boiling system: buoyancy force, surface-tension force, quasi-steady drag, added-mass force, and shear-lift force. Only limited success was achieved in comparing the prediction with the measured bubble detachment data because of the complexity of the problem and lack of sufficient information to model various forces reliably. More recent modeling efforts (Klausner et al., 1993; Zeng et al., 1993a, 1993b; Kandlikar and Stumm, 1995; van Helden, 1994; and Thorncroft, 1997) have produced better agreement with experimental data over a large range of conditions. However, significant differences exist among various models, and empirical constants are necessary to close the models.

In this chapter the bubble dynamics are considered near the point of departure and during the sliding and lift-off processes. The model of Klausner et al. (1993) is reexamined. A more rigorous derivation for the forces on the bubble is presented. Effects of the wall on some of the forces are theoretically examined. A rationale for bubble departure and lift-off criteria is provided. The model is assessed by use of more recent experimental data (Thorncroft, 1997). Necessary empiricism in the model is discussed.

## 28.2. Forces on a Growing Bubble

### 28.2.1. *Total Force*

To understand the dynamics of a growing bubble attached to a wall [Fig. 28.1(a)], it is instructive to start from an expression for the total force on the bubble. A free-body diagram (FBD) for all the forces on the vapor bubble is shown in Fig. 28.1(b), in which both the liquid and the solid walls are removed. The area denoted by $S_1$ is exposed to the liquid flow and the area $S_2$ is in contact with the heater with a contact diameter $d_w$. The pressure inside the bubble is assumed uniform and equal to the vapor pressure $p_v$. The effect of the liquid on the vapor bubble in the FBD is the stress vector $\underline{\sigma}$ and the liquid–vapor interfacial force, which is represented by the liquid–vapor surface-tension force $\underline{F}_S$. The effect of the wall is represented by the vapor pressure on $S_2$ and a reaction force $\underline{R}$. This reaction force $\underline{R}$ acts on the vapor through the three-phase common line. For simplicity, the gravitational force is assumed to be in the negative $y$-direction; the result can be easily generalized to an arbitrary orientation of the flow. The total force on the bubble is

$$\underline{F} = \rho_v \mathcal{V}_b \underline{g} + \underline{F}_S + \int_{S_1} \underline{\underline{\sigma}} \cdot \underline{n} \, dA + \int_{S_2} p_v(-\underline{n}) \, dA + \underline{R}, \tag{1}$$

where $\rho_v$ is the vapor density, $\mathcal{V}_b$ is the volume of the bubble, $\underline{g}$ is the gravitational acceleration, $\underline{\underline{\sigma}}$ is the liquid stress tensor on the bubble surface $S_1$, and $\underline{n}$ is the outward normal of the bubble.

### 28.2.2. *Surface-Tension Force*

For a given configuration the surface-tension force can be evaluated independently of the shear stress or the pressure. Assuming that the contact area is circular and the contact angle

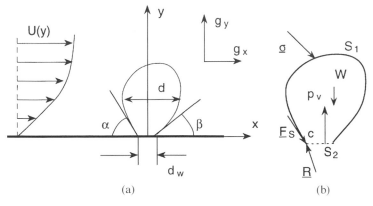

(a)                                                    (b)

Figure 28.1. (a) Schematic diagram of a bubble attached to a heater surface, (b) free-body diagram.

follows a third-order polynomial variation along the contact periphery, Klausner et al. (1993) obtained the following approximations for the $x$ and the $y$ components of the surface-tension force:

$$F_{Sx} \sim -1.25\, d_w \sigma_s \frac{\pi(\alpha - \beta)}{\pi^2 - (\alpha - \beta)^2}(\sin\alpha + \sin\beta), \tag{2}$$

$$F_{Sy} \sim -d_w \sigma_s \frac{\pi}{\alpha - \beta}(\cos\beta - \cos\alpha). \tag{3}$$

Here $\sigma_s$ is the surface-tension coefficient and $(\alpha, \beta)$ are the advancing and receding contact angles.

### 28.2.3. Buoyancy Force and Contact Force

On $S_1$, the stress of the liquid flow is

$$\underline{\underline{\sigma}} \cdot \underline{n} = -(p - \rho_l gy)\underline{n} + \underline{\underline{\tau}} \cdot \underline{n}, \tag{4}$$

where $p$ is the hydrodynamic pressure, $\rho_l$ is the liquid density, and $\underline{\underline{\tau}}$ is the deviatoric stress tensor due to viscous effects. The total force on the bubble can thus be expressed as

$$\underline{F} = \underline{R} + \underline{F}_S + \rho_v \mathcal{V}_b \underline{g} - \int_{S_1}(p - \rho_l gy)\underline{n}\, dA - \int_{S_2} p_v \underline{n}\, dA + \int_{S_1} \underline{\underline{\tau}} \cdot \underline{n}\, dA. \tag{5}$$

Since the integration on $S_1$ is not a closed integral, it is convenient to introduce a reference point $c$ [Fig. 28.1(b)] at the intersection of the common line and the $x$ axis. The last three terms in Eq. (5) are

$$-\int_{S_1}(p - p_c)\underline{n}\, dA - \int_{S_1}(p_c - \rho_l gy)\underline{n}\, dA - \int_{S_2} p_v \underline{n}\, dA + \int_{S_1} \underline{\underline{\tau}} \cdot \underline{n}\, dA$$

$$= -\int_{S_1}(p - p_c)\underline{n}\, dA + \int_{S_1} \underline{\underline{\tau}} \cdot \underline{n}\, dA - \int_{S_1 + S_2}(p_c - \rho_l gy)\underline{n}\, dA$$

$$+ \int_{S_2}(p_c - \rho_l gy - p_v)\underline{n}\, dA. \tag{6}$$

The third term on the right-hand-side of Eq. (6) gives the buoyancy force, $\rho_l \mathcal{V}_b g \underline{e}_y$; in general, this is $-\rho_l \mathcal{V}_b \underline{g}$. The last term in Eq. (6) was identified as the "contact pressure force" in Klausner et al. (1993) since $S_2$ is the contact area. The pressure difference, $p_c - p_v$, is related to the surface-tension at $r$ where $y = 0$ so that

$$\int_{S_2}(p_c - \rho_l gy - p_v)\underline{n}\, dA = \int_{S_2}(p_c - p_v)\underline{n}\, dA = -A_2(p_c - p_v)\underline{e}_y = A_2 \frac{2\sigma_s}{r_c}\underline{e}_y. \tag{7}$$

In the above, $r_c$ is the radius of curvature at the reference point $c$ and $A_2 = [(\pi/4)d_w^2]$ is the contact area. This force is in the positive $y$-direction; hence it aids the lift-off of the bubble. Thus,

$$\underline{F} = \underline{R} + \underline{F}_S - (\rho_l - \rho_v)\mathcal{V}_b \underline{g} + A_2 \frac{2\sigma_s}{r_c}\underline{e}_y - \int_{S_1}(p - p_c)\underline{n}\, dA + \int_{S_1} \underline{\underline{\tau_n}} \cdot \underline{n}\, dA. \tag{8}$$

### 28.2.4. *Hydrodynamic Force*

The last two terms in Eq. (8) contain both inviscid and viscous effects on an incomplete surface $S_1$. A better insight on these terms may be gained by first considering a growing bubble of radius $a(t)$ translating with a velocity $V(t)$ in an unbounded, unsteady, uniform flow field of velocity $U(t)$. Typically, the hydrodynamic force on such a bubble consists of (i) a quasi-steady drag, (ii) a history force, (iii) an added-mass force, and (iv) a force due to the free-stream acceleration. For a clean bubble, the history force has been found to be small (Mei and Klausner, 1992) because of the mobility of the bubble–liquid interface and is thus neglected in the present study.

For the quasi-steady drag, Mei et al. (1994) gave the following expression:

$$F_{\text{qs}} = 6\pi\rho_l\nu(U - V)a\left\{\frac{2}{3} + \left[\frac{12}{\text{Re}} + 0.75\left(1 + \frac{3.315}{\text{Re}^{1/2}}\right)\right]^{-1}\right\},\tag{9}$$

where the Reynolds number is

$$\text{Re} = 2a(t)|U(t) - V(t)|/\nu\tag{10}$$

and $\nu$ is the kinematic viscosity of the liquid. It agrees well with both numerical simulations and experimental measurements.

In the creeping flow regime the growth velocity $\dot{a}$ does not contribute to the quasi-steady drag $F_{\text{qs}}$ since the stream function for a uniform flow over a growing clean bubble is $\psi = \frac{1}{2}U(R^2 - Ra)\sin^2\theta - \dot{a}a^2\cos\theta$. The first part gives $\frac{2}{3}6\pi\rho_l\nu(U - V)a$ to $F_{\text{qs}}$. The second part is an irrotational source flow that leads to only a symmetric deviatoric normal stress on the bubble surface. It does not contribute to the pressure in the creeping flow.

The added-mass force associated with $dU/dt$, $dV/dt$, and $\dot{a}(t)$, together with the free-stream acceleration force, can be obtained by the solution of the following inviscid flow problem for the velocity potential $\phi$ in a coordinate that coincides with the sphere and moves with velocity $V(t)$:

$$\nabla^2\phi = 0;\quad \frac{\partial\phi}{\partial R}\Big|_{R=a} = \dot{a}(t) + V(t)\cos\theta;\quad \frac{\partial\phi}{\partial x}\Big|_{R\to\infty} = U(t).\tag{11}$$

The solution for $\phi$ is

$$\phi = -\dot{a}\frac{a^2}{R} + U\left(R + \frac{a^3}{2R^2}\right)\cos\theta - V\frac{a^3}{2R^2}\cos\theta.\tag{12}$$

In the above $(R, \theta)$ are the radius and the angle in spherical coordinates attached to the bubble center. The pressure distribution $p(\theta)$ around the bubble can be determined from the unsteady Bernoulli equation in a moving reference frame and the unsteady force can be obtained through integration:

$$F_{\text{FS}} + F_{\text{AM}} = \frac{4}{3}\pi\rho_l a^3\frac{dU}{dt} + \frac{1}{2}\frac{4}{3}\pi\rho_l a^3\left(\frac{dU}{dt} - \frac{dV}{dt}\right) + 2\pi\rho_l a^2(U - V)\dot{a}.\tag{13}$$

The first term is due to the free-stream acceleration, the second term is the conventional added-mass force, and the last term is the added-mass force associated with bubble growth. If the flow is nonuniform in space, $dU/dt$ in Eq. (13) must be replaced by the substantial derivative $D\underline{U}/Dt$.

In pool boiling, $U = 0$ and $V(t) = \dot{a}$. In the absence of a wall effect, the added-mass force leads to an unsteady growth force in the direction normal to the wall because of bubble growth:

$$F_{\text{growth}} = -\frac{2}{3}\pi\rho_l a^3 \ddot{a} - 2\pi\rho_l a^2 \dot{a}^2 = -\pi\rho_l a^2 \left(2\dot{a}^2 + \frac{2}{3}a\ddot{a}\right). \tag{14}$$

In Klausner et al. (1993), a hemispherical bubble was used and $F_{\text{growth}} \sim -\pi\rho_l a^2(\frac{3}{2}\dot{a}^2 + a\ddot{a})$. This is not significantly different from the present estimate based on the complete sphere model.

### 28.2.5. *Wall Effects on the Quasi-Steady Force due to the Translation of the Bubble*

After a bubble departs from the nucleation site, it often continues to slide along the heater. The wall effect on $F_{\text{qs}}$ at finite Re is difficult to evaluate exactly. However, at high Re, it may be estimated as follows. Suppose that the velocity potential $\phi$ of the inviscid flow due to a bubble at a distance $h$ above the wall moving in the $x$-direction with a velocity $V$ [Fig. 28.2(a)] is known. The drag on the bubble at high Re can be obtained by evaluation of the energy dissipation of the potential flow (Batchelor, 1967, p. 368) as

$$F_{\text{qs}} V = \rho_l \nu \int_A \underline{n} \cdot \nabla q^2 \, dA, \tag{15}$$

where $q^2 = \nabla\phi \cdot \nabla\phi$. For $(a/2h) \ll 1$, the velocity potential in the presence of the wall can be found from Lamb (1932, p. 133) through a regular perturbation analysis:

$$\phi \sim -\frac{1}{2}V\frac{a^3}{R^2}\cos\theta - \frac{1}{4}V\frac{a^6}{(2h)^3}\frac{\cos\theta}{R^2} - \frac{1}{2}V\frac{a^3}{R'^2}\cos\theta' - \frac{1}{4}V\frac{a^6}{(2h)^3}\frac{\cos\theta'}{R'^2} + O\left[\left(\frac{a}{2h}\right)^6\right]. \tag{16}$$

The integration in Eq. (15) is numerically evaluated. For $h \gg a$, the result can be fit in a simple form as

$$F_{\text{qs}}(\text{Re} \to \infty) \sim 12\rho_l \nu V a \left\{1 + \left(\frac{a}{2h}\right)^3 + O\left[\left(\frac{a}{2h}\right)^6\right]\right\}. \tag{17}$$

The presence of the wall leads to an increase in the drag by 10%–12.5% at high Re. This

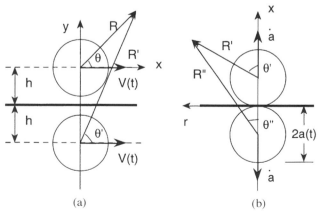

Figure 28.2. Wall effects on the liquid motion around the bubble due to (a) translation in the horizontal direction, (b) growth and motion in the vertical direction.

number may not be accurate for $a/(2h) = 0.5$ since approximation (16) is an expansion based on $a/(2h) \ll 1$. Nevertheless this result is useful since it demonstrates that the presence of the wall has limited influence on the drag of a sliding bubble.

### 28.2.6. *Wall Effects on the Growth Force*

Equation (14) gives an estimate of the growth force in pool boiling without wall effects. When the method of images is used, the wall effect on $F_{growth}$ can also be estimated. As shown in Fig. 28.2(b), the bubble is centered at $x = a$, which is moving at a velocity $\dot{a}$. The velocity potential due to the expanding bubble is

$$\phi = -\dot{a}\frac{a^2}{R'} - \dot{a}\frac{a^3}{2R'^2}\cos\theta'.$$

And

$$\phi_{images} = -\dot{a}\frac{a^2}{R''} + \dot{a}\frac{a^3}{2R''^2}\cos\theta''$$

accounts for the effect of wall images. The wall images induce an $x$-component velocity, $\dot{a}/8$, at $x = a$. This requires a doublet of velocity potential

$$\phi_{doublet} = \frac{1}{8}\dot{a}\frac{a^3}{2R'^2}\cos\theta'$$

located at $x = a$ in order to satisfy the no-penetration condition on the surface of the sphere at $x = a$. Hence,

$$\phi = -\dot{a}\frac{a^2}{R'} - \frac{7}{8}\dot{a}\frac{a^3}{2R'^2}\cos\theta' - \dot{a}\frac{a^2}{R''} + \frac{7}{8}\dot{a}\frac{a^3}{2R''^2}\cos\theta'' + \cdots. \tag{18}$$

Numerical integration of the $x$-component pressure force on the bubble surface yields

$$F_{growth} \sim -\pi\rho_l a^2(1.64\dot{a}^2 + 0.40a\ddot{a}). \tag{19}$$

Thus the wall effect on the growth force is again of limited extent.

### 28.2.7. *Effect of Shear – Shear-Lift Force*

Recent numerical results (Dandy and Dwyer, 1990) indicate that the presence of a uniform shear has very little effect on the drag. The mean flow shear in the wall region mainly contributes to the total force by generating a shear-lift force that is normal to the surface through the last two terms in Eq. (8). From Auton's (1987) result for inviscid flow with weak shear and the result of Legendre and Magnaudet (1997) for creeping flow at high shear rate, Thorncroft (1997) proposed the following approximation for the shear-lift force:

$$F_{SL} = \frac{1}{2}|U - V|(U - V)\pi\rho_l a^2\Gamma^{1/2}\left\{\left[\frac{1.146J(\sqrt{2\Gamma/Re_b})}{Re_b}\right]^2 + \left(\frac{3}{4}\Gamma^{1/2}\right)^2\right\}^{1/2}, \tag{20}$$

where

$$\Gamma = \frac{\partial U}{\partial y} \frac{a}{|U - V|} \tag{21}$$

is the dimensionless shear rate and

$$\mathrm{Re}_b = 2a|U(y = a) - V|/\nu, \tag{22}$$

$$J(\varepsilon) \approx 0.6765\{1 + \tanh[2.5(\log_{10}\varepsilon + 0.191)]\}\{0.667 + \tanh[6(\varepsilon - 0.32)]\}. \tag{23}$$

To obtain $U(y)$ and $\partial U/\partial y$ near the heater wall, the following velocity profile proposed by Reichardt for single-phase flow has been used by Klausner et al. (1993), Zeng (1993b), and Thorncroft (1997):

$$\frac{U(y)}{u^*} = \frac{1}{\kappa} \ln\left(1 + \kappa \frac{yu^*}{\nu_l}\right) + c\left[1 - \exp\left(-\frac{yu^*}{\chi \nu_l}\right) - \frac{yu^*}{\chi \nu_l}\exp\left(-0.33\frac{yu^*}{\nu_l}\right)\right], \tag{24}$$

where $\kappa = 0.4$, $\chi = 11$, and $c = 7.4$. For subcooled flow boiling, the friction velocity $u^* = u_l\sqrt{C_f/2}$ is evaluated with the approximation for $C_f$:

$$C_f/2 = (2.236 \ln \mathrm{Re}_l - 4.639)^{-2}, \tag{25}$$

where $\mathrm{Re}_l$ is the bulk liquid Reynolds number based on the duct hydraulic diameter.

### 28.2.8. *Effect of Contact Area*

The foregoing analyses pertain to a complete spherical bubble. In the case of a bubble attached to the wall with a finite contact diameter $d_w$, the effect of the contact area is somewhat uncertain. The force on $S_1$ can be expressed as

$$\underline{F}_1 = \left[-\int_{S_1+S_2}(p - p_c)\underline{n}\,dA + \int_{S_1+S_2}\underline{\underline{\tau_n}}\cdot\underline{n}\,dA\right] - \left[-\int_{S_2}(p - p_c)\underline{n}\,dA + \int_{S_2}\underline{\underline{\tau_n}}\cdot\underline{n}\,dA\right]$$

$$= \underline{F}_0 - \underline{F}_2, \tag{26}$$

where

$$\underline{F}_2 = -\int_{S_2}(p - p_c)\underline{n}\,dA + \int_{S_2}\underline{\underline{\tau_n}}\cdot\underline{n}\,dA \tag{27}$$

and $\underline{F}_0$ results from the complete surface $S_1 + S_2$. Near the point of departure the contact diameter $d_w$ is usually small compared with the bubble radius $a$. If the bubble is attached to the nucleation site and exposed to the liquid flow near the wall, the contact area $S_2$ would be exposed to a very low velocity, as shown in Fig. 28.3(a), where the wall has been removed and the flow is allowed to extend beyond $y = 0$. Little variation in pressure is expected on $S_2$ so that $(p - p_c)$ is practically zero. The deviatoric stress $\underline{\tau}$ is small since the local velocity is small. Hence the contribution from $\underline{F}_2$ to the total force is small compared with that from the complete surface integral $\underline{F}_0$. The effect of contact area may be neglected in this case.

As the bubble slides along a heater wall, it sees effectively a uniform flow in addition to a shear flow. In this case, a more quantitative estimate is needed for $\underline{F}_2$. To obtain such an estimate of $\underline{F}_2$, an inviscid flow model is used since the sliding bubble is typically of larger size. For

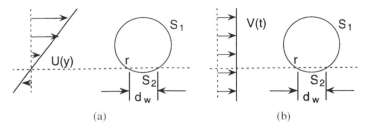

Figure 28.3. Effect of contact area on the overall force acting on a bubble (a) in a shear flow, (b) sliding in a quiescent liquid.

small contact diameter, $d_w/a \ll 1$, numerical integration of the pressure on $S_2$ based on the Bernoulli equation gives

$$-\int_{S_2}(p - p_c)\underline{n}\,dA \sim 2.65\rho_l V^2 a^2 \left(\frac{d_w}{2a}\right)^4 \underline{e}_y. \tag{28}$$

The shear stress is zero on the bubble surface. The normal stress is

$$\tau_{RR} = 2\mu\frac{\partial u_R}{\partial R}\Big|_{R=a} = -\frac{6\mu V}{a}\cos\theta.$$

Because of the symmetry of $\tau_{RR}$ near $\theta = \pi/2$, $(\int_{S_2}\underline{\underline{\tau}}\cdot\underline{n}\,dA)_y \sim 0$ and

$$\left(\int_{S_2}\underline{\underline{\tau_n}}\cdot\underline{n}\,dA\right)_x \sim -4.78\mu V a\left(\frac{d_w}{2a}\right)^4. \tag{29}$$

The effect of the contact area $S_2$ on the total force $\underline{F}$ is rather small and can be neglected.

To summarize, the main forces on the bubble are

$$\underline{F} \sim \underline{R} + \underline{F}_S - (\rho_l - \rho_v)\mathcal{V}_b\underline{g} + \frac{\pi}{4}d_w^2\frac{2\sigma_s}{r_c}\underline{e}_y + 6\pi\rho_l\nu(\underline{U} - \underline{V})a$$

$$\times\left\{\frac{2}{3} + \left[\frac{12}{\mathrm{Re}} + 0.75\left(1 + \frac{3.315}{\mathrm{Re}^{1/2}}\right)\right]^{-1}\right\} + F_{SL}\underline{e}_y + \frac{4}{3}\pi\rho_l a^3\frac{D\underline{U}}{Dt}$$

$$+ \frac{1}{2}\frac{4}{3}\pi\rho_l a^3\left(\frac{D\underline{U}}{Dt} - \frac{d\underline{V}}{dt}\right) + 2\pi\rho_l a^2(\underline{U} - \underline{V})\dot{a}, \tag{30}$$

where $\underline{F}_S$ is given by approximations (2) and (3) and $F_{SL}$ is given by Eq. (20).

## 28.3. Predicting Bubble Departure, Sliding, and Lift-Off

### 28.3.1. *Inclination Angle in Flow Boiling*

Here the $x$-direction is taken to be parallel to the heater surface while the $y$-direction is normal to the heater surface. The gravitational force may act in either the $x$-or the $y$-direction. It is important to point out that in Klausner et al. (1993), Zeng et al. (1993a and b), and Thorncroft (1997) the part of the added-mass force associated strictly with the bubble growth, $\underline{V} \sim \dot{a}\underline{e}_y$, is grouped into a growth force, $\underline{F}_{growth}$. In the present derivation,

$$\underline{F}_{growth} \sim -\frac{1}{2}\frac{4}{3}\pi\rho_l a^3\frac{d\underline{V}}{dt} - 2\pi\rho_l a^2\underline{V}\dot{a} = \left(-\frac{1}{2}\frac{4}{3}\pi\rho_l a^3\ddot{a} - 2\pi\rho_l a^2\dot{a}^2\right)\underline{e}_y. \tag{31}$$

The term $-\frac{1}{2}\frac{4}{3}\pi\rho_l a^3 \frac{d\underline{V}}{dt}$ due to the bubble sliding acceleration was labeled as the added-mass force in the traditional sense.

In flow boiling, it has been observed that bubbles at the nucleation site are typically inclined toward the downstream direction because of the liquid flow. This suggests that the bubble center moves in both $x$- and $y$-directions, $V \sim \dot{a}(\sin\theta_i \underline{e}_x + \cos\theta_i \underline{e}_y)$, where $\theta_i$ is the inclination angle (Klausner et al., 1993). To take this observation into account, $\underline{F}_{\text{growth}}$ may be expressed as

$$\underline{F}_{\text{growth}} = \left[ -\frac{1}{2}\frac{4}{3}\pi\rho_l a^3 \ddot{a} - 2\pi\rho_l a^2 \dot{a}^2 \right](\sin\theta_i \underline{e}_x + \cos\theta_i \underline{e}_y). \tag{32}$$

The inclination angle $\theta_i$ was determined from fitting the prediction with the measured bubble departure data (Klausner et al., 1993). In view of Eq. (32), the total force may be recast in the form

$$\underline{F} \sim \underline{R} + \sum{}' \underline{F}$$

$$= \underline{R} + \underline{F}_S - (\rho_l - \rho_v)V_b \underline{g} + \frac{\pi}{4}d_w^2 \frac{2\sigma_s}{R_r}\underline{e}_y + 6\pi\rho_l \nu(\underline{U} - \underline{V})a\left\{ \frac{2}{3} + \left[ \frac{12}{Re} \right.\right.$$

$$\left.\left. + 0.75\left( 1 + \frac{3.315}{Re^{1/2}} \right) \right]^{-1} \right\} + F_{SL}\underline{e}_y + 2\pi\rho_l a^3 \frac{D\underline{U}}{Dt} + 2\pi\rho_l a^2 \underline{U}\dot{a} - \frac{1}{2}\frac{4}{3}\pi\rho_l a^3 \frac{dV_x}{dt}\underline{e}_x$$

$$- 2\pi\rho_l a^2 V_x \dot{a}\underline{e}_x + \left[ -\frac{1}{2}\frac{4}{3}\pi\rho_l a^3 \ddot{a} - 2\pi\rho_l a^2 \dot{a}^2 \right](\sin\theta_i \underline{e}_x + \cos\theta_i \underline{e}_y), \tag{33}$$

where the prime $'$ after $\sum$ indicates the summation over all forces except the reaction force $\underline{R}$. In the above, it is understood that $V_x$ results from the translation velocity of the sliding bubble, not from the growth of the bubble while its base is still attached to the nucleation site.

### 28.3.2. *Horizontal Pool Boiling Lift-Off Criterion*

In pool boiling when the heater wall is placed horizontally below the liquid, the bubble lifts off the heater because of buoyancy. A commonly used criterion to predict bubble lift-off is that the summation of the forces discussed above in the $y$-direction, $\sum' F_y$, become zero. This can be rationalized as follows. While the bubble is attached to the nucleation site, the liquid–vapor surface-tension force pulls the bubble toward the wall and the growth force pushes it toward the wall while the buoyancy force aids the lift-off of the bubble. These forces, $\sum' F_y$, are not likely to balance for all the time before bubble lift-off and the wall must react through the three-phase contact line:

$$R_y + \sum{}' F_y = m_b \frac{dV_y}{dt}, \tag{34}$$

where $m_b$ is the bubble mass. Since $m_b$ is very small,

$$\sum{}' F_y \sim -R_y \le 0 \tag{35}$$

before lift-off. As the bubble size increases, the hydrostatic pressure difference between the top and the bottom of the bubble, and therefore the buoyancy force, becomes more important. There typically exists a thin liquid microlayer between the rapidly growing vapor bubble and

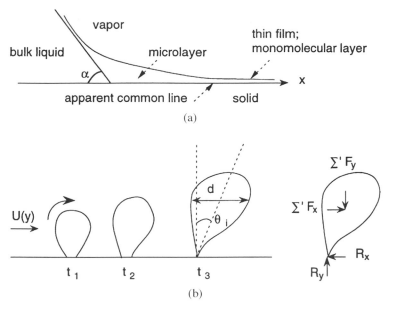

Figure 28.4. (a) Sketch of the three-phase interface near the common line, (b) sketch of the bubble at the onset of departure in flow boiling.

the heating surface. This microlayer may be described through a continuum approach since its dimension is much larger than the molecular length scales of the liquid. Toward the center of the base, there is a three-phase common line [Fig. 28.4(a)]. Around the common line, there exists a thin film whose size is on the molecular scale (Slattery, 1990). The reaction force results from the London–van der Waals force acting through the thin film. As the buoyancy force increases, the edge of the microlayer toward the bulk liquid region is pushed forward. The increase in the pressure near the common line also forces the interface of the thin film to propagate toward the center of the base so that it looks as if the apparent common line moves in. As the length of the common line decreases the thin-film region becomes smaller so that the reaction force $R_y$ diminishes. As bubble lift-off is approached, the thin film near the common line may rupture, and the bubble loses contact with the heater so that $R_y = 0$. This gives

$$\sum{}' F_y = 0 \quad \text{at lift-off.} \tag{36}$$

After lift-off, Eq. (36) remains valid since $m_b$ is practically zero. The existence of a reaction force before lift-off addresses van der Geld's (1996) criticism of using Eq. (36) as a bubble lift-off criterion.

To predict the bubble lift-off diameter in pool boiling, Zeng et al. (1993a) argued that at the point of a bubble's lifting off the wall, the contact diameter $d_w$ is small compared with the bubble radius. Hence the surface tension and the contact pressure forces are small. The dominant forces in the $y$-direction are buoyancy force and the growth force. Zeng et al. (1993a) found that close agreement between the model prediction and 190 pool boiling experimental data points in horizontal pool boiling can be obtained if the growth force term is empirically modeled as

$$F_{\text{growth}} = -\pi \rho_l a^2 \left( \frac{3}{2} C_s \dot{a}^2 + a \ddot{a} \right), \qquad C_s \sim 20/3. \tag{37}$$

The data were obtained in six different investigations (Friz and Ende, 1936; Staniszewski, 1959; Keshock and Siegel, 1964; Han and Griffith, 1965; Cole and Shulman, 1966; and van Stralen et al., 1975) over a wide range of pressure (0.02–2.8 bars), gravity (0.014–1 g), and Jacob number (4–869). The bubble lift-off radius was predicted by the solution of $a(t)$ from the following force balance equation:

$$-\rho_l \pi a^2 \left( \frac{3}{2} C_s \dot{a}^2 + a \ddot{a} \right) + \frac{4}{3} (\rho_l - \rho_v) \pi a^3 g \sim 0. \tag{38}$$

The good agreement between the model prediction given by approximation (38) and 190 data points over a wide range of conditions suggests that $F_{\text{growth}}$ is the dominant mechanism to counteract the buoyancy force. It is noted that the bubble lift-off diameters ranged from at least 0.5 mm to 10 mm. For such large bubbles in pool boiling systems, the role of the surface-tension force is very limited because of the relatively small contact diameter.

### 28.3.3. Bubble Departure and Lift-Off in Horizontal Flow Boiling

Zeng et al. (1993b) further applied the bubble dynamics to predict the bubble departure diameter in horizontal flow boiling. Again the surface-tension force was neglected by assuming that the bubble has a small contact diameter at the point of departure. The criteria for departure from the nucleation site were

$$R_x = -\sum{}' F_x = 0 \tag{39a}$$
$$R_y = -\sum{}' F_y = 0. \tag{39b}$$

These two equations were used to solve for two variables: the bubble radius $a(t)$ and the bubble inclination angle $\theta_i$ at departure. From visual observations, Zeng et al. (1993b) further postulated that after the bubble departs from the nucleation site, it rights itself during the subsequent sliding on the heater surface so that $\theta_i = 0$ after departure. The use of Eqs. (39a) and (39b) provides a systematic way of determining the inclination angle $\theta_i$. This was an improvement of the empirical determination of $\theta_i$ through fitting the data in Klausner et al. (1993). The predicted departure diameter agrees with the measured data reasonably well. Equations (39a) and (39b) have also been used by Thorncroft (1997) for determining the departure diameter, the sliding trajectory, and lift-off diameters in both vertical and horizontal flow boiling. However, a rational basis for this approach was not adequately discussed.

Here it is postulated that as the vapor bubble grows and is near the onset of departure, the part of the common line facing the upstream is being pushed toward the vapor region under increasingly larger quasi-steady drag resulting from the bulk liquid flow. With the decreasing area supporting the common line, the magnitude of $R_x$ decreases. Thus it requires a somewhat larger inclination angle $\theta_i$ to hold the bubble until the bubble finally departs the nucleation site. In the meantime, the reduction in the contact area also implies a reduction in the magnitude of $R_y$ [Fig. 28.4(b)]. As the bubble is about to leave the nucleation site, both components of the reaction force, $R_x$ and $R_y$, approach zero. Hence both Eqs. (39a) and (39b) hold at the point of departure. Once the bubble leaves the nucleation site, the bubble quickly uprights itself so that the inclination angle $\theta_i$ becomes zero, as observed experimentally. This causes the $y$-component of the growth force to increase a little so that $\sum' F_y < 0$ again and the bubble does not lift off until later.

During the sliding, the bubble continues to grow, which means that the bubble continues to withdraw energy from the heater. This indicates that the bubble is still in contact with the wall. Because the bubble is in the upright position and the inclination angle is zero, it is reasonable to expect that the receding and the advancing angles are not significantly different. Hence the $x$-component of the surface-tension force is practically zero. On the other hand, the $y$-component of the surface-tension force is not zero and is given by approximation (3). The bubble departure diameter in flow boiling is much smaller than that in pool boiling and the inclusion of the surface-tension force may become necessary for small bubbles.

Because of the bubble mirage effect discussed by Cooper et al. (1983), little experimental information is available on the three-phase contact region. Thorncroft (1997) postulated that $d_w \sim 2a \sin \beta$, in which $\beta$ is the experimentally determined receding liquid/solid contact angle. This gives the surface-tension force of the order of $F_{sy} \sim -2a\pi \sigma_s \sin^2 \beta$. With this hypothesis, the bubble sliding velocity is computed with the $x$-component of the bubble dynamic equation

$$\left( \rho_v + \frac{1}{2} \rho_l \right) \mathcal{V}_b \frac{dV_x}{dt} = 6\pi \rho_l \nu (U - V_x) a \left\{ \frac{2}{3} + \left[ \frac{12}{\mathrm{Re}} + 0.75 \left( 1 + \frac{3.315}{\mathrm{Re}^{1/2}} \right) \right]^{-1} \right\}$$
$$- 2\pi \rho_l a^2 V_x \dot{a} \underline{e}_x, \tag{40}$$

and the lift-off diameter in horizontal flow is determined with the $y$-component force balance

$$\sum{}' F_y = F_{Sy} + F_{SL} + F_{\mathrm{growth}} + (\rho_l - \rho_v) \mathcal{V}_b g \sim 0, \tag{41}$$

where the shear-lift force is based on the liquid velocity seen at the bubble center of mass and the bubble sliding velocity results from the solution of Eq. (40). Good agreement with the experimentally measured lift-off diameters for refrigerant R-113 (Zeng et al., 1993b) was

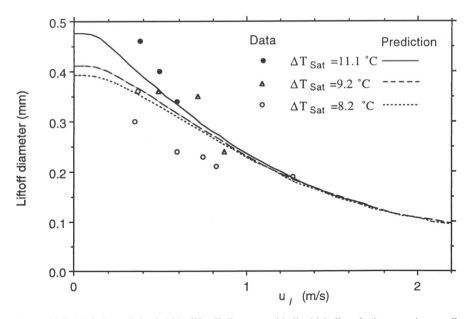

Figure 28.5. Variation of the bubble lift-off diameter with liquid bulk velocity at various wall superheats.

obtained. Since the bubble sliding velocity is not equal to the liquid flow velocity, the shear-lift force was adequately accounted for here. Figure 28.5 shows the variation of the predicted and the measured bubble lift-off diameters as functions of bulk liquid velocity $u_l$. Contrary to the model prediction of Zeng et al. (1993b), the effect of the bulk liquid velocity on the lift-off diameter can now be captured.

### 28.3.4. *Vertical Flow Boiling*

The bubble departure diameter in a vertically oriented heating surface can be predicted by the solutions of the following equations:

$$\sum{}' F_x = \pm (\rho_l - \rho_v) \mathcal{V}_b g + 6\pi \rho_l \nu U a \left\{ \frac{2}{3} + \left[ \frac{12}{\text{Re}} + 0.75 \left( 1 + \frac{3.315}{\text{Re}^{1/2}} \right) \right]^{-1} \right\}$$
$$- \rho_l \pi a^2 \left( \frac{3}{2} C_s \dot{a}^2 + a \ddot{a} \right) \sin \theta_i \sim 0, \tag{42}$$

$$\sum{}' F_y = F_{\text{SL}} - \rho_l \pi a^2 \left( \frac{3}{2} C_s \dot{a}^2 + a \ddot{a} \right) \cos \theta_i \sim 0. \tag{43}$$

The positive sign for the buoyancy force term is taken for upflow so that both the buoyancy force and the quasi-steady force promote the bubble departure while the growth force resists. For downflow, the buoyancy is in the opposite direction of $F_{\text{qs}}$. The inclination angle $\theta_i$ is positive if $F_{\text{qs}}$ is stronger than the buoyancy, and vice versa. A negative value for $\theta_i$ results if the buoyancy is larger than $F_{\text{qs}}$ so that the bubble is inclined toward the upstream of the flow; for such a case the bubble slides upward against the flow following departure as is observed experimentally.

After departure, bubbles start sliding and $\theta_i = 0$. The bubble trajectory can be obtained by the solution of the following equation:

$$\left( \rho_v + \frac{1}{2} \rho_l \right) \mathcal{V}_b \frac{dV_x}{dt} = \pm (\rho_l - \rho_v) \mathcal{V}_b g + 6\pi \rho_l \nu (U - V_x) a \left\{ \frac{2}{3} + \left[ \frac{12}{\text{Re}} \right. \right.$$
$$\left. \left. + 0.75 \left( 1 + \frac{3.315}{\text{Re}^{1/2}} \right) \right]^{-1} \right\} - 2\pi \rho_l a^2 V_x \dot{a} \underline{e}_x. \tag{44}$$

The bubble lift-off diameter in the downflow case is predicted from the $y$-component force balance:

$$\sum{}' F_y = F_{Sy} + F_{\text{SL}} - \rho_l \pi a^2 \left( \frac{3}{2} C_s \dot{a}^2 + a \ddot{a} \right) \sim 0, \tag{45}$$

where the shear-lift force $F_{\text{SL}}$ depends on the $(U - V_x)$ and $V_x$ is obtained from Eq. (44).

Figure 28.6 shows a typical comparison of the predicted bubble sliding trajectory with measured ones following bubble departure for a mass flow rate $G = 244 \text{ kg/m}^2 \text{ s}$, $\Delta T_{\text{sat}} = 0.55°\text{C}$ (Thorncroft, 1997). Good agreement can be observed. A successful prediction for the sliding trajectory indicates that the present bubble dynamics model captures the dominant forces.

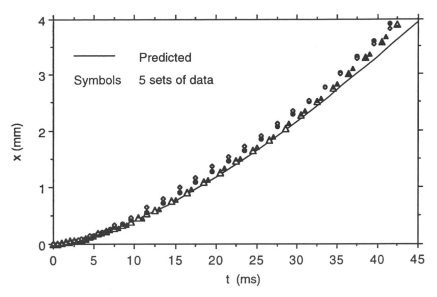

Figure 28.6. Bubble sliding trajectories after departure from the same nucleation site in a vertical upflow (mass flow rate $G = 244$ kg/m$^2$ s, $\Delta T_{\text{sat}} = 0.55\,^\circ$C).

## 28.4. Summary and Conclusion

A comprehensive derivation for the dynamic forces on a growing vapor bubble near the heater surface in flow and pool boiling is presented. A rational basis for predicting the departure of bubbles from the nucleation site and lift-off of bubbles from the heating surface is discussed. The model prediction for bubble departure diameter, sliding trajectory, and lift-off diameter in pool boiling, horizontal flow boiling, upflow boiling, and downflow boiling are in reasonable agreement with experimental measurements. The effects of wall, effects of bubble growth, gravitational (buoyancy) force, surface-tension force, quasi-steady drag, added-mass force, and shear-lift force on the dynamics of bubble departure and lift-off in various flow configurations are elucidated.

## References

Auton, T. R., 1987, "The lift force on a spherical body in a rotational flow," *J. Fluid Mech.* **183**, 190–218.

Batchelor, G. K., 1967, *An Introduction to Fluid Dynamics*, Cambridge U. Press, London/New York.

Chang, Y. P. 1963, "Some possible critical conditions in nucleate boiling," *ASME J. Heat Transfer* **85**, 89–100.

Cole, R. and Rohsenow, W. M. 1969, "Correlation of bubble departure diameters for boiling of saturated liquids," *Chem. Eng. Prog. Symp. Ser.* **65**(92), 211–213.

Cole, R. and Shulman, H. L., 1966, "Bubble departure diameters at subatmospheric pressures," *Chem. Eng. Prog. Symp. Ser.* **62**(64), 6–16.

Cooper, M. G., Mori, K., and Stone, C. R., 1983, "Behavior of vapor bubbles growing at a wall with forced flow," *Int. J. Heat and Mass Transfer* **26**, 1489–1507.

Dandy, D. S. and Dwyer, H. A., 1990, "A sphere in shear flow at finite Reynolds number: effect of shear on particle lift, drag, and heat transfer," *J. Fluid Mech.* **216**, 381–410.

Fritz, W. and Ende, W., 1936, "Über den Verdempfungsvorgang nach Kinematographischen Aufnahmen an Dampfblasen," *Phys. Z.* **37**, 391–401.

Han, C. Y. and Griffith, P., 1965, "The mechanism of heat transfer in nucleate pool boiling, part I: bubble initiation, growth and departure," *Int. J. Heat Mass Transfer* **8**, 887–904.

Kandlikar, S. G. and Stumm B. J., 1995, "A control volume approach for investigating forces on a departing bubble under subcooled flow boiling," *ASME J. Heat Transfer* **117**, 990–997.

Keshock, E. G. and R. Siegel, R., 1964, "Forces acting on bubbles in nucleate boiling under normal and reduced gravity conditions," NASA Tech. Note TN D-2299.

Klausner, J. F., Mei, R., Bernhardt, D. M., and Zeng, L. Z., 1993, "Bubble departure in forced convection boiling," *Int. J. Heat and Mass Transfer* **36**, 651–662.

Koumoutsos, N., Moissis, R., and Spyridonos, A., 1968, "A study of bubble departure in forced-convection boiling," *ASME J. Heat Transfer* **90**, 223–230.

Lamb, H., 1932, *Hydrodynamics*, Dover, New York.

Legendre, D. and Magnaudet, J., 1997, "A note on the lift force on a spherical bubble or drop in a low-Reynolds-number shear flow," *Phys. Fluids* **9**, 3572–3574.

Levy, S., 1967, "Forced convection subcooled boiling – prediction of vapor volumetric fraction," *Int. J. Heat and Mass Transfer* **10**, 951–960.

Mei, R. and Klausner, J. F., 1992, "Unsteady force on a spherical bubble at finite Reynolds number with small fluctuations in the free-stream velocity," *Phys. Fluids A* **4**, 63–70.

Mei, R., Klausner, J. F., and Lawrence, C. J., 1994, "A note on the history force on a spherical bubble at finite Reynolds number," *Phys. Fluids A* **6**, 418–420.

Slattery, J. C., 1990, *Interfacial Transport Phenomena*, Springer-Verlag, New York.

Staniszewski, B. E., 1959, "Bubble growth and departure in nucleate boiling," Tech. Rept. 16, MIT, Cambridge, Massachusetts.

Thorncroft, G. E., 1997, "Heat transfer and vapor bubble dynamics in forced convection boiling," Ph.D. dissertation, University of Florida, Gainesville, FL.

van Helden, W. G. J., 1994, "On detaching bubbles in upward flow bubble," Ph.D. thesis, Eindhoven University of Technology, Eindhoven, The Netherlands.

van der Geld, C. W. M., 1996, "Bubble detachment criteria: some criticism of 'Das Abreissen von Dampfblasen an festen Heizflachen'," *Int. J. Heat Mass Transfer* **39**, 653–657.

van Stralen, S. J. D., Cole, R., Sluyter, W. M., and Sohal, M. S., 1975, "Bubble growth rates in nucleate boiling of water at subatmospheric pressures," *Int. J. Heat Mass Transfer* **18**, 655–669.

Zeng, L. Z., Klausner, J. F., and Mei, R., 1993a, "A unified model for the prediction for bubble detachment diameters in boiling systems: Part I pool boiling," *Int. J. Heat Mass Transfer* **36**, 2261–2270.

Zeng, L. Z., Klausner, J. F., Bernhardt, D. M., and Mei, R., 1993b, "A unified model for the prediction of bubble detachment diameters in boiling systems: Part II flow boiling," *Int. J. Heat Mass Transfer* **36**, 2271–2279.

# PART FIVE

# Complex Flows

# 29

# Heat, Mass, and Momentum Exchanges between Outer Flow and Separation Bubble behind a Single Backward-facing Step with Gas Injection from One Duct Wall

TONG-MIIN LIOU AND PO-WEN HWANG

## 29.1. Introduction

In industrial burners, afterburners and ramjet combustors flames are principally stabilized by turbulent recirculation zones behind bluff bodies or backward-facing steps. The hot combustion products cause ignition by transferring heat to the cold combustible mixture across the mixing layer that exists at the outer edge of the recirculation zone. The flame-stabilization mechanism thus involves the turbulent exchange of two gas streams and appropriate residence time of the cold combustible mixture in the recirculation zone. However, there is relatively limited information in open literature on the rates of mass, momentum, and heat transfer between the recirculation zone and main flow.

Bovina (1958) first studied mass exchange behind V-shaped flame holders. A curve fit through the time record of tobacco smoke concentration decay from the equilibrium value to zero was used to infer the residence time. The mean residence time ($T_{rm}$) and turbulent exchange coefficient ($D_z$) of the recirculation zone were found to decrease and increase, respectively, with increasing upstream bulk velocity ($u_a$) such that $D_z/u_a$ remained constant. Similar trends were also found by Winterfeld (1965), who used similar techniques in investigating V-, cone-, and cylinder-shaped flame holders. He further concluded that the presence of a flame increases $T_{rm}$ considerably and, according to this result, reduces the exchange of mass, momentum, and energy behind the flame holder. Zakkay et al. (1970) experimentally studied $T_{rm}$ of a foreign gas within the recirculation zones behind a cylinder and a truncated cone placed in an axisymmetric supersonic stream by using the hot wire. They concluded that $T_{rm}$ is much shorter in turbulent flows than in laminar flows. The molar concentration of the foreign gas in the recirculation region decays exponentially for both laminar and turbulent flows. Resonance absorption of laser light by methane was performed by Baev et al. (1979) to study turbulent mass transfer between supersonic external flows and a recirculation zone in a rectangular channel with a backstep. They found that the concentration of methane has a little effect on the mass transfer process and, in turn, residence time. Adopting paraffin smoke and NaCl as tracers, Maki and Suzuki (1986) investigated mass transfer between the recirculation zone and outside flow of an annular premixed flame. $T_{rm}$ was found to decrease with increasing gas exit velocity $u_a$ for $u_a < 20$ m/s examined. The air entrainment rate into the near wake of a flat-faced cylindrical body aligned with the flow in a low-speed wind tunnel was calculated from the residence time measurements with $CO_2$ as the tracer by Roberds et al. (1989). It was concluded that 8% of the free air stream that would pass through the front area of the test body is drawn into the near wake.

All the investigations surveyed above were performed experimentally and used an Euler analysis of tracer concentration in the recirculation zone, a curve fit through the record, and a rather limited number of records (typically 8 to 10 records) for statistical average to obtain $T_{rm}$ from which the mean mass entrainment rate or exchange coefficient was calculated. There is no corresponding numerical work available from open literature. Moreover, $u_a$ is one of the important parameters that affect the transport phenomena across the boundary of the recirculation zone. The previous studies did not resolve the relevant instantaneous flow structures or particle motions near and across the zone boundary. They provided primarily zone-averaged quantities.

Below, the governing equations, the interface between the recirculation zone and outer flow, and mass, momentum, and heat transfer rate across the interface are mathematically formulated in Section 29.2. In Section 29.3 the numerical scheme, boundary and initial conditions, and grid independence test are described. The results are presented and related physical mechanisms discussed in Section 29.4 followed by concluding remarks in Section 29.5.

## 29.2. Problem/Topic Description/Formulation

### 29.2.1. *Problem*

Combustion instability is a crucial issue encountered in solid-fuel ramjet engines. It is a requirement of flame stabilization that the gas residence time in the flame-holding recirculation zone be longer than the time necessary for preparing the fresh mixture for ignition. The knowledge of residence time and turbulent transport across the outer edge of the recirculation zone is thus imperative.

### 29.2.2. *Topic Description*

Based on the preceding review of the relevant studies, the present work is undertaken to perform for the first time a numerical experiment in which a Lagrangian analysis is used to trace the fluid particles in the recirculation zone to acquire the residence time directly. From the acquired residence time the turbulent exchange coefficient across the zone boundary can be further calculated. The recirculation zone examined is one behind a confined backstep with wall injection (Fig. 29.1), a configuration simulating a solid-fuel ramjet combustor. The main parameter explored is $u_a$, whose effects on the characteristics of the recirculation zone will be revealed in terms of instantaneous vorticity field and particles' pathlines. In addition, the transport phenomena across the interface between the external flow and the recirculation flow are characterized in terms of both mean and instantaneous flow quantities.

### 29.2.3. *Formulation*

The filtering operation $\bar{g}(\mathbf{x}) = \int_\Omega G_\Delta(\mathbf{x} - \zeta)g(\zeta)\,d\zeta$, where $G$ is a filter function of volume averaging, decomposes a variable $g$ into a large-scale component $\bar{g}$ and a subgrid-scale component $g'$, which accounts for the scales not resolved by the filter width $\Delta$: $g = \bar{g} + g'$. A mass-weighted filter, which simplifies the mathematical expressions for compressible flow simulations, is defined by $\tilde{g} = \overline{\rho g}/\bar{\rho}$, where $\bar{\rho}$ is the filtered density. This implies a second decomposition of $g$: $g = \tilde{g} + g''$. The filtered forms of compressible mass, momentum, and energy equations with the Smagorinsky subgrid-scale eddy viscosity model in two-dimensional Cartesian coordinates

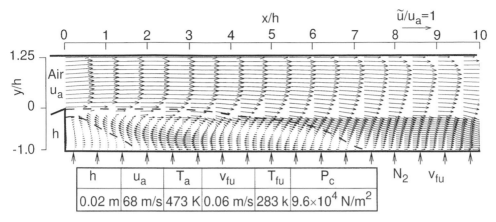

Figure 29.1. Mean flow pattern in terms of velocity vector field (dashed curve, mean dividing streamline; $p_c$ = initial chamber pressure).

(Fig. 29.1) can be respectively written as

$$\frac{\partial \bar{\rho}}{\partial t} + \frac{\partial}{\partial x}(\bar{\rho}\tilde{u}) + \frac{\partial}{\partial y}(\bar{\rho}\tilde{v}) = 0, \tag{1}$$

$$\frac{\partial}{\partial t}(\bar{\rho}\tilde{u}) + \frac{\partial}{\partial x}(\bar{\rho}\tilde{u}\tilde{u} - \bar{\tau}_{xx} - \tau s_{xx}) + \frac{\partial}{\partial y}(\bar{\rho}\tilde{u}\tilde{v} - \bar{\tau}_{xy} - \tau s_{xy}) = 0, \tag{2}$$

$$\frac{\partial}{\partial t}(\bar{\rho}\tilde{v}) + \frac{\partial}{\partial x}(\bar{\rho}\tilde{v}\tilde{u} - \bar{\tau}_{yx} - \tau s_{yx}) + \frac{\partial}{\partial y}(\bar{\rho}\tilde{v}\tilde{v} - \bar{\tau}_{yy} - \tau s_{yy}) = 0, \tag{3}$$

$$\frac{\partial \hat{E}}{\partial t} + \frac{\partial}{\partial x}(\hat{E}\tilde{u} - \overline{u\tau}_{xx} - \overline{v\tau}_{yx} - \tilde{u}\tau s_{xx} - \tilde{v}\tau s_{yx} - \kappa\partial\tilde{T}/\partial x - q_{s_x})$$

$$+ \frac{\partial}{\partial y}(\hat{E}\tilde{v} - \overline{u\tau}_{xy} - \overline{v\tau}_{yy} - \tilde{u}\tau s_{xy} - \tilde{v}\tau s_{yy} - \kappa\partial\tilde{T}/\partial y - q_{s_y}) = 0, \tag{4}$$

where viscous stresses $\tau_{ij}$ are defined and subgrid-scale stresses $\tau s_{ij}$ and heat fluxes $q_{s_j}$ are modeled as in our previous study (Liou et al., 1997). For an ideal gas, the total energy $\hat{E}$ per unit volume is related to the filtered pressure $\bar{p}$ and axial ($\tilde{u}$) and transverse ($\tilde{v}$) velocity components by means of $\hat{E} = \bar{p}/(\gamma - 1) + 0.5\bar{\rho}(\tilde{u}^2 + \tilde{v}^2)$. The filtered temperature $\tilde{T}$ is related to $\bar{\rho}$ and $\bar{p}$ by the ideal gas law $\tilde{T} = \bar{p}/(\bar{\rho}R)$. In the above $R$ and $\gamma$ are the gas constant and the specific heat ratio, respectively. Equations (1)–(4) are used to solve $\bar{\rho}$, $\tilde{u}$, $\tilde{v}$, and $\bar{p}$. $\hat{E}$ and $\tilde{T}$ can be further calculated from above-mentioned definition and ideal gas law.

Figure 29.1 depicts the typical mean flow pattern of the problem investigated. To examine the mass, momentum, and energy transport across the outer edge of the separating recirculation zone, one needs to first determine the coordinates $y_d$ of the border of the backflow zone through the definition of the dividing streamline (dashed curve in Fig. 29.1), $\int_{-1}^{y_d}\langle\bar{\rho}\rangle\langle\tilde{u}\rangle\,dy = 0$, which is a line in the flow field where the time average of the mass flow crossing the interface equals zero. The main flame-holding recirculation zone and corner recirculation zone are thus 7.2 and 1.5 h long, respectively. Each of the mass particles entering the recirculation zone will remain there for different periods of time. A mean residence time $T_{rm}$ of all mass particles in this zone can be related to the exchange mass flow rate $\dot{M}_{em}$ through $\dot{M}_{em} = \rho_{fh}V_{fh}/T_{rm}$, where $V_{fh}$ and $\rho_{fh}$ are respectively the volume and the average density of the recirculation zone.

When $\dot{M}_{em}$ is divided by $\rho_{fh}$ and the surface area of the recirculation zone $S$, a quantity has the dimension of a velocity that is designated as the average escape velocity $U_{em} = \dot{M}_{em}/(\rho_{fh}S)$ of the exchange flow crossing the interface in either direction. The momentum and heat, $\text{Mom}_{em}$ and $H_{em}$, transferred from the recirculation zone can also be determined by $\text{Mom}_{em} = \dot{M}_{em}U_{em}$ and $H_{em} = \dot{M}_{em}Cp_{fh}T_{fh}$, where $Cp_{fh}$ and $T_{fh}$ are the average values of specific heat and temperature within the recirculation zone.

### 29.3. Approach/Methodology

The code used the finite-volume technique, which involved alternating in time the second-order, explicit MacCormack's and Godunov's schemes. Reduction in phase error can be achieved by temporal switching of these two schemes since MacCormack's scheme has a lagging phase error and Godunov's scheme has a leading phase error. To improve the spatial accuracy of Godunov's original scheme, in this work the piecewise initial states to the left and the right of the cell edges were obtained by a second-order extrapolation from the cell center values. In addition, a limiting technique was applied to achieve numerical stability.

The primitive variables $(\bar{\rho}, \tilde{u}, \tilde{v}, \bar{p})$ must be specified at the boundaries because they are involved in the flux vectors. At the solid wall of a dump plane, no-slip conditions are used along with an adiabatic wall and zero normal pressure gradient. Characteristic-based boundary conditions are enforced on the inflow and the outflow boundaries for $\bar{p}$, $\tilde{u}$, $\tilde{v}$, and $\bar{p}$ (Liou et al., 1997). The specification of the initial conditions is based on the work of Krametz and Schulte (1988) and given in Fig. 29.1.

Several sets of uniform grids were examined, and the grid independence was attained for grid sizes $242 \times 47$ and $342 \times 57$ ($x \times y$). The deviations of the mean velocity profiles calculated from these two grid systems were all less than 1%. Consequently, the results presented below are based on the $242 \times 47$ grid system.

### 29.4. Results and Discussion

Validations of the present code have been performed in our previous investigations (Liou and Lien, 1995; Liou et al., 1997). Please refer to these earlier studies for more detailed information. As $u_a$ is increased gradually from 68 to 127.5 m/s (or Mach number from 0.17 to 0.29), an inlet velocity range typical of that encountered in subsonic combustion ramjet combustors, the variations of the reattachment length $x_{re}$, cross-section area, and dividing streamline length of the main recirculation zone (Fig. 29.1) are in the ranges of 5.9 to 7.4 h, 3.6 to 4.3 $h^2$, and 6.1 to 7.4 h, respectively.

### 29.4.1. *Pathlines and Instantaneous Vorticity Contours*

In applying the Lagrangian method, 120 passive particles are introduced into the numerically fully developed flow field, occurring at $t^* = $ combustor length$/(0.5u_a) = 6$ ms, from 25 traverse positions with an interval $\Delta y = 0.05$ h along the air inlet plane ($x/h = 0$) and from 95 streamwise positions with an interval $\Delta x = 0.1$ h along the porous wall ($y/h = -1$). Particles are released from these 120 positions at every $\Delta t = 0.0167$ ms until totally 309 sets of particles or $102 \times 309$ particles to attain statistical reliability. The instantaneous position of

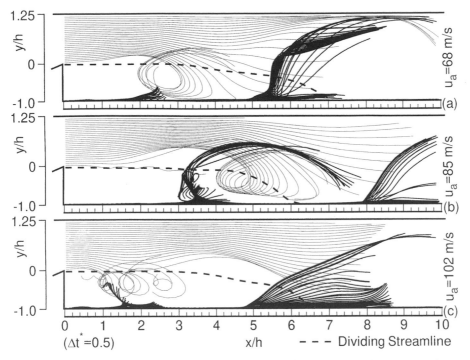

Figure 29.2. Flow structures visualized by pathlines for particle release time $t^* = 6$ ms and elapse time $\Delta t^* = 0.5$ ms: (a) $u_a = 68$ m/s, (b) 85 m/s, (c) 102 m/s (thin solid curves, air particles; thick solid curves, $N_2$ particles).

each released particle is found with the computed instantaneous velocity field and a fourth-order Runge–Kutta scheme in time. At each instant the number of particles flowing into and out of the recirculation zone can thus be monitored and the residence time of each particle in the zone can be counted. The lines traced out by the identified particles during $\Delta t^* = 0.5$ ms are plotted in Figs. 29.2(a)–(c) for various $u_a$. It is seen that the air particles released closer to the step top have more chances to be entrained into the shear layer rolled-up clockwise rotating vortices shown in Fig. 29.3 and hence depict traces of many tumbles or flow into and out of the recirculation zone many times. When $N_2$ particles are released from the porous wall, they tend to first move toward two regions located immediately behind the large clockwise rotating vortices that impinge on the $N_2$ particles' releasing surface and whose rear edges lift the $N_2$ particles up from the porous wall (Fig. 29.3). For the region closer to the reattachment point, $N_2$ particles are clockwise convected downstream directly by the clockwise rotating vortex that bounds from the porous wall and proceeds downstream (Fig. 29.3). However, for the region closer to the step, most of $N_2$ particles' traces are clustered, relatively short, and confined in the recirculation zone, indicating a longer mean residence time. Only $N_2$ particles whose traces are long enough to touch the separated shear layer have chances to escape from the backflow zone [Fig. 29.2(b)]. In general, for the range of $u_a$ examined, the $N_2$ particles (simulating fuel particles) stay longer in the backflow than the air particles, suggesting that the zone is full of fuel-rich mixture and capable of assuming flame holding. The variation of $u_a$ does not affect the basic mechanism of vortex evolution but affects the evolution time and thus the particles' pathlines.

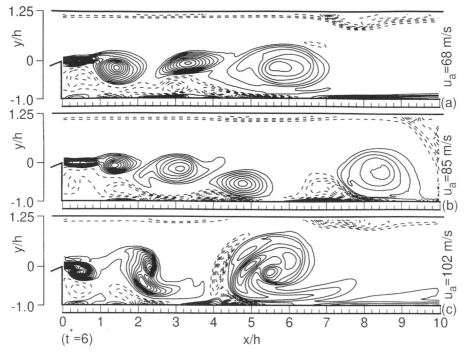

Figure 29.3. Instantaneous flow visualization of vorticity field: (a) $u_a = 68$ m/s, (b) 85 m/s, (c) 102 m/s (solid curves, clockwise rotating vortices; dashed curves, counterclockwise rotating vortices).

### 29.4.2. *Flow Properties on the Dividing Streamline*

For understanding the turbulent transport phenomena between the outer flow and recirculation zone, Fig. 29.4 shows the variations of the time-averaged axial and transverse turbulent normal stresses, $\langle u''^2 \rangle$ and $\langle v''^2 \rangle$, Reynolds shear stress, $\langle -u''v'' \rangle$, and length scale based on the eddy diffusivity concept proposed by Boussinesq and $u_a$, $l_{\text{Bou}} = \langle -u''v''/\tilde{S}_{xy} > /u_a \rangle$; along the dividing streamline. $\tilde{S}_{xy} = \partial \tilde{u}/\partial y + \partial \tilde{v}/\partial x$ in the definition of $l_{\text{Bou}}$ is the filtered strain rate. These quantities, except $l_{\text{Bou}}$, generally increase with increasing $u_a$ along most parts of the dividing streamline. $\langle u''^2 \rangle$ and $\langle v''^2 \rangle$ reveal higher values in short distances, $0.8 < x/x_{re} < 1$, upstream of the reattachment point [Figs. 29.4(a) and (b)], and in the middle portion of the dividing streamline, $0.4 < x/x_{re} < 0.6$, respectively. The observation indicates that the transverse fluctuation of the dividing streamline is largest in its middle portion and gradually vanishes toward its two ends, the separation point (step top corner) and the reattachment point. From the middle portion of the dividing streamline toward the reattachment point, $\langle v''^2 \rangle$ is gradually suppressed by the chamber wall but $\langle u''^2 \rangle$ is not. It is for this reason that $\langle u''^2 \rangle$ on the dividing streamline receives turbulent kinetic energy from $\langle v''^2 \rangle$ for $x/x_{re} > 0.6$ and reveals higher values for $0.8 < x/x_{re} < 1.0$. Values of $\langle -u''v'' \rangle$ in Fig. 29.4(c) are mostly positive, indicating that along the dividing streamline $u''$ and $v''$ are mostly different in phase and opposite in sign. $l_{\text{Bou}}$ is found to increase and be independent of $u_a$ along the dividing streamline up to $x = 0.6 x_{re}$ [Fig. 29.4(d)], a trend paralleling the vortex growth shown in Fig. 29.3. However, $l_{\text{Bou}}$ becomes negative for $u_a = 102$ m/s in the portion of $0.9 < x/x_{re} < 1.0$, an example showing the failure of Boussinesq's hypothesis due to high anisotropy of the strain.

It is informative to examine how $u_a$ affects some mean physical properties arithmetically averaged over the entire dividing streamline. The averaged turbulent kinetic energy is found

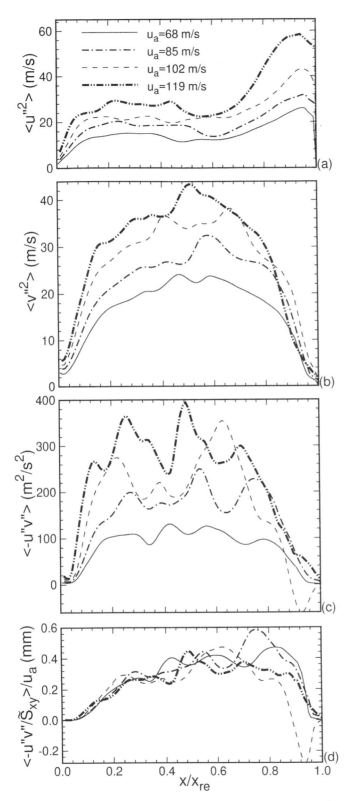

Figure 29.4. Distribution of time-averaged axial and transverse turbulent normal stresses, Reynolds shear stress, and characteristic length scale along the dividing streamline.

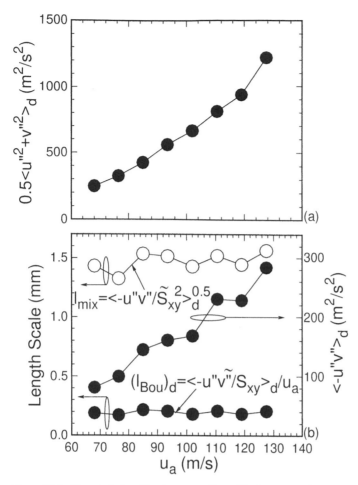

Figure 29.5. Mean turbulent kinetic energy, Reynolds shear stress, Prandtl's mixing length, and length scale based on Boussinesq's concept and $u_a$ as functions of $u_a$.

to increase with increasing $u_a$, as shown in Fig. 29.5(a), since both the strength (or velocity gradient) of the separated shear layer and $\langle -u''v'' \rangle_d$ [Fig. 29.5(b)] increase with increasing $u_a$. Figure 29.5(b) further shows that both the Prandtl's mixing length $l_{mix}$ and $(l_{Bou})_d$ are practically independent of $u_a$ with $l_{mix} \approx 7.5(l_{Bou})_d$ for the $u_a$ range examined. Note that if $u_a$ in $(l_{Bou})_d$ [Fig. 29.5(b)] is replaced by a characteristic fluctuation velocity defined from turbulent kinetic energy shown in Fig. 29.5(a), one can still obtain a constant $(l_{Bou})_d$ independently of $u_a$.

Figure 29.6 depicts the instantaneous velocity phase diagrams and axial fluctuation velocity spectra at selected points on the dividing streamline for $u_a = 68$ m/s. Near the step, $x/x_{re} = 0.052$, the fluctuation levels of $\tilde{v}/u_a$ and $\tilde{u}/u_a$ are confined to $0 \pm 0.2$ and $0.25 \pm 0.25$, respectively, by the step. The corresponding fluctuating spectrum has a peak amplitude of 250 at $f = 720$ Hz, which is the vortex shedding frequency. In the middle portion of the dividing streamline, $x/x_{re} = 0.486$, the fluctuation center moves further to the right and fluctuation levels increase to $0 \pm 0.75$ and $0.5 \pm 0.5$ for $\tilde{v}/u_a$ and $\tilde{u}/u_a$, respectively. The peak spectrum

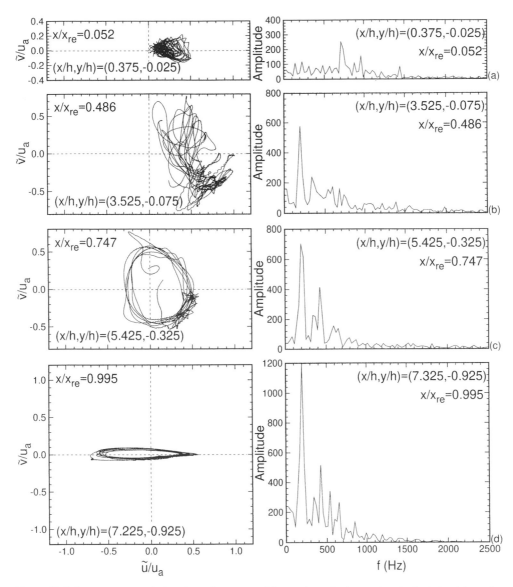

Figure 29.6. Instantaneous velocity phase diagrams and fluctuation velocity spectra at selected points on the dividing streamline for $u_a = 68$ m/s.

frequency shifts to 200 Hz, the vortex pairing frequency, with an amplitude increased to $\sim$600. In the rear portion of the dividing streamline, $x/x_{re} = 0.747$, the velocity phase diagram reveals a circlelike shape with fluctuation levels of $\tilde{v}/u_a = 0 \pm 0.5$ and $\tilde{u}/u_a = 0 \pm 0.4$. Adjacent to the reattachment point, the transverse fluctuation is suppressed by the chamber wall such that the velocity phase diagram is like an airfoil with $\tilde{v}/u_a = 0 \pm 0.1$ and $\tilde{u}/u_a = -0.15 \pm 0.6$. The peak spectrum frequency $f = 200$ Hz has an amplitude as high as 1200. When $u_a$ is varied, the above trend remains the same; nevertheless, the corresponding fluctuation centers and levels may change.

### 29.4.3. *Mass, Momentum, and Heat Transfer across the Edge of the Recirculation Zone*

$T_{rm}$ obtained directly from the present numerical experiment is compared with previous measured data by use of the curve fit method for various $u_a$ in Fig. 29.7. Although the flame-holder geometries and flow conditions are different, all the results show a decrease of $T_{rm}$ with increasing $u_a$ at nearly the same slope for $u_a > 30$ m/s. For $u_a < 30$ m/s, the slope is steeper. When $u_a$ is increased, the preceding Figs. 29.4(a), (b), and 29.5(a) indicate an increase of turbulent fluctuation or kinetic energy on the dividing streamline, resulting in a decrease of $T_{rm}$.

From the knowledge of $T_{rm}$, the mass, momentum, and heat transfer rate across the dividing streamline can be calculated. Their variations with $u_a$ are plotted in Figs. 29.8(a) and (b). The turbulent exchange diffusivity, defined as $D_z = u_{em}r_m = V_{fh}r_m/(T_{rm}S)$, with $r_m$ denoting an equivalent radius of the recirculation zone (Bovina 1958), is also plotted versus $u_a$ in Fig. 29.8(c). Both the present study and the work of Maki and Suzuki (1986) indicate that $\dot{M}_{em}/\dot{M}_0$ is essentially a weak function of $u_a$, especially for higher values of $u_a$ examined, as shown in Fig. 29.8(a). Because $\dot{M}_0$ is proportional to $u_a$, $\dot{M}_{em}$ increases approximately linearly with $u_a$. This trend is reasonable since $T_{rm}$ decreases with increasing $u_a$. Similarly, Figs. 29.8(b) and (c) show the increase of $\text{Mom}_{em}$, $H_{em}$, and $D_z$ with increasing $u_a$. The length scale $D_z/u_a$ is found to be independent of $u_a$ by the present study and Bovina's work (1958). At this point, it is interesting to point out that the three length scales, $(l_{Bou})_d$, $l_{mix}$, and $D_z/u_a$, presented above are all independent of $u_a$ and have the relative magnitude in the

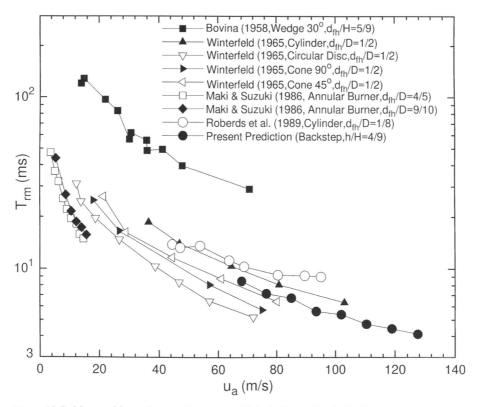

Figure 29.7. Mean residence time as a function of air inlet bulk velocity obtained by various researchers ($d_{fh}$, flame-holder size; $D$, tube diameter; $H$, chamber height).

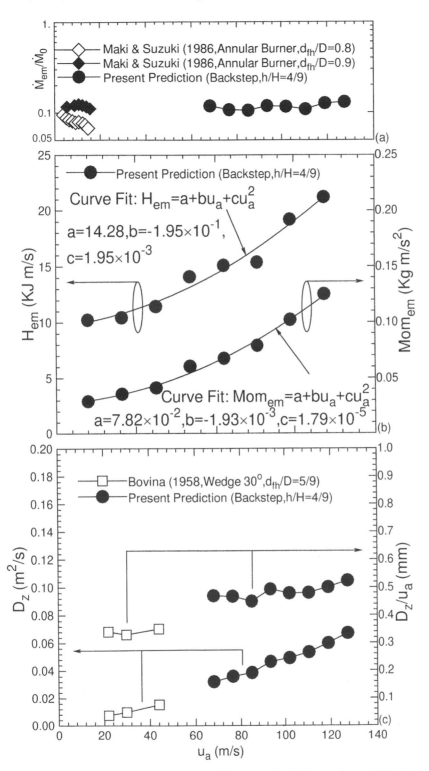

Figure 29.8. Mass, momentum and heat transfer rate, and turbulent exchange diffusivity across the dividing streamline as a function of $u_a$ ($\dot{M}_0$: the inlet air mass flow rate).

order $(l_{\text{Bou}})_d < D_z/u_a < l_{\min}$. In the present study, $(l_{\text{Bou}})_d$, $D_z/u_a$, and $l_{\text{mix}}$ are 0.2, 0.5, and 1.5 mm, respectively.

## 29.5. Concluding Remarks

For the first time a large eddy simulation together with a particle tracking method has been applied to study the effects of inlet air mass flow loading $(u_a)$ on the mean residence time $T_{rm}$ of gas in the flame-holding recirculation zone and turbulent transport across the zone boundary in a simulated solid-fuel ramjet combustor. Through numerical pathlines and vorticity contour visualizations, we are able to demonstrate the key role played by the shedding vortices in making the recirculation zone full of fuel-rich gas mixture ready for flame holding. Both turbulent normal stresses and shear stress on the zone edge are found to increase, whereas $T_{rm}$ decreases, with increasing $u_a$, suggesting the attainment of turbulent heat, mass, and momentum transfer enhancement by raising $u_a$. The decrease of $T_{rm}$ with increasing $u_a$ concluded by the present study and previous investigations for $0 < u_a < 130$ m/s seems to be independent of flame-holder geometries and reveals a decreasing slop with increasing $u_a$. A further study is thus recommended for exploring the existence of a critical $u_a$, above which $T_{rm}$ may remain a constant regardless of $u_a$. Three turbulent length scales, $(l_{\text{Bou}})_d$, $D_z/u_a$, and $l_{\text{mix}}$, on the recirculation zone boundary are examined. They are found to be rather insensitive to the variation of $u_a$ and have relative magnitudes of $(l_{\text{Bou}})_d < D_z/u_a < l_{\min}$.

## Acknowledgment

This work was supported by the National Science Council of the Republic of China under contract number NSC 85-2212-E -007-025.

## References

Baev, V. K., Vorontsov, S. S., Zabaikin, V. A., and Konstantinovskii, V. A. 1979. Determination of the time spent by gas in a recirculation zone by means of a resonance absorption technique. *Combust. Explosion Shock Waves* **15**, 758–760.

Bovina, T. A. 1958. Studies of exchange between recirculation zone behind the flame-holder and outer flow. In *Seventh Symposium (International) on Combustion*, pp. 692–696.

Krametz, E. and Schulte, G. 1988. Observation of a reacting shear layer in a backward-facing step flow. In *Proceedings of the Fourth International Symposium on Application of Laser Anemometry to Fluid Mechanics*, Paper 2.6.

Liou, T. M. and Lien, W. Y. 1995. Numerical simulations of injection-driven flows in a two-dimensional nozzless solid-rocket motor. *J. Propul. Power* **11**, 600–606.

Liou, T. M., Lien, W. Y., and Hwang, P. W. 1997. Flammability limits and probability density functions in simulated solid-fuel ramjet combustors. *J. Propul. Power* **13**, 643–650.

Maki, H. and Suzuki, K. 1986. A study on mass transfer between recirculation zone and outside flow of annular premixed flame. *Bull. JSME* **29**, 1781–1787.

Roberds, D. W., Mcgregor, W. K., Hartsfield, B. W., Schulz, R. J., and Rhodes, R. P. 1989. Measurement of residence time, air entrainment rate, and base pressure in the near wake of a cylindrical body in subsonic Flow. *AIAA J.* **27**, 1524–1529.

Winterfeld, G. 1965. On processes of turbulent exchange behind flame holders. In *Tenth Symposium (International) on Combustion*, pp. 1265–1275.

Zakkay, V., Sinha, R., and Medecki, H. 1970. Residence time within a wake recirculation region in an axisymmetric supersonic flow. *AIAA paper* 70–111.

# 30

# A Moving Boundary Problem Arising from Stratigraphic Modeling

J. MARR, J. B. SWENSON, AND V. R. VOLLER

## 30.1. Introduction

A moving boundary problem is one in which the location and the movement of one or more of the boundaries that define the problem domain need to be determined as part of the problem solution. Moving boundary problems occur in many engineering and scientific disciplines (Crank, 1984; Shyy et al., 1997). The classical example is the Stefan problem, in which the rate of melting in pure ice is driven by the discontinuity in heat flux across the isothermal solid/liquid interface. The primary contribution of this chapter is the introduction of a new class of moving boundary problems that arises from the modeling of sediment transport, deposition, and erosion over geologic time scales. The long-term objective of this work is the development of a suite of models that can be used to understand the first-order controls on the generation of stratigraphy in a variety of sedimentary basins ranging in size from small intermontaine basins ($10^1$ km) to continental margins ($10^3$ km).

A problem of interest is the formation of alluvial fans in a closed basin (e.g., Death Valley, California) subject to tectonically driven subsidence. An alluvial fan is a half-cone-shaped accumulation of sediment that forms downstream of the point where a river draining an erosional basin discharges into a depositional basin (Hooke and Rohrer, 1979). Alluvial fans are formed by channelized rivers, sheet flows, or debris flows; this study considers only fan formation by channelized flow. Alluvial fans are often located in depositional basins formed by tectonic subsidence, which creates accommodation space for sediment derived from the adjacent erosional basin (Allen and Allen, 1990). The objective of this study is to determine the morphological evolution of the fan surface and, in doing so, determine the rate of advance of the fan toe in response to sediment supply and tectonism. It is shown that the growth of the alluvial fan is in many ways analogous to the advance of a melting front within a solid and, as such, represents a moving boundary problem. Alluvial-fan sedimentation, however, possesses two features that distinguish it from traditional moving boundary problems: (a) the standard explicit relationship between flux discontinuity and the rate of moving boundary advance is absent, and (b) the value of the potential field at the moving boundary is time dependent. Solving a moving boundary problem with these conditions is challenging. In Section 30.2 the basic components of the problem are outlined. This is followed by the presentation of three alternative solution schemes: a scheme based on a deforming grid, a scheme based on a fixed grid, and a semianalytical integral solution.

## 30.2. Mathematical Model of Alluvial-Fan Sedimentation

The central ideas in this work are presented as simple one-dimensional problems. Extensions and results from more complex one- and two-dimensional examples are discussed and presented at the end of the chapter.

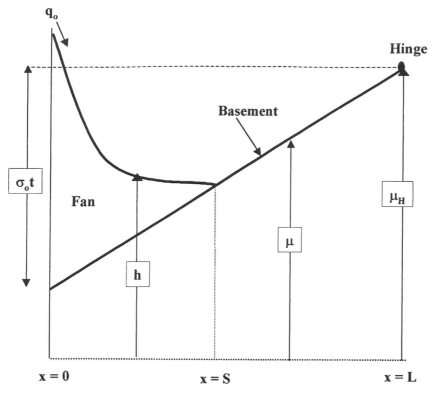

Figure 30.1. Cross section through idealized depositional basin consisting of alluvial fan, $0 < x < S(t)$, and subsiding basement, $S(t) < x < L$, atop which the fan advances in time.

The cross section of Fig. 30.1 represents an idealized alluvial-fan system that is used throughout this study. A steady flux of sediment $q_0$ is delivered to a long and narrow depositional basin $(x > 0)$ from an adjacent erosional terrane $(x < 0)$; the erosional and depositional regimes are separated by a vertical fault. The width of the basin $(L)$ is assumed to be much less than its length (normal to the page) such that the mean direction of sediment transport is in the $x$ direction. The two components of the basin are the alluvial fan, $0 < x < S(t)$, and the basement of the subsiding basin, $S(t) < x < L$, on which the alluvial fan advances.

The sediment supplied to the basin is transported and deposited on the surface of the fan system by a network of rivers. It has been shown theoretically (Paola et al., 1992) that on large spatial and time scales, the transport and deposition of sediment by rivers can be treated as a diffusional process, in which the time- and depth-averaged sediment flux $q(x, t)$ is proportional to the local slope of the alluvial-fan surface:

$$q = -\nu \frac{\partial h}{\partial x} \qquad 0 \le x \le S(t), \tag{1}$$

where $h$ is the elevation of the fan surface and $\nu$ is the diffusivity of the river system, which depends on the water discharge and the grain size of the sediment delivered to the fan.

The depositional basin that hosts the alluvial-fan system is subsiding under the influence of regional tectonism. The spatial distribution of subsidence rate, $\sigma(x)$, for many basins of the type considered in this study can be approximated by

$$\sigma(x) = \sigma_0 \left( 1 - \frac{x}{L} \right),$$ (2)

where $\sigma_0$ is the subsidence rate along the fault, measured positive downward, and $L$ is the length over which subsidence acts (i.e., distance to the hinge point). The basement elevation $\mu$ is

$$\mu(x, t) = \mu_H - \sigma_0 \left( 1 - \frac{x}{L} \right),$$ (3)

where $\mu_H$ is the elevation of the hinge point (Fig. 30.1).

Combining the sediment transport relation of Eq. (1) and the tectonic subsidence model of Eq. (2) in the Exner equation of sediment mass conservation across the alluvial fan yields a diffusion equation in the bed elevation (e.g., Paola et al., 1992):

$$\frac{\partial h}{\partial t} + \sigma_0 \left( 1 - \frac{x}{L} \right) = \nu \frac{\partial^2 h}{\partial x^2} \qquad 0 \le x \le S(t).$$ (4)

Equation (4) is subjected to the following boundary conditions. A prescribed flux of sediment is delivered to the head of the fan system:

$$-\nu \frac{\partial h}{\partial x} \bigg|_{x=0} = q_0.$$ (5)

The sediment flux at the toe of the fan system is zero, which is the mathematical representation of a closed basin:

$$-\nu \frac{\partial h}{\partial x} \bigg|_{x=S(t)} = 0.$$ (6)

The elevation of the fan toe must be equal to that of the basement:

$$h(S, t) = \mu_H - \sigma_0 \left( 1 - \frac{S}{L} \right) t.$$ (7)

The initial conditions are

$$h(x, 0) = \mu_H, \qquad S(0) = 0.$$ (8)

For later use the following auxiliary condition at the fan toe is introduced

$$\frac{\partial^2 h}{\partial x^2} \bigg|_{x=S(t)} = 0.$$ (9)

The central objective in obtaining a solution to Eqs. (4)–(8) is the determination of the rate of advance of the alluvial-fan toe across the subsiding basement of the basin. In addition, the stacking of bed-elevation profiles will form the stratigraphy in the fan deposit.

Inspection of Eqs. (4)–(8) indicates that the transport and deposition of sediment across an alluvial fan is somewhat analogous to the conduction of heat across the liquid phase in the classical Stefan problem. This analogy is, however, not exact: (a) The presence of tectonic subsidence introduces a sink term not present in the thermal analog; (b) the potential ($h$) at the fan toe is time dependent; and (c) Stefan condition (6) contains no explicit reference to the rate of fan-toe advance. These conditions present a challenge to obtaining a solution to Eqs. (4)–(8); to date, no analytical solution has been found, and numerical and approximate semianalytical techniques must be invoked.

## 30.3. A Deforming-Grid Solution

A standard numerical approach to solving Eqs. (4)–(8) involves discretization of the time-dependent solution domain into a fixed number of control volumes, the lengths of which change in accordance with the evolution of the solution domain. An appropriate scheme can be obtained by direct application of discrete versions of the underlying conservation principles on the control volumes defined by the deforming grid of nodes (e.g., Voller and Peng, 1994). An alternative approach (Lynch, 1982) is to use a transformed coordinate system to derive an equivalent governing differential equation that can be cast in discrete form in the transformed coordinate system. This latter approach is adopted in this chapter; for the one-dimensional problem of interest, the appropriate coordinate transformation is the Landau front-fixing transformation (Crank, 1984), in which a new independent variable is defined:

$$\xi = \frac{x}{S}. \tag{10}$$

With this transformation the solution domain reduces to the unit $0 < \xi < 1$, and Eq. (4) assumes the form of an diffusion–advection equation (Crank, 1984):

$$S^2 \frac{\partial h}{\partial t} + \sigma_0 \left( 1 - \frac{S\xi}{L} \right) = \nu \frac{\partial^2 h}{\partial \xi^2} + S\xi \frac{dS}{dt} \frac{\partial h}{\partial \xi} \qquad 0 \leq \xi \leq 1, \tag{11}$$

where, in discrete form, the second term on the right-hand side represents advection of mass due to deformation of the numerical grid. The boundary conditions on Eq. (11) are

$$-\frac{\nu}{S} \frac{\partial h}{\partial \xi} \bigg|_{\xi=0} = q_0, \tag{12}$$

$$\frac{\partial h}{\partial \xi} \bigg|_{\xi=1} = 0, \tag{13}$$

$$h(1, t) = \mu_H - \sigma_0 \left( 1 - \frac{S}{L} \right) t. \tag{14}$$

Integration of Eq. (11) over a time step proceeds by iteration as detailed below; a fully implicit approach is recommended. The solution domain, $0 < \xi < 1$, is covered by a fixed grid of nodes with constant spacing. At internal nodes a finite-difference representation of Eq. (11) is obtained. The resulting equation will be nonlinear since the position and the velocity of the alluvial-fan toe are unknown a priori. At the boundary nodes ($\xi = 0$ and $\xi = 1$) the finite-difference scheme is modified to account for the flux conditions of Eqs. (12) and (13).

With these conditions and estimates for the position and the velocity of the alluvial-fan toe, the linear system of finite-difference equations is solved to yield the nodal elevations of the alluvial-fan surface. If the predicted elevation of the alluvial-fan toe agrees with that of the known basement elevation of Eq. (14) to within some user-specified tolerance, then the solution for that time step is complete. If the convergence criterion is not satisfied, iteration continues, with a rearrangement of Eq. (14) used to update the position of the alluvial-fan toe and its rate of advance. Some underrelaxation is typically required for guaranteeing solution convergence.

In applying the deforming-grid technique to Eqs. (4)–(8), there are two noteworthy issues to consider: (a) The above algorithm requires initial estimates for the length, rate of growth, and surface profile of the fan. Currently, a semianalytical integral profile method (detailed below) is used to provide estimates for these quantities. (b) The solution technique described above differs from standard deforming-grid approaches in which the Dirichlet condition is typically enforced at the moving boundary and the auxiliary Neumann condition used to update its position and rate of advance. This approach fails in this study because of the lack of an explicit relationship involving the rate of advance of the moving boundary. Hence the roles of the Dirichlet and Neumann conditions are reversed; in this way the resultant numerical scheme may not be as stable as a standard deforming-grid solution. Hence the time step and amount of underrelaxation used must be chosen carefully.

### 30.4. A Fixed-Grid Solution

The objective of this section is to introduce a fixed-grid solution of Eqs. (4)–(8) in real space. The solution domain, $0 < x < L$, is discretized with a fixed number of control volumes of uniform width $\Delta x$ (Fig. 30.2). Conservation of sediment mass on a control volume fully inside the alluvial fan can be written as

$$h_P = h_P^{\text{old}} + \frac{\Delta t}{\Delta x}[q_{\text{in}} - q_{\text{out}}] - \Delta t \sigma_0 \left(1 - \frac{x_P}{L}\right), \tag{15}$$

where explicit time integration has been invoked. In Eq. (15), the subscript $P$ refers to the $P$th node, and the superscript old identifies a value from the previous point in time. Note that the last term on the right-hand side of Eq. (15) represents the rate of creation of accommodation

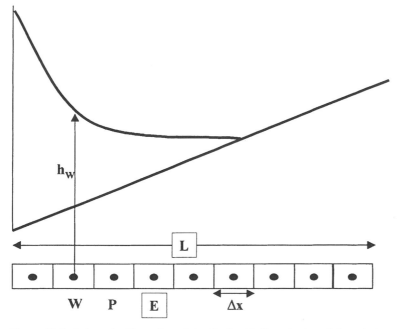

Figure 30.2. Schematic illustration of the fixed-grid discretization of the solution domain.

space in the control volume due to tectonic subsistence. The heights of nodes outside the body of the alluvial fan are set at the basement elevation, which is readily determined by Eq. (3). With use of standard finite-difference approximations, the flux terms in Eq. (15) are expressed in terms of the neighboring nodal fan elevations:

$$q_{in} = v \frac{\left( h_W^{old} - h_P^{old} \right)}{\Delta x}, \qquad q_{out} = v \frac{\left( h_P^P - h_E^{old} \right)}{\Delta x}, \tag{16}$$

where the subscripts $W$ and $E$ refer to the nodes west and east, respectively, of the $P$th node (Fig. 30.2). The Neumann condition at the sediment source is imposed by simply setting $q_{in} = q_0$ at the westernmost node. The Neumann condition of zero bed slope at the fan toe [Eq. (6)], however, can be imposed in only an average sense. With reference to Fig. 30.3, the following treatment is proposed:

$$q_{out} = \begin{cases} v \dfrac{(h_P - h_E)}{\Delta x}, & \text{if } h_P^{old} > h_E^{old} \\ 0, & \text{if } h_P^{old} < h_E^{old} \end{cases} . \tag{17}$$

This mass-conserving condition imposes the restriction that sediment cannot be transported out of a control volume until its bed elevation is greater than that of the downstream control volume. This physically intuitive restriction ensures that sediment cannot be transported out of the control volume in which the toe is situated, a condition that, in an average sense, satisfies the zero flux condition at the fan toe. Implementation of Eq. (17) is straightforward because the flux

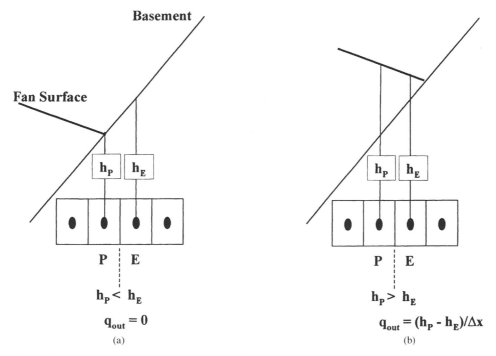

Figure 30.3. Schematic illustration of the fixed-grid treatment of sediment transport near the fan toe. (a) The fan toe is situated within the $P$th control volume (in an average sense) and the elevation of that control volume is *less* than that of the volume to its east, in which case the sediment flux leaving the $P$th volume is zero; (b) the fan toe is outboard of the $P$th control volume and the elevation of that control is volume is *greater* than that of the volume to its east, in which case the sediment flux leaving the $P$th volume is nonzero.

condition is evaluated in terms of the known (old) nodal fan elevations. This convenience offsets the restrictions placed on the time step by stability requirements. In addition, this fixed-grid scheme is easily extended to more complicated problems, such as a time-dependent sediment supply to the basin or a two-dimensional fan, which could impose severe challenges to a deforming-grid approach.

Although the above scheme conserves mass, some readers may be concerned that the treatment of sediment transport within the control volume containing the fan toe may produce oscillations in the predicted toe movement. In a subsequent section, however, it is shown that for reasonable choices of temporal and spatial discretization the proposed fixed-grid scheme yields results of similar accuracy to those obtained with the deforming-grid approach.

## 30.5. An Integral Profile Solution

Integral profile methods are commonly used to obtain semianalytical solutions to moving boundary problems. In this technique, the spatial distribution of the field variable (e.g., temperature) is approximated with a polynomial, the order of which is chosen so as to match the number of boundary conditions and auxiliary relations that characterize the problem. Enforcement of global conservation in the transported quantity (e.g., heat) yields an ordinary differential equation in the position of the moving boundary and its time derivatives that is then solved analytically or numerically. A classical example of this technique is the heat balance integral (Goodman, 1958; Crank, 1984), which is used to approximate the rate of advance of the phase front in melting and solidification problems. A similar approach, hereby defined as the mass balance integral (MBI) technique, can be used to determine the rate of advance of the alluvial-fan toe; this approach is illustrated below.

It is advantageous to introduce a new variable $\zeta$, which measures the thickness of the alluvial fan at any location:

$$\zeta \equiv h - \mu. \tag{18}$$

Global conservation of sediment mass across the alluvial fan is

$$q_0 = \frac{\mathrm{d}}{\mathrm{d}t} \int_0^s \zeta \, \mathrm{d}x. \tag{19}$$

The elevation of the fan surface is assumed to display a cubic dependence on distance from its toe:

$$\zeta(x, t) = a(S - x) + b(S - x)^2 + c(S - x)^3. \tag{20}$$

Note that this profile automatically satisfies the condition of zero fan thickness at its toe. The choice of coefficients,

$$a = \frac{\sigma_0 t}{L}, \qquad b = 0, \qquad c = \frac{q_0}{3S^2 v}, \tag{21}$$

guarantees that the Neumann conditions at the sediment source and fan toe and curvature condition (9) at the fan toe will be satisfied. Integration of Eq. (19), by use of Eqs. (20) and (21), yields the following equation for the fan length:

$$S(t) = \sqrt{\frac{q_0 t}{\frac{q_0}{12v} + \frac{\sigma_0 t}{2L}}}. \tag{22}$$

An alternative approach is to assume a quadratic profile for the fan. Such a relationship yields a similar expression for the trajectory of the fan toe:

$$S(t) = \sqrt{\frac{q_0 t}{\frac{q_0}{6\nu} + \frac{\sigma_0 t}{2L}}}.$$                          (23)

For small times, the position of the alluvial-fan toe, as given by either Eq. (22) or Eq. (23), displays the standard square-root-of-time dependence. In the limit of large time, the fan toe becomes stationary at a position that depends on the ratio of sediment supplied to the basin to the rate of creation of space, by means of tectonic subsidence, across the basin:

$$\lim_{t \to \infty} \frac{S(t)}{L} = \sqrt{\frac{2q_0}{\sigma_0 L}}.$$                          (24)

Therefore, in the large-time limit, the position of the fan toe is given by the square root of the basin's capture ratio, which is defined by Paola (1988) as

$$\frac{2q_0}{\sigma_0 L}.$$                          (25)

A capture ratio of unity implies an exact balance between the rate of sediment delivery to the basin and the rate of creation of space within it; in this limiting case, the position of the fan toe will asymptotically approach the distal edge of the basin ($L$).

### 30.6. An Initial Test Problem

The performance of the various solution techniques outlined above are evaluated on a simple example of alluvial-fan sedimentation characterized by

$$L = 2 \times 10^3 \, \text{m}; \quad q_0 = 1 \, \text{m}^2 \, \text{a}^{-1}; \quad \nu = 1 \, \text{m}^2 \, \text{a}^{-1}; \quad \sigma_0 = 10^{-3} \, \text{m} \, \text{a}^{-1}.$$

Note that a is the standard MKS-system abbreviation for a year. The fixed-grid solution uses a spatial step of 10 m and a time step of 50 years. The deforming-grid solution uses 41 nodes, a time step of 1000 years, and an initial solution generated at 1000 years. The trajectories of the alluvial-fan toe, as predicted by the three approaches, are displayed in Fig. 30.4. The fixed-grid,

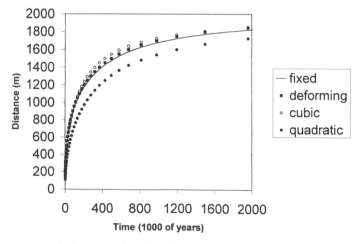

Figure 30.4. A comparative plot of fan-toe trajectories predicted by the various solution techniques outlined in the text.

deforming-grid, and MBI (with cubic polynomial representation) schemes all yield similar trajectories; the trajectory of the quadratic polynomial MBI approach is consistently less than the other solutions. Despite its simplicity, the fixed-grid approach appears to offer a comparable performance with the more complicated deforming-grid technique. When compared with the deforming-grid approach and restrictive MBI method, the fixed-grid approach appears to offer significant advantages in terms of ease of implementation, applicability to more complex problems, and overall flexibility.

## 30.7. Extensions

There is a variety of extensions to the above test problem that is of interest to the engineering and geological communities. It is possible to preserve a sequence of fan elevations and, by translating them downward to simulate the effects of tectonic subsidence, generate the stratigraphy preserved in the fan itself. The result of this exercise for the test problem given above is displayed in Fig. 30.5. It is also possible to simulate the effects of climate (precipitation) and/or variability in source rock type (erodability) by introducing time dependence to the sediment supply to the depositional basin, i.e., $q_0 = q_0(t)$. In addition, it is common for the alluvial fan to terminate in a standing body of water to form a fan delta (Nemec and Steel, 1988). A proper treatment of sediment transport and deposition in such a system requires coupling a submarine regime, with its different transport mechanisms, to the alluvial-fan system described herein. It has been shown that sediment transport in this system is analogous to heat transfer in the explicit Stefan problem (Swenson et al., 1997).

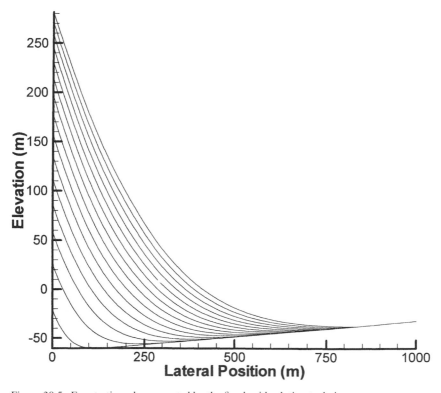

Figure 30.5. Fan stratigraphy generated by the fixed-grid solution technique.

## 30.8. Conclusions

A new class of moving boundary problems that arises from a mathematical treatment of large-scale sediment transport and deposition has been identified. The growth of an alluvial fan in a subsiding closed basin was analyzed as a moving boundary problem with two distinct features: (1) The sediment flux at the moving boundary (fan toe) vanishes, and (2) the potential at the moving boundary (the elevation of the toe) displays a time dependence. Regardless, following suitable modification, the problem is amenable to standard moving boundary techniques. Three such techniques are presented in this chapter: (1) a deforming-grid technique, (2) a fixed-grid technique, and (3) an integral profile technique. When applied to a simple test problem, all three techniques yield similar results. Current work on this class of moving boundary problems is aimed at (1) extension of the fixed-grid approach to the two-dimensional problems with multiple sediment sources and (2) extension of the deforming-grid approach to allow analysis of sediment transport across complex, multicomponent systems such as the continental margin.

## Acknowledgments

The authors wish to thank Chris Paola and Gary Parker for stimulating discussions, Mark Person for generous access to the Gibson Computational Hydrogeology Laboratory at the University of Minnesota, and the Minnesota Supercomputer Institute for a resource grant. JBS was supported through National Science Foundation grants GER 93-54936 and EAR-94-05807.

## References

Allen, P. A. and Allen, J. R. 1990. *Basin Analysis: Principles and Applications*, Blackwell Scientific.

Crank, J. 1984. *Free and Moving Boundary Problems*, Clarendon, Oxford.

Goodman, T. R. 1958. The heat balance integral and its applications to problems involving a change of phase. *Trans AIME* **80**, 335–342.

Hooke, R. L. and Rohrer, W. L. 1979. Geometry of alluvial fans: effect of discharge and sediment size. *Earth Surf. Proc. Landforms* **4**, 147–166.

Lynch, D. R. 1982. Unified approach to simulation on deforming elements with application to phase change problems. *J. Comput. Phys.* **47**, 387–411.

Nemec, W. and Steel, R. J. 1988. *Fan Deltas: Sedimentary and Tectonic Settings*, Blackie, Glasgow.

Paola, C. 1988. Subsidence and gravel transport in basins. In *New Perspectives in Basin Analysis*, K. L. Kleinspehn and C. Paola, eds., pp. 231–234, Springer-Verlag, New York.

Paola, C., Heller, P. L., and Angevine, C. L. 1992. The large-scale dynamics of grain-size variation in alluvial basins. *Basin Res.* **4**, 73–90.

Shyy, W., Thakur, S. S., Ouyang, H., Liu, J., and Blosch, E. 1997. *Computational Techniques for Complex Transport Phenomena*, Cambridge U. Press, Cambridge.

Swenson, J., Voller, V., Paola, C., and Parker, G. 1997. Modeling fluvio-deltaic sedimentation as a generalized Stefan problem. *EOS Trans.* AGU **78**.

Voller, V. R. and Peng, S. 1994. An enthalpy formulation based on an arbitrarily deforming mesh solution of the Stefan problem. *Comput. Mech.* **14**, 492–502.

# 31

# Convection Generated by Lateral Heating of a Solute Gradient: Review and Extension

C. F. CHEN

## 31.1. Introduction

When a stably stratified solute gradient contained in a vertical tank is subjected to lateral heating, a series of horizontal convecting cells is generated when the critical $\Delta T$ is exceeded. This is a double-diffusive phenomenon, which arises from the differential rates of thermal and mass diffusions. In this chapter, we first present a review of research results obtained by experimental investigations, stability analyses, and two-dimensional numerical simulations of this instability phenomenon. We then present the results of our recent experiments showing that convection in these nearly horizontal convection cells are three dimensional in nature because of the presence of salt-finger convection.

## 31.2. Review of the Literature

### 31.2.1. *Experimental Results*

The first systematic study of convection in a stably stratified fluid generated by lateral heating was carried out by Thorpe et al. (1969). However, a report on a related phenomenon was published by Brewer (1883) 86 years earlier. He noted that, under certain circumstances, the settling of fine sand particles in pure water "will not fade gradually in density from the bottom upwards through regularly diminishing opacity to the top, but will rather be disposed in successive layers, the limits of each more or less well defined." Following this discovery, Barus (1886) carried out sedimentation experiments of fine clay particles under a number of different circumstances without determining a cause for the appearance of stratified layers. Mendenhall and Mason (1923), by further experimentation, showed that the layer structure was caused by a lateral temperature gradient acting on a stably stratified suspension owing to the differential settling rates of the particles.

Thorpe et al. (1969) conducted their experiments in a linearly stratified salt solution contained in a tall, narrow tank (0.6 cm wide × 15 cm high × 7.5 cm deep). The two broad sidewalls were kept at different constant temperatures. The temperature of the hot wall was increased slowly until "two-dimensional rolls were found to develop quite rapidly and apparently simultaneously in the slot...." They also performed a linear stability analysis, assuming a basic quiescent state in which the temperature and concentrations had compensating horizontal gradients such that the density was constant on a horizontal plane. Further, because of the large solute Rayleigh numbers of the experiments, the nondiffusive condition at the wall and the boundary layer flow near the wall were neglected in the analysis. Their prediction of the critical thermal Rayleigh number for onset of layered convection was

$$R_T = 5.9(-R_S)^{5/6}/(\text{Le} - 1), \tag{1}$$

in which the thermal and the solute Rayleigh numbers are defined as $R_T = g\alpha\Delta T D^3/\kappa v$ and $R_S = g\beta(\mathrm{d}S/\mathrm{d}z)D^4/\kappa v$. In these definitions, $g$ is gravitational acceleration; $\alpha$ and $\beta$ are the thermal and solutal coefficients of expansion; $\kappa$ and $\kappa_S$ are thermal and solutal diffusivities, respectively; $\mathrm{d}S/\mathrm{d}z$ is the initial solute gradient; $\Delta T$ is the temperature difference across the slot width $D$; and $v$ is the kinematic viscosity. The Lewis number is $\text{Le} = \kappa/\kappa_S$. The prediction agreed well with their experimental results, especially at high values of $(-R_S)$.

Paliwal and Chen (1980a) extended the experimental study to an inclined slot. The marginal stability conditions were determined for the angle of inclination $\theta$ for the vertical $-75° \leq \theta \leq 75°$. The results were asymmetrical with respect to the vertical. Contrary to intuition, the system was more stable when the lower surface was heated. The results of a linear stability analysis (Paliwal and Chen, 1980b) agreed well with the experimental data. The analysis also showed that, when the lower wall was heated, the vertical solute gradient increased in the steady state before the onset of instabilities, thus enhancing stability.

Chen et al. (1971), on the other hand, considered the stability of convection in a large tank of stably stratified fluid being heated by a vertical boundary wall. Lacking an obvious length scale, they determined one by dimensional analysis. This is the natural length scale $h$, which is the height to which a heated parcel of fluid would rise in the initially stratified fluid:

$$h = \alpha\Delta T/(-\beta\,\mathrm{d}S/\mathrm{d}z). \tag{2}$$

Their experiments were carried out in a wide tank (12.7 cm wide $\times$ 30 cm high $\times$ 20 cm deep). The temperature of one sidewall was kept constant while the other was raised rapidly to a preset value, with a time constant of 3 min. By performing 15 experiments over a range of conditions, they determined the critical Rayleigh number $\tilde{R}_h$ based on $h$ to be

$$\tilde{R}_h = g\alpha\Delta T h^3/\kappa v = 15{,}000 \pm 2500. \tag{3}$$

They also determined the convection layer thickness, which varied from $0.67h$ to $0.97h$ in the range of supercritical Rayleigh numbers tested, $1.42 \times 10^4 < \tilde{R} < 5.4 \times 10^4$. In Fig. 31.1, we show the results of this "impulsive heating case" carried out in a stratified ethanol–water solution by Chen and Chen (1997). It was later found, by Huppert and Turner (1980) and Huppert et al. (1984), that the layer thickness decreased to $\sim 0.6h$ at higher Rayleigh numbers ($\tilde{R} \approx 10^6$–$10^9$). Their experiments were carried out in a number of aqueous solutions with a wide range of values of the Prandtl and Lewis numbers.

Narusawa and Suzukawa (1981) performed experiments with a constant heat flux boundary. They found that the onset of layer convection occurs when the nondimensional parameter $\pi_3$ exceeds a critical value, where

$$\pi_3 = \alpha(q/k)/\beta(-\mathrm{d}S/\mathrm{d}z), \tag{4}$$

in which $q$ is the heat flux and $k$ the thermal conductivity. Since $q/k$ is proportional to the horizontal temperature gradient at the wall, $\pi_3$ is the ratio of the horizontal density gradient due to heating to the vertical density gradient due to the initial stratification. The critical value of $\pi_3$ depends on the Lewis number of the fluid.

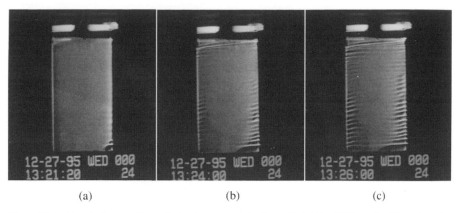

Figure 31.1. Onset of convection cells due to an impulsively applied $\Delta T = 9\,^\circ$C to a stratified ethanol–water solution at $t = 13:20$. Hot wall is on the right. (a) $t = 13:21$, (b) $t = 13:24$, (c) $t = 13:26$. Initial stratification: 100% ethanol on top and 64% ethanol at the bottom.

Tanny and Tsinober (1988) conducted experiments in wide tanks in which the temperature rise of one sidewall could be controlled to give a linear rise with time or an exponential rise with time constants varying from 1.4 to 133 min. It was found that the onset condition can be correlated by the thermal and the solutal Rayleigh numbers based on the thermal diffusion length. The condition is essentially the same as that determined by linear stability analysis for a narrow slot (Hart, 1971; Thangam and Chen, 1981). The thickness of the fully developed convection layers was found to be $\approx 0.6h$, in agreement with the results of Huppert et al. (1984).

Lee et al. (1990) performed experiments in a tank equipped with permeable walls for the top and bottom boundaries. In this manner, they were able to maintain a constant $S$ boundary condition at these boundaries. They observed four different flow regimes as the applied $\Delta T$ was increased for a given initial solute gradient. When $\Delta T$ was very small, the fluid was quiescent. At larger $\Delta T$'s, convection layers were generated successively from the top and bottom boundaries. With even larger $\Delta T$'s, the convection layers onset all along the heated wall. The latter two regimes were the focus of attention in the experiments of Chen et al. (1971). At very large $\Delta T$, the flow settled down to a unicellular convection.

Jeevaraj and Imberger (1991) studied the sideways heating of an initial density gradient stratified by both temperature and solute with a constant temperature boundary. Schladow et al. (1992) and McDonald and Koseff (1994) studied the problem with constant heat flux input.

### 31.2.2. Stability Analyses

Hart (1971) performed a linear stability analysis of the slot problem over the entire range of solute Rayleigh numbers, taking into account the boundary layer flow along the sidewalls. The results showed that the instability is shear dominated at low values of $(-R_S)$ and double-diffusive dominated at high $(-R_S)$. The critical condition [Eq. (1)] derived by Thorpe et al. (1969) becomes asymptotically correct as $(-R_S) \to \infty$. Thangam and Chen (1981) showed that the instability onsets in an oscillatory mode within the transition region between shear and double-diffusive instabilities. Recently, Young and Rosner (1998) extended the analysis to include both double-diffusive and triple-diffusive systems. The eigenvalue problem was solved by the spectral method. The marginal stability conditions for a double-diffusive system were calculated for $10^5 < -R_S < 10^{10}$, and the critical thermal Rayleigh number was found to be

$$R_t = 7.618(-R_S)^{0.8204}/(\text{Le} - 1),\tag{5}$$

which is consistent with Eq. (1). Using these results, they showed that the layers observed by Mueth et al. (1996) in decane-in-water emulsion in a vertical tube subjected to a lateral temperature gradient of $10 \pm 2$ mK are the result of double-diffusive instabilities. This is because the diffusion in the emulsion is due to Brownian motion. Its extremely low value resulted in $\text{Le} \approx 10^6$.

A weakly nonlinear analysis was subsequently carried out by Hart (1973), who showed that, if $\text{Le} > (-R_S)^{1/6}$, the onset of instability is through a subcritical bifurcation and the streamline pattern may be considerably different from that predicted by the linear theory. The nonlinear phenomenon was studied in more detail by Thangam et al. (1982), who used a Galerkin method for both vertical and inclined slots. The results show that, in all cases, the counterrotating vortices at onset as predicted by linear stability theory merge into vortices of twice their size, rotating in the same direction, as the critical Rayleigh number is exceeded.

The stability analysis for the wide-tank case is much more difficult because the basic state is time dependent. Chen (1974) carried out an initial-value stability analysis by numerically monitoring the evolution of the perturbation kinetic energy of distributed random disturbances of small magnitude. The growth or decay of the kinetic energy delineates the unstable and stable states. For a tank width of $W = 8h$, results show that the kinetic energy decays for $R_h < 10^4$ and grows for $R_h > 2 \times 10^4$. The wavelength of the faster growing wave was found to be $0.56h$. These results are consistent with our earlier experimental results.

Kerr (1989) incorporated the basic time-dependent state by nondimensionalizing the horizontal distance by the thermal diffusion length, $L = (\kappa t)^{1/2}$, and the vertical distances by the natural length scale, $h$ [Eq. (1)], suggested by Chen et al. (1971). By assuming $(h/L)^2 \ll 1$ at the time of instability onset, which was the case in all previous experiments, the eigenvalue problem was solved to determine the critical value of $Q$ of the marginal state. The nondimensional governing parameter $Q$ is related to $R_h$ and is defined as

$$Q = (1 - \text{Le}^{-1})^6 \frac{g\alpha \Delta T h^5}{\kappa_S \nu L^2} = \text{Le}(1 - \text{Le}^{-1})^6 R_h (h/L)^2.\tag{6}$$

For salt–water solutions, $Q \approx 1.48 \times 10^5$. This prediction shows good agreement with the data of Chen et al. (1971) and Tanny and Tsinober (1988). A weakly nonlinear analysis carried out by Kerr (1990) showed that, similar to the slot case, the bifurcation into instability is subcritical. Initially, the convections are counterrotating. However, they have a tendency to develop into corotating cells of twice their size.

Bifurcation studies of this problem with a given solute concentration at the top and bottom boundaries have been carried out by Tsitverblit and Kit (1993), Tsitverblit (1995), and Dijkstra and Kranenborg (1996). Results show that this problem has a complex steady bifurcation phenomenon. For a given $R_S$, unique solutions exist when the thermal Rayleigh number is either small or large. For intermediate values of $R_T$, there exist a large number of different steady flow patterns.

### 31.2.3. *Numerical Calculations*

The first numerical simulation of the problem was carried out by Wirtz et al. (1972). This and all subsequent numerical simulations were made by assuming two-dimensional motion.

The equations of motion, energy, and solute with impulsively imposed constant temperature heating from one sidewall were integrated by a finite-difference scheme. Two types of problems were considered, the finite-geometry problem and the infinite-slot problem. In the former, the rectangular boundary had an aspect ratio (height/width) of 6 and Rayleigh number $R_h = 10^5$. Results show that convection layers were generated successively, starting from the bottom boundary. For the infinite-slot case, periodic conditions were applied over a height of $2h$. Again, with $R_h = 10^5$, convection layers were generated simultaneously in the calculation domain, each with a height of $\approx 0.67h$. Both results were consistent with our previous experimental results.

Heinrich (1984) solved the equations with a finite-element method under the same conditions as those for the finite-geometry case of Wirtz et al. (1972), except that the aspect ratio was changed to 5. Results were similar to those obtained by Wirtz et al. Heinrich's calculations were carried out for a much longer time. Consequently, the convection layers were much more developed. Results were also obtained for the case in which $\Delta T$ was imposed symmetrically by increasing (decreasing) the temperature of the left (right) wall. A symmetric pattern was obtained.

Lee and Hyun (1991) simulated the experiments of Chen et al. (1971) by a finite-difference method using a parallel computing scheme. The tank had an aspect ratio of 2, and results were obtained for different values of $R_h$. Their results for the critical condition and initial layer thickness show good agreement with the experimental results of Chen et al.

Wright and Shyy (1996) performed numerical simulations of the experiments of Schladow et al. (1992) in which the stratified fluid contained in a wide tank was heated from one sidewall by a constant heat flux input. In order to obtain better resolution of the flow within the developing convection layer near the hot wall, they used a composite grid method with local refinement. They obtained remarkably detailed two-dimensional flow structures of the three types of intrusions observed by Schladow et al. in their experiments.

Chen and Liou (1997) considered convection in an inclined tank of aspect ratio 2 filled with vertically stratified fluid. Heating was at a constant temperature from one of the sidewalls. Calculations were made for $R_h = 6.4 \times 10^5$, with angles of inclination varying from $+75°$ to $-75°$. Results show that at a positive angle of inclination $\theta$ (heating from below), the onset of cellular convection is later than when the angle of inclination is $-\theta$ (heating from above). This result is consistent with the enhanced stability at the $+\theta$'s that Paliwal and Chen (1980a, 1980b) found both experimentally and theoretically for an inclined slot.

### 31.3. Recent Developments

The convection patterns are similar in all fully developed convection cells. Fluid flows upward along the heated wall, turns toward the cold wall along the top of the cell, flows downward along the cold wall, and then completes the circuit by returning along the bottom of the cell toward the hot wall. Because of the initial concentration stratification, the relatively warm fluid along the top of the cell is relatively solute rich, and the cooler fluid along the bottom of the cell is relatively solute poor. Such a vertical distribution of solute concentration and temperature can cause the onset of double-diffusive instability in the form of salt fingers. In a quiescent fluid, salt fingers are long, narrow convection cells in the vertical direction that transport both heat and solute downward in an effort to diminish the gradient. Because of this possibility, Thorpe et al. (1969) and Wirtz et al. (1972) suspected that the flow in the convection cells was three dimensional.

Recently, through a series of experiments, we detected the presence of salt-finger convection in all the vigorously convecting cells. These experiments were performed in a deep tank (Chen and Chen, 1997) and in a shallow tank (Chan and Chen, 1999) in which the convection

Table 31.1. *Experimental Solutions*

| | Stratification | | Critical $\Delta T$ | |
| | | | Gradual $\Delta T^\dagger$ | Impulsive $\Delta T^\ddagger$ |
| Solution | *Top* | *Bottom* | ($^\circ C$) | ($^\circ C$) |
|---|---|---|---|---|
| Water–salt | 2% salt | 5% salt | 6.5 | 8.4 |
| Ethanol–water | 0% water | 36% water | 6.2 | 8.6 |

$^\dagger$ According to Thorpe et al. (1969).
$^\ddagger$ According to Chen et al. (1971).

was caused either by buoyancy or by a surface-tension gradient. The presence of salt-finger convection would greatly accelerate the mixing process within each convection cell and thus affect the merging of layers in later stages of the experiment. The experiments are described in more detail below.

### 31.3.1. *Buoyancy-Driven Cases*

These experiments were carried out by Chen and Chen (1997) in a tank 5 cm wide $\times$ 9.5 cm high $\times$ 9.5 cm long. The two vertical sides (9.5 cm $\times$ 9.5 cm) were heat transfer walls made of copper, the two end walls were made of Plexiglass for visualization purposes, and the bottom of Bakelite. The experimental solutions were either water–salt or ethanol–water. Stratification was achieved by carefully admitting to the tank 21 equal layers of successively less concentrated solution. Once the filling was completed, the tank was left standing for 1 h to allow diffusion to smooth out the concentration distribution. The initial stratification of the two test fluids was chosen so that the critical $\Delta T$'s for the onset of convection cells were comparable (as shown in Table 31.1).

Since the finger convection we were trying to detect exists in the presence of shear, it should appear as longitudinal cells when viewed from above, as shown by Linden (1974) and Thangam and Chen (1981). In the former, the results were obtained in the interface between counterflowing sugar and salt solutions and, in the latter, the results were obtained from a surface discharge of warm, salty solution on a body of fresh water. To detect the presence of finger convection in the experiments, the test fluid was seeded with aluminum particles. These were added to the prepared solutions for each stratifying layer before the filling of the tank. The particles were illuminated by a horizontal light sheet generated by a 20-mW He–Ne laser with a cylindrical lens. A CCD camera was mounted directly above the test tank, and the image was displayed on a monitor and recorded for later analysis. The test tank was placed on a platform capable of vertical motion so that flow in the tank could be scanned at any horizontal plane.

The presence of finger convection was first detected in fully developed cells with vigorous convection. When viewed from above, the particles were aligned in rows spanning the entire width of the tank. A good example is shown in Fig. 31.2, which was taken from an experiment with ethanol–water solution at $\Delta T = 7^\circ$C. The hot wall is at the top and the flow is downward toward the cold wall. The light sheet was located in the bottom cell. Finger convection was detected in all cells, except a few in which the convection was very weak.

By systematic experimentation, it was found that finger convection occurs concurrently with the onset of convection. When the sidewall is heated impulsively, the finger convection pattern travels with the convection cell as it advances toward the cold wall, as shown in Fig. 31.3(a). This photo of the particles was taken in an experiment with salt–water solution at approximately

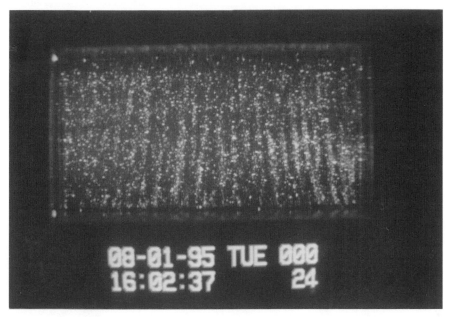

Figure 31.2. Salt-finger convection in the bottom layer of an ethanol–water experiment. The solution is seeded with aluminum particles and is illuminated by a horizontal laser light sheet. The hot wall is on top.

3 min after $\Delta T = 7.3°$C was applied. The convection that originated at the hot wall was advancing toward the cold wall. Finger convection was evident all along the span of the tank.

Chen and Chen (1997) have shown that convection cells originate at the cold wall when the $\Delta T$ is applied gradually to reach the critical state. They also showed by shadowgraph that convective motion is propagated into the bulk fluid through a series of vortices. Finger convection was detected in these vortices, as shown in Fig. 31.3(b). This experiment was also conducted with salt–water solution with gradually increased $\Delta T$. The photo was taken at 12 min after the final adjustment was made to reach $\Delta T = 6.9°$C, exceeding the critical. At that time, there was a larger convection roll in the form of a vortex along the cold wall. A second, weaker vortex was just starting along the first one. The particle trace shows a row of finger convection cells spanning the entire width of the first vortex. The second vortex was just starting and finger convection was not present yet.

These results show that the motion in these convection cells is intrinsically three dimensional because of finger convection. The finger convection is double diffusive in nature, and its onset is simultaneous with the onset of the convection cells. It persists until the end of the experiment, when the solute concentration is equilibrated.

### 31.3.2. *Surface-Tension Gradient Driven Cases*

The first set of experiments in which convection was driven by surface-tension gradients was conducted by Chen and Chen (1997) in the deep tank described in Subsection 31.3.1. The test fluid was a stratified ethanol–water solution with 100% ethanol at the free surface so that it would not be easily contaminated. Convection driven by a surface-tension gradient was generated by a small temperature difference, $\sim$1°C, set across the tank. In the main body, the fluid remained quiescent because of the stabilizing effect of the initial stable density gradient.

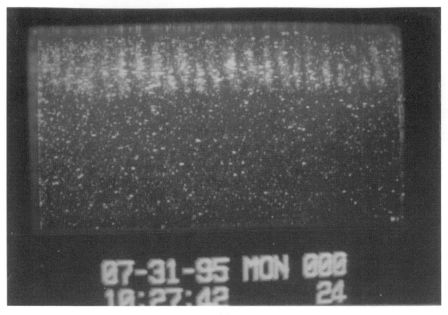

(a)

(b)

Figure 31.3. Appearance of salt-finger convection is concurrent with the onset of the horizontal convection layer. (a) In the impulsively heated case, salt-finger convection advances with the convection cell from the hot wall (top boundary) toward the cold wall. (b) In the gradually heated case, convection starts at the cold wall (bottom boundary) in a series of vortices. Salt fingers can be seen in the first vortex. The second vortex appears as the bright–dark band.

In these experiments, the $\Delta T$ was imposed by raising (lowering) the temperature of the hot (cold) wall by $\Delta T/2$. Soon after the $\Delta T$ was imposed, a small vortex was generated where the free surface intersected each of the two sidewalls. Both vortices brought solute-rich fluid to the surface. Because the presence of the solute, water in this case, reduces the surface tension, the thermal and solutal capillary effects oppose each other at the hot wall, while they reinforce each other at the cold wall. As a consequence, the convection cell grew from the cold wall into the interior and eventually spanned the whole width of the tank. This explains the fact that when the finger convection pattern was detected by Chen and Chen (1997) in a horizontal plane just beneath the free surface, it originated at the cold wall and gradually extended itself across the tank. Because of the small velocities, photographs were made with 8-s exposures in order to make the longitudinal rolls clearly visible.

Chan and Chen (1999) carried out further investigations in a shallow tank, 1 cm high $\times$ 5 cm wide $\times$ 10 cm long. They also conducted a two-dimensional numerical simulation of the early stages of the experiment. The physical experiments were carried out with an ethanol–water solution stratified from 100% ethanol at the top to 96% ethanol at the bottom. Convection driven by a surface-tension gradient was generated when a 2°C temperature difference was imposed. With this $\Delta T$ in the absence of a free surface, no motion was detected in the fluid because of the stabilizing effect of the stratification. Finger convection was observed to start at the cold wall and gradually advance toward the hot wall. Because of the limited depth of the tank, the experiment was over in approximately an hour, when the water concentration in the solution became equilibrated.

The results of the numerical simulation show steady growth of the vortex at the cold wall similar to what was observed in the experiments. At the hot wall, because of the opposing effects of thermal and solutal capillary effects, the flow is oscillatory. Consequently, the initial vortex generated there exhibits no growth at all.

## 31.4. Conclusions

The phenomenon of horizontal cellular convection generated by sideways heating of a stratified fluid has been reviewed. Experiments carried out recently show that the convection in each of these cells is three dimensional, arising from salt-finger convection in the presence of shear. The resulting convection pattern consists of longitudinal rows aligned in the direction of shear. The presence of a fingering convection greatly enhances the mixing processes of solute and heat. Two-dimensional numerical simulations of the problem could be underestimating the effect of such mixing.

## References

Barus, C. 1886. Subsidence of fine solid particles in liquids. *Bull. US Geol. Survey* **36**, 515–555.

Brewer, W. H. 1883. On the subsidence of particles in liquids. *Mem. U.S. Natl. Acad. Sci.* **2**, 165–175.

Chan, C. L. and Chen, C. F. 1999. Salt-finger convection generated by thermal and solutal capillary motion in a stratified fluid. *Int. J. Heat Mass Transfer.* **42**, 2143–2159.

Chen, C. F. 1974. Onset of cellular convection in a salinity gradient due to a lateral temperature gradient. *J. Fluid Mech.* **63**, 563–576.

Chen, C. F. and Chen, F. 1997. Salt-finger convection generated by lateral heating of a solute gradient. *J. Fluid Mech.* **352**, 161–176.

Chen, C. F., Briggs, D. G., and Wirtz, R. A. 1971. Stability of thermal convection in a salinity gradient due to lateral heating. *Int. J. Heat Mass Transfer* **14**, 57–65.

Chen, Y.-M. and Liou, J.-K. 1997. Time-dependent double-diffusive convection due to salt-stratified layer with differential heating in an inclined cavity. *Int. J. Heat Mass Transfer* **40**, 711–725.

Dijkstra, H. A. and Kranenborg, E. J. 1996. A bifurcation study of double diffusive flows in a laterally heated stably stratified liquid layer. *Int. J. Heat Mass Transfer* **39**, 2699–2710.

Hart, J. E. 1971. On sideways diffusive instability. *J. Fluid Mech.* **49**, 279–288.

Hart, J. E. 1973. Finite amplitude sideways diffusive convection. *J. Fluid Mech.* **59**, 47–64.

Heinrich, J. E. 1984. A finite element model for double diffusive convection. *Int. J. Num. Methods Eng.* **20**, 447–464.

Huppert, H. E. and Turner, J. S. 1980. Ice blocks melting into a salinity gradient. *J. Fluid Mech.* **100**, 367–384.

Huppert, H. E., Kerr, R. C., and Hallworth, M. A. 1984. Heating or cooling in stable compositional gradient from the side. *Int. J. Heat Mass Transfer* **27**, 1395–1401.

Jeevaraj, C. G. and Imberger, J. 1991. Experimental study of double-diffusive instability in sidewall heating. *J. Fluid Mech.* **222**, 565–586.

Kerr, O. S. 1989. Heating a salinity gradient from a vertical sidewall: linear theory. *J. Fluid Mech.* **207**, 323–352.

Kerr, O. S. 1990. Heating a salinity gradient from a vertical sidewall: nonlinear theory. *J. Fluid Mech.* **217**, 529–546.

Lee, J., Hyun, M. T., and Kang, Y. S. 1990. Confined natural convection due to lateral heating in a stably stratified solution. *Int. J. Heat Mass Transfer* **33**, 869–875.

Lee, J. W. and Hyun, M. T. 1991. Time-dependent double-diffusion in a stably stratified fluid under lateral heating. *Int. J. Heat Mass Transfer* **34**, 2409–2415.

Linden, P. F. 1974. Salt fingers in a steady shear flow. *Geophys. Fluid Dyn.* **6**, 1–27.

McDonald, E. T. and Koseff, J. R. 1994. The internal structure of lateral intrusions in a continuously stratified heat/salt system. *Phys. Fluids* **6**, 3870–3883.

Mendenhall, C. E. and Mason, M. 1923. The stratified subsidence of fine particles. *Proc. Natl. Acad. Sci. USA* **2**, 199–207.

Mueth, D. M., Crocker, J. C., Esipov, S. E., and Grier, D. G. 1996. Origin of stratification in creaming emulsions. *Phys. Rev. Lett.* **77**, 578–581.

Narusawa, U. and Suzukawa, Y. 1981. Experimental study of double-diffusive cellular convection due to a uniform lateral heat flux. *J. Fluid Mech.* **113**, 387–405.

Paliwal, R. C. and Chen, C. F. 1980a. Double-diffusive instability in an inclined fluid layer. Part 1. Experimental investigation. *J. Fluid Mech.* **98**, 755–768.

Paliwal, R. C. and Chen, C. F. 1980b. Double-diffusive instability in an inclined fluid layer. Part 2. Theoretical investigation. *J. Fluid Mech.* **98**, 769–785.

Schladow, S. G., Thomas, E., and Koseff, J. R. 1992. The dynamics of intrusions into a thermohaline stratification. *J. Fluid Mech.* **236**, 127–165.

Tanny, J. and Tsinober, A. B. 1988. The dynamics and structure of double-diffusive layers in sidewall-heating experiments. *J. Fluid Mech.* **196**, 135–156.

Thangam, S. and Chen, C. F. 1981. Salt finger convection in the surface discharge of heated saline jets. *Geophys. Astrophys. Fluid Dyn.* **18**, 111–146.

Thangam, S., Zebib, A., and Chen, C. F. 1982. Double-diffusive convection in an inclined fluid layer. *J. Fluid Mech.* **116**, 363–378.

Thorpe, S. A., Hutt, P. K., and Soulsby, R. 1969. The effects of horizontal gradients on thermohaline convection. *J. Fluid Mech.* **38**, 375–400.

Tsitverblit, N. 1995. Bifurcation phenomena in confined thermosolutal convection with lateral heating: commencement of double-diffusive region. *Phys. Fluids* **7**, 718–736.

Tsitverblit, N. and Kit, E. 1993. The multiplicity of steady flows in confined double-diffusive convection with lateral heating. *Phys. Fluids A* **5**, 1062–1064.

Young, Y. and Rosner, R. 1998. Linear stability analysis of doubly diffusive vertical slot convection. *Phys. Rev. E* **57**, 1183–1186.

Wirtz, R. A., Briggs, D. G., and Chen, C. F. 1972. Physical and numerical experiments on layered convection in a density-stratified fluid. *Geophys. Fluid Dyn.* **3**, 265–288.

Wright, J. and Shyy, W. 1996. Numerical simulation of unsteady convective intrusions in a thermohaline stratification. *Int. J. Heat Mass Transfer* **39**, 1183–1201.

# 32

# Heat Conduction from a Solid Particle and the Force on It in Stokes Flow in a Fluid with Position-Dependent Physical Properties

ANDREAS ACRIVOS AND YONGGUANG WANG

## 32.1. Introduction

A while back, while endeavoring to compute certain effective properties of dilute two-phase materials in which one of the phases is finely dispersed in the other, we encountered a divergent integral (such integrals generally arise when performing effective medium calculations for dilute systems – see Batchelor, 1974) that had to be renormalized by using a solution of the creeping flow equations, past a given particle, in a fluid whose viscosity varied by a large amount over distances much larger than the dimension of this particle. On failing to locate such a solution in the literature, we decided therefore to examine a class of heat conduction and creeping flow problems in fluids having position-dependent thermal conductivities and viscosities in order to determine the first-order correction to the overall rate of heat conduction from a particle, as well as that of the force on it, arising from this nonuniformity of the physical properties. Specifically, we consider the problem of heat conduction from a heated particle whose surface temperature is held constant, into a medium of infinite extent where the thermal conductivity $K$ is a function of position. General expressions are derived for the total heat flux from the particle for three classes of $K$: (1) The change of $K$ throughout the whole space is small compared with $K_0$, its typical value in the neighborhood of the particle; (2) $K$ changes slowly by an amount comparable with $K_0$ but over a distance that is much larger than the size of the particle; (3) $K$ changes exponentially but slowly along one direction from zero to infinity. These three special cases cover a wide range of heat conduction problems of interest since their contributions to the rate of heat transfer are additive. Corresponding expressions are also given for the force on a particle translating under creeping flow conditions in a fluid with a position-dependent viscosity $\mu$. These general expressions for these two bulk transport quantities are found to reduce to particularly simple forms when the particle is a sphere. It is hoped that they can be found useful in a wider context.

## 32.2. The Heat Conduction Problem

### 32.2.1. *Statement of the Problem*

Consider a particle of arbitrary shape embedded in an infinite expanse of a heat conducting medium having a prescribed nonuniform heat conductivity $K$, which is a smooth function of position. We study the rate of heat transfer from this particle to the surrounding medium, when the normalized temperature $T$ is set equal to unity on the surface of the particle and vanishes at infinity. Thus the governing equation and boundary conditions are given by

$$\nabla \cdot [K(\mathbf{r})\nabla T] = 0, \tag{1}$$

$$T = 1 \quad \text{on} \quad \partial D, \tag{2}$$

$$T \longrightarrow 0 \quad \text{as} \quad r \longrightarrow \infty, \tag{3}$$

where $\mathbf{r}$ is the dimensionless position vector relative to a reference point inside the particle normalized by $a$, the characteristic dimension of the particle. The total rate of heat transfer from the particle is then given by

$$Q = -\int_{\partial D} K(\mathbf{r})\nabla T \cdot \mathbf{n} \, dS, \tag{4}$$

where $\mathbf{n}$ is the unit vector normal to the surface of the particle pointing from the particle to the medium, and the Nusselt number Nu is defined as

$$\text{Nu} \equiv \frac{Q}{4\pi a K_0} = -\frac{1}{4\pi a K_0} \int_{\partial D} K(\mathbf{r})\nabla T \cdot \mathbf{n} \, dS, \tag{5}$$

where $K_0$ is a typical value of the thermal conductivity in the neighborhood of the particle.

We note that no solution exists satisfying Eqs. (1)–(3) if $K$ vanishes in all directions like $1/r$ or faster as $r \longrightarrow \infty$, as can easily be seen by examining Eq. (1) for the spherically symmetric form of $K$. Therefore we exclude this case in our subsequent analysis.

Another useful form of Eq. (1) is

$$\nabla^2 T + \mathbf{G} \cdot \nabla T = 0, \tag{6}$$

with

$$\mathbf{G}(\mathbf{r}) \equiv \frac{1}{K(\mathbf{r})} \nabla K(\mathbf{r}), \tag{7}$$

which is formally analogous to the familiar heat convection equation in a fluid with uniform heat conductivity (Acrivos and Taylor, 1962). Here, $\mathbf{G}$ plays a role similar to that of the velocity field in the heat convection problem. There is an important difference between the two problems however, in that $\mathbf{G}(\mathbf{r})$ is generally not divergence free in contrast to the velocity field, which is usually taken to have zero divergence. When $\mathbf{G}(\mathbf{r})$ is divergence free, the two problems are mathematically identical.

### 32.2.2. *The case $K = K_0 + \epsilon_1 K_0 f(\mathbf{r}) + \cdots$ with $f(\mathbf{r}) \sim O(1)$*

We first consider the case in which the heat conductivity $K(\mathbf{r})$ is a smooth function of position and is approximately equal to $K_0$ everywhere, i.e., when

$$K(\mathbf{r}) = K_0 + \epsilon_1 K_0 f(\mathbf{r}) + \cdots, \tag{8}$$

where $\epsilon_1 \ll 1$ is a small parameter characteristic of the magnitude of the relative variation of the heat conductivity, and $f(\mathbf{r})$ together with its gradient are both at most $O(1)$ everywhere. This problem can be treated by means of the regular perturbation expansion

$$T = T^{(0)} + T^{(1)} + \cdots, \tag{9}$$

where $T^{(1)}/T^{(0)} \longrightarrow 0$ as $\epsilon_1 \longrightarrow 0$. Then, from Eq. (6),

$$\nabla^2 T^{(0)} = 0, \tag{10}$$

$$T^{(0)} = 1 \quad \text{on} \quad \partial D, \tag{11}$$

$$T^{(0)} \longrightarrow 0 \quad \text{as} \quad r \longrightarrow \infty, \tag{12}$$

and

$$\nabla^2 T^{(1)} = -\mathbf{G} \cdot \nabla T^{(0)}, \tag{13}$$

$$T^{(1)} = 0 \quad \text{on} \quad \partial D, \tag{14}$$

$$T^{(1)} \longrightarrow 0 \quad \text{as} \quad r \longrightarrow \infty, \tag{15}$$

where $\mathbf{G} \sim O(\epsilon_1)$ and $T^{(0)}$ refers to the solution of the temperature field when the same particle is in a medium with uniform heat conductivity $K_0$ under the same boundary conditions as Eqs. (2) and (3). In what follows, we suppose that $T^{(0)}(\mathbf{r})$ is known.

Following a procedure originally introduced by Brenner and Cox (1963) in low Reynolds number flows and by Brenner (1963) and Acrivos (1980) in analogous heat convection problems, we multiply Eq. (10) by $T^{(1)}$ and Eq. (13) by $T^{(0)}$ to arrive at

$$T^{(1)} \nabla^2 T^{(0)} - T^{(0)} \nabla^2 T^{(1)} = T^{(0)} \mathbf{G} \cdot \nabla T^{(0)}$$

or

$$\nabla \cdot [T^{(1)} \nabla T^{(0)} - T^{(0)} \nabla T^{(1)}] = T^{(0)} \mathbf{G} \cdot \nabla T^{(0)}.$$

On integrating both sides of the above over the volume enclosed by the surface of the particle and a large spherical surface $|\mathbf{r}| = \mathcal{R}$ around the particle and applying the divergence theorem, we arrive at

$$\int_{\mathcal{R}} [T^{(1)} \nabla T^{(0)} - T^{(0)} \nabla T^{(1)}] \cdot \mathbf{n} \, dS - \int_{\partial D} [T^{(1)} \nabla T^{(0)} - T^{(0)} \nabla T^{(1)}] \cdot \mathbf{n} \, dS$$

$$= \int_V T^{(0)} \mathbf{G} \cdot \nabla T^{(0)} \, dV.$$

Since, as $r \longrightarrow \infty$, $T^{(0)}$ decays like $1/r$ and $T^{(1)} \longrightarrow 0$ as required by boundary condition (15), the first integral vanishes as $\mathcal{R} \longrightarrow \infty$. Thus, taking into account boundary conditions (11) and (14), we obtain that

$$\int_{\partial D} \nabla T^{(1)} \cdot \mathbf{n} \, dS = \int_V T^{(0)} \mathbf{G} \cdot \nabla T^{(0)} dV$$

$$= - \int_V \ln\left(\frac{K}{K_0}\right) |\nabla T^{(0)}|^2 dV - \int_{\partial D} \ln\left(\frac{K}{K_0}\right) \nabla T^{(0)} \cdot \mathbf{n} \, dS,$$

where the divergence theorem has been used again in the last step. Consequently, on account also of Eq. (4), the heat transfer rate can be written as

$$Q = Q^{(0)} - \int_{\partial D} (K - K_0) \nabla [T - T^{(0)}] \cdot \mathbf{n} \, dS - \int_{\partial D} K_0 \nabla [T - T^{(0)}] \cdot \mathbf{n} \, dS$$

$$- \int_{\partial D} (K - K_0) \nabla T^{(0)} \cdot \mathbf{n} \, dS,$$

or

$$Q = Q^{(0)} - \int_{\partial D} (K - K_0)\nabla[T - T^{(0)}] \cdot \mathbf{n}\, dS + K_0 \int_V \ln\left(\frac{K}{K_0}\right)|\nabla T^{(0)}|^2\, dV$$

$$+ \int_{\partial D}\left[K_0 \ln\left(\frac{K}{K_0}\right) - (K - K_0)\right]\nabla T^{(0)} \cdot \mathbf{n}\, dS + \ldots\ . \tag{16}$$

But, as was mentioned above, the perturbation is regular in this case; hence the difference between $T$ and $T^{(0)}$ will be $O(\epsilon_1)$ everywhere. Consequently, the contribution from both the first and the third integrals in Eq. (16) is $O\{\epsilon_1^2\}$, in view of the fact that $K - K_0$ is $O(\epsilon_1)$ on the surface of the particle, while the second integral is $O(\epsilon_1)$. Thus,

$$Q = Q^{(0)} + K_0 \int_V \ln\left(\frac{K}{K_0}\right)|\nabla T^{(0)}|^2 dV + \ldots, \tag{17}$$

or, from Eq. (8),

$$\mathrm{Nu} = \mathrm{Nu}^{(0)} + \frac{\epsilon_1}{4\pi}\int_V f(\mathbf{r})|\nabla T_0|^2 dV + \ldots, \tag{18}$$

where $\mathrm{Nu}^{(0)} = Q^{(0)}/4\pi a K_0$. Equation (18) further reduces to

$$\mathrm{Nu} = 1 + \frac{\varepsilon_1}{4\pi}\int_V \frac{f(\mathbf{r})}{r^4} dV + \cdots$$

for spherical particles with its characteristic dimension taken as its radius.

### 32.2.3. *The Case $K = K(\epsilon_2\mathbf{r})$ Being Bounded at Infinity*

The regular perturbation expansion presented above can also be applied to the case in which the heat conductivity $K$ changes smoothly over space by an amount $O(1)$ relative to $K_0$, but over large distances, i.e.,

$$K = K(\epsilon_2\mathbf{r}), \tag{19}$$

where $\epsilon_2 \ll 1$ is a small parameter equal to the ratio of the characteristic dimension of the particle to the macroscopic length scale over which $K$ changes significantly. Then, the governing equation (6) becomes

$$\nabla^2 T = -\mathbf{G}(\epsilon_2\mathbf{r}) \cdot \nabla T, \tag{20}$$

where $\mathbf{G}(\epsilon_2\mathbf{r}) = [\nabla K(\epsilon_2\mathbf{r})]/K(\epsilon_2\mathbf{r})$ is $O(\epsilon_2)$. Here, the ratio of the magnitude of the two terms in the above equation $|\mathbf{G}(\epsilon_2\mathbf{r}) \cdot \nabla T|/|\nabla^2 T| \sim |r\mathbf{G}| \sim O(1)$ within the region $r \geq O(1/\epsilon_2)$. This suggests that the perturbation is singular because the term on the right-hand side is not uniformly smaller than the term on the left as $\epsilon_2 \longrightarrow 0$. Nevertheless, as we can see by examining Eq. (13), this singularity in the region $r \geq O(1/\epsilon_2)$ is weak enough in this case that the leading-order correction $T^{(1)}$ can still satisfy the boundary condition (15) at infinity as long as $\mathbf{G}$ decays faster than $1/r^\delta$, where $\delta > 0$, as $r \longrightarrow \infty$. This requirement is obviously satisfied when $K$ is bounded everywhere. We should expect, however, that higher-order terms in the regular perturbation expansion would eventually fail to yield a solution and that matched expansions would have to be used.

Therefore, in the present case, we can still use Eq. (17), which, on account of the dependence of $K$ on the macroscopic variable $\mathbf{R} \equiv \epsilon_2 \mathbf{r}$, reduces to

$$
\mathrm{Nu} = \mathrm{Nu}_0 + \frac{\epsilon_2}{4\pi} \mathrm{Nu}_0^2 \lim_{\rho \to 0} \int_\rho^\infty \ln\left[\frac{K(\mathbf{R})}{K_0}\right] \frac{dV_R}{R^4}
$$

$$
+ \frac{\epsilon_2}{4\pi} \bar{\mathbf{G}}_0 \cdot \int_{r \geq r_{\min}} \mathbf{r}\left[|\nabla T^{(0)}|^2 - \frac{\mathrm{Nu}_0^2}{r^4}\right] dV + \frac{\epsilon_2}{4\pi} \bar{\mathbf{G}}_0 \cdot \int_{r \leq r_{\min}} \mathbf{r}|\nabla T^{(0)}|^2 dV + \cdots
$$

$$(21)$$

for a particle of arbitrary shape, where $\bar{\mathbf{G}}_0 \equiv (\nabla_R K)/K|_{R \to 0}$ is taken to be $O(1)$. The second integral is over the space exterior to the smallest sphere of radius $r_{\min}$ that completely encloses the particle, while the third is over the region between the surface of that sphere and the surface of the particle. In arriving at Eq. (21), the fact has been used that $T^{(0)}$ can be expanded in terms of decaying harmonics for $r \geq r_{\min}$, with its leading term being $\mathrm{Nu}^{(0)}/r$ as $r \to \infty$. Note that since all the integrals are convergent, the correction to $\mathrm{Nu}_0$ is $O(\epsilon_2)$. For a spherical particle, the second and the third integrals vanish and Eq. (21) simplifies to

$$
\mathrm{Nu} = 1 + \frac{\epsilon_2}{4\pi} \lim_{\rho \to 0} \int_\rho^\infty \ln\left[\frac{K(\mathbf{R})}{K_0}\right] \frac{dV_R}{R^4} + \cdots.
$$

### 32.2.4. *The Case $(\nabla K)/K = constant$*

In this sub-section we consider the case in which $K$ changes exponentially in a certain direction, say $\mathbf{i}_3$, i.e.,

$$
K = K_0 e^{\mathbf{G}_0 \cdot \mathbf{r}} = K_0 e^{\epsilon_3 r_3}, \tag{22}
$$

where $\mathbf{G}_0$ is a constant vector with norm $\epsilon_3 \ll 1$. Thus the thermal conductivity vanishes as $r_3 \to -\infty$ and increases without bound as $r_3 \to +\infty$. From Eqs. (6) and (7), we therefore have that

$$
\nabla^2 T + \epsilon_3 \mathbf{i}_3 \cdot \nabla T = 0, \tag{23}
$$

subject to Eqs. (2) and (3).

The above is of course identical to the equation for the convective heat transfer from a particle to a uniform velocity field, which has the well-known fundamental solution (see Acrivos and Taylor, 1962)

$$
T = \frac{1}{\rho} e^{-\frac{1}{2}(\rho - \rho_3)}, \qquad \boldsymbol{\rho} \equiv \epsilon_3 \mathbf{r}, \qquad \rho = |\boldsymbol{\rho}|,
$$

which reduces to

$$
T = \frac{1}{\rho} - \frac{\rho - \rho_3}{2\rho} + O(\rho) \quad \text{as} \quad \rho \to 0. \tag{24}
$$

Equation (23) can therefore be solved using the method of inner and outer expansions in which the inner solution is obtained by means of a regular expansion valid for $r \leq O(1/\epsilon_3)$ and is matched to a multiple of fundamental solution (24) in the overlap region $r \sim O(1/\epsilon_3)$ (Acrivos and Taylor, 1962). As shown by Brenner (1963), however, the $O(\epsilon_3)$ correction to the

Nusselt number can simply be calculated from the solution of heat conduction equation (10) subject to the conditions $T = 1$ on $\partial D$ and $T = \epsilon_3 \mathrm{Nu}^{(0)} \langle T \rangle$ as $r \longrightarrow \infty$, where $\langle T \rangle$ is the $O(1)$ term in Eq. (24) averaged over the surface of a sphere of radius $\rho$. Since clearly $\langle T \rangle = -1/2$, we are easily led to Brenner's result,

$$\frac{\mathrm{Nu}}{\mathrm{Nu}_0} = 1 + \frac{1}{2}\epsilon_3 \mathrm{Nu}_0 + \cdots,$$

which becomes

$$\mathrm{Nu} = 1 + \frac{1}{2}\epsilon_3 + \cdots$$

for a spherical particle.

We note, parenthetically, that an exact expansion for Nu valid to all orders in $\epsilon_3$ can be obtained for a sphere by using the general solution [Acrivos and Taylor, 1962, Eqs. (11)–(14)] that satisfies governing equation (23) and condition (3) and then determining the coefficients in that solution by applying Eq. (2).

We finally note that if we represent $K$ by the rather general expression

$$K = K_0[1 + \epsilon_1 f_1(\mathbf{r})] f_2(\epsilon_2 \mathbf{r}) e^{\epsilon_3 r_3},$$

we can simply add the individual contributions to the Nusselt number and thereby obtain, for a wide class of systems, the first-order correction to the overall rate of heat conduction arising from the nonuniformity in the thermal conductivity.

### 32.3. The Translational Motion of a Particle

#### 32.3.1. *The Statement of the Problem*

Consider the translational motion of a rigid particle of arbitrary shape immersed in an infinite expanse of an incompressible Newtonian fluid with nonuniform viscosity $\mu$. When the particle is small enough so that all inertia effects are negligible, the governing equations and boundary conditions are given by

$$\nabla \cdot \mathbf{T} = 0, \tag{25}$$
$$\nabla \cdot \mathbf{u} = 0, \tag{26}$$
$$\mathbf{u} = \mathbf{U} \quad \text{on} \quad \partial D, \tag{27}$$
$$\mathbf{u} \longrightarrow 0 \quad \text{as} \quad r \longrightarrow \infty, \tag{28}$$

where $\mathbf{r}$ is the dimensionless position vector relative to a reference point inside the particle, normalized by the characteristic dimension of the particle $a$, $\mathbf{u}(\mathbf{r})$ is the velocity field around the particle, $\mathbf{U}$ is the translational velocity, and $\mathbf{T}$ is the stress tensor given by

$$\mathbf{T} = -p\mathbf{I} + \mu(\mathbf{r})[\nabla\mathbf{u} + (\nabla\mathbf{u})^T], \tag{29}$$

where $p$ is the dynamic pressure.

The total force acting on the particle by the fluid is given by

$$\mathbf{F} = \int_{\partial D} \mathbf{T} \cdot \mathbf{n} \, dS, \tag{30}$$

where **n** is the unit vector normal to the surface of the particle pointing from the particle to the fluid.

Another useful form of Eq. (25) is

$$\nabla \cdot \mathbf{W} = -\mathbf{G} \cdot \mathbf{W}, \tag{31}$$

where

$$\mathbf{W}(\mathbf{r}) \equiv \frac{\mathbf{T}(\mathbf{r})}{\mu(\mathbf{r})} = -q(\mathbf{r})\mathbf{I} + [\nabla \mathbf{u} + (\nabla \mathbf{u})^T] \tag{32}$$

with

$$q(\mathbf{r}) \equiv \frac{p(\mathbf{r})}{\mu(\mathbf{r})}, \qquad \mathbf{G}(\mathbf{r}) \equiv \frac{1}{\mu(\mathbf{r})} \nabla \mu(\mathbf{r}). \tag{33}$$

### 32.3.2. *The case* $\mu = \mu_0 + \epsilon_4 \mu_0 f(\mathbf{r}) + \cdots$

When the variation of the viscosity $\mu(\mathbf{r})$ around $\mu_0$ is everywhere small compared with $\mu_0$,

$$\mu(\mathbf{r}) = \mu_0 + \epsilon_4 \mu_0 f(\mathbf{r}) + \cdots, \tag{34}$$

where $\epsilon_4 \ll 1$ is a small parameter characteristic of the magnitude of the relative variation of the viscosity, and $f(\mathbf{r})$ together with its gradient are both at most $O(1)$ everywhere, the problem can be treated by means of the regular perturbation expansion, just as in Subsection 32.2.2:

$$\mathbf{u} = \mathbf{u}^{(0)} + \mathbf{u}^{(1)} + \cdots, \tag{35}$$
$$\mathbf{W} = \mathbf{W}^{(0)} + \mathbf{W}^{(1)} + \cdots, \tag{36}$$

Then, from Eq. (31),

$$\nabla \cdot \mathbf{W}^{(0)} = 0, \tag{37}$$
$$\nabla \cdot \mathbf{u}^{(0)} = 0, \tag{38}$$
$$\mathbf{u}^{(0)} = \mathbf{U} \quad \text{on} \quad \partial D, \tag{39}$$
$$\mathbf{u}^{(0)} \longrightarrow 0 \quad \text{as} \quad r \longrightarrow \infty, \tag{40}$$

with

$$\mathbf{W}^{(0)} = -q^{(0)}\mathbf{I} + [\nabla \mathbf{u}^{(0)} + (\nabla \mathbf{u}^{(0)})^T], \tag{41}$$

which are the same as those for the case in which the same particle translates through a fluid with uniform viscosity $\mu_0$, and

$$\nabla \cdot \mathbf{W}^{(1)} = -\mathbf{G} \cdot \mathbf{W}^{(0)}, \tag{42}$$
$$\nabla \cdot \mathbf{u}^{(1)} = 0, \tag{43}$$
$$\mathbf{u}^{(1)} = 0 \quad \text{on} \quad \partial D, \tag{44}$$
$$\mathbf{u}^{(1)} \longrightarrow 0 \quad \text{as} \quad r \longrightarrow \infty. \tag{45}$$

In the following, we suppose that $\mathbf{u}^{(0)}$ and the corresponding force $\mathbf{F}^{(0)}$ are known. Because of the linear dependence of Eqs. $(37) --(41)$ on $\mathbf{U}$, these can be represented as,

$$\mathbf{u}^{(0)} = \mathcal{U}^{(0)} \cdot \mathbf{U}, \qquad \mathbf{F}^{(0)} = \mathcal{F}^{(0)} \cdot \mathbf{U}, \tag{46}$$

where $\mathcal{U}^{(0)}$ and $\mathcal{F}^{(0)}$ are taken to be known tensors that, for a spherical particle, for example, are given by

$$\mathcal{U}^{(0)} = \frac{3}{4} \left( \frac{\mathbf{rr}}{r^3} + \frac{\mathbf{I}}{r} \right) - \frac{3}{4} \left( \frac{\mathbf{rr}}{r^5} - \frac{\mathbf{I}}{3r^3} \right) \qquad \mathcal{F}^{(0)} = -6\pi \mu_0 \mathbf{I}. \tag{47}$$

Again, following the procedure originally introduced by Brenner and Cox (1963) in low Reynolds number flows and by Brenner (1963) and Acrivos (1980) in analogous heat convection problems, we multiply Eq. (37) by $\mathcal{U}^{(1)}$ and Eq. (42) by $\mathcal{U}^{(0)}$ to arrive at

$$[\nabla \cdot \mathbf{W}^{(0)}] \cdot \mathcal{U}^{(1)} - [\nabla \cdot \mathbf{W}^{(1)}] \cdot \mathcal{U}^{(0)} = \nabla \cdot [\mathbf{W}^{(0)} \cdot \mathcal{U}^{(1)} - \mathbf{W}^{(1)} \cdot \mathcal{U}^{(0)}] = \mathbf{G} \cdot \mathbf{W}^{(0)} \cdot \mathcal{U}^{(0)},$$

where $\mathcal{U}^{(1)}$ is defined by $\mathbf{u}^{(1)} = \mathcal{U}^{(1)} \cdot \mathbf{U}$. Integrating both sides of the above equation over the volume enclosed by the surface of the particle and a large spherical surface $|\mathbf{r}| = \mathcal{R}$ around the particle and applying the divergence theorem, we obtain

$$\int_{\mathcal{R}} \mathbf{n} \cdot [\mathbf{W}^{(0)} \cdot \mathcal{U}^{(1)} - \mathbf{W}^{(1)} \cdot \mathcal{U}^{(0)}] \, \mathrm{d}S - \int_{\partial D} \mathbf{n} \cdot [\mathbf{W}^{(0)} \cdot \mathcal{U}^{(1)} - \mathbf{W}^{(1)} \cdot \mathcal{U}^{(0)}] \, \mathrm{d}S$$

$$= \int_{V} \mathbf{G} \cdot \mathbf{W}^{(0)} \cdot \mathcal{U}^{(0)} \mathrm{d}V. \tag{48}$$

Since, as $r \longrightarrow \infty$, $\mathcal{U}^{(0)}$ and $\mathcal{W}^{(0)}$ decay like $1/r$ and $1/r^2$, respectively, while $\mathcal{U}^{(1)} \longrightarrow 0$ and $\mathbf{W}^{(1)} \sim o(1/r)$ as required by boundary condition (45), the first integral vanishes as $\mathcal{R} \longrightarrow \infty$. Therefore, taking into account boundary conditions (39) and (44), we have that

$$\int_{\partial D} \mathbf{n} \cdot \mathbf{W}^{(1)} \, \mathrm{d}S = \int_{V} \mathbf{G} \cdot \mathbf{W}^{(0)} \cdot \mathcal{U}^{(0)} \mathrm{d}V$$

$$= - \int_{\partial D} \left( \ln \frac{\mu}{\mu_0} \right) \mathbf{n} \cdot \mathbf{W}^{(0)} \cdot \mathcal{U}^{(0)} \mathrm{d}S - \frac{1}{2} \int_{V} \left( \ln \frac{\mu}{\mu_0} \right) \{\nabla \mathcal{U}^{(0)}$$

$$+ [\nabla \mathcal{U}^{(0)}]^T\} : \{\nabla \mathcal{U}^{(0)} + [\nabla \mathcal{U}^{(0)}]^T\} \cdot \mathbf{U} \, \mathrm{d}V,$$

where, once again, the divergence theorem has been applied in the last step, and the transpose and double inner product both refer to operation on the first two indices of the corresponding tensors. This expression, together with Eqs. (30), (32), and (36), can be used to determine the leading-order correction to the force, yielding

$$\mathbf{F} = \int_{\partial D} [\mu_0 + (\mu - \mu_0)] \mathbf{n} \cdot [\mathbf{W}_0 + (\mathbf{W} - \mathbf{W}_0)] \, \mathrm{d}S$$

$$= \mathbf{F}^{(0)} - \frac{\mu_0}{2} \int_{V} \left( \ln \frac{\mu}{\mu_0} \right) \{\nabla \mathcal{U}^{(0)} + [\nabla \mathcal{U}^{(0)}]^T\} : \{\nabla \mathcal{U}^{(0)} + [\nabla \mathcal{U}^{(0)}]^T\} \cdot \mathbf{U} \, \mathrm{d}V + \cdots. \tag{49}$$

When $\mu(\mathbf{r})$ is given by Eq. (34), Eq. (49) reduces to

$$\mathbf{F} = \mathbf{F}^{(0)} - \frac{1}{2} \epsilon_4 \mu_0 \mathbf{U} \cdot \int_{V} f(\mathbf{r}) \{\nabla \mathcal{U}^{(0)} + [\nabla \mathcal{U}^{(0)}]^T\} : \{\nabla \mathcal{U}^{(0)} + [\nabla \mathcal{U}^{(0)}]^T\} \mathrm{d}V + \cdots,$$

which, for a spherical particle, simplifies to

$$\mathbf{F} = -6\pi\mu_0\mathbf{U} \cdot \left\{\mathbf{I} + \frac{3}{8\pi}\epsilon_4 \int_V f(\mathbf{r}) \left[\frac{\mathbf{I}}{r^8} + \frac{\mathbf{rr}}{r^2}\left(\frac{3}{r^4} - \frac{6}{r^6} + \frac{2}{r^8}\right)\right] dV + \cdots\right\}.$$

### 32.3.3. *The Case $\mu = \mu(\epsilon_5\mathbf{r})$ Being Bounded at Infinity*

Similarly to the corresponding heat conduction problem in Subsection 32.2.3, the regular perturbation expansion presented in Subsection 32.3.1 can also be applied to the case in which the viscosity $\mu$ changes smoothly over space by an amount $O(1)$ relative to $\mu_0$, but over large distances, i.e.,

$$\mu = \mu(\epsilon_5\mathbf{r}), \tag{50}$$

where $\epsilon_5 \ll 1$ is a small parameter equal to the ratio of the characteristic dimension of the particle to the macroscopic length scale over which $\mu$ changes significantly. Then, the governing equation (31) becomes

$$\nabla \cdot \mathbf{W} = -\mathbf{G}(\epsilon_5\mathbf{r}) \cdot \mathbf{W}, \tag{51}$$

where $\mathbf{G}(\epsilon_5\mathbf{r}) = [\nabla\mu(\epsilon_5\mathbf{r})]/\mu(\epsilon_5\mathbf{r})$ is $O(\epsilon_5)$. To determine the leading-order correction to the force, we use the same method as that in Subsection 32.2.3 and find again that, although the regular perturbation expansion fails when $r \geq O(1/\epsilon_5)$, this singularity is weak enough that the leading-order corrections $\mathbf{u}^{(1)}$ and $\mathbf{W}^{(1)}$ can still satisfy boundary condition (45) as long as $\mathbf{G}$ decays faster than $1/r^\delta$, where $\delta > 0$. This can be seen by examining Eq. (42). This requirement is obviously satisfied when $\mu$ is bounded everywhere. Therefore, in the present case, we can still use the regular perturbation expansion and obtain the leading term correction to $\mathbf{F}^{(0)}$, as given by Eq. (49), which on account of the dependence of $\mu$ on the macroscopic variable $\mathbf{R} \equiv \epsilon_5\mathbf{r}$, reduces to

$$\mathbf{F} = \mathbf{F}^{(0)} - \frac{9}{8\pi}\epsilon_5\mathbf{F}^{(0)} \cdot \left(\lim_{\rho\to 0}\int_\rho^\infty \left\{\ln\left[\frac{\mu(\mathbf{R})}{\mu_0}\right]\right\}\frac{\mathbf{RR}}{R^6}dV_R\right) \cdot \frac{\mathcal{F}^{(0)}}{6\pi\mu_0}$$
$$- \frac{1}{2}\epsilon_5\mu_0\int_{r\geq r_{\min}}(\bar{\mathbf{G}}_0 \cdot \mathbf{r})\left(\{\nabla\mathcal{U}^{(0)} + [\nabla\mathcal{U}^{(0)}]^T\} : \{\nabla\mathcal{U}^{(0)} + [\nabla\mathcal{U}^{(0)}]^T\} \cdot \mathbf{U}\right.$$
$$\left. - \frac{27}{2}\frac{\mathbf{F}^{(0)}}{6\pi\mu_0} \cdot \frac{\mathbf{rr}}{r^6} \cdot \frac{\mathcal{F}^{(0)}}{6\pi\mu_0}\right)dV - \frac{1}{2}\epsilon_5\mu_0\mathbf{U} \cdot \int_{r\leq r_{\min}}(\bar{\mathbf{G}}_0 \cdot \mathbf{r})\{\nabla\mathcal{U}^{(0)}$$
$$+ [\nabla\mathcal{U}^{(0)}]^T\} : \{\nabla\mathcal{U}^{(0)} + [\nabla\mathcal{U}^{(0)}]^T\}dV + \cdots \tag{52}$$

for a particle of arbitrary shape, where $\bar{\mathbf{G}}_0 \equiv (\nabla_R\mu)/\mu|_{R\to 0}$ is supposed to be $O(1)$. The second integral is over the space exterior to the smallest sphere of radius $r_{\min}$ that completely encloses the particle, while the third is over the region between the surface of that sphere and the surface of the particle. Note that since all the integrals are convergent, the correction to $\mathbf{F}^{(0)}$ is $O(\epsilon_5)$. In arriving at Eq. (52), the fact has been used that $\mathcal{U}^{(0)}$ can be expressed as Lamb's expansion for $r \geq r_{\min}$ with its leading term being the Stokeslet term as $r \longrightarrow \infty$. For a spherical particle, the second and the third integrals vanish and Eq. (52) simplifies to

$$\mathbf{F} = \mathbf{F}^{(0)} + \frac{9}{8\pi}\epsilon_5\mathbf{F}^{(0)} \cdot \left\{\lim_{\rho\to 0}\int_\rho^\infty \left[\ln\frac{\mu(\mathbf{R})}{\mu_0}\right]\frac{\mathbf{RR}}{R^6}dV_R\right\} + \cdots.$$

### 32.3.4. *The Case* $(\nabla \mu)/\mu = constant$

In this subsection we consider the case in which $\mu$ changes exponentially along a certain direction, say $\mathbf{i}_3$, i.e.,

$$\mu = \mu_0 e^{\mathbf{G}\cdot\mathbf{r}} = \mu_0 e^{\epsilon_6 r_3}, \tag{53}$$

where $\mathbf{G}$ is a constant vector with norm $\epsilon_6 \ll 1$. Then the governing equation is still given by Eq. (31), with $\mathbf{G}$ being a constant vector.

To find the leading-order correction to the force in this case, we follow the same singular perturbation approach as given in Subsection 32.2.4 for the analogous heat conduction problem. First, the point force solution of the governing equation is found by the method of Fourier transform, and then its inner expansion is used to determine the leading-order correction to the force.

Denoting by $\boldsymbol{\alpha}(\mathbf{r})$ the $O(1)$ term in the inner expansion of the velocity expression of the point force solution as $r \longrightarrow 0$, we show in what follows that, for the present purpose, we need only $\langle \boldsymbol{\alpha} \rangle$, i.e, the average of $\boldsymbol{\alpha}(\mathbf{r})$ over the solid angle $\Omega$. First, from the matching requirement, the term $\mathbf{u}^{(1)}$ in the inner expansion must match $\boldsymbol{\alpha}(\mathbf{r})$ and satisfy Eqs. (42)–(44). Obviously the homogeneous solution of Eq. (42) can only contribute a constant as $r \longrightarrow \infty$ and the $\mathbf{r}$ dependent part of $\boldsymbol{\alpha}(\mathbf{r})$ must be matched by the particular solution of Eq. (42). But, since the leading term of $\mathbf{W}^{(0)}$, which is $O(1/r^2)$ as $r \longrightarrow \infty$, is odd in $\mathbf{r}$, we have from Eq. (42) that

$$\mathbf{u}^{(1)} \longrightarrow \mathbf{C} + \bar{\mathbf{u}}_p^{(1)} \quad \text{as} \quad r \longrightarrow \infty, \tag{54}$$

where $\mathbf{C}$ is a constant and $\bar{\mathbf{u}}_p^{(1)}$, the leading term of the particular solution of Eq. (42) as $r \longrightarrow \infty$, which is $O(1)$ with $\langle \bar{\mathbf{u}}_p^{(1)} \rangle = 0$, must also be odd in $\mathbf{r}$. By matching the above form of $\mathbf{u}^{(1)}$ with $\boldsymbol{\alpha}(\mathbf{r})$ we obtain that $\mathbf{C} = \langle \boldsymbol{\alpha} \rangle$ and $\bar{\mathbf{u}}_p^{(1)} = \boldsymbol{\alpha}(\mathbf{r}) - \langle \boldsymbol{\alpha} \rangle$. Substituting the form of $\mathbf{u}^{(1)}$ given by condition (54) into Eq. (48), which is valid as long as $\mathcal{R} \leq O(1/\epsilon_6)$, and choosing $\mathcal{R} = O(1/\epsilon_6)$, we find that the leading term of the first integral equals $\mathcal{F}^{(0)} \cdot \langle \boldsymbol{\alpha} \rangle / \mu_0$ as $\epsilon_6 \longrightarrow 0$. Also, the leading term of the second integral equals $\mathbf{F}^{(1)}/\mu_0$, on account of Eqs. (39) and (44), and that of the last, although vanishing for a spherical particle, becomes, for a particle of arbitrary shape,

$$\mathbf{G} \cdot \int_{r \geq r_{\min}} \left[ \mathbf{W}^{(0)} \cdot \mathcal{U}^{(0)} + \frac{27}{4} \frac{\mathbf{F}^{(0)}}{6\pi\mu_0} \cdot \frac{\mathbf{rrr}}{r^6} \cdot \frac{\mathcal{F}^{(0)}}{6\pi\mu_0} \right] dV + \mathbf{G} \cdot \int_{r \leq r_{\min}} \mathbf{W}^{(0)} \cdot \mathcal{U}^{(0)} dV,$$

where, again, the first integral of the above is over the space exterior to the smallest sphere of radius $r_{\min}$ that completely encloses the particle, while the second is over the region between the surface of that sphere and the surface of the particle. Note that the first integral converges since the integrand decays like $1/r^4$ as $r \longrightarrow \infty$, in view of the fact that $\mathcal{U}^{(0)}$ can be expressed in terms of Lamb's expansion for $r \geq r_{\min}$ and that the leading term of $\mathbf{W}^{(0)} \cdot \mathcal{U}^{(0)}$ is canceled by the next term in the integrand.

Therefore we obtain that

$$\mathbf{F}^{(1)} = -\langle \boldsymbol{\alpha} \rangle \cdot \mathcal{F}^{(0)} + \mu_0 \mathbf{G} \cdot \int_{r \leq r_{\min}} \mathbf{W}^{(0)} \cdot \mathcal{U}^{(0)} dV$$

$$+ \mu_0 \mathbf{G} \cdot \int_{r \geq r_{\min}} \left[ \mathbf{W}^{(0)} \cdot \mathcal{U}^{(0)} + \frac{27}{4} \frac{\mathbf{F}^{(0)}}{6\pi\mu_0} \cdot \frac{\mathbf{rrr}}{r^6} \cdot \frac{\mathcal{F}^{(0)}}{6\pi\mu_0} \right] dV + \dots . \tag{55}$$

Next, to obtain $\langle\alpha\rangle$, we note that the governing equations in terms of the variable $\rho = \epsilon_6 \mathbf{r}$ for the flow due to a point force with strength $\mathbf{F}$, acting by the fluid at the point $\rho = 0$, are given by

$$\nabla_\rho \cdot \mathbf{T} = \mathbf{F}\delta(\rho), \qquad \nabla_\rho \cdot \mathbf{u} = 0, \tag{56}$$

and therefore, on account of Eqs. (31), (33) and (53), the momentum equation can be written as

$$\nabla_\rho \cdot \mathbf{W} + \mathbf{i}_3 \cdot \mathbf{W} = \frac{\mathbf{F}}{\mu_0}\delta(\rho), \tag{57}$$

On taking the Fourier transform of Eqs. (56) and (57) with $\mathbf{W}$ given by Eq. (32), we obtain

$$i\mathbf{k} \cdot \hat{\mathbf{u}} = \mathbf{0}, \tag{58}$$

$$(i\mathbf{k} + \mathbf{i}_3) \cdot (-\hat{q}\mathbf{I} + i\mathbf{k}\hat{\mathbf{u}} + i\hat{\mathbf{u}}\mathbf{k}) = \frac{\mathbf{F}}{\mu_0},$$

which, on account of Eq. (58), becomes

$$-(i\mathbf{k} + \mathbf{i}_3)\hat{q} - (k^2 - ik_3)\hat{\mathbf{u}} + i\hat{u}_3\mathbf{k} = \frac{\mathbf{F}}{\mu_0}, \tag{59}$$

where

$$\hat{q}(\mathbf{k}) = \int q(\rho)e^{-i\mathbf{k}\cdot\rho}\,\mathrm{d}^3\rho,$$

and similarly for $\hat{\mathbf{u}}$.

Next, on taking the inner product of both sides of Eq. (59) with $\mathbf{k}$, we obtain

$$\hat{q}(\mathbf{k}) = \frac{\hat{u}_3 + i\frac{\mathbf{F}}{\mu_0}\cdot\frac{\mathbf{k}}{k^2}}{1 - i\frac{k_3}{k^2}}, \tag{60}$$

which, when substituted into Eq. (59), allows us to solve $\hat{\mathbf{u}}$ for any given $\mathbf{F}$. We consider two separate cases in the following subsections.

**32.3.4.1.** *The Case $\mathbf{F}^{(0)}$ parallel to $\mathbf{G}$*

When $\mathbf{F}^{(0)} = F^{(0)}\mathbf{e}_3$, we have from Eqs. (59) and (60) that

$$\hat{u}_3(\mathbf{k}) = -\frac{k^2 - k_3^2}{k^4 + k^2 - 2k_3^2 - 2ik^2k_3}\frac{F^{(0)}}{\mu_0}, \tag{61}$$

and therefore

$$u_3(\rho) = \frac{1}{(2\pi)^3}\int \hat{u}_3(\mathbf{k})e^{i\mathbf{k}\cdot\rho}\,\mathrm{d}^3\mathbf{k}. \tag{62}$$

But, since we are interested in only the average velocity $\hat{u}_3$ over the solid angle $\Omega$ for fixed $\rho$ in $\rho$ space, we have that, on account of Eq. (62),

$$\langle u_3(\rho)\rangle = \frac{1}{(2\pi)^3}\int \hat{u}_3(\mathbf{k})\frac{\sin(k\rho)}{k\rho}\,\mathrm{d}^3\mathbf{k}. \tag{63}$$

Substituting Eq. (61) into Eq. (63), we can then obtain the asymptotic expression for $\langle u_3(\rho) \rangle$ as $\rho \longrightarrow 0$:

$$\langle u_3(\rho) \rangle = -\frac{F^{(0)}}{6\pi\mu_0\rho} + \langle\alpha\rangle_{\parallel} + \cdots,$$

where

$$\langle\alpha\rangle_{\parallel} = \frac{1}{(2\pi)^3} \int \left[ \hat{u}_3(\mathbf{k}) - \frac{k_3^2 - k^2}{k^4} \frac{F^{(0)}}{\mu_0} \right] d^3\mathbf{k} = \frac{3(\pi+3)}{16} \frac{F^{(0)}}{6\pi\mu_0} = 1.152 \frac{F^{(0)}}{6\pi\mu_0},$$

with $\hat{u}_3(\mathbf{k})$ given by Eq. (61).

Therefore, from Eq. (55),

$$\mathbf{F} - \mathbf{F}^{(0)} = -1.152 G \frac{\mathcal{F}^{(0)}}{6\pi\mu_0} \cdot \mathbf{F}^{(0)} + \mu_0\mathbf{G} \cdot \int_{r \leq r_{\min}} \mathbf{W}^{(0)} \cdot \mathcal{U}^{(0)} \, dV$$

$$+ \mu_0\mathbf{G} \cdot \int_{r \geq r_{\min}} \left[ \mathbf{W}^{(0)} \cdot \mathcal{U}^{(0)} + \frac{27}{4} \frac{\mathbf{F}^{(0)}}{6\pi\mu_0} \cdot \frac{\mathbf{rrr}}{r^6} \cdot \frac{\mathcal{F}^{(0)}}{6\pi\mu_0} \right] dV + \cdots, \tag{64}$$

where $G \equiv |\mathbf{G}|$.

For spherical particles, Eq. (64) reduces to

$$\mathbf{F} = -6\pi\mu_0\mathbf{U}(1 + 1.152 G + \cdots) \cdot \tag{65}$$

### 32.3.4.2. *The Case $\mathbf{F}^{(0)}$ perpendicular to $\mathbf{G}$*

Using the technique as described above, we can also obtain the leading-order correction to the force due to the variation of the viscosity when $\mathbf{F}^{(0)}$ is perpendicular to $\mathbf{G}$, say $\mathbf{F}^{(0)} = F^{(0)}\mathbf{e}_1$ and $\mathbf{G} = G\mathbf{e}_3$.

In this case, Eq. (60) reduces to

$$\hat{q}(\mathbf{k}) = \frac{\hat{u}_3 + i\frac{F^{(0)}}{\mu_0}\frac{k_1}{k^2}}{1 - i\frac{k_3}{k^2}},$$

while the projection of Eq. (59) in the directions $\mathbf{e}_3$ and $\mathbf{e}_1$ gives

$$\hat{u}_3(\mathbf{k}) = \frac{-k_1k_3 + ik_1}{k^4 + k^2 - 2k_3^2 - 2ik^2k_3} \frac{F^{(0)}}{\mu_0}$$

$$\hat{u}_1(\mathbf{k}) = \frac{-\frac{F^{(0)}}{\mu_0} - ik_1\hat{q} + ik_1\hat{u}_3}{k^2 - ik_3}, \tag{66}$$

respectively. Consequently, the asymptotic expression for the average of $u_1$ over the solid angle $\Omega$ for fixed $\rho$ in $\rho$ space,

$$\langle u_1(\rho) \rangle = \frac{1}{(2\pi)^3} \int \hat{u}_1(\mathbf{k}) \frac{\sin(k\rho)}{k\rho} \, d^3\mathbf{k},$$

becomes, as $\rho \longrightarrow 0$,

$$\langle u_1(\rho) \rangle = -\frac{F^{(0)}}{6\pi \mu_0 \rho} + \langle \alpha \rangle_\perp + \cdots,$$

where

$$\langle \alpha \rangle_\perp = \frac{1}{(2\pi)^3} \int \left[ \hat{u}_1(\mathbf{k}) - \frac{k_1^2 - k^2}{k^4} \frac{F^{(0)}}{\mu_0} \right] d^3\mathbf{k} = 0.469G \frac{F^{(0)}}{6\pi \mu_0},$$

with $\hat{u}_1(\mathbf{k})$ given by the second equation of Eqs. (66). Therefore the total force on the particle is found in the same way as in Subsection 32.4.2.1 to equal

$$\mathbf{F} - \mathbf{F}^{(0)} = -0.469G \frac{\mathcal{F}^{(0)}}{6\pi \mu_0} \cdot \mathbf{F}^{(0)} + \mu_0 \mathbf{G} \cdot \int_{r \leq r_{\min}} \mathbf{W}^{(0)} \cdot \mathcal{U}^{(0)} \, dV$$

$$+ \mu_0 \mathbf{G} \cdot \int_{r \geq r_{\min}} \left[ \mathbf{W}^{(0)} \cdot \mathcal{U}^{(0)} + \frac{27}{4} \frac{\mathbf{F}^{(0)}}{6\pi \mu_0} \cdot \frac{\mathbf{rrr}}{r^6} \cdot \frac{\mathcal{F}^{(0)}}{6\pi \mu_0} \right] dV + \cdots. \tag{67}$$

For spherical particles, Eq. (67) reduces to

$$\mathbf{F} = -6\pi \mu_0 \mathbf{U}^{(0)} (1 + 0.469G + \cdots). \tag{68}$$

The fact that the correction to the force for the case $\mathbf{F}^{(0)} \parallel \mathbf{G}$ given by Eq. (65) is larger than that when $\mathbf{F}^{(0)} \perp \mathbf{G}$ given by Eq. (68) is consistent with the results for the corrections due to the presence of a solid wall on the force of a sphere translating in a uniform fluid far away from the wall [Happel and Brenner, 1973, Eqs. (7-4.28) and (7-4.39)], in which case the correction for the motion of the sphere toward the wall is twice that for the motion parallel to the wall.

In conclusion, we were able to determine the leading-order correction to the force on a particle translating through a fluid with its viscosity given by Eq. (53) when $\mathbf{F}^{(0)}$ is either parallel or perpendicular to $\mathbf{G}$. These expressions can therefore be combined to give the correction for the case when $\mathbf{U}$ and therefore $\mathbf{F}^{(0)}$ is along an arbitrary direction, because of the linearity of the problem.

We note, as was the case with the heat conduction problem, that if we represent $\mu$ by the rather general expression

$$\mu = \mu_0 [1 + \epsilon_4 f_3(\mathbf{r})] f_4(\epsilon_5 \mathbf{r}) e^{\epsilon_6 r_3},$$

we can simply add the three individual contributions to the force and thereby obtain, for a wide class of systems, the first-order correction to the force arising from the nonuniformity in the viscosity.

## Acknowledgment

This work was supported by the U.S. National Science Foundation grant CTS-9012937.

## References

Acrivos, A. 1980. A note on the rate of heat or mass transfer from a small particle freely suspended in a linear shear field. *J. Fluid Mech.* **98**, 299.

Acrivos, A. and Taylor, T. D. 1962. Heat and mass transfer from single spheres in Stokes flow. *Phys. Fluids* **5**, 387.

Batchelor, G. K. 1974. Transport properties of two phase materials with random structure. *Ann. Rev. Fluid Mech.* **6**, 227.

Batchelor, G. K. 1979. Mass transfer from a small particle suspended in a fluid with a steady linear ambient velocity distribution. *J. Fluid Mech.* **95**, 369.

Brenner, H. 1963. Forced convection heat and mass transfer at small Peclet number from a particle of arbitrary shape. *Chem. Eng. Sci.* **18**, 109.

Brenner, H. and Cox, R. G. 1963. The resistance to a particle of arbitrary shape in translational motion at small Reynolds number. *J. Fluid Mech.* **17**, 561.

Happel, J. and Brenner, H. 1973. *Low Reynolds Number Hydrodynamics.* Noordhoff International, Groningen, The Netherlands.

# 33

# Radiation-Affected Ignition Phenomena with Solid–Gas Interaction

SEUNG WOOK BAEK AND JAE HYUN PARK

*Korea Advanced Ins. of Sci. & Tech. Aerospace Engineering Department 373–1 Gusungdong, Yusungku Taejon, South Korea*

## Nomenclature

| | |
|---|---|
| $C$ | specific heat |
| $d_p$ | particle diameter |
| $G$ | incident radiation |
| $I$ | radiative intensity |
| $L$ | characteristic length |
| $n_p$ | particle number density |
| $\vec{q}^C$ | conductive heat flux |
| $\vec{q}^R$ | radiative heat flux |
| $Q_{abs}$ | absorption efficiency |
| $Q_{ext}$ | extinction efficiency |
| $Q_{sca}$ | scattering efficiency |
| $T$ | temperature |

### *Greek Symbols*

| | |
|---|---|
| $\beta$ | extinction coefficient ($=\kappa + \sigma_s$) |
| $\varepsilon_p$ | solid particle emissivity |
| $\varepsilon_w$ | wall emissivity |
| $\theta$ | polar angle |
| $\kappa$ | absorption coefficient |
| $\lambda$ | thermal conductivity |
| $\mu$ | direction cosine ($=\cos\theta$) |
| $\sigma$ | Stefan–Boltzmann constant $= 5.670 \times 10^{-8}$ W/m$^2$ K$^4$ |
| $\sigma_g$ | gas concentration |
| $\sigma_p$ | particle concentration |
| $\sigma_s$ | scattering coefficient |
| $\tau_L$ | optical thickness ($=\beta L$) |
| $\Phi$ | scattering phase function |
| $\phi$ | equivalence ratio |
| $\omega_s$ | scattering albedo ($=\kappa/\beta$) |

### *Subscripts*

| | |
|---|---|
| $b$ | blackbody |
| $g$ | gas |
| $p$ | particle |

## 33.1. Introduction

In the area of fire safety, the physical state under which a reactive two-phase mixture (solid particles/gas) can ignite and generate a heat release is one of critical problems still remaining to be elucidated in further detail. Basic phenomena involved are controlled by the balance between heat release and heat losses to the surroundings through conduction, convection, and radiation. Given a suitable ignition source, the reactive solid particles or gases can react exothermically with an oxidant if the heat removal is not sufficient.

Thermal ignition of a reactive material has been well reviewed by Merzhanov and Averson (1971). The thermal ignition by conduction has been examined (Shouman et al., 1974; Kordylewski, 1979). As far as the radiative heat transfer is concerned, Khalil et al. (1983) developed a model predicting the stationary thermal ignition of a heat-generating particle suspension bounded by emitting and diffusely reflecting walls. Smith et al. (1988) examined the thermal ignition phenomenon with a single homogeneous absorbing and emitting medium. Generally speaking, when the ignition heat source is the external radiation, the existing theoretical models for the ignition of a two-phase mixture can be divided into two types, solid or gas phase ignition, depending on the key exothermic reaction leading to ignition, which were investigated in detail by Baek (1990, 1992, 1994, 1996).

From this viewpoint, in this chapter the past works on the radiation-affected ignition phenomena with solid–gas interaction are presented and then further research work to be done is discussed.

## 33.2. Basic Formulation

In order to take account of radiation in the modeling of ignition, the divergence of radiative heat flux is required to be added into the conventional energy conservation equation (Viskanta and Menguc, 1987):

$$\rho \frac{\mathrm{D}h}{\mathrm{D}t} = \frac{\mathrm{D}p}{\mathrm{D}t} - \nabla \cdot \vec{q} + \Phi + \dot{Q} + \rho \sum_{k=1}^{N} Y_k \vec{f}_k \cdot \vec{V}_k, \tag{1}$$

$$\nabla \cdot \vec{q} = \nabla \cdot (\vec{q}_C + \vec{q}_R)$$
$$= -\lambda \nabla T + \kappa \left( 4\sigma T^4 - \int_{4\pi} I \, \mathrm{d}\Omega \right), \tag{2}$$

where $\Phi$ is dissipation by viscous stress, $\dot{Q}$ is the heat generation, and the fifth term on the right-hand side represents the body-force work.

To derive the radiative heat flux and its divergence, it is inevitable to solve the radiative transfer equation (RTE), which is given by highly nonlinear integro-differential form with spectrally varying properties:

$$\frac{\mathrm{d}I}{\mathrm{d}s} = -(\kappa + \sigma_s)I + \kappa I_b + \frac{\sigma_s}{4\pi} \int_{4\pi} \Phi(\hat{s}, \hat{s}_i) I(\hat{s}_i) \, \mathrm{d}\Omega_i, \tag{3}$$

where $s$ is the path length along the pencil of rays. The terms on the right-hand side of Eq. (3) represent the attenuation of radiative intensity due to absorption and out-scattering and the augmentation through emission and in-scattering, respectively.

There have been many attempts to solve the RTE for application to various industrial high-temperature systems. Among others they are the Monte Carlo method, zone method, spherical harmonics method ($P_N$), discrete ordinates method, finite-volume method (Raithby and Chui, 1990), etc. In one-dimensional geometry, the exact formulation or a more simplified approach such as differential approximation and two-flux method are used. Methods other than the finite-volume method are summarized in the classical text about radiation in detail (Modest, 1993).

Particles are the principal contributors to the absorption and scattering of radiation. These effects can be well accounted for in the limiting cases of very small or very large concentrations. The radiative properties of a cloud consisting of the uniform-sized particles can be described by (Modest, 1993; Siegel and Howell, 1992)

$$\sigma_s = \pi d_p^2 n_p Q_{\text{sca}}, \qquad \kappa = \pi d_p^2 n_p Q_{\text{abs}}, \qquad \beta = \kappa + \sigma_s + = \pi d_p^2 n_p Q_{\text{ext}}, \qquad (4)$$

where $\sigma_s$, $\kappa$, and $\beta$ are scattering, absorption, and extinction coefficient, respectively. If the particles are mainly diffusely reflecting spheres, the scattering, absorbing, and extinction efficiencies in Eqs. (4) can be simplified to

$$Q_{\text{sca}} = 1 - \varepsilon_p, \qquad Q_{\text{ext}} = 1, \qquad Q_{\text{abs}} = \varepsilon_p. \qquad (5)$$

### 33.3. Ignition of Solid Particles due to Absorption of Radiation

As schematically shown in Fig. 33.1, a mixture of carbon particles and air is contained between two parallel walls. The system is one dimensional, and the mixture is assumed to be quiescent. Two walls held at different temperatures $T_1$ and $T_2$ are assumed to be diffuse reflectors and emitters. The carbon particles are assumed to be spherical and uniformly monodispersed in the gaseous phase. The particle volume is neglected in comparison with the suspension volume. The particles are supposed to absorb and emit as well as isotropically scatter radiation and are also assumed to have constant monochromatic radiative properties. The inner particle temperature is considered to be uniform.

Under the above assumptions, the energy conservation equations for gas and particles can be represented by

$$\rho_g C_g \frac{\partial T_g}{\partial t} = \lambda_g \frac{\partial^2 T_g}{\partial x^2} - n_p Q, \qquad (6)$$

$$\sigma_p C_p \frac{\partial T_p}{\partial t} = n_p Q - \frac{\partial q_R}{\partial x} + H_p \Gamma_p, \qquad (7)$$

where $H_p$ is the heat of combustion per unit mass of fuel particle and $\rho_g$ and $\sigma_p$ are the gas density and the particle concentration, respectively. The amount of heat transferred between gas and one particle, $Q$, is expressed as

$$Q = \pi d_p^2 h (T_g - T_p), \qquad (8)$$

where the heat transfer coefficient $h$ can be calculated from the Ranz–Marshall correlation for the Nusselt number (Ranz and Marshall, 1952). The particle burning rate $\Gamma_p$ is given

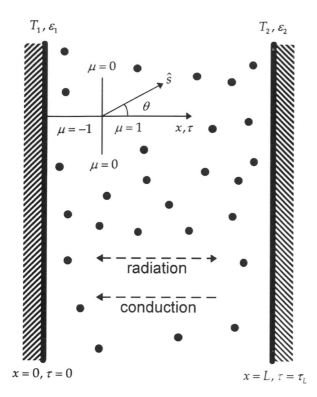

Figure 33.1. Schematic diagram for the ignition of the carbon particle suspensions.

by

$$\Gamma_p = n_p \pi d_p q, \tag{9}$$

where $q$ is the carbon burning rate per unit external geometric surface (Field et al., 1967).

In a one-dimensional geometry, the exact forms for the incident radiation, the radiative heat flux, and its divergence can be obtained as follows (Sparrow and Cess, 1970):

$$G(\tau) = 2B_1 E_2(\tau) + 2B_2 E_2(\tau_L - \tau)$$
$$+ 2\int_0^{\tau_L} \left[ (1 - \omega_s)\sigma T_p^4(\tau') + \frac{\omega_s}{4}G(\tau') \right] \cdot E_1(|\tau - \tau'|)\,d\tau', \tag{10}$$

$$q_R(\tau) = 2B_1 E_3(\tau) - 2B_2 E_3(\tau_L - \tau)$$
$$+ 2\int_0^{\tau} \left[ (1 - \omega_s)\sigma T_p^4(\tau') + \frac{\omega_s}{4}G(\tau') \right] \cdot E_2(\tau - \tau')\,d\tau'$$
$$- 2\int_{\tau}^{\tau_L} \left[ (1 - \omega_s)\sigma T_p^4(\tau') + \frac{\omega_s}{4}G(\tau') \right] \cdot E_2(\tau' - \tau)\,d\tau', \tag{11}$$

$$-\frac{dq_R(\tau)}{d\tau} = 2B_1 E_2(\tau) + 2B_2 E_2(\tau_L - \tau) - 4(1 - \omega_s)\sigma T_p^4 - \omega_s G(\tau)$$
$$+ 2\int_0^{\tau_L} \left[ (1 - \omega_s)\sigma T_p^4(\tau') + \frac{\omega_s}{4}G(\tau') \right] \cdot E_1(|\tau - \tau'|)\,d\tau', \tag{12}$$

$$B_1 = \varepsilon_1 \sigma T_1^4 + 2(1 - \varepsilon_1)$$

$$\times \left\{ B_2 E_3(\tau_L) + \int_0^{\tau_L} \left[ (1 - \omega_s)\sigma T_p^4(\tau') + \frac{\omega_s}{4} G(\tau') \right] \cdot E_2(\tau') \, d\tau' \right\}, \tag{13}$$

$$B_2 = \varepsilon_2 \sigma T_2^4 + 2(1 - \varepsilon_2)$$

$$\times \left\{ B_1 E_3(\tau_L) + \int_0^{\tau_L} \left[ (1 - \omega_s)\sigma T_p^4(\tau') + \frac{\omega_s}{4} G(\tau') \right] \cdot E_2(\tau_L - \tau') \, d\tau' \right\}, \tag{14}$$

where the exponential integral function $E_n(x)$ is defined by

$$E_n(x) = \int_0^1 \mu^{n-2} \exp\left( -\frac{x}{\mu} \right) d\mu. \tag{15}$$

### 33.3.1. *Ignition Delay*

The energy equation with the divergence of radiative heat flux is numerically solved with or without radiation. Figure 33.2 illustrates the temperature distributions near two walls. When the radiation is included, the carbon particle temperature rapidly increases with the temperature of radiatively transparent air almost unchanged. Evidently it is due to the radiative absorption of carbon particles. The increase in total heat flux due to radiation makes the ignition delay shorter, as shown in Fig. 33.3. Here, ignition delay is defined by the time interval from an initial exposure of the mixture to the radiation to the onset of ignition.

Temporal variations of gas and particle temperatures are presented in Fig. 33.4. In the vicinity of the hot wall, the conduction is more dominant than radiation because of the large gas temperature gradient; therefore, the air is always hotter than the particles. Contrary to this, the particle temperature is higher than air in the region far from the hot wall because of the far-reaching effect of radiation. This is a noteworthy characteristic of radiation.

The influence by scattering albedo was found to be negligible since the scattering redistributes only the radiative intensity, while the change in the hot wall emissivity $\varepsilon_2$ was more influential in predicting the ignition delays (Baek, 1992).

The effect of the particle size on the ignition delay is shown in Fig. 33.5. When the particle size increases at a fixed loading ratio, the particle number density decreases. This in turn leads to a reduction of extinction coefficient as well as a decrease in the total particle surface area available to the convective heat transfer between gas and particles. Consequently the particles are rendered less radiatively active and the increasing rate of particle temperature becomes much slower as the particle size increases. Finally, this results in longer ignition delay.

However, the carbon mass loading is known to affect the ignition delays negligibly because of two counterbalancing effects on the particle temperature. An appreciable change in the lower wall temperature makes no noticeable difference in ignition delays either.

### 33.3.2. *Influence of Nonuniform Particle Temperature*

The effect of nonuniform inner particle temperature was also taken into account (Baek, 1992). This can be accounted for by adopting the energy equation for a single particle, which replaces the energy equation for a particle cloud. The ignition delay was found to become shorter compared with the uniform particle temperature case. As the particle size increased, the ignition delay also increased because of the large volume-to-surface-area ratio.

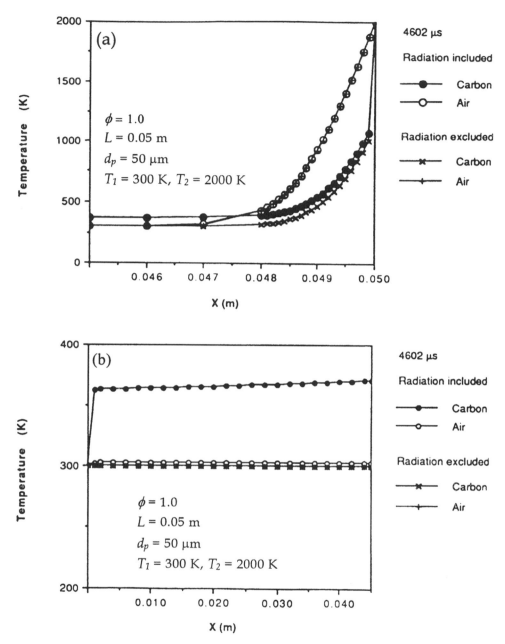

Figure 33.2. (a) Effect of radiation on temperature variation of carbon particles and air (magnified) adjacent to the hot wall for $\phi = 1.0$, $L = 0.05$, $d_p = 50\,\mu$m, $T_1 = 300$ K, and $T_2 = 2000$ K; (b) magnified regime adjacent to the cold wall.

## 33.4. Ignition of Combustible Gas by Radiatively Absorbing Inert Particles

The combustible gas with a sufficient amount of inert particles, which is exposed to external radiation, also poses a serious safety problem because of its inherent ignition risk. Its risk results from the fact that the inert particles absorbing the radiation have the possibility of directly

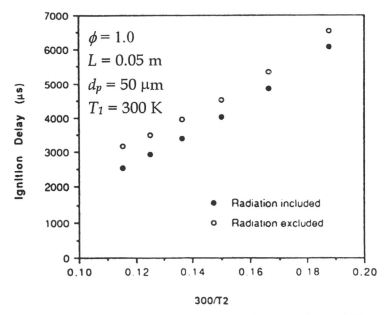

Figure 33.3. Effect of radiation on ignition delays for $\phi = 1.0$, $L = 0.05$ m, $d_p = 50\,\mu$ m, and $T_1 = 300$ K.

igniting the combustible gas by conduction and convection, even if the gas is not in immediate contact with the ignition source.

Hill et al. (1992) numerically and experimentally investigated the ignition of a hydrogen/air mixture by laser-heated coal particles. However, a very simplified radiation model has been adopted. Using a more rigorous radiation model and detailed chemistry, Baek (1994, 1996) explored the ignition of a suspension comprising inert particles and combustible gas ($H_2$/air, $C_3H_8$/air) in slab geometry.

### 33.4.1. *Mathematical Modeling*

As schematically shown in Fig. 33.6, a mixture of inert aluminum oxide particles and combustible gas is contained between two transparent walls that are heated by the external blackbody radiative heat source maintained at a high temperature. A one-dimensional open system is assumed, neglecting the gas expansion. The assumptions taken in Section 33.3 still apply here.

However, different from the model in Section 33.3, the reaction term appears in the gaseous energy equation, not in the particle energy equation:

$$\rho_g C_g \frac{\partial T_g}{\partial t} = \lambda_g \frac{\partial^2 T_g}{\partial x^2} - n_p Q - \sum_{i=1}^{N} \dot{\omega}_i \Delta H_{g,i}, \tag{16}$$

where $\dot{\omega}_i$ is the reaction rate for the $i$th species.

In this problem, the ignition source is the external radiation so that the two-flux radiation model is preferred. The net radiative heat flux $q^R$ can be expressed in terms of $q^+$ and $q^-$ in the forward and the backward directions. The governing equations for $q^+$ and $q^-$ can be derived by the integrating RTE over a hemisphere (Modest, 1993; Siegel and Howell, 1992):

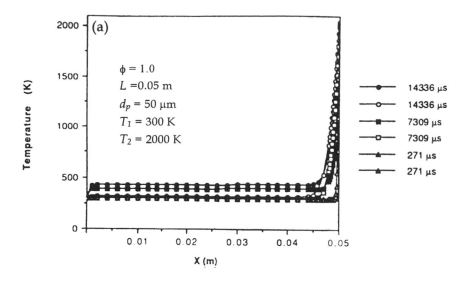

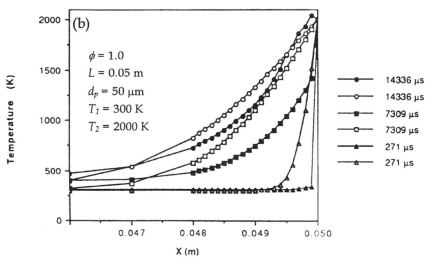

Figure 33.4. (a) Temperature variation of carbon particles and air for $\phi = 1.0$, $L = 0.05\,\mathrm{m}$, $d_p = 50\,\mathrm{mm}$, $T_1 = 300\,\mathrm{K}$, and $T_2 = 2000\,\mathrm{K}$; (b) magnified regime adjacent to the hot wall.

$$q^R = q^+ - q^-, \tag{17}$$

$$\frac{\mathrm{d}q^+}{\mathrm{d}x} = -2(\kappa + \sigma_s)q^+ + 2\sigma_s q^- + 2\kappa\sigma T_p^4, \tag{18}$$

$$\frac{\mathrm{d}q^-}{\mathrm{d}x} = 2(\kappa + \sigma_s)q^- + 2\sigma_s q^+ - 2\kappa\sigma T_p^4. \tag{19}$$

### 33.4.2. *Discussions*

In order to predict the ignition delay of a gas mixture accurately, a precise estimation of the reaction rate of each gas species is definitely required so that usage of the multiple-step chemical

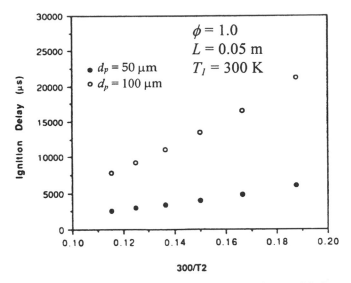

Figure 33.5. Effect of particle size on ignition delays for $\phi = 1.0$, $L = 0.05$ m, and $T_1 = 300$ K.

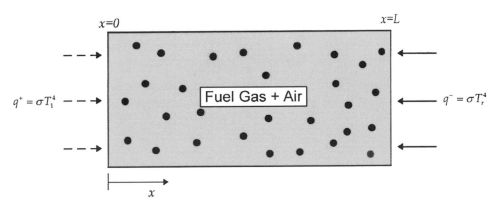

Figure 33.6. Schematic diagram for the ignition of the mixture of combustible gas and inert particles.

kinetics is highly demanded. While for a mixture of $H_2$/air a system of 10 species ($H_2$, $O_2$, $N_2$, H, O, N, OH, $HO_2$, $H_2O$, NO) with 32 elementary reaction steps has been adopted (Baek, 1994), for a mixture of $C_3H_8$/air a system of 31 species ($C_3H_8$, $O_2$, OH, H, O, $N_2$, $H_2$, $H_2O$, $HO_2$, CO, $CO_2$, $CH_4$, $CH_3$, $CH_2O$, HCO, $C_2H_6$, $C_2H_5$, $C_2H_4$, $C_2H_3$, $C_2H_2$, $CH_2CO$, $CH_2$, CH, $C_2H$, HCCO, *iso*-$C_3H_7$, *n*-$C_3H_7$, $C_3H_6$, $CH_3HCO$, $H_2O_2$, $CH_3CO$) with 123 elementary reaction steps has been chosen (Baek, 1996). A mixture of $CH_4$/air has also been examined with a system of 16 species ($CH_4$, $O_2$, H, O, OH, $H_2$, $H_2O$, $N_2$, $HO_2$, CO, $CO_2$, $CH_3$, $CH_2O$, HCO, $CH_3O$, $H_2O_2$) and 35 elementary reaction steps. The reaction rate $\dot{\omega}_i$ for the $i$th species is calculated from the CHEMKIN subroutines (Kee et al., 1980).

As shown in Fig. 33.7, the particles warm up first because of the absorption of radiation, and the particle temperature is always higher than the gas temperature before ignition. The onset of ignition is therefore defined as the time required for two temperature curves to cross each other, as revealed in the figure, in which $T_r$ is the temperature of external radiative thermal source. In Fig. 33.8 the ignition delays for hydrogen, methane, and propane with air are plotted on a linear scale against the inverse of the external source temperature. In general, the ignition

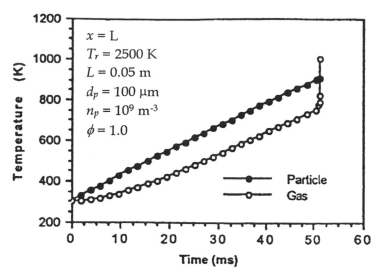

Figure 33.7. Temporal variation of particle and gas mixture temperature at $x = 0.05$ m for $\phi = 1.0$, $L = 0.05$ m, $d_p = 100\,\mu$m, $n_p = 10^9$ m$^{-3}$, $T_r = 2500$ K.

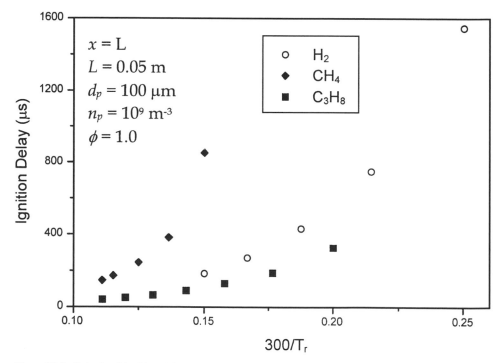

Figure 33.8. Calculated ignition delays.

delay time increases with decreasing $T_r$. The ignition delay is shown to be the shortest in the order of propane, hydrogen, and methane.

The change of particle number density at a fixed particle size exerts a strong influence such that the smaller the particle number density, the longer the ignition delay. The decrease in particle size at a fixed particle number density leads to a decrease in the extinction coefficient as well as the particle concentration. The particles then become less radiatively active so that their temperature increase slower. This finally leads to a substantial increase in ignition delay.

## 33.5. Conclusions and Recommendations

This chapter presented an ignition phenomenon of a two-phase mixture that is exposed to the external radiation. The absorption of radiation by inert particles is found to play a significant role in directly igniting the combustible gas mixture. The combustible solid particles are also discovered to be very hazardous, even if no heat source other than external radiation is available. In deriving the results above, we made several assumptions so that several recommendations for further research can be made.

First, the radiation effects by gases as well as particles are recognized. In the previous results only the radiation effects by particles were assumed, neglecting gas radiation. However, gases such as CO, $CO_2$, $O_2$, $NO_x$, $SO_x$, and $CH_4$ are known to be involved in thermal radiation so that the deliberation of two-phase radiation (both gas and particle radiation) deserves further exploration.

For a gray mixture in which for the particles emission, absorption, and scattering are considered while the scattering for the gas is neglected, the RTE can be written as (Park et al., 1998; Denison and Webb, 1993)

$$\frac{dI}{ds} + (\kappa_g + \kappa_p + \sigma_s)I = \kappa_g I_{bg} + \kappa_p I_{bp} + \frac{\sigma_s}{4\pi} \int_{4\pi} I(\hat{s}_i)\Phi(\hat{s}, \hat{s}_i)\,d\Omega_i,$$

where $I_{bg}$ and $I_{bp}$ are the blackbody emissive powers corresponding to local temperatures of the gas and the particles, respectively. Now, the divergences of the radiative heat flux, $\nabla \cdot \vec{q}_g^R$ for gas and $\nabla \cdot \vec{q}_p^R$ for particles, which are required for gas and particle energy equations, can be denoted by

$$\nabla \cdot \vec{q}_g^R = \kappa_g \left( 4\pi I_{bg} - \int_{4\pi} I\,d\Omega \right),$$

$$\nabla \cdot \vec{q}_p^R = \kappa_p \left( 4\pi I_{bp} - \int_{4\pi} I\,d\Omega \right).$$

Thereby the two-phase radiation effects can be examined in the future. Park et al. (1998) have already shown that the gas radiation together with particle radiation can play a significant role in determining the gas and particle temperature variations compared with the case of particle radiation only.

Second, the nongray behavior is an intrinsic characteristic of a material (Modest, 1993). Buckius and his colleagues have investigated the nongray behavior of a solid–gas mixture by experiments (Skocypec et al., 1987; Walters and Buckius, 1991). For a gas-only medium, many reliable theoretical models, such as the narrow-band model, wide-band model, weighted sum of gray gases model, etc., were reported (Modest, 1993). In contrast, the concrete theoretical nongray model of a solid–gas two-phase medium has not been proposed yet.

Third, when particles were dealt with in previous works they were assumed to be homogeneous and spherical. While Hottel and Sarofim (1967) have shown that the effects of particle shape on radiative energy transfer are negligible in systems involving a distribution of particle sizes, the shape of a particle may severely affect the ignition characteristics (Vorsteveld and Hermance, 1987). For this reason, it is necessary to explore further the effects of particle size and polydispersion of particles. Even though the radiative properties alone for the polydispersed mixture have been extensively studied (Menguc and Viskanta, 1985), their effects in practical applications are yet to be reported.

Finally, a careful exploration and consideration of changes in geometry and boundary conditions also remain to be identified.

## References

Baek, S. W., 1990, "Ignition of particle suspensions in slab geometry," *Combust. Flame* **81**, 366–377.

Baek, S. W., 1992, "Influence of nonuniform particle temperature on the radiatively active particle ignition," *AIAA J. Thermophys. Heat Transfer* **6**, 382–384.

Baek, S. W., 1994, "Ignition of combustible gases by radiative heating of inert particles," *Combust. Flame* **97**, 418–422.

Baek, S. W., 1996, "Ignition of propane-air mixture by radiatively heated small particles," *AIAA J. Thermophys. Heat Transfer* **10**, 539–542.

Denison, M. K. and Webb, B. W., 1993, "Modeling of radiative transfer in pulverized coal-fired furnace: effect of differing particle and gas temperature," in *Proceedings of the Sixth International Symposium on Transport Phenomena in Thermal Engineering*, Seoul, Korea, Vol. 1, 187–192.

Field, M. A., Gill, D. W., Morgan, B. B., and Hawksley, P. G. W., 1967, *Combustion of Pulverized Coal*, The British Coal Utilization Research Association, Leatherhead, Surrey, England.

Hill, P. C., Zhang, D. K., Samson, P. J., and Wall, T. F., 1992, "Laser ignition of combustible gases by radiative heating of small particles," *Combust. Flame* **91**, 399–412.

Hottel, H. C. and Sarofim, A. F., 1967, *Radiative Transfer*, McGraw-Hill, New York.

Kee, R. J., Miller, J. A., and Jefferson, T. H., 1980, "CHEMKIN: a general-purpose, problem-independent, transportable, Fortran chemical kinetics code package," Sandia Rep. 80-8003.

Khalil, H., Shultis, J. K., and Lester, T. W., 1983, "Stationary thermal ignition of particle suspensions," *ASME J. Heat Transfer* **105**, 288–294.

Kordylewski, W., 1979, "Critical parameters of thermal ignition," *Combust. Flame* **34**, 109–117.

Menguc, M. P. and Viskanta, R., 1985, "On the radiative properties of polydispersions: a simplified approach," *Combust. Sci. Tech.* **44**, 143–159.

Merzhanov, A. G. and Averson, A. E., 1971, "The present state of the thermal ignition theory: an invited review," *Combust. Flame* **16**, 89–124.

Modest, M. F., 1993, *Radiative Heat Transfer*, McGraw-Hill, New York.

Park, J. H., Baek, S. W., and Kwon, S. J., 1998, "Analysis of a gas-particle direct-contact heat exchanger with two-phase radiation effect," *Num. Heat Transfer A*, **33**, 701–721.

Raithby, G. D. and Chui, E. H., 1990, "A finite-volume method for predicting a heat transfer in enclosures with participating media," *ASME J. Heat Transfer* **112**, 415–423.

Ranz, W. E. and Marshall, W. R., 1952, "Evaporation from drops," *Chem. Eng. Prog.* **48**, 141–173.

Shouman, A. R., Donaldson, A. B., and Tsao, H. Y., 1974, "Exact solution to the one-dimensional stationary energy equation for a self heating slab," *Combust. Flame* **23**, 17–28.

Siegel, R. and Howell, J. R. 1992, *Thermal Radiation Heat Transfer (3rd ed.)*, Hemisphere Publishing, Washington, D.C.

Skocypec, R. D., Walters, D. V., and Buckius, R. O., 1987, "Spectral emission measurements from planar mixtures of gas and particulates," *ASME J. Heat Transfer* **109**, 151–158.

Smith, T. F., Byun, K.-H., and Chen, L.-D., 1988, "Effects of radiative and conductive transfer on thermal ignition," *Combust. Flame* **73**, 67–74.

Sparrow, E. M. and Cess, R. D., 1970, *Radiation Heat Transfer*, Brooks-Cole, Belmont, MA.

Viskanta, R. and Menguc, M. P., 1987, "Radiation heat transfer in combustion system," *Prog. Energy Combust. Sci.* **13**, 97–160.

Vorsteveld, L. G., and Hermance, C. E., 1987, "Effect of geometry on ignition of a reactive solid: acute angle," *AIAA J. Propul. Power* **5**, 26–31.

Walters, D. V. and Buckius, R. O., 1991, "Normal spectral emission from nonhomogeneous mixtures of $CO_2$ gas and $Al_2O_3$ particulate," *AMSE J. Heat Transfer* **113**, 174–184.

# 34

# Biomagnetic Fluid Dynamics

YOUSEF HAIK, VINAY M. PAI, AND CHING-JEN CHEN

## 34.1. Introduction

Electromagnetic fields have become a part of daily life in the 20th century. Over the years, there has always been a persistent concern among the scientific community about the induced effects of electromagnetic fields on the biological activity of human beings. However, a large number of studies carried out over the past few decades have not been able to identify conclusively the possible effects of such electromagnetic fields. A recent report (Oak Ridge-Associated Universities, 1992), published by the Committee on Interagency Radiation Research and Policy Coordination in 1992 found no convincing evidence available either to support or deny the contention that exposure to low-frequency (because of alternating electric fields) magnetic fields would be hazardous to health.

On the other hand, many studies were also made on the effect of a direct magnetic field, as opposed to an alternative magnetic field, on humans; many researchers reported that a steady magnetic field has some beneficial effects for humans. Probably the first notable medical application of magnetism was made by Mesmer in 1774 (Richard, 1986). In Paris, his Magnetic Institute attracted hundreds of the idle rich to his work of animal magnetism. Approximately 100 years later, a French scientist, Durval (Battocletti, 1976), reported on the results of an experiment on 100 men, for which he used what was called a "magnetic bracelet": two thirds of these men felt some kind of sensation or irritation or itching. Other scientists also began to report effects of magnetic bracelet. Thus the effect of magnetism on humans was recognized. However, no scientific studies of biomagnetic fluid dynamics were available.

Red blood cells have the characteristics of a paramagnetic fluid when deoxygenated (in veins) and diamagnetic when oxygenated (in arteries) (Shaylgin et al., 1983). This leads to the belief that there will be an effect of the magnetic field on the biological systems. For example, patients exposed to the magnetic resonance image (MRI) diagnostic machine or a passenger in the magnetic field generated from a magnetic levitation device will experience the effect of magnetization. Therefore many studies have been initiated to study the effect of the magnetic fields on the human body. Higashi et al. (1993) studied the orientation of the erythrocytes in strong magnetic field, from 1 to 8 T, and reported that the erythrocytes are found to orient with their disk plane parallel to the magnetic field. In this study, experiments were conducted to measure the whole-blood magnetic susceptibility in both the oxygenated and the deoxygenated states. The effect of a high magnetic field on blood flow is also reported in this study.

Haik (1997) proposed a mathematical model to describe the dynamics of the blood under the influence of a high magnetic field. In the formulation of the mathematical model the Navier–Stokes equations are retained if the applied magnetic field is zero. However, in the presence of the magnetic field, the angular momentum equations become nonlinear partial differential equations. In the limit in which the applied magnetic field reaches zero, the magnetic torque reaches zero, and hence the algebraic statement of the stress is reached ($\tau_{ij} = \tau_{ji}$). The magnetic torque is due to the difference in the orientation between the magnetization vector and the applied magnetic field. It reaches zero when the magnetization vector is aligned with the applied field.

The use of magnetic fluids other than blood has existed for more than 30 years (Haik, 1997). In industrial applications, magnetic fluids, which possess a ferromagnetic property, had been used often as lubricants and seals. A ferromagnetic fluid is a fluid mixed with particles susceptible to the magnetic field and are stable in colloidal suspensions (Rosensweig, 1985). Ferromagnetic particles suspended in the blood, for various medical applications, have also been proposed in recent years (Haik, 1997). A device that is used to separate red blood cells from whole blood is being developed. The device utilizes ferromagnetic particles that are prepared to link safely with the red blood cells. Magnetic force is applied to attract and thus retain the coupled red blood cells. The retardation of the red blood cells by the magnetic attraction leads to separation of red cells from the whole blood. The white cells are then directed toward an ultraviolet light for further treatment. This device is capable of providing a continuous, on-line separation of blood cells.

In this chapter, static and dynamic experiments of the effect of high magnetic field on human blood are presented. Mathematical formulations of the biomagnetic fluid dynamics are also presented. The development of a new device that can be used to separate red blood cells from whole blood based on biomagnetic fluid principles is also reported.

## 34.2. Human Blood in a High Magnetic Field

Several experiments were conducted to study the effect of a high magnetic field on human blood (Haik, 1997; Pai, 1997). In this study the results of two experiments that study the effect of a high magnetic field on human blood are presented. The first experiment studies the effect of a high magnetic field on human blood in the static condition. The second experiment studies the effect of a high magnetic field on human blood in the dynamic condition. The static experiment reports the magnetic susceptibility of the whole blood. The dynamic experiment reports the effect of a high magnetic field on human blood in the flow condition. Details of the experiments are given in the subsequent subsections.

### 34.2.1. *Magnetic Susceptibility*

The magnetic susceptibility of oxygenated and deoxygenated whole blood was determined with a superconducting quantum interference device (SQUID) magnetometer at different magnetic field strengths. This research was done at the Florida State University Center for Materials Research and Technology. For the oxygenated blood the sample was sealed in a plastic tubing of 5-mm height and 3-mm diameter. A Teruflex tube sealer (ACS-152) was used for sealing the tubing to ensure a perfect seal. For the deoxygenated blood, the sample was place in a coarse glass tube of 3-mm diameter and 30-mm height. The sample was placed at the bottom of the tube. The tube was subjected to nitrogen pressure for ~5 min. The nitrogen was introduced as bubbles at a small rate. The change of color was noticed instantaneously. In order to ensure deoxygenating of the 1-ml sample, the bubbling was kept for 5 min. Then the glass tube was sealed with an Oxy-Acton burner. The sealed sample of blood was then placed inside a plastic straw of 6-mm diameter. The magnetometery study was done with a Quantum Design MPMS2

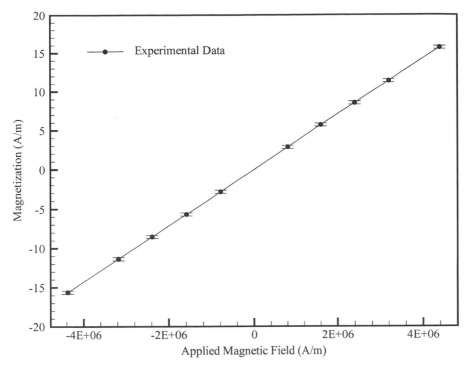

Figure 34.1. Magnetic susceptibility of oxygenated whole blood.

SQUID magnetometer. The magnetometer temperature was raised to 310 K (the temperature of blood inside the human body) before the sample was introduced into the SQUID chamber. A small field (100 G) was initially applied to center the specimen in the SQUID.

The field was then varied from +5 to −5 T, with steps of 0.5 T. Three measurements were taken at each magnetic field strength and averaged. This process was repeated at least twice. The plots of the magnetization versus the applied magnetic field for the oxygenated and the deoxygenated whole blood are shown in Figs. 34.1 and 34.2, respectively. As can be seen, the plots exhibit linear behavior. For the deoxygenated blood the susceptibility that represents the slope of the line is positive and equal to $3.5 \times 10^{-6}$, while for the oxygenated blood the susceptibility is negative and equal to $-6.6 \times 10^{-7}$. The linear behavior of the magnetization implies that the equilibrium magnetization was achieved at the time measurements were taken.

### 34.2.2. *Blood Pipe Flow*

Experiments were carried out at the National High Magnetic Field Laboratory to study the dynamic behavior of blood flow. The experimental setup shown in Fig. 34.3 was used to study the effect of the magnetic field on vertically downward blood flow due to gravity. The blood was drawn from the vein of a healthy male volunteer at the Leon County Blood Bank. The blood was tested for infections and diseases and found clear. The hematocrit value of the blood was 0.51. The blood that was collected in a 300-cc bag was allowed to flow in a plastic tubing of 3-mm internal diameter and 800-mm total length, while the tube was placed inside the 50-mm core of a 20-T resistive magnet. The blood from the upper bag was allowed to flow by gravity and was collected in the lower blood bag. The blood was in the oxygenated state during the experiment.

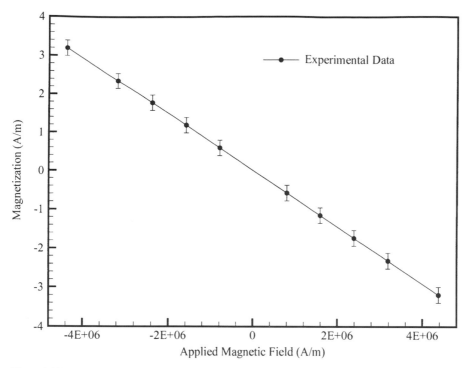

Figure 34.2. Magnetic susceptibility of deoxygenated whole blood.

The temperature in the setup was monitored by means of thermocouples to ensure that the temperature remained constant at 20°C throughout the experiment. The applied magnetic field was antiparallel to the blood flow direction.

The blood flow was directed downward while the applied field was directed upward. The blood flow rate in the magnetic coil was measured first in the absence of magnetic fields. The Reynolds number in this case was approximately 390 with an average blood velocity of 0.37 m/s in the 3-mm tube. The experiment was repeated at various magnetic field strengths up to 10 T.

At different field strengths, the time required for the flow of the blood from the blood bag above the magnet to the bag below the magnet was carefully measured. Figure 34.4 shows the results of the experiment. This series of experiments clearly showed that the blood flow was retarded when the magnetic field was applied. For example, without the magnetic field, the time of flow is 115 s, but at a 10-T field the time of flow is 148 s. The static and dynamic experiments on the human blood under the influence of a high magnetic field clearly show that human blood is affected by the high magnetic field.

In some medical applications, the low magnetic susceptibility of human blood calls for an enhancement of the magnetic susceptibility by coupling the red cells with magnetic particles. Lubbe et al. (1996) developed a magnetic fluid to which drugs and other molecules can be chemically bonded to enable those agents to be directed within an organism by a strong magnetic field. Ruuge and Ruestski (1993) suggested that the red blood cells can be regarded as a biocompatible container that can be loaded with a drug that is a chemical or magnetic compound so that the drug can be derived by a magnet for certain aspects. In this study magnetic microspheres are used to enhance the magnetic susceptibility of human red blood cells to achieve on-line, continuous separation of the red blood cells from the whole blood.

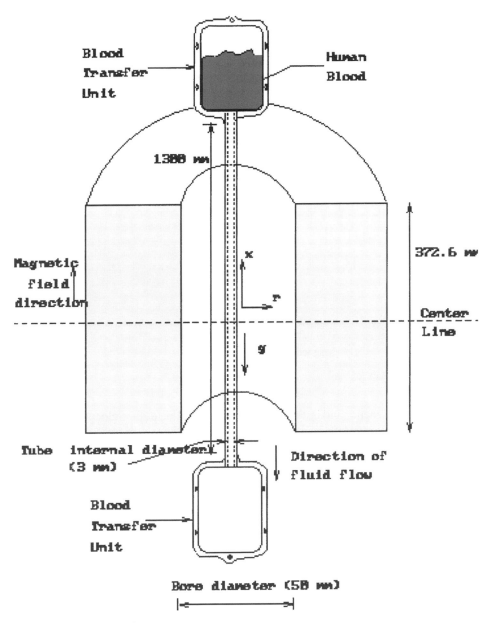

Figure 34.3. Experimental setup.

Few results that point to the effect of a high magnetic field on human blood are known. However, the details of fluid dynamics of blood, biomagnetic fluid, are not well known. In the following, a mathematical model of biomagnetic fluid motion is presented.

## 34.3. Mathematical Formulation

The mathematical model for biomagnetic fluid dynamics is based on the modified Stokes principles stated below (Haik, 1997):

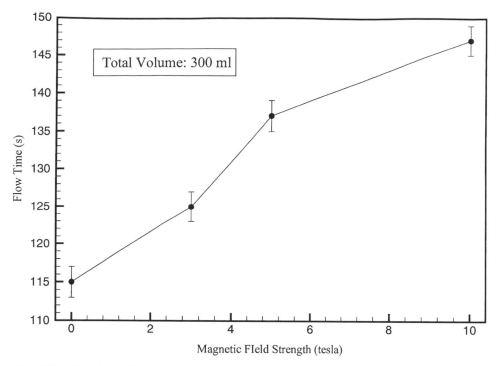

Figure 34.4. Experimental results.

1. The stress in the magnetic fluid is a continuous, linear function of the velocity gradient, the pressure, density, temperature, magnetization, space, and time (stress rate of deformation relation).
2. The magnetic fluid is homogeneous (homogeneous fluid postulation).
3. The stress in the magnetic fluid is an addition of isotropic stress and the anisotropic stress created by the magnetization (modified isotropic postulation).
4. In the presence of the magnetic field the stress of the magnetic fluid at rest is equal to the hydrostatic and magnetic pressure (stress–pressure relation).
5. If the deformation is purely dilatation the coupled stress, because of the rate of rotation, vanishes, and the normal stress is equal to the hydrostatic and magnetic pressure (modified Stokes hypothesis).

The governing equations under the above-modified Stokes principle are given below

**Continuity Equation**

$$\nabla \cdot \bar{V} = 0. \tag{1}$$

**Linear Momentum**

$$\rho \frac{D\bar{V}}{Dt} = -\nabla P + \rho \bar{F} + \eta \nabla^2 \bar{V} + \zeta(\nabla^2 \bar{V} + 2\nabla \times \bar{\omega}) + \mu_0(\bar{M} \cdot \nabla)\bar{H}. \tag{2}$$

Equation (2) states that, for the fluid element, the time rate change of momentum on the left-hand side of the equation is balanced by the right-hand terms:

- $-\nabla P$ is the force due to the pressure gradient on the flow field.
- $\rho \bar{F}$ is the body force due to the Earth's gravitational field.
- $\eta \nabla^2 \bar{V}$ is the force acting on the element because of the viscous shearing of the fluid.
- $\zeta(\nabla^2 \bar{V} + 2\nabla \times \bar{\omega})$ is the force arising because of the relative difference in the velocity of the fluid element and its particulate matter because of the applied magnetic field. Here $\zeta$ is the rotational viscosity caused by the magnetic field, while $\omega$ is the spinning rate for the particulate matter.
- $\mu_0(\bar{M} \cdot \nabla)\bar{H}$ represents the influence of the magnetic force as a body force on the fluid element. Here $\mu_0$ is the absolute permeability and is equal to $4\pi \times 10^{-7} \mathrm{Am}^2$ in SI units. $\bar{M}$ is the magnetization of the material, while $\bar{H}$ is the applied magnetic excitation.

**Angular Momentum**

$$\rho I \frac{\mathrm{D}\bar{\omega}}{\mathrm{D}t} = \mu_0 \bar{M} \times \bar{H} + \eta' \nabla^2 \bar{\omega} + 2\zeta(\nabla \times \bar{V} - 2\bar{\omega}). \tag{3}$$

$I$ is the moment of inertia of the element under consideration. Equation (3) states that the internal angular momentum for the fluid element is balanced by the right-hand-side terms:

- $\mu_0 \bar{M} \times \bar{H}$ is the body couple term and is analogous to a body couple on a small magnetic body in the presence of a local magnetic field.
- $\eta' \nabla^2 \bar{\omega}$ is the surface couple term and is dependent on the internal strain within the element, analogous to the viscous stress force that is a function of the strain rate of the fluid element. $\eta'$ is known as the spin diffusion viscosity.
- $2\zeta(\nabla \times \bar{V} - 2\bar{\omega})$ is the angular momentum exchange term and arises because of the lack of synchronization between the rate of rotation of the fluid element and the rate of internal spin of the matter making up the fluid element.

**Magnetization**

$$\frac{\mathrm{D}\bar{M}}{\mathrm{D}t} = \bar{\omega} \times \bar{M} - \frac{1}{\tau_r}(\bar{M} - \bar{M}_0),$$

$$\bar{M}_0 = M_0 \frac{\bar{H}}{\|\bar{H}\|},$$

$$M_0 = \frac{\bar{M} \cdot \bar{H}}{\|\bar{H}\|}. \tag{4}$$

The magnetization $\bar{M}$ represents the perturbation of the magnetization from the equilibrium value $\bar{M}_0$. The equilibrium magnetization $\bar{M}_0$ is defined as the magnetization attained by the stationary fluid in a steady applied magnetic field $\bar{H}$. For strong paramagnets, the direction of $\bar{M}_0$ tends to be in the direction of the applied excitation. However, for weak paramagnetic materials such as deoxygenated blood, a larger strength of excitation is required for obtaining the same degree of orientation of the equilibrium magnetization.

**Langevin Equation**

$$M_0 = M_s \left[ \coth\left(\frac{\mu_0 m H}{\kappa T}\right) - \frac{\kappa T}{\mu_0 m H} \right]. \tag{5}$$

The above equation can be considered as the equation of state of a magnetic fluid in the magnetic field. Equation (5) physically states that the magnetization tends toward saturation magnetization as $H$, the magnitude of the applied magnetic field, becomes very large. Once saturation is attained, any further increase in $H$ will not have any additional influence on the magnetic behavior of the fluid under consideration.

### Maxwell Equation

$$\nabla \times \bar{H} = 0. \tag{6}$$

The set of equations (1)–(6) represents 14 equations for 14 variables [$p$, Eq. (1); $\bar{V}$, Eq. (3); $\bar{\omega}$, Eq. (3); $\bar{M}$, Eq. (3); $M_0$, Eq. (1); and $\bar{H}$, Eq. (3)]. The viscosity $\eta$, rotational viscosity $\zeta$, density $\rho$, body force $\bar{F}$, angular spin rate $\eta'$, relaxation time $\tau_r$, temperature $T$, and the saturation magnetization $M_s$ are considered as known and constants. $\mu_0$ is the magnetic permeability in free space and is equal to $4\pi \times 10^{-7}$ H/m, $\kappa = 13.805 \times 10^{-24}$ J/K is the Boltzmann constant, and $I$ in Eq. (3) is the moment of inertia of the magnetic fluid particle or blood cells.

**Compatibility Equations.** Two additional equations are needed as a compatibility condition for Eq. (6):

$$\nabla \cdot \bar{B} = 0, \tag{7}$$
$$\bar{B} = \mu \bar{H}, \tag{8}$$

where $\mu$ is the permeability of the magnetic fluid. In summary, assuming that the body force is the gravitational force and known, the above model consists of 13 equations with 13 unknowns [pressure $P$, Eq. (1); velocity $\bar{V}$, Eq. (3); intrinsic angular velocity $\bar{\omega}$, Eq. (3); magnetization $\bar{M}$, Eq. (3); and magnetic field $\bar{H}$, Eq. (3)]. In order to complete the mathematical modeling, two additional compatibility conditions, Eqs. (7) and (8), are used. The variables and the constants that appears in the equations are listed in Table 34.1.

### 34.4. Biomagnetic Fluid Dynamics Applications

In the 21st century biological, medical, and environmental technologies will bring new challenges and improve our lives. Among these new technologies is the use of biomagnetic fluid dynamics to develop new therapeutical and clinical techniques. Several investigators reported the use of magnetic particles in medical applications. Table 34.2 lists some of the medical applications that have been cited in the literature proposing the use of magnetic fluid or particles in blood. Table 34.1 illustrates that the study of biomagnetic dynamics is a relatively new emerging field.

In this chapter the separation of red blood cells from the whole blood by a magnetic technique is discussed. Certain cancer treatments require clean separation of white cells from whole blood. At present, the separation depends on centrifugal techniques. During the centrifugal separation, red blood cells, which comprise 45% of the whole blood, often trap white cells. In general the efficiency of the centrifugal cell separation is ∼50%. A device that utilizes the biomagnetic fluid principles is being developed to separate red blood cells from the whole blood.

Red blood cells were found to be a paramagnetic material when deoxygenated and a diamagnetic material when oxygenated (Haik, 1997; Pai, 1997; Shaylgan, 1983). The magnetic

Table 34.1. *Definitions of Variables in the Governing Equations*

| Name | Symbol | Units |
|---|---|---|
| Velocity vector | $\vec{V}$ | m/s |
| Density | $\rho$ | kg/m$^3$ |
| Gravity vector | $\vec{F}$ | N/kg |
| Pressure | $P$ | N/m$^2$ |
| Viscosity | $\eta$ | kg/ms |
| Rotational viscosity | $\zeta$ | kg/ms |
| Angular spinning vector | $\vec{\omega}$ | rad/s |
| Permeability of vacuum | $\mu_0 = 4\pi \times 10^{-7}$ | Henry/m or mT/A |
| Magnetization | $\vec{M}$ | A/m |
| Magnetic field intensity | $\vec{H}$ | A/m |
| Moment of inertia | $I$ | kg m$^2$ |
| Shear spin viscosity | $\eta'$ | kg/m$^3$s |
| Relaxation time | $\tau$ | s |
| Boltzmann constant | $\kappa$ | J/K |
| Temperature | $T$ | K |
| Particle magnetization | $m$ | A/m |
| Saturation magnetization | $M_s$ | A/m |
| Equilibrium magnetization | $M_0$ | A/m |

Table 34.2. *Medical Applications of Biomagnetic Fluid Dynamics*

| Medical Applications | Biomagnetic Dynamics | Reference |
|---|---|---|
| Seals | Blocks blood flow near the surgery areas | Perry, 1977 [?] Ohara and Makinouchi, 1994 [?] |
| Drug carriers | Carries drugs to designated areas | Ruuge and Rusetski, 1993 [?] |
| Thrombus stabilizers | Strengthens blood vessels by sedimentation | Roth, 1969 [?] Engelhard and Petruska, 1992 [?] |
| X-ray contrast | Derives substances useful for the x-Ray | Fertman, 1990 [?] |
| Separators | Separates the red blood cell from the blood | Plavins and Iauva, 1993 [?] |
| Guidance | Guides devices | Senyei et al., 1978 [?] Buckingham et al., 1994 [?] |
| Tracer | Measures blood flow | Newbower, 1973 [?] |
| Therapy | Accelerates blood flow | Nikken Co., 1994 [?] |

susceptibility of the red blood cell when oxygenated or deoxygenated is of the order of $10^{-6}$. Thus in order to enhance the magnetic susceptibility of red blood cells they are coupled to a protein-coated magnetic microsphere. The magnetic microspheres are 1 $\mu$m in size. The magnetic microspheres are coated with protein to protect the cell and the body from direct interaction with the magnetic microspheres. The coupling of cells with magnetic microspheres enhances the magnetic susceptibility of the red blood cells by ~4 orders of magnitude.

The coated microspheres can be easily decoupled from the red cells surface by the introduction of a small concentration of sugar to the red blood cells that are coupled with the coated microspheres.

The flow of red blood cells now can be manipulated, in addition to pressure, gravitational, and viscous force, by the magnetic force from the externally generated magnetic field. Electromagnetic fields or permanent magnets can create the magnetic field. Although the permanent magnet's field strength is limited by the existing technology of material science, permanent magnets of 1.2 T are commercially available and relatively inexpensive. A suitable distribution of magnetic force can be designed by arranging different patterns of permanent magnets.

Microscopic visualization of red cell separation from whole blood was experimentally demonstrated. In the experiment 40-mg of magnetic microspheres were added to 30 ml of

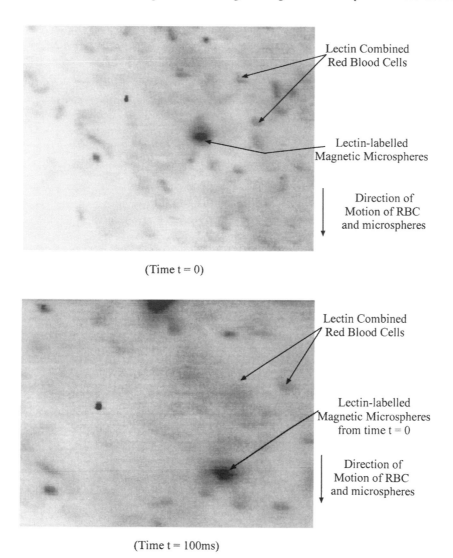

Figure 34.5. Coupled red blood cells in the absence of a magnetic field.

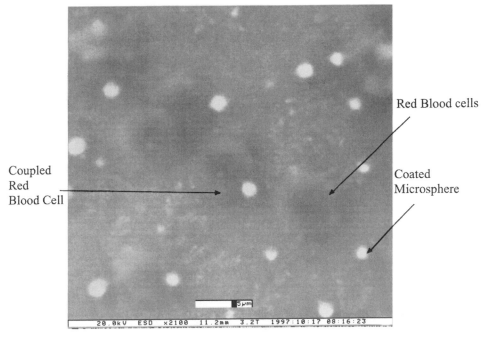

Figure 34.6. Electron microscopic view of microspheres coupled to red blood cells.

pH 7.4 phosphate-buffered saline solution. Whole blood weighing 63 mg was added to the solution. The mixture was agitated manually. 4 ml of the solution was placed a plastic covet. The covet was placed in the sample holder. The system was then observed with a microscopic flow visualization system. The magnification was set at 100×. Figure 34.5 shows the behavior of the red cells in the absence of an applied magnetic field. The images are taken at approximately 100-ms intervals. The figures show magnetic microspheres (dark spots) coupled with the red blood cells. The figure also shows that after 100 ms, the microspheres coupled to red cells have moved down because of gravitational force. It should be remarked that not all microspheres are coupled to red cells. Figure 34.6 is the electron microscopic view of microspheres (white spots) coupled to the red blood cell surface. The picture was obtained with a scanning electron microscope with 2100×. A magnet of 1.2 cm diameter was placed next to the plastic covet. A highly accelerated flow of red cells toward the magnetic field is observed from the video sequence. Two pictures are shown in Fig. 34.7. In these figures, the magnetic field is applied along the left wall so that the coupled red cells move left toward the applied magnetic field. The technique then can be utilized to separate the red cells from the whole blood. One of the greatest advantages of this technique is the ability to separate the cells from whole blood in a continuous fashion.

A device that utilizes the magnetic field for red blood cell separation is being developed. The device is shown in Fig. 34.8. The device consists of a mixing chamber in which the whole blood is mixed with the magnetic microsphere. In the mixing chamber the red blood cells will couple with the magnetic microspheres. The whole blood is then directed toward a channel compartment. Specially arranged permanent magnets to achieve highest attraction force are placed under the channel compartment. When the whole blood flows in the channel,

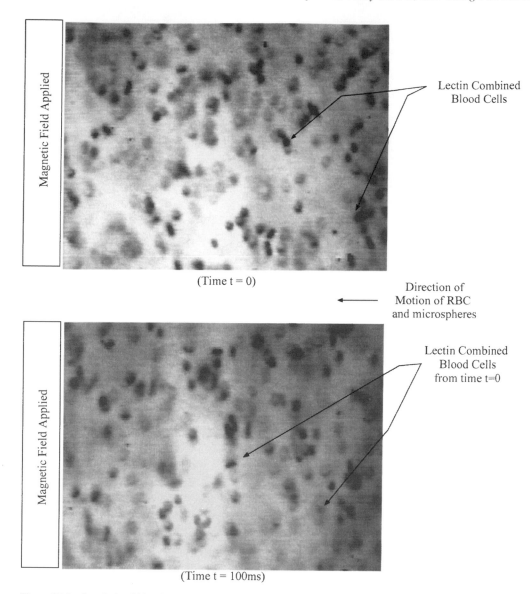

(Time t = 0)

Direction of
Motion of RBC
and microspheres

(Time t = 100ms)

Figure 34.7. Coupled red blood cells in the presence of a magnetic field.

an attraction force applied by the magnets will retain the coupled red blood cells. The plasma, white cells, and platelets leave the channel to go to the cancer treatment unit while the red cells remain in the channel. Once the plasma, white cells, and platelets leave the channel a fluid that contains a small amount of sugar will be used to flush the channel. Because of the small sugar concentration, the red blood cells will decouple from the magnetic microspheres. Because of thee difference in susceptibility between the magnetic microspheres and the red blood cells, the red blood cells will be able to leave the channel while keeping the magnetic microspheres retained because of the magnetic attraction. The red blood cells will then be returned to the patient with the treated white cells, plasma, and platelets.

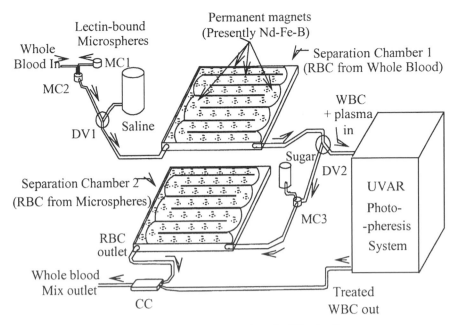

Figure 34.8. Magnetic device for continuous separation of red blood cells.

## 34.5. Conclusions

In this chapter the effect of a high magnetic field on human blood is presented. The magnetic susceptibility of whole oxygenated and deoxygenated blood with a SQUID is reported. The susceptibility of the deoxygenated whole blood is $3.5 \times 10^{-6}$ while for the oxygenated blood the susceptibility is negative and equal to $-6.6 \times 10^{-7}$. The flow of blood in a tube, because of the gravitational effect under the influence of a high magnetic field, was slowed down by 25% when the applied magnetic field reached 10 T. The mathematical formulation of the biomagnetic fluid dynamics is presented. A device that utilizes the biomagnetic fluid principles to separate red blood cells from whole blood on a continuous basis is also presented.

## References

Battocletti, J. H., *Electromagnetism, Man and Environment*, Westview Press, 1976.

Buckingham, T., Patil, V., Willoughby, P., and Szeverenyi, N. M., "Magnetic orotracheal intubation: a new technique," *Anesth. Analg.* **78**, 749–752, 1994.

Engelhard, H. and Petruska, D., "Imaging and movement of iron oxides bound antibody microparticles in brain and cerebrospinal fluid," *J. Cancer Biochem. Biophys.* **13**, 1–12, 1992.

Fertman, V. E., *Magnetic Fluids Guidebook*, Hemisphere, 1990.

Haik, Y., "Development and mathematical modeling of biomagnetic fluid dynamics," Ph.D. dissertation, Florida State University, 1997.

Higashi, T., Yamagishi, T., Takeuchi, A., Kawaguchi, N., Sagawa, S., Onishi, S., and Date, M., "Orientation of erythrocytes in strong static magnetic fields," *J. Blood* **82**, 1328–1334, 1993.

Jenson, R. and DiRusso, E., "Passive magnetic bearing with ferrofluid stabilization," NASA Tech. Memo. 107154, 1997.

Lubbe, A. S., Bergemann, C., Huhnt, W., Fricke, T., and Riess, H., "Preclinical experience with magnetic drug targeting: tolerance and efficacy," *Cancer Res.* **56**, 4694–4701, 1996.

Newbower, R. S., "Magnetic fluids in the blood," *IEEE Trans. Magn.* **MAG-9**, 447–450, 1973.

Nikken Co., Product literature, California, 1994.

Oak Ridge Associated Universities, "Health effects of low frequency electric and magnetic fields," ORAU 92/F8, 1992.

Ohara, Y. and Makinouchi, K., "An ultimate compact, seal-less centrifugal ventricular assist device," *J. Artif. Organism*, 17–24, 1994.

Pai, V. M., "Theoretical and experimental study of biomagnetic fluid dynamics," Ph.D. dissertation, Florida State University, 1997.

Palvins, J. and Lauva M., "Study of colloidal magnetic binding erythrocytes: prospects for cell separation," *J. Magn. Magn. Mater.* **122**, 349–353, 1993.

Perry, M. P., "A survey of ferromagnetic liquid applications," in *Proceedings of the International Advanced Course and Workshop on Thermomechanics of Magnetic Fluids*, pp. 219–230, 1977.

Raj, K. and Moskowitz B. D., "Ferrofluids step-up motor precision," *J. Mach. Des.* pp. 57–60, 1995.

Richard, B. F., "Biological effects of static magnetic fields," in Charles P. and Elliot P., eds., *Editors, Handbook of Biological Effects of Electromagnetic Fields*, pp. 169–225, CRC, 1986.

Rosensweig, R. E., *Ferrohydodynamics*, Cambridge University Press, 1985.

Roth, A. D., "Occlusion of interacranial aneurysms by ferromagnetic thrombi," *J. Appl. Phys.* **40**, 1044–1045, 1969.

Ruuge, E. K. and Rusetski, A. N., "Magnetic fluids as drug carriers: targeted transport of drugs by a magnetic field," *J. Magn. Magn. Mater.* **122**, 335–339, 1993.

Senyei, A., Widder, K., and Gzerlinski, G., "Magnetic guidance of drug carrying microspheres," *J. Appl. Phys.* **49**, 3578–3583, 1978.

Shaylgan, A. N., Norina, S. B., and Kondorsky, E. I., "Behavior of erythrocytes in high gradient magnetic field," *J. Magn. Magn. Mater.* **31**, 555–556, 1983.

# The Man I Know: Chia-Shun Yih, July 25, 1918 – April 25, 1997

YUAN-CHENG FUNG

Chia-Shun loved fluid mechanics. He was a fluid dynamicist par excellence. His work touched many other fluid dynamicists. We remember him. In memory of him we congregate here to talk about fluid mechanics and to dedicate our best thoughts to him. Knowing his love of friends, I am sure this symposium would please him greatly. I myself, however, have devoted most of my life to biomechanics and solid mechanics. I have only one joint paper with Chia-Shun, published in the *Journal of Applied Mechanics* in 1968, on the subject of peristatic transport, of which I will say a few words later. I am keenly aware of the fact that I am not active in the mainstream of this symposium. However, I have the distinction of being Chia-Shun's oldest friend in this gathering. It was sixty-four years ago that Chia-Shun and I met for the first time as two new students at a high school. We became classmates and pretty soon became best friends. Then we went through college together. We won China's National Scholarship Exam to study abroad in the same year. We came to the United States on the same ship. There was never a single year that our families did not spend some time together. Hence I know his early days. Maybe I can sketch a portrait of a young Chia-Shun for you.

Chia-Shun Yih was born on July 25, 1918, in Kweiyang City, Kweizhou Province, which lies in the southern midwest part of China. Kweizhou is a beautiful mountainous country. Chia-Shun was born into a scholarly family, the son of Yih Ding-Jan and Hsiao Wan-Lan. His father was a specialist on silk and silkworm culture. His father's profession was fortunate for me, because it made his father come to work in Kiangsu Province, which lies on the east coast of China, where I was born, where the silk industry was flourishing, where every family raised silkworms in the spring, and where the improvement of the genetic makeup of the silkworm eggs was a big business of the Provincial Government. Chia-Shun attended junior middle school (grades 7–9) in Zhengkiang, the provincial capital of Kiangsu. Then he and I both passed the entrance exam of the Soozhou Senior Middle School (grades 10–12) and entered in 1934. Soozhou is a city with a long history. Two thousand years ago it was the capital of the Kingdom of Woo. Our school ground was old and beautiful. The oldest hall of the school, where we often took examinations, was call the Purple Sun Hall, in honor of Master Zhu Hsi (1130–1200 AD), founder of new Confucianism in the Sung Dynasty. The walls of the Hall were lined with plaques of black stone on which Zhu Hsi's poems and lectures were carved. We were not sure whether Master Zhu actually lectured there, but the halo of tradition was real and palpable, and we were fortunate that the education we received at the school was worthy of the hallowed site.

In our high school class, Chia-Shun distinguished himself by his talent for language and mathematics. One day, our English teacher picked on him about his homework. Chia-Shun talked back in English. The whole class was surprised and impressed: talking back in classroom was never done in China at that time, and no one in our class was aware that somebody among us could carry on an argument in English with our teacher. Later, I saw this sort of demonstration again and again. For example, Chia-Shun and I started learning German together in college. By the time I could read some German math books, he was able to converse in German with a German missionary in Chongqing.

While we were in high school, the storm of war was gathering in China. Japan had occupied Manchuria in 1931 and invaded Shanghai in 1932. Full-scale war between China and Japan finally broke out on July 7, 1937, soon after our graduation from high school. We managed to take the entrance examination of the National Central University and got accepted. The University was located originally in Nanjing, the capital of China at that time. Before we could enter, however, it was moved to Chongqing, in Sichuan Province in the central midwest of China, by a quick and farsighted decision of Dr. Chia-Luen Lo, its President. Thus we entered the University in Chongqing and luckily escaped the rape of Nanjing by the Japanese army in December, 1937.

At our university, Chia-Shun studied, among other things, mathematics and the theory and design of bridges. Our college years were spent in makeshift classrooms and laboratories, classes at the crack of dawn to avoid air raids, long hours in the dugouts, military training, and an endless stream of exciting or sad news. One wintry day, Japanese planes came and bombed out our simple shower hut, and for weeks afterwards some of us had to bathe in the emerald water of the nearby Chia-Ling River, beautiful but cold. Chia-Shun liked to say that the dominant sensation when one jumps into icy water is an immediate headache. "Afterwards, the shiver in the sun seemed almost pleasant by comparison."

After graduation, Chia-Shun worked first in the National Hydraulics Laboratory in Guanshien, Sichuan. There he studied the work of Li Bing who, more than 2300 years ago, had invented a system of constructing and reconstructing control dikes every year, which works to this day. There he learned how a superior engineering design could make the Chengdu plain one of the richest areas in China for 2300 years. Then he worked for the National Bureau of Bridge Design in Kweiyang, his hometown. At that stage, we both intended to be practicing engineers. Then he was married to Loh Hung-Kwei, who gave birth to his first son, Yiu Yo Yih. But the marriage lasted only a few years.

Then something changed our lives. A group of American professors visited China and on their return raised forty-some graduate scholarships from various American Universities and offered them to the Chinese Ministry of Education. By nationwide examinations, the Ministry of Education chose forty-two scholars to study in the United States. We were among the forty-two, and after the War came to the United States via India. At Calcutta we boarded the American troop ship "General Hase," which took a month crossing the Red Sea, the Mediterranean, and the Atlantic, before reaching New York City on December 28, 1945. I then went to Caltech, and Chia-Shun, after a very brief stay at Purdue University, went to the University of Iowa (which was called the State University of Iowa before 1970) to study fluid mechanics with Hunter Rouse and John McNown, with whom he maintained a warm friendship throughout his life.

The research atmosphere at the University of Iowa suited Chia-Shun very well. He was inspired, worked hard, and was getting outstanding results that won Rouse's approval. He signed up for courses in music appreciation and French conversation. The young instructor of French conversation was Shirley Ashman from Maine. Chia-Shun and Shirley soon fell in love and were married in 1949.

In the summer of 1947, Chia-Shun went to Brown University and listened to C. C. Lin's lectures on fluid dynamics. On returning to Iowa, he made rapid progress on his dissertation and completed his Ph.D. requirements in 1948. He explained the motivation of his dissertation to me by way of observing the smoke from a lighted cigarette. He wanted to describe the curling and waving of the plume mathematically. He succeeded in solving some idealized aspects of this problem and realized that the solution could be extended to deal with many geophysical and atmospheric problems. He knew its relevance to problems of air and water pollution by car and by industrial heat waste. His solution was published in two papers: one in the *Journal of Applied Mechanics* (1950) under the title "Temperature Distribution in a Steady, Laminar, Preheated Air Jet," and another in the *Proceedings of the First U.S. National Congress of Applied Mechanics* (1950), under the title "Free Convection due to a Point Source of Heat." His formulation of the problem and his solutions were really elegant. The laminar flow solution was exact, and it was accompanied by a systematic experimental investigation of the transition from laminar to turbulent flow. He conducted the experiment himself. These studies were followed by a series of papers dealing with atmospheric diffusion, gravitational convection from a boundary source, turbulent buoyant plumes, buoyant plumes in transverse wind, etc. His characteristic approach was to find closed-form exact solutions as far as possible and to check with experimental results. From his first paper to the one hundred thirtieth, the spirit was the same.

From 1948 to 1955, Chia-Shun taught and conducted research at the University of Wisconsin, University of British Columbia, Colorado A & M University, the University of Nancy in France, and the State University of Iowa. He finally settled down at the University of Michigan in Ann Arbor. For sabbatical leaves he went to Europe. He spent a year (1959–1960) at Cambridge University, England, a year (1964) in Geneva, another year (1970–1971) in the Universities of Paris and Grenoble in France, and a year (1977–1978) at Chatoux Lab in Paris and the Technische Hochschule Karlsruhe, in Germany. After he retired in 1988, he served as a Graduate Research Professor of the University of Florida in Gainesville for three years. Then Shirley and he made Gainesville their winter home and stayed in Ann Arbor the rest of the year. Chia-Shun never stopped researching. Important papers flowed out at a regular rate. He maintained his good health. Two days before his death, he himself planted five young flowering trees in his garden. Friends watching him digging the holes asked him, why he must dig the holes so big and so deep? He answered, "At my age, I want to make sure that every sapling gets its full share of endowment, to grow up big and strong! None should be shortchanged."

Honors followed Chia-Shun's achievements. In 1968, the University of Michigan celebrated its sesquicentennial and chose to give special honors to a few outstanding professors among its faculty. Chia-Shun was among them and was given the permanent title of "Stephen P. Timoshenko Distinguished University Professor of Fluid Mechanics of the University of Michigan." In 1970, he was elected a member of Academia Sinica. In 1980, he was elected a member of the United States National Academy of Engineering. He was honored by the Chinese Institute of Engineers with the 1968 Achievement Award and by the Chinese Engineers and Scientists Association of Southern California with the 1973 Achievement Award. In 1974 he was the University of Michigan's "Henry Russel Lecturer." In 1981, he was given the Theodore von Kármán Medal by the American Society of Civil Engineers. The American Physical Society gave him the Fluid-Dynamics Prize in 1985, and the Otto Laporte Award in 1989. In 1992, he had the honor to present the "Sir Geoffrey Taylor Lecture" at the University of Florida. Chia-Shun was a great admirer of Sir Geoffrey. Earlier, in 1976, Chia-Shun had dedicated a volume of *Advances in Applied Mechanics* that he edited to Sir Geoffrey. In the Preface, Chia-Shun said of G.I., "His work was always marked by an originality of thought and a freshness of approach that continue to delight his readers, and a characteristic welding of analysis to experiments that is rarely attempted, let alone attained, by others." My feeling is that this describes Chia-Shun himself very well.

Chia-Shun's scientific papers published between 1950 and the early part of 1990 have been collected in a two-volume set called *Selected Papers by Chia-Shun Yih*, published by World Scientific, Singapore, 1991. It contains ninety-seven articles divided into five categories:

(1) stratified flows and internal waves, (2) theory of hydrodynamic stability, (3) gravity waves, (4) jets, plumes and diffusion, and (5) general. Chia-Shun's papers that appeared after the publication of the *Selected Papers* are listed at the end of this chapter. His important contributions of the last period, from 1990 to 1997, include his theory of colliding solitons, his theory of infinitely many superposable solutions of the Navier–Stokes equations, his theory of added mass, the kinetic-energy mass, momentum mass, and drift mass in steady irrotational subsonic flows, and in periodic water waves and sound waves, and his theory of the instability due to viscosity stratification. The last paper is entitled "Tornado-like Flows," and is a most remarkable contribution.

In addition to journal articles, Chia-Shun published two books on stratified flow and one on the whole field of fluid mechanics. The former grew out of the seeds planted in his paper, "On the Flow of a Stratified Fluid," presented at the Third U.S. National Congress of Applied Mechanics in 1958 (*Proceedings of the Congress*, pp. 857–861). In this paper, he introduced a transformation of variables that simplifies the equations describing the two-dimensional stratified flow. Using this transformation, he obtained exact solutions of a number of large amplitude motions. Germination of these seeds developed into his book *Dynamics of Nonhomogeneous Fluids* (MacMillan, 1965). The second edition of this book, which contains a great deal of new material, was given a new title, *Stratified Flows* (Academic Press, 1980). The Yih style of fresh and concise writing shines through. This style is particularly evident in his third book, *Fluid Mechanics, A Concise Introduction to the Theory* (McGraw-Hill, 1969). When this book went out of print in 1979, Chia-Shun issued an improved edition through the West River Press in order to reduce its price for the benefit of students. This fluid mechanics book was translated into Chinese by Zhang Kebun, Zhang Diming, Chen Qiqian, and Tsai Zhungxi, and published by Higher Education Press, Beijing, 1982. The rendering was excellent, and Chia-Shun was very grateful to the translators.

Chia-Shun was interested in biomechanics also. In 1968, he and I published a paper together, entitled "Peristaltic Transport" (*J. Appl. Mech.*, 1968, 669–675). We were aiming to understand a disease called "hydroureter," in which the ureter becomes enlarged, the peristaltic transport becomes ineffective, and the kidney may be injured. The ureteral peristalsis was described by Aristotle twenty-three centuries ago. We solved the two-dimensional, small-amplitude, low Reynolds number case. Frank Yin, then a graduate student working with me at UCSD, (now a famous biomedical engineering professor at Washington University in St. Louis) performed exhaustive model experiments and extended the theory to the axisymmetric tubular case. Later, our attention was focused on the biological study of the ureteral smooth muscle and we learned a great deal. In the meantime, we realized also that the hydroureter problem is best solved surgically by sewing up the ureter to a smaller diameter.

Chia-Shun did not work much further on biological problems. But he laid out a plan to study the blood flow in large arteries by means of the Orr–Sommerfeld equation. When he solved the colliding soliton problem in 1993, we discussed extensively to aim further research on the arterial blood flow problem. There is no doubt that solitons can exist in arteries because of the nonlinear characteristics of the elasticity of the blood vessel wall. But the arterial tree is characterized by its branching pattern, each branch is not very long, and the flow is characterized by the forward and reflected waves. Could these solitons really resolve the secret of pulse-wave diagnosis? Could the method used by Chinese traditional medicine, which was rooted in the book *Nei Jin* written approximately twenty-three centuries ago, be understood through a study of these solitons? How do these solitons transmit messages about the health and disease of internal organs to the radial artery on the wrist? How to interpret the signals detected by the fingers pressing on the radial arteries? If the relationship between the solitons and the boundary conditions at the branching points of the arteries were understood clearly, perhaps all these questions could be answered. How I wish Chia-Shun were here to solve these mysteries!

The *Selected Papers* includes only Chia-Shun's mathematical and physical articles. His other writings were omitted. I am glad that his "Remembrance of G. I. Taylor" remained in the *Selected Papers* (pp. 1005–1009). But I wish you could read his literary piece, "Old China Remembered," published in *The Ohio Review* **18** (1977), 67–77, (Bibliography No. 84, *Selected Papers*, p. 1020). It consists of five short stories: The Slate Court, Crepuschule, Mulberries, Silk from Wild Cocoons, and Winter-Sweet. Through them you would really understand young Chia-Shun. Donald Hall, the poet, in his introduction to this article, said, "When I think of Chia-Shun now in his absence, he smiles with a wild enthusiasm – and it may be enthusiasm over a poem a thousand years old, or over a problem he is solving, . . . , or over the petal of a flower in front of us. He delights in the world. . . but unlike most humans – scientist or poet or salesman or factory worker – his world moves far outside the borders of his work; it is wide with things to be loved and cherished."

An illustration of Chia-Shun's seeing poetry in fluid mechanics and fluid mechanics in poetry can be found in the frontispieces of his books. He chose a 1946 photograph of a wheat field in western Kansas to illustrate the dynamics of nonhomogeneous fluids, a 13th century Chinese painting of a tidal bore to illustrate fluid mechanics, and quoted the poems of La Fontaine, Li Chong Chu, and Fung Yen Chi to introduce various topics in fluid mechanics.

In daily life the Yih family is warm, relaxed, and somewhat idealistic. Son Yiu Yo is a computer expert, son David is a Ph.D. musician, and daughter Katherine is an ecological biologist working on public health. Chia-Shun himself played flute and painted with oil in the style of the French impressionists. He was gregarious and a wonderful storyteller. He loved to eat and often cooked for friends. He was a true gardener and could name many plants by their Latin names. He loved students and treated them as family members. Inspiration could come to him at any time, in any place. During the garden wedding ceremony of his daughter, he whispered to me that he had suddenly found the solution of a solitary wave.

Last year there were an unusually large number of tornadoes in the U.S. One day, after their long drive from Gainesville to Ann Arbor, he called to tell me that he and Shirley had arrived home safely and that while Shirley was driving, he mentally added to a swirling horizontal flow of a fluid a core of another fluid of different density and temperature, and suddenly he got a model of the tornado. In his head, he worked out the mathematical facets of how a core leads the weather condition at a higher altitude to the ground, how the horizontal swirling will generate the maximum speed at the surface of the core at the ground level, how the cyclonic action would cause the tornado to spin counterclockwise looking down toward the earth on the northern hemisphere, but clockwise on the southern hemisphere, and how sometimes a $\lambda$-shaped tornado can be formed. He could even explain why the debris of a tornado is always thrown to the left in the northern hemisphere. Thus a two-fluid theory of tornadoes was born. It remained only to check the literature, ask the experts about the facts, do some numerical calculations, and write it up. As his friend, you got this kind of enjoyment!

On April 24, 1997, Chia-Shun and I left for Taipei to participate in a workshop on "Mechanics and Modern Science and Technology." This workshop was initiated by the Chinese Academy (Academia Sinica) to study how the Academy could contribute to the advancement of mechanics through the establishment of a new Institute of Mechanics. I started my trip from San Diego. Chia-Shun started from Detroit. Both of us were supposed to arrive early in the evening. I waited and waited for him. At midnight I called home. Then I learned that when his plane was preparing to stop at Tokyo on the way, the stewardess went to wake him up before landing and found it impossible to awake him. A passenger sitting next to him said he did not notice any signs of distress. So he left, so peacefully, so elegantly!

## Bibliography

A list of papers published before 1990 is given in *Selected Papers by Chia-Shun Yih*. Those published in 1990–1997 are presented below.

Yih, C. S. 1990. Wave formation on a liquid layer for de-icing airplane wings. *J. Fluid Mech.* **212**: 41–53.

Yih, C. S. 1990. Infinitely many superposable solutions of the Navier–Stokes equations: damped Beltrami flows. In *Of Fluid Mechanics and Related Matters*, Proceedings of a Symposium Honoring John Miles on his 70th Birthday, Dec. 1990.

Yih, C. S. 1993. General solution for interaction of solitary waves including head-on collisions. *Acta Mech. Sin.* **9**: 97–101, Science Press, Beijing.

Yih, C. S. 1993. Solitary waves in stratified fluids and their interaction. *Acta Mech. Sin.* **9**: 193–209, Science Press, Beijing.

Yih, C. S. 1994. Solitary waves in Poiseuille flow of a rotating fluid. *Q. Appl. Math.* **52**: 739–752.

Yih, C. S. 1994. Intermodal interaction of internal solitary waves. *Q. Appl. Math.* **52**: 753–758.

Yih, C. S. 1995. Kinetic-energy mass, momentum mass, and drift mass in steady irrotational subsonic flows. *J. Fluid Mech.* **297**: 29–36.

Yih, C. S. and Wu, T. Y.-T. 1995. General solution for interaction of solitary waves including head-on collisions. *Acta Mech. Sin.* **11**: 193–199.

Yih, C. S. and Zhu, S. 1996. Selective withdrawal from stratified streams. *J. Austral. Math. Soc. Ser. B* **38**: 26–40.

Yih, C. S. 1996. Added mass. *Chin. J. Mech.* **12**: 9–14.

Yih, C. S. 1997. The role of drift mass in the kinetic energy and momentum of periodic water waves and sound waves. *J. Fluid Mech.* **331**: 429–438.

Yih, C. S. 1997. Tornado-like flows.

# Index

Anode evolution, 331

Benjamin-Feir instability, 99
Bernard convection, 31
Bidirectional waves, 177
Biofluids, 439
Biot number, 19, 36, 223
Boll wave, 130
Bond number, 44
Bubble detachment in boiling, 364
Bubble rise, 279

Capillary number, 5, 223
Casting, 306
Coating, 112
Continuum surface-tension models, 294
Convection
    bilayer, 15, 26
    evaporation in, 3, 32
    experiments, 26, 317
    Marangoni, 15
    Meniscus characteristics, 273
    pattern, 287
    Rayleigh, 15
    salt-finger, 403
    solidification and, 313, 317
    in stably stratified fluid, 403
    thermosolutal, 285
Core-annular flows, 258
Coriolis instabilities, 67
Couette flow, 107, 166
Coupled ocean atmosphere system, 179
Crispation number, 19
Crystal growth from melts, 306
CST (continuum surface tension) models, 294
Curvature number, 223
Cylindrical bridge, 62

Damping rates of waves, 246
Deicing fluids, 158
Deicing of wings, 156
Density stratification, and instability, 163

Eightfold patterns, 215
Elasticity number, 266

Electromagnetic fields, in biofluids, 439
El Niño Southern oscillation (ENSO)
    phenomemon, 179
Eötvös number, 329
Eulerian methods, 331, 382
Evaporation in convection, 3, 32
Evaporation number, 5

Faraday waves, 174, 211, 246
Films
    liquid flow, 73
    rupture of, 221
    thin-liquid, 3, 221
    three-film flow, 120
    two-film flow, 120
Fixed-grid solution of moving boundary, 397
Flame-stabilization, 381
Flexible structures, and fluid flows, 265
Flowing water, waves in, 234
Fluid flows, and flexible structures, 265
FNFD (fully nonlinear fully dispersive)
    waves, 171, 176
Front-track algorithm, 287
Frost heave, and ice lenses, 347
Froude number, 74, 158, 200, 234

Galileo number, 44, 85
Gas-liquid interfacial waves, 129
Gas-liquid packed bed, 130, 139
Gravity-capillary waves, 246

Hexagonal patterns, 215

Ice lenses, and frost heave, 347
Ignition, 428
Immersed boundary technique, 265
Immiscible fluid interface, 142
Instability
    Benjamin-Feir, 99
    control of, 73
    in coupled ocean atmosphere system, 179
    of a cylindrical bridge, 62
    density stratification and, 163
    interfacial, 107, 156, 158, 166
    Kelvin–Helmholtz, 142

Instability (*cont.*)
  of liquid film flow, 73
  Marangoni type, 3
  of parallel flow on an inclined plane, 74
  in a pycnocline, 198
  Rayleigh–Taylor, 356
  soft mode of, 156
  subharmonic resonance in, 213
  thermocapillary, 43, 57
  Tollmien–Schlichting mode of, 156
  two-layer Couette flow, 107, 166
  viscosity stratification, 161
  window of stabilization, 80
  Zebiak–Cane model, 182
Interface tracking, 287
Internal solitary waves, 198

KdV (Kortweg de Vries)
  equation, 89, 234
  first order wave, 203
  model, 177
Kelvin–Helmholtz instability, 142
Knudsen number, 321, 336
Kuramoto–Sivashinski models, 95, 100

Lagrangian method, 331, 382
Landau front-fixing, 396
Lattice Boltzmann methods, 352
Layered fluids, waves in, 171
Level-set numerical method, 267
Linear waves, 118, 175, 234
Liquid films
  flow in, 73
  thin, 3, 221

Marangoni convection, 15
Marangoni effect, 3, 43
Marangoni number, 6, 19, 36, 58, 223
Marangoni stress, 43
Marangoni type instability, 3
Meniscus characteristics, and convection, 273
Moving boundary problem, 393

Non-Newtonian fluid, 158
Nonlinear waves in shear flows, 2399

Oblique waves, 91
Oceanic internal waves, 171
Ohnesorge number, 330
Oscillating waves, 132
Oscillation, and El Nino, 179
Oscillation frequencies of surface waves, 246

Parallel flow on an inclined plane, 74
Parametrically driven surface waves, 211
Patterns, 214–219
Peclet number, 325, 326
Phase-change processes, 278

Phase-field numerical method, 267
Photographic technology, 112
Pinned-edge waves, 246
Plasma heating, 325
Power law fluid, 158
Prandtl number, 36, 44, 58, 309
Premelting, and Van der Waals forces, 340
Prony fitting of extraction, 113
Pycnocline instability, 198

Quasi-periodic patterns, 215

Radiation, 433
Rayleigh convection, 15
Rayleigh number, 19, 44, 309
Rayleigh–Taylor instability, 356
Regelation, 339
Reynolds number, 74, 148, 158, 200, 223, 265, 269
Richardson number, 207

Salt-finger convection, 403
Schmidt number, 200
Shear flows, 239
SIMPLE algorithm, 281
SIMPLER algorithm, 332
Solidification, and convection, 313, 317
Solidification problems, 282, 285
Stability. *See* Instability
Stefan problem, 393, 401
Stokes flow, 413
Stratigraphic modeling, 393
Strouhal number, 74
Subharmonic resonance in instability, 213
Surface tension gradients, 43

Taylor number, 57
Thermocapillarity
  Coriolis instabilities and, 57
  instability and, 43, 57
  vs. regelation, 339
  *see also* Marangoni phenomena
Thermocapillary instability, 57
Thermocapillary waves, and instability, 43
Thermosolutal convection, 285
Thin film waves, model, 100
Thin liquid films, 85
Three-dimensional waves, 85
Tollmien–Schlichting instability, 156
Two-fluid-layer flow, 156

Van der Waals forces, 340
Viscosity stratification, 161
VOF (volume of fluid) numerical technique, 267

Waves
  anisotropic Kuramoto–Sivashinsky equation, 95
  bidirectional, 177
  boll, 130

in core-annular flows, 258
damping rates of, 246
Faraday, 174, 211, 246
FNFD (fully nonlinear fully dispersive), 171, 176
gas-liquid interfacial, 129
generated by surface tension gradients, 43
generation of, 85
gravity-capillary, 246
interfacial, in gas liquid flows, 129
interfacial, in layered fluids, 171
internal solitary, in pycnoclines, 198
Kuramoto–Sivashinski model of thin
    film, 100
linear, 118, 175, 234
long-wave theory, 5

nonlinear, in shear flows, 239
oblique, 91
oceanic internal, 171
oscillating, 132, 246
parametrically driven surface, 211
thin film, model, 100
Wave-wall collision, 205
Wave-wave interaction, 206
Weber number, 19, 144, 269, 270
Welding, 306

Young Laplace equation, 266

Zakharov–Kuznetsov (ZK) equation, 92
Zebiak–Cane model of instability, 182